SURFACE COATINGS

VOL 2-PAINTS AND THEIR APPLICATIONS

Prepared by the
Oil and Colour Chemists' Association, Australia

Springer-Science+Business Media, B.V.

Softcover reprint of the hardcover 2nd edition 1974

Publishing history
First published in 1974
Second revised edition in two volumes
Volume 1 1983
Volume 2 1984

Both editions first published by
New South Wales University Press
Box 1 PO Kensington NSW Australia 2033

ISBN 978-94-010-9812-0 ISBN 978-94-010-9810-6 (eBook)
DOI 10.1007/978-94-010-9810-6

Macarthur Press, Parramatta, NSW

Those members of OCCAA who have contributed to this book could not have done so without the understanding, forbearance, advice and encouragement of their companies, colleagues and families, and it is to them that we dedicate this work.

TEXTBOOK COMMITTEE

MEMBERS

J M Waldie* *Chairman; editorial revision*
D D Bonney *Biographies*
R Drummond* *Editorial revision*
J A Foxton *Editorial revision*
K Freeman *Technical*
W W Gallagher *Technical; major reviews*
J Hale *Technical*
D S Howie *N S W University Press*
G Liepins *Index preparation*
B J Lourey *Technical*
D M D Stewart *Technical*
R E Walton* *Editorial revision*
I R Waples *SI units*
R J Willis* *Technical; Chairman, Technical Education Sub-committee*

* Volume 2 Editorial Committee

CONTENTS

PREFACE

Volume 2 of *Surface Coatings* provides a comprehensive overview of the technology and utilisation of decorative and industrial paints in Australia.

It is a companion to Volume 1 of this work, in which the diverse range of raw materials employed in the surface coating and allied industries is examined in considerable detail. Readers of Volume 2 will find this introductory text invaluable; cross references to appropriate chapters in Volume 1 are provided and the index covers both volumes.

The Editorial Committee decided, as with Volume 1, to omit references unless the nature of the relevant chapter was such that direction of the reader to additional information sources was essential—for example, where an overview of a subject not directly connected with paint technology was included for completeness.

Note that proprietary trade names in this volume are indicated by an asterisk; attempts have been made, however, to replace these as far as possible by chemical descriptions.

Suggested amendments and improvements should be directed to the NSW University Press, for the attention of Mr D Howie, who will ensure that these receive the attention of the Editorial Committee.

J M Waldie *chairman* Textbook Committee
SYDNEY

31 RHEOLOGY

31.1 INTRODUCTION

There have been many attempts to describe, in a single word such as sticky, long, buttery, ropey, rubbery and thin, the complex rheological characteristics of liquids and semi-solids. The following section attempts to unravel some of the mysteries associated with viscosity and rheology.

Rheology is *the science of flow and deformation of matter* or, more simply, the study of viscosity over a wide range of conditions. This definition embodies another term, viscosity, which is *the resistance to flow offered by a fluid.*

The above definitions form the basis of the study of rheology, not only in paints and surface coatings but in many other fields.

31.2 SHEAR FORCES AND VISCOSITY

Before considering rheology in detail it is necessary to look into the effects of forces on a body. When a body (of any shape) is placed under a force (strain), an alteration to the shape of the body will occur; for example, a force or strain imposed on one end of a fixed wire will cause elongation. However, if the strain is applied over the surface of a flat body while a parallel flat surface is kept rigidly in contact with it, the two surfaces will slide relative to each other in a direction parallel to the two surfaces. This process is called shearing and the force applied is called a shear force.

To understand the concept of viscosity consider what happens in a liquid when it is stirred, or in other words, when a shear force is applied. Work is applied to the stirrer and into the liquid as a shear force. This force causes the liquid to move at a speed which is dependent on both the rheology of the system (i.e. how thick or thin it is) and on the size of the shear force applied (i.e. effort put into stirring).

Imagine the liquid as a small cube made up of a number of parallel layers of planes (see figure 31.1). The following symbols can be used to denote the various factors involved:

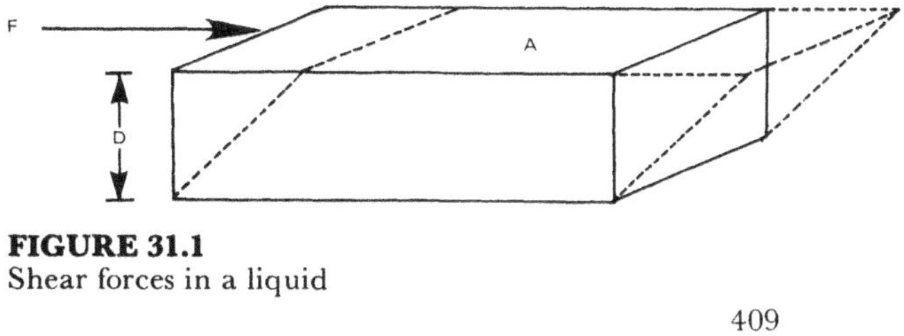

FIGURE 31.1
Shear forces in a liquid

F = force applied
v = velocity at which the top plane moves
A = area of top plane
D = distance between top and bottom plane

This leads to some new definitions:

Shear rate is defined as the velocity gradient and is the change of velocity with distance in the system, i.e.

$$\text{Shear rate } (\nu) = \frac{\text{v}}{\text{D}}$$

Shear stress is defined as the force per unit area acting on a plane, i.e.

$$\text{Shear stress } (\tau) = \frac{\text{F}}{\text{A}}$$

Thus *viscosity* may now be redefined as the shear stress required to produce unit shear rate in a liquid.

$$\text{Viscosity } (\eta) = \frac{\text{shear stress}}{\text{shear rate}}$$

Historically, the unit usually applied to viscosity was the *poise* (dynes. s/cm^2). The poise was the unit of *dynamic viscosity* and was related to the *kinematic viscosity* unit (the *stoke*) by the following equation:

$$\text{(stokes)} = \frac{\eta \text{ (poise)}}{\rho \text{ (g/cm}^3)}$$

where ρ was the density of the liquid.

Within the International System of Units (SI), the unit of dynamic viscosity is the pascal second (Pa.s). The use of the centipoise (1 cP = mPa.s) is permitted, but not in conjunction with SI units. The use of the poise is not permitted.

The unit of kinematic viscosity is mm^2/s. The use of the centistoke (1 cSt = 1 mm^2/s) is permitted, but not in conjunction with SI units. As with the poise, the use of the stoke is not permitted.

31.3 NEWTONIAN AND NON-NEWTONIAN LIQUIDS

Liquids are characterised by the type of flow properties they exhibit. A short discussion of the types of flow commonly encountered follows.

31.3.1 Newtonian (or Simple Flow)

Newton postulated that the shear stress required was proportional to the shear rate produced by the stress. Unfortunately this is not always the case. Changes in temperature, shear stress, shear rate and time may alter a given viscosity value. However, at a constant temperature a number of liquids do exhibit behaviour that complies with Newton's postulate—these liquids have *Newtonian flow*, are known as *Newtonian liquids* and include water, organic solvents, mineral oils and some resin solutions.

A linear plot of shear rate against shear stress for a Newtonian liquid results in a straight line passing through the origin with a slope corresponding to the viscosity (see figure 31.2).

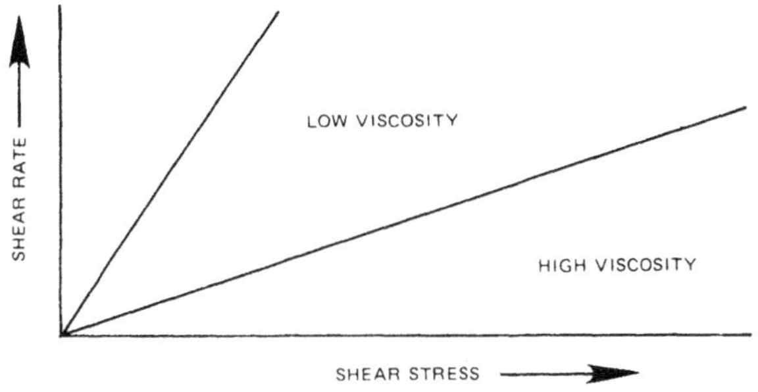

FIGURE 31.2
Newtonian liquid characteristics

A liquid with Newtonian flow displays a constant viscosity irrespective of the shear rate at constant temperature.

31.3.2 Non-Newtonian Flow

Liquids that do not follow Newton's postulate exhibit non-Newtonian flow and may be divided as follows:

Liquids with plastic flow
Liquids with pseudoplastic flow
Liquids with dilatant flow
Liquids with thixotropic flow

31.3.2.1 Plastic Flow

This occurs when there is a certain minimum shear stress that must be applied before any flow occurs. This minimum value is called the *yield value*. Above this value the liquid behaves as a Newtonian liquid. A linear plot for *plastic flow* results in a straight line that intercepts the shear stress axis at the yield point (see figure 31.3).

Liquids that exhibit this flow behaviour are called *Bingham liquids*.

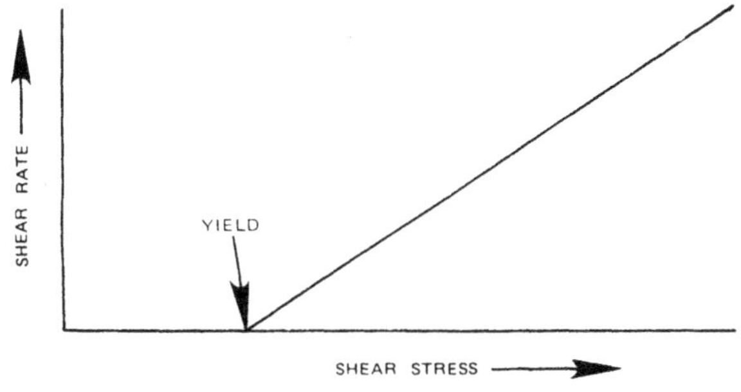

FIGURE 31.3
Plastic flow

31.3.2.2 Pseudoplastic Flow

Pseudoplastic flow may be regarded as a hybrid of plastic and Newtonian flows, with the fluid exhibiting plastic flow above moderate shear rates and Newtonian flow at low shear rates. In systems with this type of flow the viscosity decreases with increasing shear rate.

A linear plot of shear stress against shear rate results in a gentle curve convex to the shear stress axis (see figure 31.4).

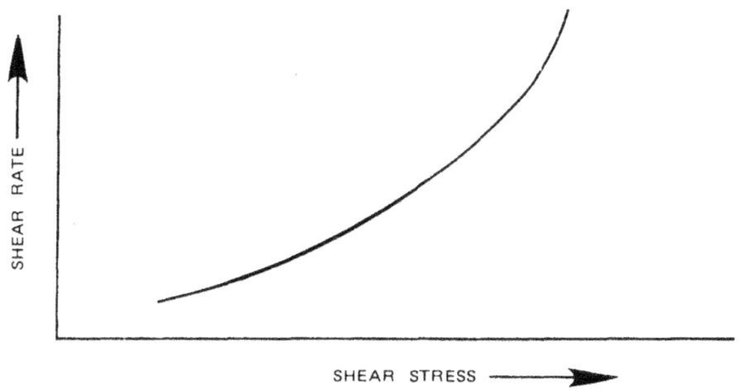

FIGURE 31.4
Pseudoplastic flow

The shear stress is, therefore, related to the shear rate according to the following formula:

Shear stress = $(k \times \text{shear rate})^n$

where k and n are constants.

This curve relationship with its constantly changing slope means that a liquid with this type of behaviour exhibits a series of viscosities, each one depending on the shear stress applied.

31.3.2.3 Dilatant Flow

This type of flow may be thought of as the opposite of pseudoplastic flow. In a system exhibiting *dilatant flow*, the viscosity increases as the shear rate increases. The formula quoted above also applies but the values of k and n differ.

Here a linear plot of shear stress against shear rate results in a smooth curve concave to the shear stress axis (see figure 31.5).

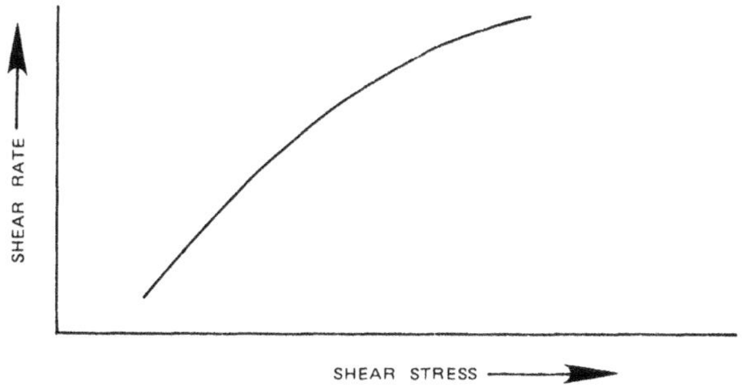

FIGURE 31.5
Dilatant flow

31.3.2.4 Thixotropic Systems

A *thixotropic* system is a special case of the pseudoplastic system which occurs frequently in the surface coating and related industries. The thixotropic system follows the characteristic viscosity curves of the pseudoplastic system (that is, a drop in viscosity with increasing shear rate) but also has a time dependency factor. It is worth noting here that with thixotropic systems at a constant shear rate, the viscosity also diminishes with the increased time of application of shear.

The time dependency makes measurement of thixotropic systems quite difficult. Because most systems are thixotropic to some extent, when carrying out rheological studies the rheological history of samples being tested or compared should be identical, otherwise invalid results could be obtained.

The plot of shear stress against shear rate for thixotropic systems gives a continuous curve convex to the shear stress axis as the shear rate is increased to the maximum. If shear rate is decreased, a *hysteresis loop* is created with another curve generated, different to that obtained when increasing the shear rate (see figure 31.6).

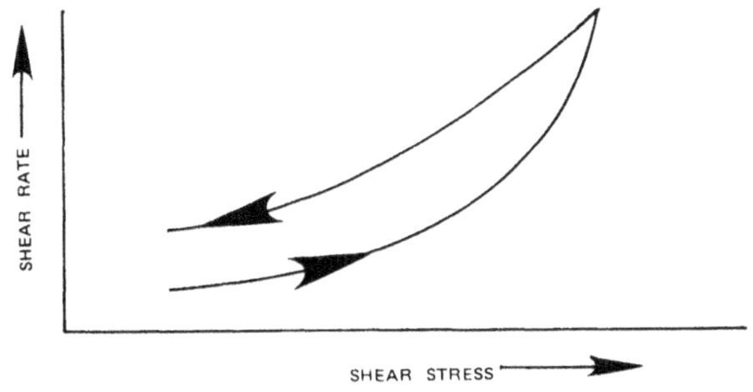

FIGURE 31.6
Thixotropic behaviour

When carried out under controlled conditions, the degree of thixotropy is a function of the area contained within the two curves—that is, within the hysteresis loop.

Flow behaviour under pseudoplastic and thixotropic systems is very similar but the difference is significant.

Thixotropy may be defined as a reversible change of viscosity accompanying the application or removal of a shear. A thixotropic liquid possesses a structure that breaks down with time when sheared until an equilibrium condition is reached, at which point the internal forces tend to rebuild the structure as fast as the shearing force breaks it down. Further reduction in this structure can only be induced by an increase in the rate of energy input (that is, increasing the shear rate). If the shearing (or energy) is stopped, the structure starts to rebuild to its initial viscosity. In addition, if the shearing forces are partially reduced, the structure is partially rebuilt. The inherent strength of the thixotropic structure determines the rate of breakdown and the 'rate of recovery'. This 'recovery' always takes a finite time varying from almost instantaneous to many hours. Thus a thixotropic system satisfies the following:

(a) the viscosity decreases at a constant shear rate;
(b) the viscosity is time dependent; and
(c) the reduced viscosity approaches a minimum value (which depends on the magnitude of the shear rate).

In contrast, pseudoplastic flow is characterised by a viscosity decrease when the shear rate is increased and vice versa. This viscosity, however, is not time dependent at a constant shear rate. Thus, at constant temperature, a pseudoplastic system is also characterised by a constant viscosity at a given shear rate no matter how long that shear rate is applied.

The rheological behaviour of a thixotropic system depends on the 'shear history' of the coating. Many difficulties in practical measurement of the viscosity of the thixotropic system are the result of the neglect of the time-dependent structure-rebuilding behaviour. For this reason thixotropic systems including paints should be kept undisturbed for a long period (2 to 3 weeks) before testing.

31.4 METHODS OF MEASUREMENT OF RHEOLOGY

The previous sections considered shear rate and shear stress, the ratio of which gives viscosity. Thus the measurement of rheological behaviour is the measurement of shear stress produced by a shearing force, and all rheology machines are based on this. Some of the more well-known *viscometers*, as these machines are known, are discussed below.

31.4.1 Bubble Viscometers

The best known of the bubble viscometers is that of Gardner Holdt. This consists of a range of tubes containing fluids ranging in viscosity from 0.5 mm^2/s through to 15 × 10^3 mm^2/s intervals such that each successive tube is about 25 per cent higher in viscosity than the preceeding one. These bubble tubes, designated alphanumerically, are calibrated to be correct at 25° C.

In use, a sample tube is filled to a level with the liquid under test so that the air space is of the same dimensions as the standard tubes. The sample tube is sealed and immersed in a water bath at 25° C, together with selected standard tubes. The viscosity is determined by removing the sample bubble tube and the appropriate standard bubble tube(s) from the water bath, inverting them, and judging the speed at which the bubble rises in the tube. The value of the Gardner tube most closely matching the sample is recorded as the sample's *Gardner viscosity*. The main advantage of this procedure is that the sample is always compared against a standard, the test is simple, and the operator does not require a great degree of skill. Obviously one of the problems with this type of equipment is that it is confined largely to transparent or translucent materials such as varnishes, oils, etc.

31.4.2 Falling Sphere Viscometers

These are not often used in the surface coatings industry. Here the sample under test is brought to a constant temperature, usually 25° C, and a standard ball is allowed to fall through the liquid for a given distance; the time taken is recorded. Applying complicated mathematics the viscosity expressed in Pa.s may be obtained, but this is beyond the scope of this chapter.

The falling sphere viscometer has the advantage of simplicity and cheapness. Provided it is used to measure viscosity of a Newtonian liquid, good results can be obtained. It is not recommended for non-Newtonian or opaque liquids.

31.4.3 Efflux Cups

Various efflux cups are available to the surface coatings industry. The types most commonly encountered are:

BS cups (to BS 1733, BS 3900)
'Ford' cups (to ASTM D1200)
Zahn cups

All three types are available in a range of sizes with varying orifice diameters. All operate similarly, with the time taken to empty via the orifice being recorded as the 'viscosity'.

Within Australia, the efflux cup most commonly used is the BS 1733 B4 cup, which has been made the basis of the Australian Standard test method: AS 1580 method 214.2.

In an attempt to rationalise the situation, the International Standards Organisation (ISO) in 1972 produced an international efflux cup (ISO 2431), which was intended to replace the variety of cups then in use throughout the world. There is little evidence to date of the widespread acceptance of the ISO cup or of its success in superseding any existing types of cups.

Realistic flow rates with these cups are between 20 and 100 seconds, and therefore the cup should be selected accordingly. The principal defect with this type of instrument is that the shear rate and shear stress change as the determination is being carried out; the rate at which the sample passes through the orifice is much greater at the beginning of the determination than it is at the end because of the decreasing volume in the cup. The shear rate will in most cases be a maximum of $10^1 s^{-1}$.

Efflux cups are not suitable for non-Newtonian liquids.

31.4.4 The Stormer Viscometer

The Stormer Viscometer consists of a paddle attached to a rotating spindle. The energy is applied by a weight operating through a gear train to the spindle.

Original models required the weight to be adjusted until the time required for 100 revolutions was 30 seconds, but recent models are fitted with a stroboscopic head which reduces the time taken for each determination.

The standard temperature used for a determination is 25° C. The result is obtained by recording weight required for 200 revolutions/min. The value, in grams, is referred to as *Krebs* units, which cannot be related to other viscosity units.

This instrument can handle both Newtonian and non-Newtonian liquids as some attempt is made to maintain a constant shear rate. However, the design of the paddle is such that the shear rate varies throughout the sample while the determination is being carried out and therefore the actual shear rate cannot be quoted. The other defect with this system is that the Krebs unit is not a standard unit. However, the Stormer Viscometer is a simple instrument to use; it is robust, gives information concerning package viscosity and will differentiate between samples of different rheology. This instrument has an advantage in that a yield point can be determined on an undisturbed sample by gradually adding weights and recording the weight at which the first movement of the paddle is detected. It was probably the first simple rotation viscometer to be used by the surface coatings industry.

31.4.5 The Rotothinner

The Rotothinner is an electrical instrument incorporating a motor-driven paddle constantly rotating at 570 rpm. The torque produced when the paddle is rotated in the sample under test is transferred to the base, via the sample container, which revolves against a torsion spring giving a direct reading in poise.

The spindle is disc shaped with a number of holes bored at angles through it. Because of the design of the instrument it is very useful for control work, and it can also be used for laboratory thinning of material to a required viscosity, thus facilitating predictions for batch correction. The Rotothinner is robust, quick and simple to use and is easily cleaned. A flame-proof version (air driven) is also available.

With this instrument there is an attempt to maintain a constant shear rate and produce a direct reading in poise. However, because of the design of the spindle, the shear rate is not constant throughout the sample and cannot be quoted; this must, therefore, place some doubt on the validity of the reading. It is, however, an excellent instrument for routine comparison work.

31.4.6 The Brookfield Viscometer

The Brookfield Viscometer is probably the best attempt, in a simple instrument, to combine validity of results, simplicity of operation and overall robustness.

The instrument incorporates a rotating spindle, immersed in the material under test, revolving at constant speed. The reading is indicated on a rotating dial by a pointer attached to the spindle shaft. The pointer deflection is carried through a torsion spring.

Depending on the model, various rotational speeds are possible, and different sized spindles are available, which makes the instrument quite useful for the study of viscosity characteristics within the shear rate range in which it operates.

This instrument does not suffer from the defects encountered in the other instruments discussed earlier, as it generates a constant shear rate (dependent on speed and spindle selected). Provided the side wall effect from the container is small, it is possible to relate shear stress to shear rate. The major problem is that high shear rate ranges cannot be obtained, but shear rates in the range 10^0 to $10^2 s^{-1}$ are possible.

31.4.7 Cone and Plate Viscometers

Perhaps the most sophisticated of the currently available instruments are the cone and plate viscometers. There are many types available and all are based on the same principle.

Consider two contra-rotating discs, as in figure 31.7.

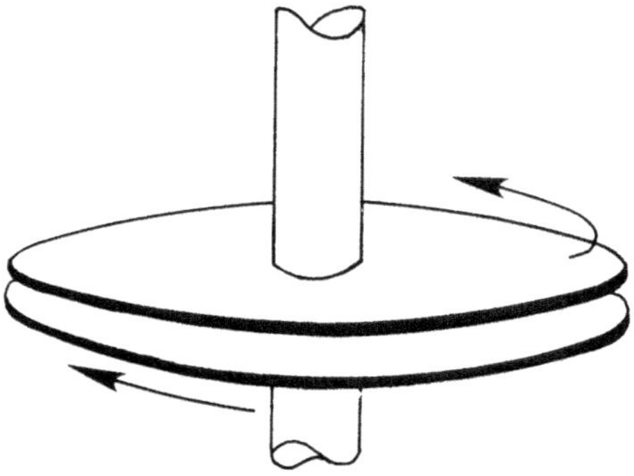

FIGURE 31.7
Rotating discs

Examination of this diagram will show that the portion closest to the circumference of the plates is under greater shear than that portion of the sample in the centre. This is a similar situation to the other viscometers, involving varying shear rate. Now consider the use of two conical shaped plates, as in figure 31.8.

In this situation it is obvious that the shear rate differences between the centre and outer edges have been compensated for. It is also obvious that there is a plane (X–Y in figure 31.8) where, since the two cones are rotating in different directions, the sample under test remains (theoretically, at least), stationary.

It is possible to take advantage of this fact, and use one surface flat and the other conical, as in figure 31.9.

This then highlights another problem: since the cone and plate are in contact at A, any attempt to measure shear stress/shear rate would give erroneous results because of friction at the point of contact. To overcome this, the cone is truncated at O–P. Another point is that there is no need to rotate both plate and cone—one may be kept stationary (usually the plate).

Some of the machines available are discussed below.

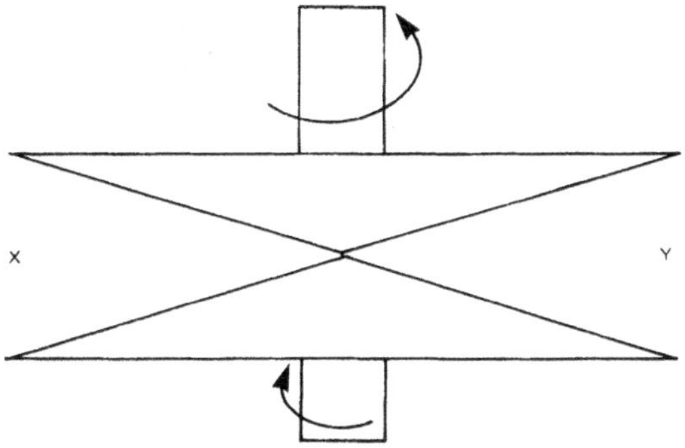

FIGURE 31.8
Rotating cones

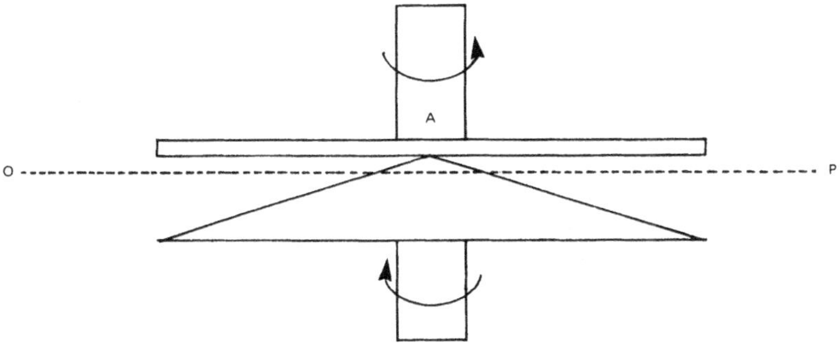

FIGURE 31.9
Rotating cone and disc

31.4.7.1 ICI Brushometer

The simplest cone and plate viscometer is the ICI brushometer, which is a cone and plate viscometer which operates at a fixed shear of between 10^3 and $10^4 s^{-1}$. This is a constant speed–constant shear rate machine which operates in the region of brushing shear.

31.4.7.2 The Weissenberg Rheogoniometer

The Weissenberg is one of the instruments that covers the shear rate range experienced in practice in the surface coatings industry. Minimum shear rate obtainable is in the order of $2 \times 10^{-3} s^{-1}$ and the maximum is in the order of $1 \times 10^4 s^{-1}$. Basically the Weissenberg consists of a cone and plate viscometer with the cone rotatable via a fixed-speed motor and multi-speed gear box. The induced shear in a sample is transmitted to the static plate. This plate is held in place by a frictionless air-bearing and torsion bar. The bar indicates the deflection by altering the field inside a transducer, producing a very small change in current. This change is amplified and recorded.

In operation the sample is applied to the lower platen (the cone) and the static platen/torsion bar/air-bearing assembly is lowered onto the sample. The gap required to compensate for the removal of the top of the cone is set using a transducer. The sample is then pre-sheared at a given shear rate, the test shear rate selected and the bottom platen set in motion, and the deflection measured and calculated to shear stress. Shear stress obtained for a given shear rate enables viscosity to be calculated at that shear rate. By increasing the speed of rotation, the shear rate can be increased in steps and a full plot from approximately 10^{-3} to $10^{4}s^{-1}$ can be carried out.

Earlier, in the sections dealing with pseudoplastic and dilatant flow, the following formula was established:

$$\text{shear stress} = (k \times \text{shear rate})^{n}$$

Taking the logarithm of both sides we get:

$$\begin{aligned} \log (SS) &= \log (k \times SR)^{n} \\ &= n \log (k) + n \log (SR) \\ &= \text{constant} + n \log (SR) \end{aligned}$$

giving an equation of the form:

$$y = a + bx$$

where a and b are constants.

This is a straight line of slope b with intercept a. Thus plotting shear stress and shear rate on log axes, a straight line should be (and in fact is) obtained.

With Newtonian liquids, the value of $n = 1$ and thus a slope of 45° is obtained. With dilatant and pseudoplastic systems, the value of n is such that the slope is greater than and less than 45°, respectively.

This use of log–log co-ordinates then makes the concept of rheology comprehensible, because any changes in rheology that occur with changes in shear rate can be seen easily and the point at which these changes occur can be accurately defined. It also enables the calculation of n to be carried out either graphically or mathematically. The calculation of n can be carried out by relatively unskilled people who need to measure the shear stress co-ordinate and the shear rate co-ordinate and divide accordingly.

32 FORMULATING PRINCIPLES

32.1 INTRODUCTION

Chapters elsewhere in these books have dealt in detail with the physical and chemical properties of the raw materials used in paint formulations. This chapter will indicate practical techniques used by the formulator in the selection of raw materials and their correct combination to develop a new product. With the vast and increasing number and complexity of raw materials suitable for use in paints, it is difficult to guide the reader in such a way as to make him or her an immediately competent paint formulator; paint formulating is a matter of experience, inspired guesswork and, above all, logical thought, coupled with a thorough knowledge of raw materials. The latter has been supplied, and experience is a product of time; the *logic* of paint formulation is therefore the subject of this chapter.

In order to illustrate the various development stages, an example of the principles involved in the formulation of a simple gloss enamel will be given. Multiple simultaneous formulation changes applied to as complex a product as paint can lead to confusion; the key phrase in any development should always be 'one step at a time', particularly when experience is lacking.

32.2 ESSENTIAL PREREQUISITES

32.2.1 Discussion

Before any thought can be given to the formulation and its ingredients, it is essential that the market requirements of the product are thoroughly understood by the formulator. The 'market' in this context could range from a single specialist applicator to the marketing division of the paint company itself.

32.2.2 Requirements

Information on the following parameters is essential in most cases. The requirements are not given in any order of importance, as this will change from one formulation to another. More specialised criteria may need to be satisfied from time to time.

32.2.2.1 Substrate

The substrate, or substrates, to which the product is to be applied will dictate many of the properties of the paint (for example, flexibility, alkali resistance, permeability). Thus a product for use over concrete as a paving paint will have very different properties from one intended for application to timber floors. The two products may even be based on the same generic type of vehicle, but the formulations will be quite different in their final composition. The appropriate substrate or substrates should always be used in service or application tests during development.

32.2.2.2 Pretreatment

The term *pretreatment* is used in its broadest sense to mean the previous history of the substrate before application of the paint, and can vary from a simple light sanding (before application of a timber primer) to the use of the most sophisticated degreasing and descaling techniques, followed by a priming and undercoating system (before application of a final topcoat). As any pretreatment of the substrate will modify its paint-holding properties, the same technique should be used during application/service testing.

32.2.2.3 Physical Properties

The physical requirements are generally detailed for a mixture of decorative and protective properties and usually include performance specifications. In this classification are such properties as viscosity, corrosion resistance, gloss, colour, film hardness and flexibility. The appearance, performance and properties of both the liquid paint and the dry film must be detailed precisely, and they are the most likely area of significant discussion and argument concerning the relative importance of each of the various properties specified. This single most important review will decide the compromise of properties achieved in the final formulation and, for the formulator, the boundaries of his or her flexibility to engineer the formulation.

32.2.2.4 Application Conditions and Equipment

A knowledge of the application equipment and conditions is essential if the customer is to be able to use the product effectively. A paint formulated to a performance specification for use by brush at low temperatures would be significantly different, if only in solvent composition, to one formulated to the same specification but to be spray applied at elevated temperatures. Equally importantly during application trials, conditions and equipment as close as possible to those of the final applicator should be used. Fine details (for example, tip size and pressure for spray application) should be obtained, as these parameters can have significant influence on the properties and characteristics that need to be built into the formulation. In critical cases it may be necessary to arrange for the application trials to be carried out by the customer, and to formulate the product specifically for the customer's equipment and application techniques.

32.2.2.5 Drying Time/Curing Conditions

The specification of drying time or curing conditions has a major bearing on the selection of the paint vehicle (resin), as well as the solvent to be used. The conditions must be specified precisely; for example, there would be no point producing a product based on a long oil alkyd using conventional driers and solvents if it were intended to be cured by baking for 10 minutes at 150° C.

32.2.2.6 Durability/Service Conditions

These two factors are related and should always be considered together. The term 'durability' must be quantified not only by time but also by such factors as the aggressiveness of the atmosphere, the liability to physical damage, any special needs such as acid resistance, and the film properties that are required to remain adequate throughout the service life of the film.

32.2.2.7 Toxicity

Many raw materials used in paint are actively or passively toxic; some traditional ingredients have been shown over the years to be significantly toxic and are currently very strictly controlled.

When formulating a product, its end use must be borne in mind during raw-material selection; every paint formulator should have at hand a copy of the legislation and poisons regulations current throughout the country. It must also be remembered that minor constituents of the formulation, although harmless when incorporated into the final product, may constitute significant health hazards during paint manufacture. It is therefore the duty of the formulator to ensure that factory staff are adequately warned of the possible hazards of any material, despite the fact the final paint is well within the limits of any toxicity legislation. Naturally, any finished product must comply with the poisons legislation in all respects. Details of some aspects of the NHMRC recommendations are provided in chapter 58.

32.2.2.8 Cost

Most cost calculations are made using one of two criteria: cost per litre of paint or cost per finished job. The choice of criterion may significantly alter the type of material selected and modify the formulating techniques. The market defines the acceptable price per litre of the formulation, and the formulator must engineer the product accordingly.

32.2.2.9 Existing Specifications

An increasing quantity of paint products is being sold to specifications which may be written by a number of specifying authorities, either governmental or industrial. These specifications will usually be either:

(a) composition, or
(b) performance.

In the first case the formulator's latitude is usually restricted to close limits and is frequently only the ability to meet the specification at the lowest possible cost, or to produce the most cost-effective formula, calculated by whichever method is agreed with the customer. The performance type of specification gives the formulator almost complete freedom of choice of raw materials and formulating parameters; providing the performance requirements (frequently weatherability, opacity and various resistance tests) are met, the formulator is able to use his or her skills to achieve the best, most cost effective formula, to the benefit of both the manufacturer and the customer.

32.2.2.10 Special Factors

Under this heading are the unclassified or special requirements of the customer; they could include such factors as re-coat time, special performance requirements or possible viscosity restrictions which may vary from customer to customer. Any special requirements should be noted from the commencement of development, as they may affect the initial raw material/formulation parameter selection. It is often the special requirements that cause the formulator the greatest problem: if they are not completely understood from the outset, they may be in conflict with the formulator's original decisions.

32.2.3 Example

Throughout this chapter it is intended to give as an example the application of recommended formulating techniques to the development of a formulation for an exterior gloss alkyd paint, and the way in which the logic steps are applied to this formulation. Whilst it is realised that this type of product is readily available and an integral part of the inventory of most paint companies, the basic simplicity of the formulation can be used to demonstrate the 'creativity' which is so necessary a part of formulating skills.

TABLE 32.1
Main requirements for formulating a white gloss alkyd paint

	Parameter	*Requirement*
32.2.2.1	Substrate	Timber or other general building material
32.2.2.2	Pretreatment	Primer and undercoat (or just undercoat)
32.2.2.3	Physical properties	
	(a) Opacity	High (one-coat coverage over undercoat)
	(b) Gloss	High
	(c) Viscosity	Ready for use
	(d) Colour	White plus selected range
	(e) Solids content	Minimum 50% by volume
	(f) Spreading rate	14–16 m^2/litre average
	(g) Appearance in can	Well bodied
	(h) Stability	24 months
32.2.2.4	Application conditions and equipment	Ambient temperatures; brush, roller or spray
32.2.2.5	Drying time/curing conditions	Maximum 6 h touch-dry, maximum 16 h hard-dry under ambient conditions
32.2.2.6	Durability/service conditions	(a) High durability (b) Residential atmosphere
32.2.2.7	Toxicity	Non-toxic
32.2.2.8	Cost	$1.50/litre raw material cost
32.2.2.9	Existing specifications	Nil
32.2.2.10	Special factors	(a) Series of tint bases of equal quality required (b) Equipment clean-up with mineral turps

32.3 BASIC PLANNING

32.3.1 Discussion

Having obtained as much information as possible about the characteristics of the product to be formulated, it is essential that a clear plan should be laid down before commencing the practical aspects of the development programme. The initial plan should be a framework upon which to hang the detailed evaluation programme and should be flexible. With no plan to follow, it is relatively easy to be misled by interesting but irrelevant results, although these might be followed up either as a sub-programme or a separate exercise if warranted. If the framework is too rigid, important points may be overlooked in a short-sighted attempt to stay strictly within the narrow guidelines. This is basically the first stage of the project during which the experience of the formulator plays a major part in saving time and effort. Such skill is of assistance in acquiring the information outlined in section 32.2, but at this later stage definite savings can be made with experienced planning.

32.3.2 Collection of Information

With the objectives firmly established—namely the characteristics of the product to be formulated—the first planning step is to assemble as much relevant information as is available. This could come from any number of diverse sources, which will include journals, text-books (including this one) and manufacturers' literature; two sources that should never be overlooked are the current product inventory and reports of past projects. With this information to hand and thoroughly researched, the programme then requires a number of decisions.

32.3.3 Test Schedule

The product requirements have been defined at this stage and the relevant background information assessed, but before the practical work is commenced it is valuable to set down a rough outline of the test procedures to be adopted during the development programme. This stage is often critical to the success and duration of the programme, as too little testing can cause the neglect of some major problem leading to failure at the field-testing stage. Even worse, in the initial stage of service, too much testing will slow down the development programme beyond its deadline.

Testing should be confined to those properties that pertain to the end use of the product; for example, testing a paint for use on concrete by measuring its adhesion to steel is usually irrelevant. Similarly, testing the abrasion resistance of a ceiling paint will be unnecessary in most cases.

The testing programme should develop with the formulation. In the initial stages of development, only a few tests should be applied. As more detail is fitted into the product formula, then further tests can be added; upon reaching the final formulation, a comprehensive testing programme is undertaken to ensure that no unwanted side-effects have occurred during the development period. The accuracy of the test methods must be taken into account when selecting tests; if there are standard industry tests or standards issued by government bodies (for example, the Standards Association of Australia), these should be used if possible. However, on occasion, a quick in-house test can be valuable and may be used at the initial stages, gradually phasing in more accurate tests later.

There are many test methods available, details of which are beyond the scope of this chapter; the selection of the tests is dependent upon experience and good planning. Relevant methods should be reviewed at the planning stage.

32.4 INITIAL DECISIONS

32.4.1 The Basic Paint

The basic paint consists of only two major components:

(a) The *resin* (vehicle, medium), and
(b) the *pigment*.

Two further groups of components should be added for preliminary review:

(c) the *solvents*, and
(d) the *additives*.

32.4.2 The Resin

The resin, or resin combination, is the key component of any paint. Chapters in volume 1 have provided detailed information concerning the main resin types, but at this stage it is useful to review the parameters controlled by the resin in the context of paint formulation. These are shown below and are not in any special order of importance.

32.4.2.1 Basic Film Properties

Under this heading are such properties as hardness, flexibility, abrasion resistance, alkali resistance and adhesion. In effect, the resin system is the sole governor of these and similar properties. Some minor variations can be made by modifying the other basic components, but the resin system has the greatest influence.

32.4.2.2 Curing Conditions

Similar comments apply as in section 32.4.2.1. Although modifications to conditions and rates can be made by the use of specific solvent blends or additives, the chosen resin system governs the basic drying and curing performance.

32.4.2.3 Durability

Although pigment suppliers sell 'high-durability' grades of pigment, this is assessed on a pigment-to-pigment basis, and such grades of pigment may not to any significant degree improve the 'weathering' of a low-durability resin system. Durability in the more general sense is also governed by the resin system.

32.4.2.4 Application Properties

Solvent selection can significantly modify the application properties of a product, but it is the resin system itself that determines application characteristics in the widest sense; for example, can the product be brushed, or must it be sprayed?

32.4.2.5 Compatibility with Pretreatments

The vehicle has the sole influence upon the specific adhesion of the product to the various substrates or pretreatments to which the product is to be applied.

32.4.2.6 Cost

The resin system is likely to be the major component of the formulation, and the cost of the vehicle must have a major impact on the cost of the finished formulation.

32.4.3 The Pigment

The pigment is intended to control only two major properties of a paint: *colour* and *opacity*. It has no other *basic* function, but there are a number of ancillary influences that can be exerted by the pigment. One example is anti-corrosive properties: zinc dust, zinc chromate and zinc phosphate have become known as 'anti-corrosive pigments', although whether they are pigments in the true sense of the word is debatable.

The pigment can also influence the weatherability of a paint, as well as its more general durability such as alkali resistance; the application and manufacturing properties are affected by the rheological performance of the pigment. It is also, together with the resin, a major contributor to the cost of the formulation.

Thus the choice of the pigment is governed initially by the colour and opacity required, but the particular grade or chemical type is then decided on the secondary properties of that pigmentation system. For example, the choice between a phthalocyanine blue or a Prussian blue is made on light and chemical fastness requirements plus cost, rather than the colour of the two pigments.

32.4.4 The Solvent System

The solvent system is required to perform only one basic function: to enable the paint to be satisfactorily applied. However, in a similar manner to the secondary influences of pigment selection, selection of solvent affects a number of other properties of the paint.

The evaporation rate of the solvent can influence the flow and sag resistance of a wet film, as well as the degree of cure of a baking finish. A low evaporation rate will improve the flow but increase the sagging of an applied finish, and may result in excessive solvent retention in the cured film with subsequent rapid breakdown of the coating in service). In a number of cases, evaporation rate can be modified by blending solvents to form azeotropes and improve the evaporation rate of the mixture.

The 'strength' or 'cutting power' of a solvent can be successfully used to control to some extent the cost of the product. A 'weak' solvent reduces viscosity more slowly than a 'strong' solvent to achieve a given viscosity from a particular vehicle. Hence by using a 'weak' solvent the solids content of the paint is reduced, usually with a saving in cost. The solvent system can also have a significant effect upon the manufacturing process by its influence on the flow of the resin (which is a function of vehicle solubility). A secondary solvent is often used to achieve specific formulation characteristics. The solvent type used will depend to a large extent on the vehicle system, and resin compatibility should always be the main consideration.

32.4.5 Additives

The term '*additives*' covers a wide range of ancillary raw materials used in the formulation of a paint. They can vary from the extender pigments to anti-skinning agents, fungicides and other materials which are used in minor quantities, as discussed in chapter 30 (Volume 1).

32.4.6 Example

The above paragraphs serve to show how the formulator's choice, from the multitude of raw materials available, is restricted by the requirements of the finished paint. It is now possible to apply these principles to the preparation of the paint described in table 32.1.

32.4.6.1 The Resin System

The requirements of DIY (do-it-yourself market) application, drying conditions and cost significantly reduce the range of basic vehicles for the product, and three appear to be the most likely starting points:

(a) a long oil alkyd;
(b) a urethane-modified oil or alkyd; or
(c) an acrylic emulsion.

The acrylic emulsion can be discarded immediately, because there are:

(a) the requirement for mineral turps 'clean-up'; and
(b) a solids content of 50% solids by volume.

The latter parameter is difficult to achieve with current acrylic latex vehicles.

The dry time does not necessarily rule out the use of latex vehicles, as those identified are maximum times not specified times; the shorter times of latex products may well be more acceptable if the other requirements can be met.

The choice between a urethane oil (or alkyd) and a long oil alkyd is rather more complex, as all could be used as a satisfactory basis for formulation. It may be necessary to run parallel developments until one alternative is selected or ultimately to give the customer a choice at some point in the development when accurate performance can be predicted.

Assuming the long oil alkyd approach has been selected (experience also confirms this), a particular vehicle or choice of vehicles must now be decided. In many cases formulators will have one or two suitable 'work-horse' vehicles; if not, then preliminary trials of alkyds selected from manufacturers' data or other available information must be run. For simplicity it is assumed that one vehicle has been selected for preliminary development, but modifying vehicles or a completely different type might eventually be used.

32.4.6.2 The Pigment

The initial trials will be run on the white product, although the need for tint bases must always be borne in mind; pigment selection is reduced to titanium dioxide of a 'high-durability' grade (this is a requirement of the product).

32.4.6.3 The Solvent

As mineral turps clean-up is required, mineral turps will be used as the sole solvent unless a change is shown as being necessary during development.

32.4.6.4 Additives

Because the system is required to be high gloss, no extender will be used initially. The only additives to be used at this stage will be the drier system required by the alkyd.

Thus preliminary trials will be run using the long oil alkyd, a high-durability grade of titanium dioxide, mineral turps and driers, giving a total of four basic variables (although a combination of up to four different drier types may be required by the alkyd, giving overall up to seven variables).

32.5 THE COMPROMISE

32.5.1 Discussion

Assuming that the basic formulation has seven variables, it is obviously confusing to change all seven at one time. Changes should be made only to the parameter that most significantly influences the performance of the dry film, and it is generally recognised that this is the pigment volume concentration.

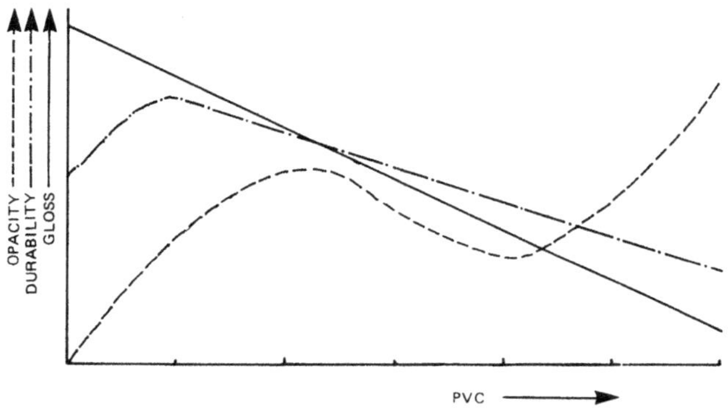

FIGURE 32.1
The effect of pigment volume concentration (PVC) on the gloss, durability and opacity of a white alkyd paint

32.5.2 Pigment Volume Concentration (PVC)

The PVC is defined as the percentage of pigment volume in the total volume *solids* of the paint.

The effect of PVC on three important properties of a paint is shown in figure 32.1. As the paint to be formulated initially is white, the influences shown are those for titanium dioxide only.

$$\text{PVC\%} = \frac{\text{volume of pigment} \times 100}{\text{volume of pigment} + \text{volume of binder non-volatile}}$$

Note. No units have been put on the axes of the graphs, as the actual values achieved will depend upon the grade of pigment and the specific vehicle used. The degree of dispersion may also significantly shift the curves within the axes. The positioning of the curves has no significance—only their general shapes.

It can be seen from figure 32.1 that the requirements of high gloss, high opacity and high durability are in conflict: maximum gloss is given at very low PVC, maximum durability at low

PVC and maximum opacity at either moderate or very high PVC. Obviously the very high PVC section of the curve is unsatisfactory when gloss and durability requirements are considered, as the latter factors are approaching minimum values in this region. The point of optimisation of gloss and durability is of no interest, as inadequate opacity is developed. Hence the section of the curves of interest is that corresponding to the initial recurving of the opacity graph, and the best compromise of properties will be achieved by working within these fairly narrow limits, which usually lie between 15 and 25 per cent PVC. Formulations varying only PVC within these limits (a PVC 'ladder') should be evaluated to obtain the point of optimum compromise.

Superimposed upon these three factors may be additional needs (for example, tinctorial strength alignment with other products and, naturally, cost considerations). A few preliminary trials will normally be sufficient to achieve the optimum compromise, but the PVC affects other properties such as:

(a) film flexibility,
(b) adhesion,
(c) impact resistance,
(d) sandability,
(e) moisture permeability.

These are minor considerations in the majority of cases, and they assume importance only in critical formulations; they can be ignored until it becomes apparent that attention is needed. Within the PVC limits set by the main factors involved, the influence of PVC on these secondary factors is minimal, and other techniques are usually more rewarding (when manipulating, for example, impact resistance). The use of extenders is a valuable method of achieving high PVC's without the necessity of using high levels of expensive prime pigments, and extenders are included in the PVC calculation as part of the pigment when formulating other than high gloss finishes. Thus a flat enamel based on the raw materials used to obtain the graphs in figure 32.1 could have used titanium dioxide to achieve optimum opacity then included an extender to reduce the gloss to the required level. The selection of the correct extender is in itself a complex subject. Full descriptions and uses of generic types of extender are covered in chapter 28 (volume 1).

32.5.3 Critical Pigment Volume Concentration (CPVC)

The critical pigment volume concentration is usually described as the PVC at which there is just sufficient binder available to completely wet the pigment and fill the voids between the particles at a particular degree of dispersion. The CPVC of each system will depend upon the surface area of the pigments and extenders as well as the 'binding power' of the vehicle—'binding power' being the ease with which the vehicle system will wet the pigment/extender combination. Hence CPVC is very specific to the total system. Changing a simple extender within a system may change the CPVC by an appreciable amount.

As the CPVC is basically the PVC at which the observation of voids within the dry film commences (the onset of significant porosity), it has appreciable effects upon the properties of the film. The direct calculation of CPVC is not possible, but its value can be estimated by experiment with reasonable accuracy by plotting the change of a specific property such as corrosion resistance, scrub resistance, opacity or moisture permeability against PVC. An inflection will occur in the curve at a point that is approximately equivalent to the CPVC, but which will vary by a few per cent depending upon the particular property being studied. A good-quality paint should be formulated at a PVC at least 5 per cent above or below the CPVC, depending on the properties required. Use can be made of 'underbinding' (that is, a PVC greater than CPVC), particularly in latex paints, as the voids in the film act as points of multiple reflection of incident light and reinforce the opacity. Just as the refractive index difference between the white pigment particles and the binder confers opacity, so does the greater difference between the air in the voids and the pigment, extender and binder, allowing reduced binder and pigment costs. 'Underbinding' causes some reduction in the overall integrity of the paint film, but this technique is very useful for some types of product. The second opacity inflection in figure 32.1 indicates the position of the CPVC in the system illustrated.

32.6 PRELIMINARY TRIALS

32.6.1 Discussion

To produce sufficient information to plot the curves in figure 32.1 requires a series of paints at varying PVC (a PVC ladder) and the measurement of sufficient properties to assure the formulator that his formulations will satisfy the customer's requirements. To do this, the formulator must have a knowledge of the dispersing properties of the pigments in the selected vehicle. If a 'work-horse' combination is being used (as in the modification of an existing paint), the formulator will in most cases have sufficient knowledge of the system to predict quite accurately efficient grinding formulations; however, with new systems this knowledge is not available and must be generated. The following technique is useful in predicting dispersion formulations.

32.6.2 Daniel's Flowpoint

Daniel's Flowpoint estimation is a well-documented technique from which it is possible to calculate millbase formulations for a variety of dispersing equipment. However, the technique (which consists of making flowpoint titrations for pigment/vehicle combinations at varying vehicle solids) produces information which, in its raw state, is rather difficult to interpret. The results consist of a series of volumes of vehicle solutions which will 'flow' a given weight of pigment and will usually appear in the form given in figure 32.2 (the first graph).

resin concentration %	*titre mL resin solution*	*PVC (f)* %
0	20.0	19.3
2	11.0	30.2
5	8.0	37.4
10	6.5	42.2
15	6.5	42.2
20	6.7	41.5
25	7.2	40.0
30	8.0	37.4
40	9.5	33.7
50	12.0	28.5

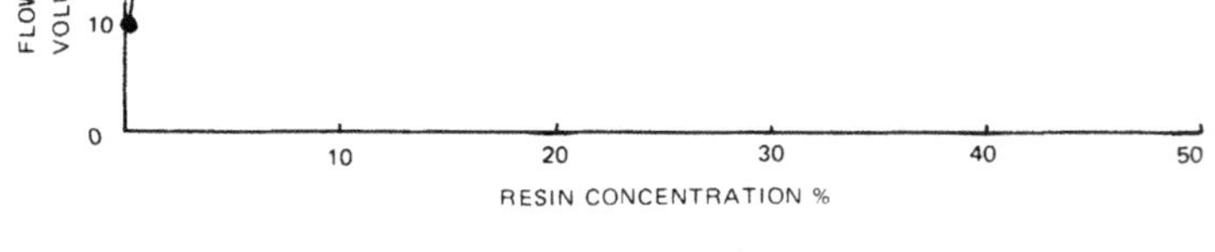

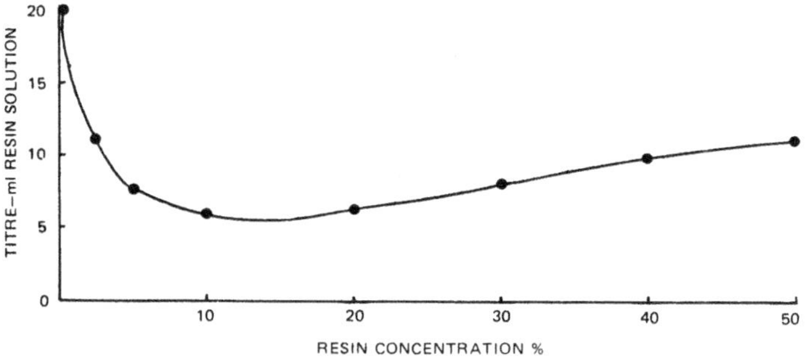

FIGURE 32.2
Transposition of flowpoint data. *Figures 32.2 to 32.6 courtesy Tioxide Australia Ltd.*

It can be seen that this curve shows a minimum at approximately 15 per cent resin concentration, which is the point of maximum dispersing power of the system. For the purpose of obtaining more valuable information, the pigment volume content of the flowpoint paste PVC(f) is a more reasonable parameter to study; this is expressed as:

$$\text{PVC(f)\%} = \frac{\text{volume of pigment} \times 100}{\text{volume (pigment + resin + solvent)}}$$

The data used to develop the first curve in figure 32.2 are transposed using the above relationship to produce the lower curve.

32.6.3 Classification of Flowpoint Curves

To select dispersion equipment, an arbitrary allocation of the flowpoint curves is useful, and figure 32.3 shows the classification found from experience to yield the most reliable information.

These can be summarised as:

(a) *Good dispersions.* The PVC(f) of the flowpoint curve maximum is greater than 38 per cent. The rheology of such pastes is normally slightly dilatant.
(b) *Average dispersions.* The PVC(f) of the flowpoint curve maximum lies between 30 and 38 per cent.
(c) *Poor dispersions.* The PVC(f) of the flowpoint curve maximum is less than 30 per cent. The rheology of such pastes is normally pseudoplastic.

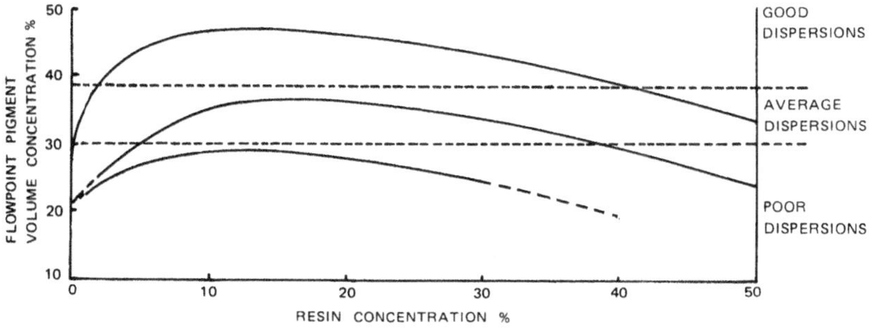

FIGURE 32.3
Arbitrary classification of flowpoint curves

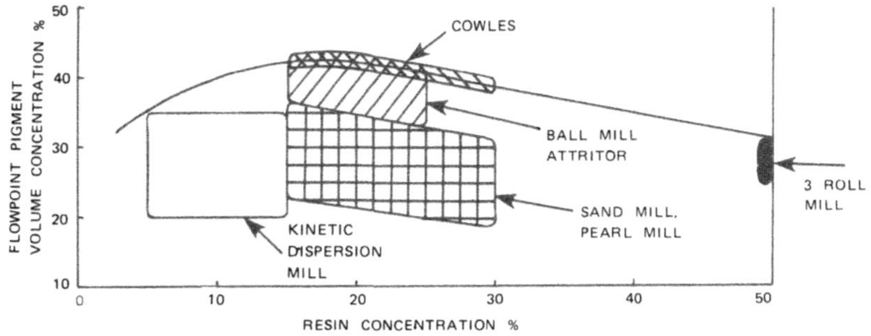

FIGURE 32.4
Suggested millbase formulae for 'good dispersions'

32.6.4 Selection of Dispersion Equipment

Figures 32.4 to 32.6 show the recommended limits of PVC(f) and resin solids concentration for various dispersion equipment, and the influence of the flowpoint classification upon the type of equipment which is suitable. Thus 'good dispersions' should mill quite satisfactorily on all modern equipment, but with a 'poor dispersion' it is possible that only the ball mill or triple roll mill will give acceptable results.

It can be readily seen that the 'average dispersion' curve has already discounted the use of simple high-speed equipment. Dispersion by such equipment is very sensitive to change in the PVC(f) and the flowpoint rheology. The point must also be made that the use of additives in the dispersion stage can critically alter the dispersion classification of the system and should be studied carefully before use.

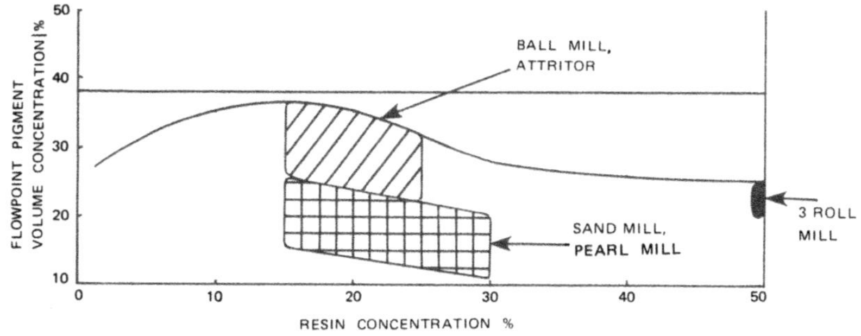

FIGURE 32.5
Suggested millbase formulae for 'average dispersions'

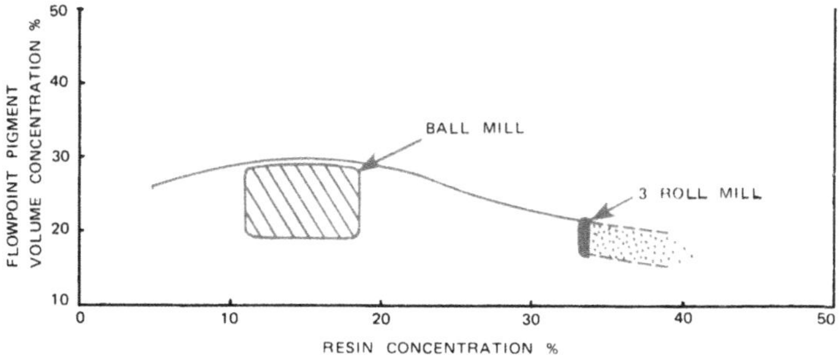

FIGURE 32.6
Suggested millbase for 'poor dispersions'

32.7 TRIAL BATCH MANUFACTURE

32.7.1 Discussion

The techniques described in sections 32.5 and 32.6 generate sufficient information to commence laboratory trial production of paints, as starting points to meet the requirements of the initial specification. However, it is important to understand exactly what is meant by the customer's descriptions of the properties required. These should be discussed completely with the customer at the time of specification. As in table 32.1, the requirements are often subjective ('high' gloss, 'high' opacity). Before commencing any trial batch, these requirements should be quantified wherever possible. This information can frequently be obtained by reference to specifications where limits are applied to the performance requirements of the product. Thus 'high gloss' will often mean greater than 85 units at 60° incidence; 'high opacity' could mean a contrast ratio of 94 per cent at a given spreading rate (usually 16 m^2/L); 'high durability' may be interpreted as a series of ratings of, for example, discoloration, colour change, chalking and dirt retention after a specified exposure period either in an artificial weathering machine or, more likely, on exterior exposure testing.

For these preliminary assessment, only a restricted number of trials need be carried out, but they should include a test of the application properties of the product.

32.7.2 Batch Preparation

At this stage it is essential to have obtained a reliable formulation and production method of manufacture. All too frequently, time is lost in attempting to solve problems that occur in the final film but which are caused by faulty manufacture. Careful control over each stage of manufacture should be maintained, as described below.

32.7.2.1 Dispersion

Having used the Daniel's Flowpoint technique to generate the millbase formulation, little difficulty should be encountered in achieving the optimum time of dispersion on the particular equipment selected. However, the introduction of an additive to the millbase can negate all the careful pre-planning, and it is bad practice to introduce additives at the dispersion stage unless absolutely essential. They should be added at the stabilisation (or millbase letdown) stage where practicable. One drawback of the flowpoint formulating technique is that it tends to incorporate the maximum possible amount of pigment in the minimum quantity of vehicle (highest PVC(f)), and bases so formulated *always* require stabilisation before either full letdown or storage prior to completion.

32.7.2.2 Stabilisation

During stabilisation it is important to ensure that no pigment flocculation takes place. Constant checks on flocculation should be included at all stages of the stabilisation step, particularly if an additive is to be incorporated. These flocculation checks should continue from batch to batch until the formulator is sure that no tendency to flocculate exists. It is good practice to maintain a single flocculation check in production testing to detect any unforeseen or inadvertent 'shock' that may occur.

32.7.2.3 Millwash

It is desirable that sufficient solvent is available for efficient mill washing and viscosity adjustment. The quantity will depend upon the final viscosity, the 'cutting power' of the solvent, and the resin viscosity, but inadequate millwash is a significant source of loss during manufacture. At the preliminary stages of formulation the quantity of solvent available is not always accurately known, but at later stages of development millwash maximisation can take place.

32.7.2.4 Letdown (Completion)

At letdown (completion), the remaining raw materials needed to 'dilute' the reduced millbase to the final paint formulation are added. The order of addition can have an influence on the final paint properties and should be carefully monitored during the preliminary trials. It is important to ensure that at no stage do two or more raw materials come together which may be incompatible (although quite compatible in the final product).

32.7.2.5 Viscosity Adjustment

In the preliminary trials an accurate knowledge of the viscosity may not be available and therefore a viscosity adjusting quantity of solvent cannot be predicted. It is frequently useful at this stage to divide the first batch into a number of aliquots and adjust the viscosity to a different figure for each (a viscosity 'ladder'). Application trials will indicate the best range for viscosity, and the bulk of the preliminary trial batch can then be adjusted to this.

32.7.3 Testing

At this stage, only testing of significant parameters should take place. If, for example, the gloss is inadequate, it is of little relevance to know that the product spreads at the correct rate, and 'tuning' testing should be omitted. It is frequently of more value to use quick checks, which yield slightly less accurate results than precise time-consuming tests, in the initial stages. Quick reliable answers are required in order that development is not held up while awaiting test results. It is most useful at this stage to have a standard product against which to compare the test paint. This standard may be a competitive material intended for the same end use, or a purely arbitrary product of known quality (which may be either better than or inferior to the required standard of the test product). Application and opacity trials, particularly, yield very quick results against a reference standard.

32.7.4 Example

For the example, the requirements of which are listed in table 32.1, it is intended to test five properties:

(a) application (viscosity ladder testing has already been carried out),
(b) gloss,
(c) opacity,
(d) colour, and
(e) drying.

These can of course all be carried out by one application of the product. As the final paint is to be applied over an undercoated surface, it is intended that for initial trials the test will simply be a brush application to undercoated paperfaced plasterboard in a vertical position. Over the white undercoat are added two grey stripes (also of undercoat) to show opacity.

Assuming the application trials have indicated that the performance of the test sample (or one of the test samples if a series has been evaluated) is satisfactory, then more testing of the same properties should be carried out on the same test sample to obtain a more accurate assessment of properties. A test report at this stage may appear as in table 32.2.

TABLE 32.2
Test results of preliminary batch

Test	*Method*	*Result*
Application	Brush on undercoated paperfaced board	Sags and slightly sticky
Gloss (60° head)	Draw-down on glass	82 per cent
Opacity	Draw-down on Morest Chart	Better than standard
Colour	Brush on white card vs standard	Better (bluer) than standard
Drying	Dry recorder at 25° C 50 per cent RH	Touch 4 h. Through 10 h.

In addition to the above tests, it is appropriate at this stage to prepare samples of the test batch for stability trials (at both elevated and ambient temperatures) to be examined at a later date according to the stability testing schedule. As the preliminary test batches are usually the simplest formulation, early stability trials indicate the inherent properties of the system in case of problems in future development batches. Similarly a costing of a satisfactory starting point will yield a useful guide to whether or not it is 'over-formulated'.

32.8 ADJUSTMENT OF FORMULATION

The next step in the development of the product is to undertake a study of the results obtained in the preliminary stages. If necessary, further tests should be carried out to identify the problems indicated in the preliminary testing and to isolate their causes.

Once the problems and causes are identified, possible areas of adjustment should be sought, and modifications made to the product by manipulation of the formulation (for example, change of PVC; adjustment of solids/viscosity). Additional or different materials may be introduced where necessary to assist in overcoming problems (possibly a change in vehicle or a blend of vehicles).

32.8.1 Example

32.8.1.1 Fault Diagnosis

Evaluation of the results might indicate the following problems and possible causes (the causes suggested are not intended to be comprehensive, but simply for demonstration purposes).

(a) low gloss: gloss appears hazy—excessive pigment, incorrect vehicle, poor drier balance, poor grind, defective vehicle;
(b) application: sticky—viscosity incorrect, high solids;
(c) cost: high—excessive pigment or vehicle price or quantity, insufficient solvent;
(d) sagging: excessive flow—excessive solvent, low viscosity.

The low gloss and high cost together may indicate that the product is over-pigmented; coupling these results with the fact that the colour is a rather bluer-white than the standard and the opacity is higher than standard tends to bear out this theory. One change to be made is therefore to reduce the pigment volume concentration of the product. This will allow a greater proportion of vehicle to be used, reducing the pigment cost and the density of the product; it may also allow more solvent to be introduced.

There are conflicting requirements indicated from the application trials: the product is sticky on application which could imply that the viscosity is too high, yet it sags, which is an indication that viscosity may be low.

As the cost is high it may be necessary to change the solvent to one of lower 'cutting power'. However, at this stage a change of solvent is undesirable, as from further testing the appearance of the product in the can is very thin (a consumer quality judgement). Therefore it is intended to make an addition of thixotrope (in this case, a thixotropic alkyd); the proportion of this resin should be determined by testing various addition levels.

32.8.1.2 Trial Batches

It is unrealistic to make all of the changes suggested in section 32.8.1.1 at one time and test the resulting product. A series of formulations should be developed to isolate each change and obtain the maximum effect from the changes. Once determined by more searching testing than in the preliminary batches, the optimised changes can then be combined in one formulation, or a small range of formulations, for trial again in the laboratory.

In the example, at least two sub-series should be planned: The first would study a restricted PVC ladder to isolate the optimum pigment level for the improvement required in gloss and cost, whilst maintaining the best achievable level of opacity. This trial is in fact a further refining of the original PVC compromise selected during preliminary work. The second series would study the influence of increasing quantities of the selected thixotropic alkyd and the most efficient method of incorporation.

The two experiments should then be combined to give a restricted range of alternative formulations for test. During the evaluation of this series, the test schedule should be expanded to a stage where a comprehensive test regime is applied to the final samples.

32.8.1.3 Test Results

At this point it will be assumed that an apparently satisfactory product has been developed. It is necessary to review all previous test results, including stability (which should have been constantly monitored) and exposure series prepared to ensure that the durability of the product is satisfactory. It is of considerable assistance to include artificial weathering tests against a known standard, as in many cases it may be impractical to wait the required exposure period before releasing the new product. Some specifying authorities will require exterior exposure data for the full period, because of the occasionally misleading results obtained from artificial weathering tests. The tint bases required can now be developed, based on the vehicle system used in the white paint; the system will probably need some modification to correct changes in properties due to adjustments in pigmentation level.

32.9 INDEPENDENT TESTING

It is important that the proposed product is evaluated by a third party, to ensure that the formulator has not overlooked some problem (by becoming too closely involved with the development programme).

It is the duty of this tester to highlight potential causes of complaint against the product in field use. This type of testing can reveal faults that the formulator, who has been intimately concerned with the development programme, has not considered. This is no criticism of the formulator, whose brief it was to produce a product that met the specification under conditions of normal application and use; the formulator may fail to recognise that field conditions are not always 'normal' and will have a resistance to mistreating a product so carefully developed.

If at any stage during this test schedule some fault is shown to exist, modification and adjustment should take place along the lines indicated in section 32.8. This procedure may be repeated at any time during the development programme when a problem is indicated.

Until both the formulator and the initial independent tester are satisfied that the product meets the specification under all conditions, no sample should be submitted to outside testing except in unusual circumstances.

32.10 FIELD TESTING

Field testing is the stage at which the product is sampled to an outside tester. In the case of a specialised product, field testing will normally be carried out by the applicator requesting the product. For a more general type of decorative product, testing is best carried out by a 'master painter' customer of the company. However, prior to submission to field testing a further review of the product should be made.

32.10.1 Product Review before Field Testing

This review should be broken down into four parts, in the order: package condition, application properties, initial appearance and serviceability.

32.10.1.1 Package Condition

Does the product appear to be of high quality in the can? Is it of well-bodied appearance? Is separation (syneresis) absent? Are there any aspects of the product in the can that may cause customer resistance? Has it gelled, settled, skinned or discoloured?

32.10.1.2 Application Properties

Does the product apply readily by the specified techniques, or is one or more methods likely to give trouble? Difficulty in application can often cause customer resistance to the product, no matter how well formulated it may be.

32.10.1.3 Initial Appearance

Does the product when first applied appear to be of the required quality? This is an extremely subjective judgement and will vary from customer to customer, particularly in the less obvious areas of performance. All immediate apparent causes of concern should be carefully reconsidered.

32.10.1.4 Serviceability

This aspect is really the heart of the matter, although considered last. The formulator should question whether the product meets the specification in all respects. It is not intended to deny the importance of serviceability by placing it last in this review, but the formulator should by now be satisfied that the specification has been fully met. It is a truism that 'first impressions count most', and any adverse comments in the first three review points will build consumer resistance to a product no matter how well it meets the specification in the long run. An unimpressive-looking product which is difficult to apply will certainly be questioned, as these points are apparent immediately the customer applies the paint, whereas a two-year or more durability rarely presents problems directly and solely attributable to the product.

32.10.2 The Field Trial

Where possible, the initial field testing of a new product should be attended by the formulator and carried out by an experienced, trusted customer. The tests should be carried out according to the points raised in the paragraphs above on package condition, application properties and initial appearance—serviceability cannot adequately be tested. The 'trusted customer' should be encouraged to criticise any aspect of product performance deemed inadequate. The formulator should have prepared a list of performance aspects for comment: these will mainly be application and initial properties. Much value can be obtained from unsolicited comments during field trial testing, and the formulator, if he or she is present at the tests, should be prepared to listen without trying to justify any adverse aspects of product performance.

Where possible or relevant, multiple trials should be arranged with independent testers under as many test conditions as possible, and the results and conclusions correlated when all the trials are completed.

32.10.3 Staff Testing

With satisfactory replies from 'trusted customer' testing, the members of the sales and marketing forces should be encouraged to carry out trials of the product themselves, particularly in the case of a product intended for the DIY market. They should be thoroughly briefed upon the results of both laboratory tests and field trials. Comprehensive performance demonstrations should be arranged and full written reports provided if a major product launch is anticipated. The competitive advantages of the product should be stated, and it is equally important that any potential disadvantages are discussed; even at this stage it is not too late to modify the formulation if the sales staff believe the disadvantages to be significant.

32.11 INITIAL PRODUCTION

During production of the initial batches of a new product, close liaison should take place between the formulator and the production staff. Where possible, the formulator should be present during manufacture of first few batches of each formulation, to ensure that no problems arise during scaling-up from the laboratory tó full-scale manufacture. No amount of detailed reporting by production staff is as valuable as the formulator witnessing the problems at firsthand. Repeated supervision should take place of any problem formulation until the formulator is satisfied that the procedures and techniques of manufacture have been made as efficient and simple as possible. Only when this point has been reached should full routine production of the product be allowed.

32.12 MONITORING OF SERVICE PERFORMANCE

After the product has been on sale for six to twelve months, it is a useful exercise for the formulator to carry out investigations into its acceptance. Over protracted periods, faults or problems may arise which were not foreseen from either laboratory or field testing. This information should be canvassed from as many sources as possible, then collated by the formulator to ensure that no definite pattern of complaint has arisen; if it has, the formulator must be prepared to modify the product to eliminate the fault. Continual checking of this type is a valuable source of information and encouragement to a formulator, in ensuring that the products for which he or she is responsible are performing satisfactorily.

32.13 RECORDS

Good records are the most valuable asset any formulating laboratory can possess. Although formulators move from company to company, the original formulating laboratory will 'always' exist. The successors to the formulator will require complete records of all successes and *failures* on any given project. A simple final formulation filed with all the rest of the company's product formulations is inadequate: both the company and the formulator will benefit greatly from good, complete records.

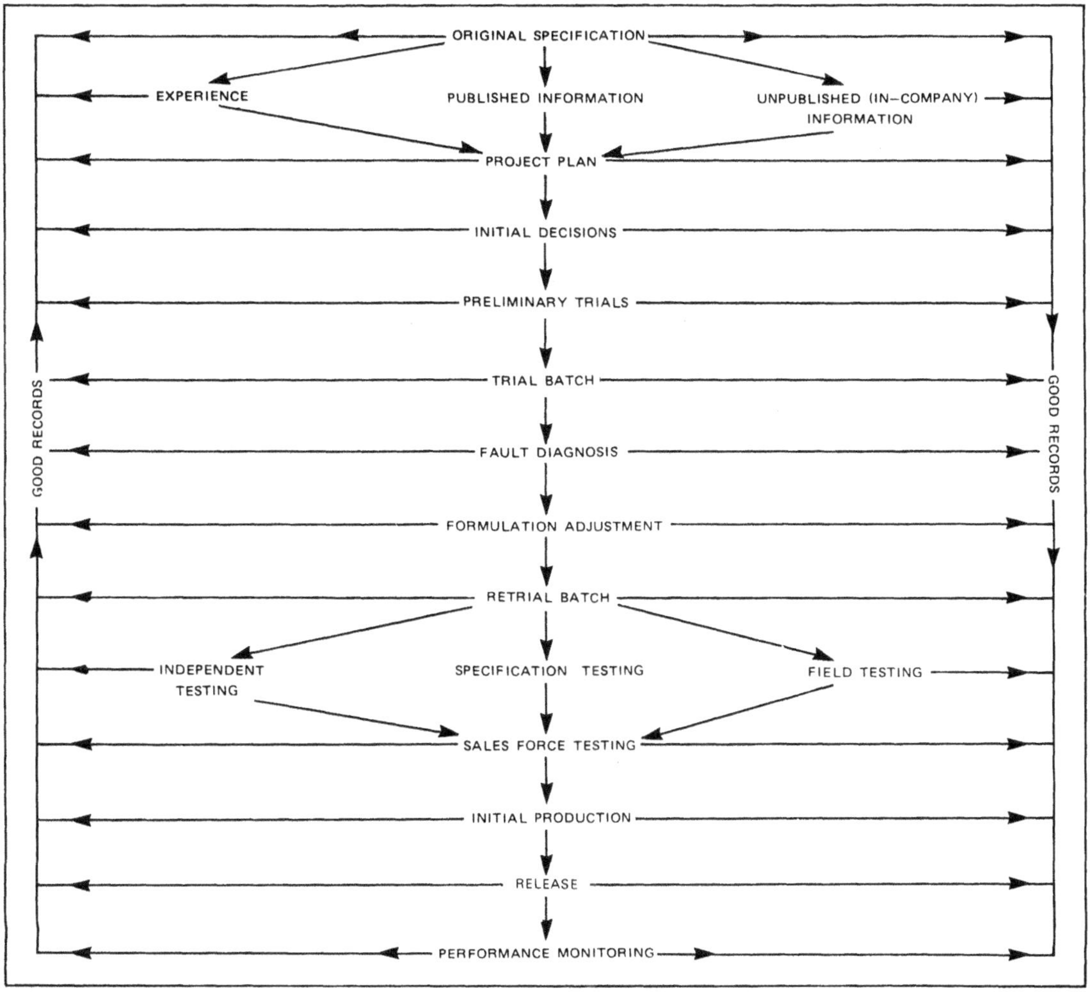

FIGURE 32.7
Project planning network

32.14 SUMMARY

An essential part of any project is to know and understand the logical stages in the plan which will lead, with a minimum of side-tracking, to a satisfactory conclusion. Figure 32.7 identifies the basic steps of a formulation development project, the central vertical progression being the major development route.

However, flexibility in planning is essential; therefore the right-hand vertical progression indicates that any step may be omitted if not completely necessary. It is, for example, quite in order to proceed from the original specification to release without intermediate steps if a product that will meet the specification is already available on the company's inventory. Similarly, having reached the field testing stage and finding the sample to be unsatisfactory, a return to any previous stage may be necessary; this reversal of direction is indicated by the left-hand vertical regression.

The requirements of a successful formulator include experience, knowledge of raw materials and good information sources. One more essential prerequisite remains to be identified—common sense—the ability to recognise the flaws in the plan and realise that the particular approach may not achieve the goal. Without this flexibility in a project leader, many faulty designs are pressed forward and eventually fail; it is highly desirable that corrective modifications are made as early as possible in the development process and that the probable need for such changes is accepted as being unavoidable in a real-life context.

33 PAINT MANUFACTURE: PIGMENT DISPERSION

33.1 INTRODUCTION

The commonly used and incorrect term *pigment grinding* refers to the dispersion of minute pigment particles in a polymeric vehicle; the actual grinding of the pigment is carried out by the pigment manufacturer.

The process of preparing a pigment dispersion in a liquid takes place in steps which, although separate and identifiable, usually occur concurrently. Firstly the vehicle or the liquid phase, wets the pigment, the solid phase. This liquid, as it is wetting the pigment, must firstly displace the entrapped air and must satisfy the *oil absorption* of the solid: that is, the amount of vehicle required to just fill the voids within each of these particles and between the particles. Secondly, the excess vehicle, and there should always be some excess, will wet and coat the surface of the agglomerated particles. At this stage, if nothing else happened, there would be a mass of agglomerated particles wetted with vehicle, all suspended in the excess vehicle.

The second stage is to mechanically separate the prime particles by means of the dispersion equipment (ball mill, bead mill or other high energy milling equipment). Once separated, further extended mechanical mixing, or milling, is required to effect *dispersion*, which is best described as 'the permanent particle separation of the wetted pigment particles in the liquid vehicle'.

To achieve the fine dispersion required for modern surface coatings, a number of different types of equipment have been devised. These will be discussed later in this chapter.

33.2 DISPERSING EQUIPMENT

33.2.1 General Considerations

Almost all mechanical dispersing equipment achieves dispersion by one or more forces: shear; impact, with other pigment particles and/or the grinding medium; or attrition. Generally all dispersion techniques depend on the use of large amounts of energy, either at low levels for long periods of time or at high levels for shorter times, but in either event, of significant importance in the costing of surface coatings formulations.

The efficiency of any dispersing equipment is dependent upon the energy that it can impart to the vehicle/pigment mixture; and it is important to recognise that vehicle/pigment systems must be formulated to meet the requirements of a particular piece of dispersing equipment. In the selection of dispersing equipment there are certain factors which should be defined:

(a) degree of dispersion required,
(b) output (volume in a given time),
(c) manpower requirements,
(d) capital, running and installation costs,
(e) maintenance costs,
(f) size,
(g) noise levels,
(h) power requirements.

Some types of dispersing equipment currently available are:

Ball and pebble mills
Roll mills
Sand, bead or pearl mills
Cavitation mixers
Attritors
Heavy-duty paste mixers
Microflow mills

The aim during the dispersing process is to direct as much power as possible into the wetting and dispersion process and waste as little power as possible in stirring up the millbase, moving the grinding charge and generating heat.

The major factor, under the control of the formulator, which affects power wastage is vehicle viscosity. As the vehicle viscosity is increased, for example by increasing the vehicle solids, more power is used to move the millbase around the mill. This results in the generation of heat and leaves less power available for wetting and dispersion.

When considering the requirements for a stable dispersion, there is therefore a compromise to be struck between high vehicle viscosity, and excessive power utilisation, since a reduction in vehicle viscosity, which is desirable to reduce power wastage, also reduces dispersion stability.

The power of the motor used by the equipment supplier is related to the high 'start-up' load and is not necessarily related to the power used during the actual dispersing process. This usually means that there is excess power available once the dispersing machine is operating.

33.2.2 Ball and Pebble Mills

Historically the most commonly used piece of dispersing equipment was the ball or pebble mill. Basically the ball mill consists of a large steel cylinder, closed at the ends and internally lined with flint, porcelain, rubber or other inert material. Ball mills achieve dispersion by rotating about their longitudinal axis and cause the *charge* (porcelain or other grinding balls, and the pigment/vehicle mix) to cascade over itself for a long period of time.

There are a number of factors to be controlled to efficiently use a ball mill. These are:

(a) ball charge (type, quantity and size),
(b) speed of rotation,
(c) volume of millbase,
(d) millbase and resin solids,
(e) millbase rheology.

Brief notes on each aspect are shown below.

33.2.2.1 Ball Charge

As mentioned earlier the dispersing medium may be selected from a large range of materials; the density is important. The higher the density, the smaller the ball may be and thus the more points of contact (for impact) that are present in the system. The dispersing charge should occupy about 35–55 per cent of the mill volume.

33.2.2.2 Speed of Rotation

The mill should rotate at a speed such that the balls cascade over each other and are not held on the wall of the mill by centrifugal force.

33.2.2.3 Volume of Millbase

This is easily controlled and should be sufficient to just fill the voids in the dispersing charge. As a 'rule of thumb' about 25 per cent of the volume of the mill is usually regarded as appropriate.

33.2.2.4 Millbase and Resin Solids

It is very important, to obtain optimum output, to control both the millbase solids and the solids of the resin in the millbase. Using the *slurry grinding technique* of very low resin solids and high total solids, a rapid dispersion may be obtained. Stabilisation stages must be very tightly controlled in order to prevent flocculation of the pigment. The more normal technique is to maintain a higher resin solids level, with a similar or slightly lower pigment content, and accept a slightly longer dispersing time. Pigment flocculation during let-down is less likely to occur.

33.2.2.5 Millbase Rheology

The rheology of the millbase is important; it should be of low viscosity with Newtonian or near Newtonian flow. Some thixotropy may be tolerated, in fact it could help dispersion, but generally increases the difficulty of unloading. Dilatancy cannot be tolerated as a dilatant system results in poor milling, 'balling up' in the mill and excessive generation of heat. This heat often may cause resin gellation to occur which then makes unloading difficult or even impossible.

33.2.3 Attritors

This class of dispersing machine is a development from ball mills and the attritor may be regarded as a ball mill standing with its axis vertical. The mill charge is moved about in the mill by the use of a vertical shaft carrying 'fingers' at right angles. Rotation of the shaft at about 100 rpm causes movement within the mill charge. The attritor handles millbases similar to those employed in ball mills with slightly higher pigment loadings being tolerated; short grinding times are normal.

As with ball mills, attritors require Newtonian or near Newtonian rheology of the millbases but higher viscosities may be used. Dilatancy cannot be tolerated but slight thixotropy may be an advantage except when unloading.

33.2.4 Roll Mills

Single and triple roll mills are common these days, with the single roll mill used mainly for refining poorly dispersed millbases and the triple roll mill used to disperse 'difficult to disperse' pigments. These are more common in the ink industry.

The single roll mill consists of a roller rotating past a bar such that very fine clearances are available for the pigment paste to pass through. The action of passing from the hopper through the gap into the recovery tank causes very high shear which leads to grinding and dispersion.

The triple roll mill consists of three rollers rotating in alternate opposite directions such that the mill paste is drawn into a very fine gap between the rollers. This occurs twice during a single pass and creates dispersion by the high shear and impact that takes place. A high degree of dispersion may be obtained by multiple passes.

Both these machines handle heavy-bodied pigment pastes with high total and resin solids. Ideally Newtonian flow is best but thixotropic bases may be easily handled. Once again, dilatancy cannot be tolerated because of the high shear involved. For best results, mill-bases should have a degree of 'tackiness'. Roll mills demand a high level of operator skill.

33.2.5 Sand and Bead Mills

Sand and bead mills are logical developments from the ball mill and attritor. The mill consists of a cylindrical pot with a vertical axis and a multi-disc impeller blade. The pre-mixed base is pumped into the bottom of the mill at a rate such that the emerging base at the top is dispersed to the required degree of fineness.

As the names imply, the grinding medium is either sand (usually Ottawa sand), glass beads or porcelain balls, and the machines are designed for either continuous or batch production.

The base forced out at the top of the mill passes through a screen which holds back the grinding medium. The pot is generally cooled with a water jacket (see figure 33.1).

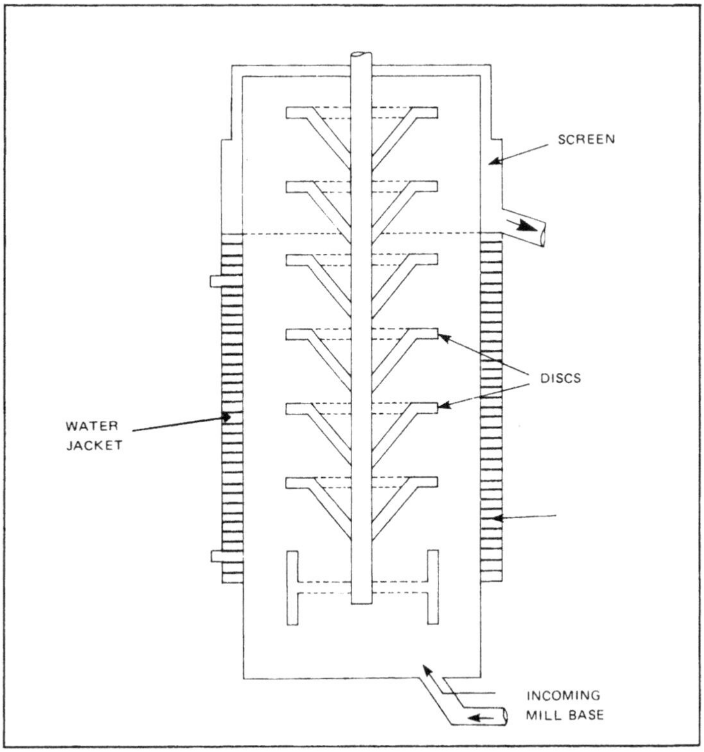

FIGURE 33.1
Cross-section of a typical sand mill

Dispersion occurs due to adjacent layers of the dispersing medium having high differential velocities. The medium travels in both horizontal and vertical layers, as shown in figure 33.2. The shearing force on each pigment particle in the mill paste is an inverse function of the size of the agglomerate involved; that is, the smaller the pigment particle the greater the shearing force applied. This leads to a more efficient dispersing system.

Resin solids, pigment : binder ratio, and grinding medium : millbase ratio are the main variables to be controlled. Generally low to medium levels of resin solids are employed, with high to very high pigment to binder ratios (8 : 1 to 20 : 1 are typical).

The mass of grinding medium in the mill should be at least 80% of the mass of the mill base solids. In general, good dispersing is obtained with a 1 : 1 (by volume) sand to millbase ratio.

Millbase rheology has very little effect on the efficiency of this type of mill except that dilatant millbase should be avoided. Dilatant millbases result in a throwing action in the mill with an accompanying rotational speed variation. Near Newtonian rheology is preferred.

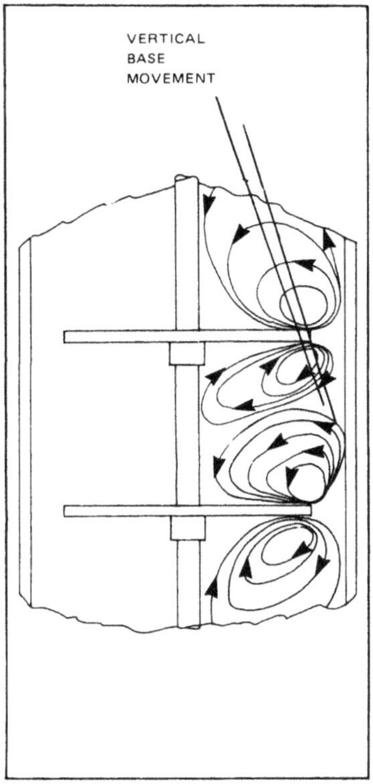

FIGURE 33.2
Movements within a typical sand mill

A recent innovation in this area has been the introduction of what might be termed a 'horizontal bead mill'. This mill constructed with its axis horizontal (as with ball mills) offers the following advantages over the vertical axis bead/sand mills:

(a) the working system is completely enclosed (minimises solvent losses and maximises environmental protection);
(b) higher throughout than comparable size vertical mill;
(c) interchangeable liners;
(d) easy cleaning and short colour change-over time;
(e) high-viscosity millbases can be handled; and

(f) wear rate is considerably less and thus thinner walls may be used resulting in better cooling and heat transfer.

33.2.6 Cavitation Dispersers

These types of mixers are also known as *high speed mixers* or *dissolvers*, and consist of a serrated disc, or rotor, which rotate at high speed in containers of millbase. Dispersion and wetting of the pigment is caused by the rotor imparting a high velocity to particles which, in turn, collide with similar particles moving at slower velocities. This results in a shearing action and impact at the points of contact as well as at the rotor edges.

Control of the dimensions of the various mechanical components is very important with cavitation mixers. Guidelines have been published on a number of occasions which indicate the following:

Impeller diameter	D
Pot diameter	2 to 3 D
Height of impeller above pot base	$\frac{1}{2}$D
Maximum millbase depth when stationary	2D

Rotor centrally placed and rotated such that 1500 m/min peripheral speed is obtained.

The cavitation mixer performs well with a large range of rheologies. Near Newtonian millbases are best, with very slight dilatancy sometimes being an advantage. High dilatancy induces very high impeller wear and heat generation, while high thixotropy reduces efficiency because of the lack of movement near the edge of the pot, away from the shearing forces of the impeller.

Heavy duty paste cavitation mixers have developed from the cavitation mixer. Here higher pigment loadings are handled by offsetting the impeller towards one edge of the pot which is free to rotate about the central axis. This allows complete mixing of very much heavier pastes.

33.2.7 Microflow Mills

Microflow mills achieve dispersion by the action of rods and ploughs rotating against a stationary liner. The dispersion chamber is a stationary, horizontal cylinder. The rotor, inside, carries a series of ploughs around its periphery. These ploughs act as scrapers and move the pigment paste into the rotating chamber where the rods, moving at a different speed to the chamber, disperse the pigment.

The rotational velocity differential between the rods and the ploughs creates high shear conditions. By varying the 'dwell time' in the mill the required degree of dispersion may be controlled.

The millbase generally should be near Newtonian for microflow mill operation. Some slight dilatancy under high shear conditions may be an advantage since heat dissipation is not a problem.

33.2.8 Miscellaneous Mixers

There are many other types of dispersing machines available, and some of them are listed below:

pug mixers,
turbine mixers,
screw mixers,
vibratory ball mills,
colloid mills,
planetary ball mills, and
rotary mixers.

However, these are less common and are being replaced by more efficient, less power-consuming mixers.

34 PAINT MANUFACTURE: PROCESSING OPERATIONS

34.1 GENERAL OBJECTIVES

The manufacture of paint is an increasingly complex and challenging activity. Basically, it is a step-by-step process which must be accomplished within demanding time schedules. The production sequence aims at the efficient conversion of a wide range of raw materials—many themselves the fruits of specialised branches of the chemical industry—into an increasing range of finished products to meet consumers' sophisticated needs. Paint factories therefore need to be organised, with effective communication as well as with systematic procedures, for successful results.

Typical Paint Production Sequence: Schema

Plant:	*Laboratory:*
Raw material delivery	Quality check
Store	
Select	
Weigh	
Mill	Dispersion check
Let down/mix	
Tint	Colour check
	Dry check
Adjust dry time	Gloss check
Adjust gloss	
Solvent addition	Viscosity check
	Density check
	Approval to fill
Strain	Cleanliness check
Fill	Sample retained
	Filing manufacturing card
Label/batch mark	Release of filled material
Warehouse	
Dispatch	

Typical Paint Production Sequence: Detail

(a) Raw materials received and identified; samples passed to the laboratory to be assessed for compliance with specifications. A typical factory would take in more than 1000 individual raw materials. Pigments would probably be the largest single group, comprising approximately 25% of a typical inventory.

(b) Approved materials coded and stacked.
(c) Raw materials, as required, selected from stock against manufacturing card and weighed or dispensed volumetrically in the case of liquids.
(d) Weighed items transferred to mill area by fork lift or trolley, or by hand.
(e) Raw material items loaded into the mill and simultaneously checked against the manufacturing formulation. Frequently materials need to be added slowly in order to achieve full wetting down, particularly if highly pigmented products are used.
(f) After an appropriate grind cycle the mill is sampled and checked, usually by means of a Hegman gauge reading.
(g) If this reading meets the specification, additional grinding medium will be added to the mill. Then the mill will be spun for a period to stabilise the dispersion.
(h) The mill is finally dropped through a coarse grating which retains the ball charge.
(i) The drained charge is pumped or drained to large mixing tanks where additional medium, driers or other additives are incorporated under mechanical agitation and brought to uniformity.
(j) Tinting carried out where required.
(k) When the batch is close to shade, the tank contents will be sampled for colour evaluation, with dry film checks for hardness, gloss, flow.
(l) Additional solvents and/or driers may be added at this stage to meet viscosity and drying specifications.
(m) Final check for viscosity, density, opacity.
(n) The material is strained or filtered immediately before filling.
(o) Cleanliness check for freedom from oversize or foreign matter.
(p) The product is gravimetrically or volumetrically filled into cans, then labelled and warehoused.

34.2 PRODUCTION MANUFACTURING CARDS

An important function of the laboratory is to issue manufacturing cards to the plant. Each batch of paint manufactured is covered by a batch number and a manufacturing card, which forms a permanent record of the details of the materials used, the various processes in the manufacture and the results of the tests carried out before releasing the batch for filling.

34.2.1 Issue

The usual system for issuing a manufacturing card to the plant involves the transfer of a 'master formula' to the manufacturing card. The master formula may need adjustment before issue to the plant; some of the matters that might need to be taken into consideration include:

(a) The suitability of the batch size for the milling equipment, pot or tank available. In a plant making industrial paints, the size of a batch is often dictated by the size of the customer's order, and considerable adjustment may be necessary to suit the equipment available.
(b) Adjustment of the formula to cover variations in raw materials. These might be variations detected during raw material testing, or could be substitutions due, for example, to shortages or strikes, and will often be temporary changes.
(c) Adjustment of the formula to allow the re-work of unwanted or slow-moving stock or the re-work of out-of-specification material.
(d) When scaling up a laboratory or pilot scale formulation to plant production, an attempt should be made to rationalise the formula to reduce time and wastage. This can be done

by using multiples of full bags of powders, or full drums of resins, or by taking account of the weighing or volumetric measuring equipment available. In general, the simpler the operation, the less the chance of costly errors.

34.2.2 Manufacturing Instructions

As well as details of the formula to be used, the manufacturing card should contain full manufacturing instructions. These will usually include:

(a) the type of mill or mixer to be used;
(b) pre-wash instructions for mill or other equipment, taking into account the previous batch manufactured;
(c) the required milling or mixing time, or degree of dispersion required;
(d) millbase stabilising requirements;
(e) size or type of pot or tank to be used for the let-down and tinting stages;
(f) mill, pump and line wash instructions;
(g) special batch requirements—for example, lead-free tinters, and
(h) health and safety precautions which need to be observed in handling any of the raw materials used.

During manufacture, the manufacturing card is used to record the addition of all items added to the batch. The results of testing carried out before approval for filling of the batch are also recorded on the manufacturing card.

34.2.3 After Manufacture

After filling the batch, the actual filled yield is compared with the theoretical yield of the manufacturing card and any discrepancies are reconciled, if possible. An assessment is then made of how the 'actual' formula compares with the 'master' formula and any necessary changes can be made to the master formula.

The manufacturing card is then used for stock control and batch costing, to be filed in the laboratory for reference for future production or in investigation of complaints.

34.3 RAW MATERIALS

34.3.1 Classification

The very large range of raw materials used in a typical paint manufacturing operation can be subdivided numerically as shown by the following list.

It can be seen that in a typical operation pigments will usually form the largest group of individual items because of the variety of colours and types needed to manufacture a comprehensive range of modern surface coatings.

Pigments	27%
Extenders	8%
Binders (resins and emulsions)	14%
Additives	17%
Solvents	9%
Chemicals	14%
Oils	9%
Others	2%

34.3.2 Storage

Pigments are usually stacked in a dry area on pallets with access to fork lift trucks. The main problems with storage of pigments are associated with dust from part or broken bags and the quantity of the items to be handled. It is essential to provide for clear, accurate identification by permanent marking of bags or stacks.

Generally, it is good housekeeping practice to store coloured pigments quite separately from white pigments or extenders to reduce the risk of contamination. Even quite minor contamination with certain pigments, such as Toluidine red, can cause serious off-colour or bleeding problems in coatings subsequently manufactured from the contaminated white or extender pigments.

Resins are usually stored in bulk tanks of some tens of thousands of litres, with pipelines and valving, so that the resin can be delivered to the mill or weighing areas within the plant. Care must be taken if volumetric metering is used to ensure that this equipment is maintained within accurate calibration. It is also important to ensure that any resins to be handled in this manner will remain fluid at the lower temperatures likely to be experienced or, alternatively, to provide some form of trace heating on the exterior of the pipe work to hold the temperature of the resin at a point above ambient during cold periods. Usually, bulk storage tends to even batch-to-batch resin manufacturing variations, to some extent because of the automatic blending of batches within this system.

Smaller-volume items are usually maintained in drum stocks. These have the disadvantages of higher handling costs and potential damage due to leaks; consequent spillage, or entrapment of water or other foreign material in the stocked resin can occur. Drum contents must be positively identified, bearing in mind that such containers may be roughly handled. There is greater risk in drum storage than in bulk, necessitating particular care to avoid errors of identity or contamination.

Emulsion polymers require special care in storage because their stability can be adversely affected by storage at either high or low temperatures. Also certain chemicals will set to a solid at low ambient temperatures, becoming difficult to discharge from drums. Ideally, all fluids are best stored at temperatures of about 16–20° C for paint-making purposes.

Solvents are also usually handled in bulk tanks and circulated through pipe systems, both underground and overhead, for ease of dispensing from special manifolds either volumetrically or by mass. Special precautions are required in earthing such systems to avoid the mechanical generation of sparks in areas where solvents are dispensed. Small quantities of solvents held in drums, particularly those of a highly flammable nature, are best stored in racks above ground with a suitable bunding surrounding the base of the racks to contain spillages.

34.4 MANUFACTURING PROCEDURES

Manufacturing instructions must be followed carefully if laboratory or pilot plant formulations are to be reproduced in the factory with consistent success.

Accurate selection of the exact grade of raw material and the specified quality will be more likely if distinctly displayed through good housekeeping, clear identification codes and adequate lighting.

After the right item is selected, an accurate measurement in the factory environment should be in the order of $\frac{1}{2}$–1%.

34.4.1 Measurement by Volume

Goods may be dispensed by volume. In the case of liquids, this method is used for bulk resins and solvents in conjunction with flow meters. Volume is often used in the addition of tinters from variously sized calibrated dippers or buckets. The accuracy of the latter approach is not usually high and becomes even less so if liquid viscosities are high or the liquids aerated.

34.4.2 Measurement by Weighing

Weighing is the most generally applicable and reliable method, provided that scales are well maintained and attention is paid to correct taring.

34.4.3 Measurement of Small Items

Small items needed in the manufacture of typical paint products are often also the most critical and, because they are so small, are apt to be regarded as trivial additions. Therefore it is advisable to provide some means of checking their identification, weighing and addition.

Typical examples are catalytic agents required for polymerisation processes and small silicone additions such as flow promoters. Drier additions need special care, and usually small scales with a high degree of accuracy should be provided or appropriate volume measuring devices that will maintain maximum accuracy with this type of raw material.

34.4.4 Sequence and Rate of Addition

Often the success of the manufacturing operation depends as much on the sequence of raw material addition as on any other single factor. This particularly applies to the addition, and order of addition, of ingredients used in the preparation of synthetic vehicles. It is also important in the manufacture of latex paints. Certainly rates of resin addition, if excessively fast, will have disastrous effects on the condition of dispersion of typical alkyd paints.

34.4.5 Compatibility

As our paint systems grow more complex and the variety of polymers used as binders increases, the risk of materials becoming inadvertently intermixed is increased, with consequences in rejection and material wastage.

Thus in modern factories, care must be taken in scheduling to avoid using mills and pipelines successively with incompatible materials. In this context, extreme importance is attached to the thoroughness of cleaning and flushing between batches milled or pumped.

'Cocktails' are expensive and difficult to deal with in the paint factory. Of course, when paints were virtually all oil-based, some intermingling of items was not of such significance as it is today.

34.5 PRODUCTION COLOUR CONTROL

Colour control can be exercised in several ways and at various stages during the manufacturing operation. It must always be borne in mind that the colour of a paint is arrived at by a subtractive process where the individual pigments absorb portions of the light spectrum falling upon them, reflecting the remainder of the wave lengths. Therefore, the intermixing of pigments will not conform to the type of pattern that is arrived at from the intermixing of coloured lights. Mixing complementary colours will result in greys rather than attractive shades. A full explanation of this phenomenon can be found in any standard text on colour, but it is surprising how often these elementary facts are overlooked when first approaching colour control work. Efficient colour control is achieved using the following techniques:

(a) Controlling the quality, tint strength and consistency of liquid pre-ground tinters. This is the first step that can be taken towards maintaining colour uniformity in finished products.
(b) Enforcing a logical and consistent selection of tinters to be used from the total range of tinters available to achieve each shade economically and with due regard to light fastness and other requirements such as freedom from lead or good alkali resistance.
(c) Avoiding the use of bleeding pigments wherever possible. Such pigments tend to complicate the picture and can produce disastrous failures in certain types of pigmentary mixtures.

(d) Avoiding the use of complementary colours such as, for example, bright red and bright green, which (as observed above) tend to cancel out, making very expensive greys.
(e) Accurately recording all tinter additions. This is useful as a guide in subsequent production of the same shade.

Colour control must, at all times, be appropriate to the end use for the product; the degree of accuracy achieved in matching the standard should be related to this use.

34.5.1 Visual Colour Control Techniques

In order to use fully the amazing discriminatory powers of the human eye, lighting conditions must be adjusted for eye comfort and for maximum visibility. Careful selection of the operator in control of this task must also be ensured.

Such colour selection assessment techniques as Ishihara and others should be used to screen potential colour operators. It is equally important to use adequate size colour standards and panels of the trial colours during checking to avoid eye fatigue and to improve overall results. It is necessary in visual colour control to watch for metameric effects by checking the material being tested under different lighting conditions.

Today the practice of wet colour matching is not regarded seriously, as it does not produce precise matches when used visually. Normally it is essential to dry the material, and a variety of forced drying techniques are used. Some techniques such as the use of infra-red heating can, if used excessively on air-dry products, produce colour drift during drying which will not be duplicated when the product is dried under normal ambient conditions. Drying must not be accelerated in a way that will impair the consistent reproduction of colour.

34.5.2 Electronic Measurement

Colorimeters such as the IDL Color-Eye, Hunterlab and Gardner, and reflectometers such as the G E Hardy are widely used.

Advantages:
(a) These units present numerical results capable of direct comparison, recording and, if necessary, statistical treatment.
(b) The instruments are completely objective.

Disadvantages:
(a) The equipment is fairly expensive and difficult to repair in the event of electronic failures.
(b) The procedure is relatively time consuming compared with a visual examination, especially if no computer equipment is available. But, with a computer, the total colour difference ΔE can be computed in 1 minute, a fact that largely removes this objection.
(c) The method is difficult to apply in certain circumstances. For example, colour standards on curved surfaces will not be readily scanned by most electronic colour measuring equipment, and certain types of paints—metallic paints and paints with glitter pigments—are difficult to apply to this type of equipment. In some instances they are virtually impossible to control accurately other than by visual examination.

34.5.3 Colour Standards

Colour standards should be at least 100 mm square and preferably larger to allow the eye to accommodate comfortably. They should be uniform, flat cards or plates for easy handling and filing. Ideally, colour standards should be prepared in the same gloss level as the product that is to be controlled. They should be prepared in a medium with minimum yellowing characteristics, such as aliphatic two-pack polyurethanes or acrylic latex products; or, alternatively, special non-yellowing laquers are suitable.

Standards should be handled with care by the edges. A domestic refrigerator, for example, can form an ideal filing cabinet, and the lower temperatures tend to retard colour drift, particularly yellowing. Colour standards should also be stored as far as possible from chemical fumes. Storage near any source of heat must be avoided. Frequently, working colour standards should be checked against a master stored for this purpose; they should be prepared in sufficient number so that they may be discarded when they become damaged or soiled. Finally, it is of importance to ensure that a full and solid coat of material is applied when preparing representative panels of the batch so that any colour judgements will be made on an obliterated background.

Ideally, materials should be prepared for comparison in a manner comparable to that which will be employed when the product is used. For example, enamels designed primarily for spray application should be sprayed so that the degree of flooding and other colour changes inherent in the application technique will not be overlooked when final judgement is made.

34.6 QUALITY CONTROL OF PRODUCTION

This topic is covered from a technical point of view in some detail in chapter 47. Here, the subject will be examined from a slightly different, more general viewpoint.

34.6.1 The Need for Testing

Testing must be carried out in order to detect and correct any errors that may occur during manufacture, or deficiencies in formulations that only become apparent when material is manufactured on the large scale. These errors may be among the following:

(a) Weighing or counting errors—for example, in the handling of bags.
(b) Identification errors. A worker may select the wrong grade or type of raw material.
(c) Errors in manufacturing technique—for example, the use of incorrect type or size of stirrer, faulty ball mill charging, the use of damaged filter media, or the selection of the wrong grade of filter.
(d) The use of unspecified tinting material. Any formulation should specify precisely what grade and range of tinters are permitted to be used in the product. If other tinters are employed, the batch will probably be found to be metameric with other batches of the same product, or metameric with colour standards. This can result in a number of field problems, and may give bleeding, poor light fastness or excessive cost in production.

34.6.2 Quality Control as a Preventative of Quality Drift

Unless most positive and careful control of all procedures is instituted, quality will drift. It may become too high, resulting in excessive manufacturing time to meet unnecessarily demanding standards and consequently high costs of production.

Or the standard may drift too low, when a marketable material is no longer produced on a consistent basis. Such drift can be checked by constant re-calibration of quality-control equipment such as viscometers and balances, against reliable standards kept for this purpose alone. Standards at times do, however, need deliberate revision to suit changes in the market or raw materials supplies. Such changes must be given careful consideration before being implemented.

34.6.3 The Control of Cleanliness in the End Product

Test procedures. Many tests call for a draw-down or flow-out on glass, or a practical brush-out, under dust-free conditions. Such tests are difficult to standardise and maintain; also some judgement must come into the final decision. It is often preferred to flow the material down after thinning it slightly. Such a procedure avoids using any instrument on the paint which can either introduce contamination or wipe away the very particles one wants to find.

Panels should be force dried under dust-free conditions before inspection and may be compared with previously dried standards. Such procedures need be applied only to quality finishing coats.

Paint filtration. Many procedures are employed, varying from the use of 355 μm (44 mesh) screens for certain types of priming materials or undercoats of low quality, through to heavy felt and cartridge filters for ultra-fine straining of quality enamels. Passing the material through at least two filter elements is of advantage; the use of mechanical aids for forcing paint through filter bags and similar equipment is mostly deleterious to the finished product. Fine wire meshes are important and with care can handle all but the highest quality finishes.

Contamination of batches by skins, pieces of bags, and dust should be avoided. Lastly, head pressure is used to advantage in filtration but, of course, this helpful pressure is not used in the typical sieve-type strainers.

34.7 FILLING PROCEDURES

The Weights and Measures Act calls for the achievement of certain *standard volumes* (for example, 1 litre). All filling operations must therefore be designed to deliver an accurate volume to the can and, ultimately, to the customer. This can be achieved in several ways.

34.7.1 Volumetric Procedure

The simplest procedure simply fills the container. This may be refined by the use of a small gauge to indicate the correct ullage for a particular-sized container. While this method has the merit of simplicity it lacks potential for precision combined with speed.

Automatic filling by adjustable piston displacement, mechanically driven, is highly accurate and fast if it is properly calibrated by mass when initially set up. The Elgin filler, for example, operates at many times the speed achieved by even the most expert hand filler.

34.7.2 Mass Basis—Filling by Mass

Simple hand filling with counterbalanced weight by the can on the scales is an accurate method which makes allowances for such factors as aeration but is not sufficiently rapid for most applications.

Automatic and semi-automatic filling using an electrical cut-off from a mass principle is a system triggered by the movement of the scale when the correct mass has entered the can. This method is both fast and exceedingly accurate.

34.7.3 Manual versus Automatic Filling

In deciding how a particular batch should be filled, one has to consider both the size of the batch (obviously the larger the batch the more suited to automatic filling) and the number of containers into which the batch will be filled. A batch with a large number of small sizes involved suggests the need for automatic filling. The time saved in automatic filling must be weighed against the additional time normally required to set up automatic machinery before filling can start and also provision made for the additional clean-up time on larger, more complex equipment at the end.

It is essential to provide rapid straining of the batch if automatic equipment is to be used; where straining speeds must be low because of the nature of the material, it is unlikely that automatic equipment can be employed satisfactorily.

34.7.4 Filling Problems and Errors

There is a common tendency to overfill. This can often be to a surprisingly high level, as most cans have appreciable ullages which allow for activities such as tinting in the field.

Underfilling can result from serious aeration of material when presented to filling point, so it is desirable to carry out a weight check to ensure that upon de-aeration the quantity in the can will comply with the normal weights and measures requirements for volume.

Filling errors can result from a lack of batch uniformity, if for some reason the batch is incompletely stirred and not quite homogenous before or during filling. The latter result can be due to unforeseen delays in the filling operation; it may extend over a considerable period of time in large batches.

A failure to seal containers adequately, especially in the case of drums which may be difficult to seal perfectly, can result in either leakage of material or, more often, the passage of air into the head space. This will cause premature skinning, viscosity increase, or even product gellation.

34.8 CONTAINERS

The common container sizes used in the paint and allied industries range from the 200 litre drum, 20, 10, 4 and 1 litre, and then the sub-litre sizes 500, 250 and 100 mL. It is general practice to refer to 20 and 10 L sizes as 'drums' or 'pails', and from 4 L down as 'tins' or 'cans'. There are regulations in transport systems, such as the railways, for the gauge of metal to be used on drums for safe transit and handling of the heavy weights involved. Tins are normally produced in a very light gauge metal.

The tin can be lithographed, with multiple print colours: it may be plain or lined internally. Epoxy-phenolic baked linings are quite popular for the storage of latex products and other materials that tend to attack tinplate on prolonged exposure. The closure itself may be either 'single' or 'double' tight, or screw cap in the case of cans used for such materials as solvents and non-pigmented products; these screw caps may either be in metal or plastic. Drums may have open tops with crimp or clincher-type lids and gaskets, or they may have screw-cap bungs for non-pigmented materials. Some plastic containers are used, particularly for latex paints, although to date they have not been very popular in this country except for paste products.

34.9 STORAGE OF FINISHED PRODUCTS

34.9.1 Temperatures of Storage

For the maximum life of the product, low temperatures above freezing point are desirable. Such temperatures reduce the risk of unwanted side reactions which can lead to poor dry, seediness or viscosity increase; settling tends to be minimised at low temperatures also. Paint viscosity is higher under these conditions, giving better suspension. Very low temperatures, at or below freezing, can adversely affect the condition of paints containing water. Such temperatures frequently result in the emulsion breaking, although steps can be taken by the formulator to reduce this risk, should action be warranted by the intended service conditions. Elevated temperatures (i.e. above 25° C) promote reactions that result in some loss of drying potential with alkyd paints; the risks of settling, skinning or separation are heightened under these conditions.

34.9.2 Stock Rotation

Any well-organised warehousing system will provide for stock rotation to prevent the build-up of pockets of aged stock, otherwise overlooked while fresh stocks pass into the field. Paint has a limited shelf life, and alterations of factors such as can design, label directions or up-dating of formulae can only be fed into the market in the correct sequence if stock rotation is practised. Most paint products are batch numbered or coded for rapid identification and reference.

34.9.3 Precautions against Fire Risks

Many paints constitute a fire risk during manufacture, storage and application. Paint warehouses need to be well set out for easy access. This assists operations and is helpful should an emergency arise. Adequate supplies of extinguishers, fire alarms and trained personnel are essential to the safe warehousing of paint products.

Cleanliness is desirable not only for workers' health and aesthetic aspects but also in the reduction of fire risk from spontaneous combustion. Paints with low flash points need to be labelled according to the relevant Dangerous Goods Act. Such action will protect both customers and those who handle the material in liquid form.

35 ARCHITECTURAL COATINGS

Architectural coatings in this context include those products used by the retail or do-it-yourself (DIY) market and by the trade or master painter.

35.1 COATING TYPES

There is a logical and basic way of dividing up the coating types under the title of architectural coatings:

Solvent-based
Water-based

Each of these can in turn be divided into the components of the final paint system, namely:

Primers
Undercoats
Finish coats

35.2 SOME FORMULATION CONSIDERATIONS FOR ARCHITECTURAL COATINGS

There are some fundamental considerations that should be noted:

(a) Paint technology is not the study of individual products but rather of systems. The formulator must therefore consider the effects of formulation changes on all components of the paint system and not just on one particular product.
(b) The considerations in formulating architectural coatings are different from those for industrial coatings. Because the industrial customer has known plant locations, equipment and stock control, the formulator through the technical representative can have access to application conditions and stock turnover. The architectural coatings formulator is designing products for the trade and the person in the street. There is very little control over stock rotation because of the diversity of paint outlets. One consequence of this is that architectural formulators must design products for longer shelf lives than their industrial counterparts.
(c) There is not just one correct formulation to meet a particular set of requirements. This is what makes paint formulation so interesting and challenging. Different formulators may take different routes and yet still develop a product that meets the original requirements. Naturally, the balance of properties may be different.
(d) It is the formulator's job to develop the best product that meets the requirements at the lowest raw material cost.

(e) The formulator's products must also be designed to be conveniently and reproducibly manufactured using equipment that is available in the plant.

Chapter 32 provides further insight into the philosophy of paint formulation.

35.3 SOLVENT-BASED PRIMERS

A *primer* is the first coat of paint to be applied to a surface. Its main functions include:

(a) providing adhesion to the substrate,
(b) providing good intercoat adhesion for subsequent coats,
(c) regulating moisture movement,
(d) providing corrosion resistance, in the case of primers for metals.

35.3.1 Wood Primers

Wood is a porous, hydrophilic material of relatively poor dimensional stability. It moves substantially with changes in moisture content, and so primers are required which regulate the rate at which moisture enters and leaves the wood and which are able to follow its movements with the minimum of cracking.

Traditionally wood primers were formulated to a pigment volume concentration (PVC) of 35–45 per cent with drying oil as the binder. In more recent years, wood primers utilise mixtures of oil and alkyd resins or, in some cases, alkyd resins as the sole binder(s). The use of alkyd resins substantially speeds up the drying time and thereby permits quicker overcoating. The formulation for a typical wood primer is given in formula 35.1.

FORMULA 35.1
Wood primer

	Kilograms	*Litres*
Rutile titanium dioxide	100.0	24.6
Calcium carbonate	700.0	260.0
Lecithin solution (50%)	16.0	17.5
Anti-settling agent }	3.0	4.3
Methylated spirits }	1.0	
Raw linseed oil	150.0	161.0
Thixotropic alkyd resin (40% NV)	150.0	170.0
Mineral turps	40.0	49.0
Raw linseed oil	65.0	70.0
Mineral turps	24.0	30.0
Medium oil length alkyd resin (55% NV)	150.0	158.0
Cobalt drier (6% Co)	4.0	4.1
Manganese drier (6% Mn)	2.0	2.1
Lead drier (24% Pb)	17.0	14.4
Red oxide tinter	20.0	9.5
Anti-skin solution (25%)	6.0	7.5
Mineral turps	14.7	18.0
	1462.7	1000.0

Method of manufacture. Grind the first section in a porcelain ball mill or alternatively disperse under HSD to 50 μm. Stabilise and wash with the second section. Add the remainder under an efficient stirrer. Adjust viscosity with the final solvent.

Characteristics

Density	1.463 kg/L
Stormer viscosity	64–68 KU
Mass solids	64.4%
Volume solids	31.5%
PVC	68.7%

This formulation should be used only as a guide. If formulating for a tropical climate, for example, one might incorporate fungicidal pigments such as zinc oxide or barium metaborate. The use of zinc oxide in the primer assists in conveying fungal resistance to the total paint system; the level needed to give adequate protection is in the order of 300 kg/1000 L.

The small quantity of red iron oxide is used in the formula to provide a pink colour. This colour has been maintained over the years to keep faith with the older breed of painter who believed a good primer needed to be mixed by oneself and needed to contain a blend of white and red lead pigments. Nowadays no white lead is used for health and ecology reasons. Special primers using red lead are still sold but these must contain special label warnings and are mainly used on ferrous substrates. Some claim that these red lead primers are superior on wood and that this is because of a tight film being formed as a result of soap formation between the red lead and the oil. It is generally found that modern alkyd/oil wood primers, free of lead, perform as well or better on Australian timbers than any red lead primer. The formulation for a typical red lead primer is given in formula 35.2.

FORMULA 35.2
Red lead primer

	Kilograms	*Litres*
Premium red lead	850.0	97.7
Calcium carbonate	400.0	148.1
Clay	150.0	57.7
Lecithin solution (50%)	30.0	33.0
Anti-settling agent	10.0	9.3
Medium oil alkyd resin (55% NV)	120.0	127.0
Raw linseed oil	350.0	375.0
Mineral turps	20.0	24.5
Medium oil alkyd resin (55% NV)	90.0	95.3
Cobalt drier (6% Co)	3.0	3.1
Manganese drier (6% Mn)	3.0	3.1
Lead drier (24% Pb)	14.0	12.0
Anti-skin solution (25%)	6.0	7.1
Mineral turps	5.8	7.1
	2051.8	1000.0

Method of manufacture. Load the first section into a clean porcelain ball mill. Grind to a grind reading of 45–55 μm. Empty the mill and wash with solvent. Add the remainder under efficient stirrer. Adjust viscosity with the final solvent.

Characteristics

Density	2.052 kg/L
Stormer viscosity	78–82 KU
Mass solids	74.8%
Volume solids	42.6%
PVC	71.2%

35.3.2 Metal Primers

Metal primers tend to be formulated for specific metals. In the case of steel the most common primers are referred to as 'red oxide zinc chromate'. The resin systems used are either alkyd or phenolic modified oils. The formulation for a typical metal primer is given in formula 35.3.

FORMULA 35.3
Metal primer

	Kilograms	*Litres*
Natural iron oxide red	200.0	43.6
Zinc chromate	200.0	60.3
Talc	150.0	54.1
Anti-settling agent }	6.0	7.0
Methylated spirits }	3.0	
Calcium drier (6% Ca)	5.0	5.5
Lecithin solution (50%)	6.0	6.6
Phenolic modified alkyd resin (73% NV)	120.0	126.3
Mineral turps	200.0	246.0
Phenolic modified alkyd resin (73% NV)	300.0	316.0
Mineral turps	60.0	73.0
Cobalt drier (6% Co)	2.5	2.6
Lead drier (24% Pb)	12.0	10.1
Mineral turps	29.8	36.4
Anti-skin solution (25%)	10.0	12.5
	1304.3	1000.0

Method of manufacture. Grind the first section in a clean porcelain mill to 40–50 μm. Stabilise in mill with second section resin. Drain. Wash out with solvent. Add driers under stirrer. Adjust viscosity with the solvent. Add anti-skin before filling.

Formula characteristics

Density	1.304 kg/L
Stormer viscosity	70–78 KU
Mass solids	67%
Volume solids	48%
PVC	34.5%

The level of rust-inhibitive pigments such as zinc chromate is important; good primers contain 150 kg/1000 L. Zinc chromate is a reactive pigment, and care must be taken to ensure adequate viscosity stability with the resin system. Another common primer group is referred to as 'yellow zinc chromate'. These tend to be similar to the red oxide versions, without the red oxide pigment, and can be used on steel. They are usually recommended for use on aluminium, where the red oxide type is not recommended because the iron oxide can cause pit corrosion when in contact with aluminium.

Both of the above primers are distinctly coloured and may give opacity problems on overcoating; white or pale-coloured primers are sometimes preferred. White pigments such as zinc phosphate, barium metaborate and calcium molybdate are sometimes used in place of zinc chromate.

For hot-dipped galvanised steel, special primers must be used. These include one- and two-pack etch primers and two-pack zinc oxide–zinc dust products. The formulation for a typical one-pack etch primer is given in formula 35.4.

FORMULA 35.4
One-pack etch primer

	Kilograms	*Litres*
Rutile titanium dioxide	100.0	25.1
Polyvinyl butyral (100%)	100.0	90.9
Methylated spirits	120.0	148.2
Butanol	80.0	99.3
Methylated spirits	200.0	246.5
Butanol	50.0	62.1
Methylated spirits	250.0	308.6
Phosphoric acid (81%) }	12.0	7.3
Water }	12.0	12.0
	924.0	1000.0

Method of manufacture. Load the first section into a clean porcelain mill and grind to less than 12 μm. Unload and wash the mill with solvents. Pre-mix acid and water in a plastic or stainless steel vessel. Add the pre-mix slowly to the millbase with thorough mixing. Do not store in mild steel vessels, as acid would be consumed and poor adhesion of primer may result.

Characteristics

Density	0.924 kg/L
Ford 4 viscosity	40–45 s
Mass solids	21.6%
Volume solids	11.6%
PVC	21.6%

The formulation for a typical zinc oxide–zinc dust type is given in formula 35.5.

Method of manufacture. Grind the first section in a clean porcelain mill to 50 μm maximum. Unload and wash out with solvent. Add the remainder in a pot or tank while stirring. Add anti-skin just before filling and adjust viscosity with the final solvent.

FORMULA 35.5
Two-pack zinc oxide/zinc dust primer

Part A	*Kilograms*	*Litres*
Zinc oxide	450.0	86.7
Barytes	150.0	34.5
Lecithin solution (50%)	10.0	11.0
Blown and bodied linseed oil to give required visc./solids	300.0	320.0
Mineral turps	130.0	160.5
Mineral turps	120.0	148.1
Mineral turps	150.0	185.2
Cobalt drier (6% Co)	2.0	2.1
Lead drier (24% Pb)	8.0	6.6
Manganese drier (6% Mn)	1.5	1.5
Anti-skin solution (25%)	6.0	7.5
Mineral turps	29.4	36.3
	1356.9	1000.0

Characteristics
Density 1.357 kg/L
Viscosity 18–24 s BS B4 cup

Part B
Zinc dust

Finished product. Mix 2 parts Part A with 1 part Part B by mass.

35.4 SOLVENT-BASED UNDERCOATS

It used to be fairly common to market different undercoats specifically for interior use or exterior use. This is not the case today, and most undercoats are designed to be interior/exterior.

Interior applications on timber are usually adequately served by applying undercoat as the first coat, without using wood primer. Interior applications also need good drying, good flow, very good gloss holdout, free sanding and low odour. These properties are obvious, because interior applications are on critical-view areas such as cupboards and doors where gloss and flow of the paint system are important; low odour is required because of the possible need to live in the dwelling while painting is being undertaken.

Exterior applications, on the other hand, need a slightly different balance of properties. Outside, the undercoat must be applied over primer and it needs slower solvent so that brushing properties can be maintained when the sun and wind evaporate the solvent more quickly. It must have good bridging properties across cracks, and above all it must give good durability to the system.

The formulator must balance all these considerations in the interior/exterior undercoat. A typical formulation is given in formula 35.6.

Method of manufacture. Disperse the first section under HSD to 40–50 μm. Add the remainder under stirrer. Add anti-skin just before filling and adjust viscosity with the final solvent.

FORMULA 35.6
Interior/exterior undercoat

	Kilograms	*Litres*
Thixotropic alkyd resin (40% NV)	80.0	90.9
Long oil soya alkyd resin (70% NV)	150.0	156.1
Lecithin solution (50%)	8.0	8.7
Anti-sag gel (8%)	40.0	48.0
Clay	100.0	38.5
Calcium carbonate	400.0	148.2
Rutile titanium dioxide	380.0	95.0
Mineral turps	140.0	172.8
Long oil soya alkyd resin (70% NV)	200.0	208.3
Cobalt drier (6% Co)	2.5	2.6
Lead drier (24% Pb)	12.0	10.1
Calcium drier (6% Ca)	5.5	6.0
Anti-skin solution (25%)	8.0	10.0
White spirit	3.9	4.8
	1529.9	1000.0

Characteristics

Density	1.530 kg/L
Stormer viscosity	73–78 KU
Mass solids	77%
Volume solids	56.3%
PVC	51.8%

Many variations to this formulation are possible, especially with respect to the type and degree of fineness of the extenders.

The trend over the past few years has been to undercoats that do not contain free oil. These undercoats offer much better drying, sanding and overcoating properties. Durability studies show no adverse effects by not using free oil, although there are slight reservations when applied over heavily chalked old finishes.

As discussed for wood primers, special tropical undercoats can be formulated using pigments such as zinc oxide. Zinc oxide is a reactive pigment, and care must always be taken to ensure that the formula does not contain resins with high acid values that can lead to rapid soap formation and excessive viscosity increases on storage.

35.5 SOLVENT-BASED FINISH COATS

35.5.1 Full Gloss Paints

These materials make up a very large and important section of the architectural paint market. It is still customary to market specific interior and specific exterior products, although it is well within the scope of the formulator to provide interior/exterior formulations. White full gloss finishes are usually pigmented to a PVC range of 15–20%. Considerable advances have been made with titanium dioxide pigments, and it is possible to formulate paints with high gloss, good opacity and good exterior durability. Dispersion capabilities are also good, allowing the paints to be manufactured using a high-speed disperser. The formulation for a typical white paint is given in formula 35.7.

FORMULA 35.7
White full gloss paint

	Kilograms	*Litres*
Rutile titanium dioxide	350.0	86.0
Anti-sag gel (8% NV)	35.0	42.0
Lecithin solution (50%)	4.0	4.4
Long oil alkyd resin (70% NV)	75.0	78.9
White spirit	62.7	79.0
Long oil alkyd resin (70% NV)	620.0	645.8
Cobalt drier (6% Co)	2.5	2.5
Lead drier (24% Pb)	14.0	11.7
Calcium drier (6% Ca)	6.0	6.2
Anti-skin solution (25%)	5.0	6.0
White spirit	30.4	37.5
	1204.6	1000.0

Method of manufacture. Pre-mix the vehicle in a clean vessel under HSD. Add the pigment and disperse to less than 12 μm, using solvent as required to maintain a suitable consistency. Add the let-down under an efficient stirrer. Adjust viscosity with the final solvent.

Characteristics

Density	1.26 kg/L
Stormer viscosity	65–68 KU
Mass solids	70.2%
Volume solids	55.9%
PVC	15.6%

The choice of alkyd resin will depend on several factors. Vegetable oil prices will determine whether the resin is based on linseed, soya or sunflower, or blends of these. Straight linseed is not suitable for white interior finishes as it yellows more than the semi-drying types when not directly exposed to light. Sunflower and tall oil may be blended with linseed to balance dry, cost and colour retention.

It is possible to produce lower-cost, lower-quality finishes by using less titanium dioxide and lower resin solids. The latter can be achieved by using alkyd resins with higher reduced viscosities.

Where regulatory laws or natural conditions (such as high sulfur-containing atmospheres) demand either low levels or no lead drier, enamels can be formulated using zirconium as a through-drier. These formulations usually require higher cobalt drier levels to maintain overall drying performance.

The colour ranges available today are extensive. The need to provide a colour service to the customer and yet control stock items has led to the use of tinting systems in preference to large numbers of ready-mix colours. Full gloss enamels form part of these tinting systems, and so the formulator must produce formulae for white and coloured bases.

The white bases are usually identified as *light*, *deep* and *accent*. The titanium dioxide level reduces from light to accent so that deeper colours can be obtained. Tint bases are tint-strength adjusted to a consistent level so that reproducible colours can be obtained. This is done by including a tint-strength test. The formulation for a typical full gloss deep base is given in formula 35.8.

FORMULA 35.8
White full gloss deep tint base

	Kilograms	*Litres*
Rutile titanium dioxide	140.0	35.0
Lecithin solution (50%)	5.0	5.5
Long oil alkyd resin (70% NV)	25.0	26.0
Anti-sag gel (8%)	50.0	59.7
White spirit	20.0	25.5
Long oil alkyd resin (70% NV)	680.0	708.3
Cobalt drier (6% Co)	4.0	4.1
Lead drier (24% Pb)	17.0	14.2
Calcium drier (6% Ca)	8.0	8.5
Anti-skin solution (25%)	6.0	7.5
White base, *or*	33.2	20.0
Long oil alkyd resin (70% NV)	or 19.2	
White spirit	66.9	85.7
	1041.1	1000.0

Method of manufacture. Disperse the first section under HSD or alternatively in porcelain ball mill to less than 12 μm. Let down under efficient stirrer with the second section. Adjust tint strength as shown. Adjust viscosity with the final solvent.

Characteristics (using Long oil alkyd resin)

Density	1.051 kg/L
Stormer viscosity	63–67 KU
Mass solids	64.3%
Volume solids	53.2%
PVC	9.6%

Coloured bases will vary from system to system, but they frequently include ochre or yellow, red, blue and neutral. The formulation for a typical full gloss yellow base is given below.

Method of manufacture. Grind the first section in a clean porcelain ball mill or bead mill. Stabilise and wash out with the second section. Add let-down under an efficient stirrer. Adjust the tinting strength and viscosity with the final section.

Characteristics (using White base)

Density	1.105 kg/L
Stormer viscosity	64–66 KU
Mass solids	57.2%
Volume solids	51.4%
PVC	17.5%

FORMULA 35.9
Yellow full gloss tint base

	Kilograms	*Litres*
Rutile titanium dioxide	140.0	35.0
Organic yellow pigment	35.0	2.3
Calcium drier (6% Ca)	8.0	8.5
Lecithin solution (50%)	5.0	5.5
Long oil alkyd resin (70% NV)	90.0	94.0
White spirit	50.0	61.0
Long oil alkyd resin (70% NV)	70.0	73.0
White spirits	35.0	40.0
Long oil alkyd resin (65% NV)	550.0	573.0
Cobalt drier (6% Co)	4.0	4.1
Lead drier (24% Pb)	17.0	14.2
Calcium drier (6% Ca)	5.0	5.3
Anti-skin solution (25%)	10.0	12.0
White base *or*	33.2	20.0
Yellow base *or*	*or* 17.8	
Long oil alkyd resin (70% NV)	*or* 19.2	
White spirits	42.6	52.1
	1094.8	1000.0

The brightness demanded of the yellow and red bases often requires the use of organic pigments.

The type and quality of pigments used in the yellow and red bases will determine whether they can be used both interior and exterior or interior only. Most organic yellows and reds have high oil absorption and this requires special formulator attention to ensure that the pigment dispersion section works efficiently and that driers are not adsorbed on storage to the point where aged wet paint has seriously impaired drying performance. This latter problem is sometimes compensated for by including a quantity of synergistic drier in the pigment dispersion section. This drier is preferentially adsorbed as a wetting agent and so leaves the active surface and through driers (such as cobalt and lead, respectively) to carry out their drying functions.

35.5.2 Semi-gloss/Satin Enamels

Semi-gloss and satin enamels are basically full gloss finishes with a lower gloss level. They usually have a gloss range of 25–50% at 60°.

It is not suggested that the formulation of these products is very simple. Unless the correct type and level of flatting agent or extender is used, a product that gives a patchy gloss paint on application will result. A typical formulation is given in formula 35.10.

Method of manufacture. Grind the first section in a clean porcelain ball mill or bead mill to a grind reading of 20–30 μm. *Note:* Reproducibility of gloss of product will depend on the consistency of mill loadings and running times. Wash mill and add let-down under stirrer. Adjust gloss and viscosity as shown.

Characteristics (using Long oil alkyd resin)

Density	1.47 kg/L
Stormer viscosity	64–68 KU
Mass solids	75.3%
Volume solids	53.8%
PVC	29.8%

These finishes are used in tinting systems, and the same previous comments apply.

FORMULA 35.10
White semi-gloss/satin paint

	Kilograms	*Litres*
Rutile titanium dioxide	400.0	100.0
Calcium carbonate	350.0	129.6
Thixotropic alkyd resin (42% NV)	75.0	85.0
Long oil alkyd resin (70% NV)	230.0	239.6
Lecithin solution (50%)	20.0	21.8
White spirits	46.1	57.0
White spirits	23.7	30.0
Medium oil alkyd resin (50% NV)	252.0	264.0
Cobalt drier (6% Co)	2.6	2.6
Lead drier (24% Pb)	14.0	11.7
Calcium drier (6% Ca)	5.6	6.0
Long oil alkyd resin *or*	32.6	33.0
Flattening paste (23%)	*or* 37.6	
White spirits	9.1	11.2
Anti-skin solution	7.0	8.5
	1467.7	1000.0

FORMULA 35.11
White flat paint

	Kilograms	*Litres*
Medium oil length alkyd resin (55% NV)	300.0	341.0
Thixotropic alkyd resin (40% NV)	100.0	114.0
Lecithin solution (50%)	10.0	11.0
Anti-sag gel (8%)	50.0	59.7
Calcium drier (6% Ca)	5.0	5.5
Calcium carbonate	200.0	74.1
Rutile titanium dioxide	400.0	100.0
Diatomaceous silica	80.0	34.4
White spirits	30.0	38.2
Medium oil length alkyd resin (55% NV)	126.3	143.6
Cobalt drier (6% Co)	1.5	1.5
Anti-skin solution (25%)	6.0	7.1
White spirit	54.5	69.9
	1363.3	1000.0

35.5.3 Flat Paints

The importance of this group has declined dramatically in the past 15 years, as flat water-based products have grown in popularity. However, they are still used in situations where ambient temperatures are not conductive to the use of latex paints. An example of such a situation is where the ambient temperature is at or below 10° C and the dwelling is unoccupied and unheated. The formulation for a typical flat white paint is given in formula 35.11.

Method of manufacture. Load the vehicle and run HSD at low speed; gradually add pigments. Use all or part of the solvent to maintain correct consistency. Disperse to 55–65 μm. Let down and adjust viscosity.

Characteristics

Density	1.363 kg/L
Stormer viscosity	73–80 KU
Mass solids	70.4%
Volume solids	47.3%
PVC	44.2%

35.5.4 Solvent-based Interior Wood Finishes

The range of products is many and varied; here are a few of the more common clear and semi-transparent types:

Clear finishes. These are available in gloss, satin and matt. The resin is usually alkyd, urethane oil or two-pack urethane. Moisture curing urethane resins are also available. The urethane oils offer the best finish for interior timber but are not as wear resistant as the two-pack urethanes for heavy traffic areas such as floors and stairs. The formulation for a typical clear, gloss wood finish is shown in formula 35.12.

FORMULA 35.12
Clear gloss interior wood finish

	Kilograms	*Litres*
Urethane modified oil (60% NV)	660.0	720.0
White spirit	102.0	130.0
Cobalt drier (6% Co)	2.0	2.2
Lead drier (24% Pb)	3.1	3.3
Anti-skin solution (25%)	10.0	12.0
White spirit	104.0	132.5
	881.1	1000.0

Method of manufacture. Blend. Adjust viscosity with the final solvent.

Characteristics

Density	0.881 kg/L
Ford 4 cup viscosity	16–20 s
Mass solids	44.9%
Volume solids	39.6%

The selection of driers in these formulae is critical to the prevention of wrinkling on overcoating.

Satin and matt versions are usually made by dispersing fine silica into the gloss type under high-speed dispersion. To maintain viscosity and yield, it is not uncommon to use a lower solids urethane oil as the basis for the lower gloss versions.

Semi-transparent stains. These can be of the 'wipe on–wipe off' type or of the type that is applied as a self-finish or overcoated with a clear finish. The pigmentation is chosen so that both the grain and the texture of the timber are apparent. The formulation for a typical self-finish/overcoatable type is given in formula 35.13.

FORMULA 35.13
Semi-transparent interior stain

	Kilograms	*Litres*
Synthetic yellow iron oxide	38.0	9.5
Natural red iron oxide	24.0	5.2
Anti-settling agent	12.0	6.5
Lecithin solution (50%)	1.0	1.1
Suitable long oil alkyd resin (70% NV)	115.0	116.0
Zirconium drier (6% Zr)	1.0	1.2
Mineral turps	120.0	146.0
Mineral turps	20.0	24.4
Mineral turps	20.0	24.4
Suitable alkyd resin (non-yellowing oil) (55% NV)	110.0	116.2
Mineral turps	300.0	366.0
Cobalt drier (6% Co)	1.0	1.1
Manganese drier (6% Mn)	1.0	1.1
Anti-skin solution (25%)	5.0	5.5
Mineral turps	144.2	175.8
Red oxide tinter	trace	trace
Black tinter	trace	trace
	912.2	1000.0

Method of manufacture. Grind the first section in a porcelain ball mill to less than 20 μm. Empty mill and wash out with solvent. Add let-down with thorough stirring. Adjust viscosity, and tint to colour and depth as shown.

Characteristics

Density	0.913 kg/L
Ford 4 cup viscosity	15–18 s
Mass solids	22.2%
Volume solids	15.2%
PVC	9.7%

35.5.5 Solvent-based Exterior Wood Finishes

There are three main categories, namely clear, semi-transparent and opaque.

Clear finishes. These are usually only in gloss for exterior applications. The resin is either alkyd, urethane oil or phenolic-modified oil. The main difference between clear finishes for interior use and those for exterior applications is related to the destructive effects of ultraviolet radiation which degrades both the clear varnish and the surface layer of the wood substrate. It is well established that exterior clears perform much better on a dimensionally and ultraviolet-stable substrate such as stone than they do on timber. In Australia, a clear finish on timber exposed to full sunlight lasts only one or two summers. This life can be extended by the incorporation of UV absorbers which are usually incorporated at a low level (1.5–2%). Care should be taken with their selection because, apart from high cost, some types deactivate the driers, leading to poor drying after storage. Different UV absorbers preferentially absorb different wavelengths of ultraviolet radiation. It is for this reason that the UV absorber should be matched to the particular resin so that it can be protected from the ultraviolet wavelengths to which it is most vulnerable. This procedure involves practical exposure assessment on local timbers. The formulation for a typical exterior clear finish is given in formula 35.14.

FORMULA 35.14
Clear gloss exterior wood finish

	Kilograms	*Litres*
Long oil linseed alkyd resin (70% NV)	664.8	707.2
UV absorber	17.3	13.9
White spirits	163.2	208.2
White spirits	15.2	19.4
Cobalt drier (6% Co)	4.0	4.2
Lead drier (24% Pb)	16.0	13.6
Calcium drier (6% Ca)	5.0	5.5
Anti-skin solution (25%)	5.5	6.5
White spirits	16.8	21.5
	907.8	1000.0

Method of manufacture. Disperse the first section in a thoroughly clean bead mill or porcelain ball mill to less than 5 μm. Wash out equipment, and let down under stirrer. Adjust viscosity with the final solvent.

Formula characteristics

Density	0.908 kg/L
BS B4 cup viscosity	70–80 s
Mass solids	51.7%
Volume solids	46.4%

Semi-transparent stains. Like their interior counterparts, these stains are formulated to reveal both the grain and the texture of the timber. The pigment level, although low, acts as a UV screen and so these finishes do not suffer from the total film delamination problems on exposure, which is characteristic of clear finishes.

Stains based on unmodified vegetable oils as the binder are quite susceptible to mould growth, but they display less gloss variation on timber than stains on straight alkyd resins. For this reason, many stains are based on a blend of oil/alkyd to balance the best and worst features of each. Some stains of this type contain paraffin waxes to provide extra water resistance, which helps to prevent the wooden substrate becoming damp and thus lowers the incidence of mould growth. On the negative side, the use of wax slows drying, makes recoating with conventional paint difficult and can contribute to pigment settling. Besides water resistance, waxes are sometimes used as matting agents to ensure uniform gloss on substrates of different porosity.

Preservatives are often incorporated but these are to prevent mould growing on the resin in the stain, not to prevent it attacking the wood cells. The stains protect the wood by screening out UV radiation and regulating moisture movement, not by carrying wood-protecting preservatives.

The formulation for a typical exterior semi-transparent wood stain finish is given in formula 35.15.

FORMULA 35.15
Semi-transparent exterior stain

	Kilograms	*Litres*
Vegetable black	20.0	9.09
Lecithin solution (50%)	2.0	2.20
Long oil alkyd resin (70% NV)	50.0	52.08
Anti-sag gel (8%)	50.0	59.52
Mineral turps	30.0	36.59
Boiled linseed oil	160.0	170.27
Medium oil alkyd resin (60% NV)	100.0	104.17
Fungicide	4.5	3.87
Cobalt drier (6% Co)	4.0	4.17
Lead drier (24% Pb)	16.0	13.56
Anti-skin solution (25%)	6.0	7.14
Mineral turps	234.0	300.0
Mineral turps	185.12	237.34
	861.62	1000.0

Method of manufacture. Grind the first section in a bead mill to 10–15 μm. Rinse equipment with portions of oil and solvent, and let down under efficient stirring. Add the final solvent and record viscosity.

Characteristics

Density	0.862 kg/L
Ford 4 cup viscosity	15–18 s
Mass solids	30.8%
Volume solids	28.5%
PVC	3.3%

Opaque stains. These are fairly heavily pigmented and allow the texture of the timber to show but not the grain. They can be described as a low-viscosity, low-solids version of a flat paint and are formulated similarly. This group has been substantially replaced by the water-based opaque stains.

35.5.6 Aerosol Finishes

Many of the primers, undercoats and finishes discussed above are also available in aerosol or spray packs. Special formulation considerations are necessary for the aerosol version.

In an aerosol can, it is necessary to incorporate a significant amount of propellant in order that the paint be expelled from the can and atomised. In the can, the propellant is a liquid and is a very weak solvent for most resin systems. The formulator must select resins that have good tolerance to aliphatic solvents and should dilute the basic paint concentrate with very strong solvents (toluol, ketones) so that the overall solvent strength in the aerosol can, after addition of the weak solvent propellant, will be as high as possible.

The relatively low viscosity of the paint demands that the formula have good pigment suspension and minimal settling, because settling will only be redispersed by shaking with the glass or metal balls that are included in the aerosol can.

The formulation for a typical aerosol undercoat is given in formula 35.16.

FORMULA 35.16
Aerosol undercoat

	Kilograms	*Litres*
Rutile titanium dioxide	250.0	65.0
Talc	150.0	52.0
Whiting	130.0	47.0
Lecithin solution (50%)	16.2	16.9
Anti-settling agent	5.5	3.0
Calcium drier (5% Ca)	3.0	3.2
Suitable short oil quick-drying alkyd resin (50% NV)	145.0	156.0
Toluol	200.0	230.0
Suitable short oil quick-drying alkyd resin (50% NV)	100.0	107.6
Toluol	20.0	22.9
Suitable short oil quick-drying alkyd resin (50% NV)	200.0	185.0
Cobalt drier (6% Co)	2.5	2.6
Lead drier (24% Pb)	3.0	2.8
Anti-skin solution (25%)	5.0	6.0
Toluol	87.1	100.0
	1317.3	1000.0

Method of manufacture. Load the first section into a clean porcelain ball mill and disperse to 26–40 μm. Wash and let down as shown. Adjust viscosity with the final solvent.

Formula characteristics

Density	1.317 kg/L
Ford 4 cup viscosity	25–30 s
Mass solids	57.2%
Volume solids	35.7%
PVC	45.9%

35.6 WATER-BASED PRIMERS

35.6.1 Wood Primers

(The reader should refer to the initial comments in section 35.3.1, which include the considerations and requirements of a wood primer.)

Water-based wood primers usually have an acrylic emulsion as their principal binder, and there are acrylic emulsions that are ideally suited for timber because of their adhesion and excellent flexibility. Emulsions are thermoplastic and do not dry by oxidation. This is in contrast to the solvent-based oils and alkyd resins which dry by oxidation and continue to oxidise for the rest of their life, leading to embrittlement and loss of flexibility. Thus the water-based primers are more able to follow the movement of a dimensionally unstable substrate such as timber, with a minimum of cracking.

Some timbers contain natural tannins which are water soluble. This is a problem for water-based systems (using a water-based wood primer), as unsightly brown stains will appear in all coats. Special additives are needed in the primer. The current materials that suppress tannin staining are zinc oxide, barium metaborate and special acrylic emulsions which fix the tannin in the first coat. There are newer emulsions being developed which suppress the staining physically by forming very tight films.

Some formulators choose to include some vegetable oil or alkyd resin into the formulation to provide better adhesion, particularly to chalky or powdery surfaces. The formulation for a water-based wood primer is given in formula 35.17

FORMULA 35.17
Water-based wood primer

	Kilograms	*Litres*
Water	50.0	50.0
Anionic dispersant (25% NV)	18.0	16.98
Non-ionic wetting agent	2.0	1.94
Antifoam/defoamer	2.0	2.25
Rutile titanium dioxide	150.0	37.50
Calcium carbonate	50.0	18.52
Zinc oxide	50.0	9.52
Water	20.0	20.0
Long oil alkyd resin (70% NV)	40.0	40.0
Acrylic emulsion (46%) with suitable tannin fixing or suppression	615.0	580.0
Antifoam	3.0	3.37
Propylene glycol	30.0	28.85
Coalescent	10.0	10.53
Mercurial preservative (10% Hg)	2.0	2.05
Cellulosic thickener solution (3.5% NV)	100.0	98.04
Amine for pH adjustment	2.0	2.22
Cellulosic thickener solution or water	78.23	78.23
	1222.23	1000.0

Method of manufacture. Disperse the first section to 40–50 μm. Empty mill and wash out. Add alkyd resin under a high-speed disperser in order to emulsify. Add the remainder in order with efficient stirring. Adjust pH and viscosity.

Characteristics

Density	1.222 kg/L
Stormer viscosity	65–70 KU
Mass solids	45.9%
Volume solids	34.1%
PVC	17.2%
pH	9

FORMULA 35.18
Water-based red oxide/zinc chromate primer

	Kilograms	*Litres*
Synthetic iron oxide	210.0	45.0
Extender (barytes)	280.0	63.0
Zinc chromate	60.0	18.0
Surface active agent (non-ionic)	4.0	3.8
Sodium nitrite	2.0	1.0
Defoamer	1.0	1.2
Preservative (mercurial type)	0.5	0.2
Antifoam	2.0	1.7
Dispersing agent	3.0	2.3
Wetting agent	1.0	0.4
Water	160.0	160.0
Water	13.6	13.5
Suitable acrylic emulsion (40% NV)	620.0	580.0
Defoamer	1.0	1.2
Thickener (3% NV)	60.0	58.0
Surface active agent (anionic)	1.5	1.7
Thickener (3%) or water	49.0	49.0
	1468.6	1000.0

Method of manufacture. Grind the first section in a porcelain ball mill to 20–30 μm. Empty mill and wash out with water. Add let-down slowly while stirring. Adjust viscosity.

Characteristics

Density	1.469 kg/L
Stormer viscosity	69–75 KU
Mass solids	56.9%
Volume solids	37.1%
PVC	34%

35.6.2 Metal Primers

Water-based metal primers have been growing in popularity over the past decade and the majority are now based on acrylic or styrene acrylic emulsions. One of the most difficult problems with these formulations is to obtain consistently good adhesion, especially where greasy or oily metal is involved.

Other problems such as flash rusting have been overcome by emulsion modification or the use of additives (such as sodium nitrite) or slower evaporating amines (rather than ammonia).

Inhibitive pigments are necessary on steel to prevent rusting and preferred on galvanised substrates to minimise formation of white corrosion product and blistering. Zinc chromate is still probably the best pigment for corrosion inhibition, but it does have toxicity and colour disadvantages. Other white inhibitive pigments such as zinc phosphate, barium metaborate and calcium molybdate are growing in popularity.

Water-based metal primers can exhibit several total paint system adhesion problems. Some acrylic-based primers display poor adhesion to metal when overcoated with solvent-based alkyd enamels. Some styrene-acrylic-based primers have good metal adhesion but very poor intercoat adhesion to solvent-based alkyd enamels.

The formulations of two examples of water-based metal primers are given in formulas 35.18 and 35.19.

FORMULA 35.19
Water-based galvanised iron primer

	Kilograms	*Litres*
Zinc phosphate	67.0	22.5
Rutile titanium dioxide	200.0	50.0
Extender (barytes)	273.0	63.0
Antifoam	2.0	2.2
Dispersing agent	7.7	7.0
Ammonia solution	2.6	1.9
Preservative (mercurial type)	0.5	0.3
Surface active agent (non-ionic)	3.2	2.9
Propylene glycol	17.0	16.2
Wetting agent	1.0	0.5
Sodium nitrite	1.2	0.7
Water	162.0	162.0
Styrene acrylic emulsion (50% NV)	586.0	571.4
Ammonia solution	2.5	2.8
Thickener solution (3% NV)	51.0	50.0
Coalescing agent	15.7	16.2
Propylene glycol	17.0	16.2
Water or thickener (3% NV)	10.0	14.2
	1419.4	1000.0

Method of manufacture. Grind the first section in a porcelain ball mill to 25–35 μm. Unload and wash with water. Add the remainder slowly with mixing. Adjust viscosity and pH.

Characteristics

Density	1.419 kg/L
Stormer viscosity	64–68 KU
Mass solids	58.7%
Volume solids	41.6%
PVC	32.6%

35.7 WATER-BASED UNDERCOATS

(Readers are referred to section 35.4 for the general requirements of an undercoat.)

Emulsion-based undercoats have substantial advantages over their solvent-based counterparts, in that they are touch-dry in less than 1 hour (compared to 2 to 8 hours), they are recoatable in 2 hours (compared to 16 to 24 hours), application equipment can be cleaned up with water instead of turps, and they are non-yellowing. Their drawbacks for interior use are poorer flow and gloss holdout, and for exterior use, inferior adhesion to chalky or powdery surfaces. Sanding properties may also be poorer.

Newer water-solubilised resins, which in many cases are colloidal dispersions as opposed to emulsions, offer better flow but slower drying and recoating times. Formulators should consider blending emulsions with the water-solubilised resins to optimise performance. Long oil alkyds are often emulsified into emulsion systems to provide better adhesion to chalky and powdery surfaces.

To optimise gloss holdout, the pigment and extenders used must be ground or dispersed to a fine state; it is usual to specify a grind reading of 12–20 μm. Finer grind readings mean that there is more surface area of pigment of extender to satisfy with dispersing agent; the formulator must allow for this to ensure good stability on storage. Undercoats must be formulated below critical PVC to reduce film porosity.

The most commonly used emulsions are vinyl/acrylic and acrylic. The acrylics offer advantages under certain conditions but raw material cost is higher on a solids basis.

The formulation for a typical water-based undercoat is given in formula 35.20.

FORMULA 35.20
Water-based undercoat

	Kilograms	*Litres*
Rutile titanium dioxide	300.0	75.0
Calcium carbonate	280.0	103.7
Non-ionic wetting agent	3.5	3.4
Anionic dispersant	3.0	2.27
Polyphosphate dispersant	2.0	0.80
Preservative/fungicide (mercurial type)	0.4	0.15
Antifoam/defoamer	1.5	1.65
Ammonia solution	1.0	1.11
Water	200.0	200.0
Water	40.0	40.0
Propylene glycol	30.0	28.85
Coalescent	12.0	12.63
Suitable acrylic emulsion (46% NV)	424.0	400.0
Antifoam	1.0	1.12
Cellulosic thickener (3.5% NV)	50.0	49.02
Ammonia solution	1.0	1.11
Cellulosic thickener or water	79.19	79.19
	1428.59	1000.0

Method of manufacture. Disperse the first section in a clean porcelain ball mill or bead mill to 15–20 μm. Empty mill and wash out with water. Add the remainder slowly under efficient stirring. Adjust pH and viscosity.

Characteristics

Density	1.429 kg/L
Stormer viscosity	75–80 KU
Mass solids	54.3%
Volume solids	35.1%
PVC	50.9%

35.8 WATER-BASED FINISH COATS

35.8.1 Interior Flats Finishes

In the retail market, water-based flat paints have almost entirely ousted their solvent-based counterparts. They are based on a range of emulsions—PVA homopolymer, PVA-acrylic, acrylic and styrene-acrylic. There are other combinations such as styrene-butadiene, and PVA-VeoVa$^+$ which are not very common in this country. Newer emulsions utilising vinyl acetate with other internally plasticising monomers are being examined for economic reasons.

The properties of the various emulsion binders are reviewed in detail in vol. 1, chapters 17 to 21. In general terms:

(a) PVA-acrylics are the most efficient, all-purpose type.
(b) Acrylics have better scrub resistance and wet adhesion.
(c) Styrene-acrylics have better pigment binding capacity and the best alkali resistance.
(d) Paints based on PVA homopolymers have poor scrub resistance, excellent brushing, and are usually restricted to finishes such as ceiling paints.

Formulations for interior flat latex paints have changed considerably over the past decade from a PVC of 45–50 per cent, to so-called 'dry hide' formulations with PVC's in the order of 60 to 75 per cent. These higher PVC's give flatter finishes with less side-sheen; they give dramatically improved opacity without significant increase in raw material cost but generally at the expense of washability. Along with this trend to 'dry hide' formulae, manufacturers have developed special grades of rutile titanium dioxide which have much higher levels of inorganic material encapsulating the base particle, typically TiO_2–84 per cent, coating 16 per cent. This heavier layer spaces the pigment particles more efficiently in the paint film and leads to improved opacity.

These 'dry hide' formulae must have pigment loadings that exceed the critical PVC of the paint. Air is entrapped into the dry film and this substantially adds to the opacity of the dry film by creating additional interfaces of high refractive index difference.

The formulation for a typical PVA-acrylic emulsion-based interior wall paint is given in formula 35.21.

Method of manufacture. Grind the first section under HSD to 50–60 μm. Add the remainder in order with efficient stirring. Adjust viscosity and pH.

Formula characteristics

Density	1.492 kg/L
Stormer viscosity	75–85 KU
Mass solids	52.5%
Volume solids	32.8%
PVC	57%

FORMULA 35.21
Water-based interior flat paint

	Kilograms	*Litres*
Water	280.0	280.0
Cellulosic thickener	1.5	1.0
Anionic dispersant	3.0	2.2
Wetting agent	2.5	2.4
Antifoam	1.0	1.1
Ammonia solution	1.5	1.6
Preservative (mercurial type)	0.5	0.2
Talc	200.0	73.5
Diatomaceous silica	75.0	32.0
Rutile titanium dioxide (special grade for high PVC latex paints)	300.0	81.0
Coalescing agent	10.0	10.5
PVA acrylic emulsion (55% NV)	308.0	280.3
Antifoam	1.5	1.6
Ammonia solution	1.5	1.6
Cellulosic thickener solution (3% NV)	150.0	148.0
Cellulosic thickener solution or water	83	83
	1419.0	1000.0

An acrylic emulsion could be substituted in this formulation on a solids basis. If a styrene-acrylic emulsion were considered, however, it could not be substituted on a solid–solid basis and maintain opacity. The greater binding capacity of the styrene acrylic resin reduces the 'dry hide' effect, and it would not be uncommon to use only 60 per cent of the PVA-acrylic solids as styrene acrylic solids to obtain comparable opacity. Other checks on the dry film such as scrubbability would then need to be carried out to ensure comparable performance.

Interior flat latex paints for interior decorating are part of the established tinting systems.

The dry-hide nature of these formulations makes line-to-line colour matching more difficult (this refers to the match between suitably adjusted flat and gloss bases when using the same tint formula). A colour tinted from a dry-hide formula has a very milky look when viewed at low angles (85°) and appears to have an excessive TiO_2 content.

Conventional dry-hide flats can suffer poor scrub and burnish resistance. In heavy wear areas, products with a higher gloss level such as low-sheen 'vinyls' and satins are often preferred. Special extenders are also used in the more expensive paints of this type.

35.8.2 Exterior Flat Finishes

The 'dry hide' formulations are not ideally suited to full exterior applications. They have PVC's above the critical PVC and this means that the dry paint film is under-bound. Full exterior exposure of a dry-hide style formula results in rapid chalking and frequently in film checking. These formulae are satisfactory for sheltered exterior exposures such as under eaves and in situations where a self-cleaning action in desired.

The preferred PVC for an exterior flat latex paint is 45–50 per cent. Conventional premium titanium dioxide is used in conjunction with extenders such as talc, calcium carbonate and mica. As a general rule the finer the particle size of the extender, the more rapid the chalking. Mica is sometimes used because of its plate-like, interlocking structure, usually in conjunction with other extenders. Both talc and mica have been shown to increase the chalking rate over a similar formula based on calcium carbonate as sole extender; each formulation must be carefully tailored to the PVC and emulsion type being used.

If special conditions warrant the use of zinc oxide as a fungicide, the dispersing system must be adjusted to ensure can stability.

The formulation for a typical exterior flat finish is given in formula 35.22.

FORMULA 35.22
Water-based exterior flat paint

	Kilograms	*Litres*
Water	130.0	130.0
Anionic dispersant	4.0	3.03
Non-ionic wetting agent	3.0	2.90
Polyphosphate dispersant	2.5	1.00
Preservative/fungicide (10% Hg)	0.6	0.22
Antifoam/defoamer	1.0	1.12
Ammonia solution	1.0	1.11
Calcium carbonate	50.0	18.52
Mica	40.0	14.29
Rutile titanium dioxide	350.0	87.50
Suitable acrylic emulsion (46% NV)	583.0	550.0
Antifoam/defoamer	1.0	1.12
Propylene glycol	30.0	28.85
Coalescent	15.0	15.79
Cellulosic thickener solution (3.5% NV)	100.0	98.04
Water or cellulosic thickener	45.4	45.40
Ammonia solution	1.0	1.11
	1357.5	1000.0

Method of manufacture. Load the vehicle into clean vessel. Add pigments slowly under HSD. Disperse to 40–50 μm. Add the remainder in order with stirring. Adjust viscosity and pH.

Characteristics

Density	1.357 kg/L
Stormer viscosity	75–80 KU
Mass solids	52.2%
Volume solids	35.7%
PVC	33.7%

The choice of emulsion will vary with the required balance of cost and performance. A PVA-acrylic emulsion will give reasonable performance although an acrylic will probably perform better. The formulator must ascertain whether the extra cost of the acrylic is justified in the particular situation. These finishes are formulated on emulsions that have minimum film forming temperatures of about 7–10° C.

Exterior flat finishes also form part of a total tint system and suitable white and coloured bases must be formulated. Coloured pigments chosen for use in such tint bases must have suitable light-fastness properties. The formulation for typical exterior ochre base is given in formula 35.23.

FORMULA 35.23
Water-based ochre tint base

	Kilograms	*Litres*
Synthetic yellow iron oxide	100.0	24.39
Mica	40.0	14.29
Calcium carbonate	150.0	55.56
Anionic dispersant	4.0	3.03
Non-ionic wetting agent	3.0	2.90
Antifoam/defoamer	1.0	1.12
Polyphosphate dispersant	1.5	0.60
Ammonia solution	1.0	1.11
Water	160.0	160.0
Water	20.0	20.0
Suitable acrylic emulsion (46% NV)	551.0	520.0
Antifoam/defoamer	1.0	1.12
Propylene glycol	30.0	28.85
Coalescent	15.0	15.79
Cellulosic thickener solution (3.5% NV)	100.0	98.04
Acrylic emulsion *or* Yellow oxide stainer	trace	trace
Ammonia solution	1.0	1.11
Water or cellulosic thickener	52.09	52.09
	1230.59	1000.00

Method of manufacture. Load the first section into a clean porcelain ball mill or a bead mill. Disperse to 40–50 μm. Empty mill and wash out with water. Add the remainder slowly with efficient stirring. Adjust tint strength, pH and viscosity.

Characteristics

Density	1.231 kg/L
Stormer viscosity	62–66 KU
Mass solids	44.2%
Volume solids	30.2%
PVC	31.2%
pH	9

35.8.3 Interior and Exterior Low-sheen Vinyls, Satins and Semi-gloss Finishes

Such products cover the gloss range from 10 to 50 per cent (60°). The interior and exterior formulations vary only in the colours available, and this aspect is dependent on the light fastness of the organic pigments used, either in the tint base or the colorant.

These products have formulations where the PVC range is 25–40 per cent. Conventional, premium titanium dioxides are used, and high levels of pigment are needed to achieve an opacity that is still below that provided by the interior 'dry hide' flat type. The formulation for a typical white formula is given in formula 35.24.

FORMULA 35.24
Water-based interior satin paint

	Kilograms	*Litres*
Rutile titanium dioxide	350.0	87.50
Calcium carbonate	100.0	37.04
Anionic dispersant	4.0	3.03
Non-ionic wetting agent	3.0	2.91
Polyphosphate dispersant	2.0	0.81
Antifoam/defoamer	1.0	1.12
Preservative/fungicide (10% Hg)	0.4	0.15
Ammonia solution	1.0	1.11
Water	100.0	100.00
PVA-acrylic emulsion (55% NV)	550.0	499.64
Propylene glycol	30.0	28.85
Coalescent	15.0	15.79
Antifoam/defoamer	2.0	2.24
Cellulosic thickener solution (3.5% NV)	102.0	100.00
PVA-acrylic emulsion *or*	35.20	19.74
Flattening base	or 21.50	
Cellulosic thickener solution or water	98.96	98.96
Ammonia solution	1.0	1.11
	1395.56	1000.00

Method of manufacture. Disperse the first section under HSD or grind in porcelain ball mill or bead mill to 20–25 μm. Add the remainder with careful, efficient agitation. Adjust gloss, viscosity and pH as shown.

Characteristics (using PVA acrylic emulsion)

Density	1.395 kg/L
Stormer viscosity	70–75 KU
Mass solids	55.3%
Volume solids	38.7%
PVC	32.2%
Gloss (60°)	20–25%

As the gloss level increases, poor flow-out of the product becomes more noticeable. There are some newer emulsions which will provide better flow than the standard PVA-acrylic type.

Higher gloss products are used in preference to the dry-hide interior flats when better washability is required or in 'water' areas such as laundries and bathrooms. Emulsions that contain a wet adhesion promoter can be useful for repaint work in these situations.

As the gloss level of the paint film rises, it may be necessary to formulate on a harder emulsion, since this usually involves a decrease in the level of the reinforcing pigment/extender blend. The dry film becomes more thermoplastic, which gives a greater tendency to exterior dirt retention, sticking of sash window frames, and similar problems.

35.8.4 Interior Gloss Finishes

This is a market area where in the past water-based products have not been able to match the gloss and flow characteristics of the solvent-borne alkyd finish.

Interior gloss finishes are usually required for critical viewing areas such as kitchen cupboards and doors. These areas have traditionally been expected to have deep gloss and excellent flow. Water-based gloss paints using the best flowing emulsions and thickeners and with the highest gloss-producing pigments do not quite match the alkyd-solvent enamel performance. In addition the emulsion-based products are thermoplastic, and so matching surfaces painted with these materials may stick together under hot ambient conditions. Special harder acrylic emulsions are now becoming available for this application, because the interior situation allows the flexibility aspect to be of secondary importance. Development of novel polymeric thickeners has also upgraded gloss and flow considerably.

One approach has been to use water-solubilised resins, either on their own or in conjunction with suitable emulsions. The water-solubilised resins are usually based on short oil alkyds. A high residual acid value is designed into the polymer which can be solubilised in water with the aid of a suitable amine. In many cases these resins are not truly solubilised but are present as very fine or colloidal dispersions. Because these products use vegetable oils, they require driers to convert or oxidise the wet film to a dry, usable condition. Gloss enamels based on water-solubilised resins have good flow and gloss and allow water clean-up but they have drying characteristics more like solvent-borne alkyd paints than the usual emulsion-based product.

Modification of the water-solubilised products with emulsion resins to speed up drying and recoatability is under evaluation. The main problem is to obtain formulations where the materials being blended are compatible, since blends that are not compatible can cause lower gloss in the dry film and/or can stability problems.

A starting formula for an interior gloss paint is given in formula 35.25.

FORMULA 35.25
Water-based interior gloss paint

	Kilograms	*Litres*
Water-reducible soya resin (40% NV)	316.0	295.5
Glycol ether solvent	62.3	67.0
Dispersant	7.5	7.2
Defoamer	5.0	5.9
Rutile titanium dioxide	360.0	89.0
Water-reducible soya resin (40% NV)	137.0	128.5
Propylene glycol	37.4	36.0
White spirit	10.0	11.5
Urethane-modified water-reducible soya resin (42% NV)	362.0	339.0
Water-dispersible cobalt drier (6% Co)	8.0	8.4
Preservative	2.1	2.0
Cellulosic thickener Solution (3% NV)	12.0	10.0
	1319.3	1000.0

Method of manufacture. Disperse the first section to less than 10 μm. Let down with resin, and discharge mill. Wash out mill with solvents. Add the remainder as a pre-mix. Adjust viscosity.

Characteristics

Density	1.319 kg/L
Stormer viscosity	75–80 KU
Mass solids	52.5%
Volume solids	38.2%
PVC	23.3%

FORMULA 35.26
Water-based exterior gloss paint

	Kilograms	*Litres*
Propylene glycol	60.0	57.0
Water	60.0	60.0
Dispersant	3.0	2.8
Antifoam	2.0	2.2
Ammonia solution	1.0	1.3
Rutile titanium dioxide	300.0	79.0
Coalescent	20.0	22.0
Suitable acrylic emulsion (46% NV)	650.0	590.0
Cellulosic thickener solution (3% NV) (grade to optimise flow and gloss)	100.0	97.0
Antifoam	2.0	2.2
Water or thickener solution	86.5	86.5
Ammonia solution	trace	trace
	1284.5	1000.0

Method of manufacture. Disperse the first section under HSD or alternatively ball mill to a grind reading of 12 μm maximum. Add the remainder with efficient but minimum stirring. Adjust viscosity and pH.

Characteristics

Density	1.285 kg/L
Stormer viscosity	63–67 KU
Mass solids	46.6%
Volume solids	32.8%
PVC	24.1%
pH	9

35.8.5 Exterior Gloss Finishes

Water-based gloss finishes for exterior use have achieved good acceptance. These products have superior chalk resistance, flexibility and gloss retention in comparison to their alkyd solvent-borne counterparts. The initial gloss and flow aspects, discussed above, are not as important for exterior applications; in most instances the painted surfaces are not in a 'critical viewing area'. They frequently return useful film life of double that achieved with conventional systems, provided they are applied to suitably prepared/primed surfaces.

Exterior water-based gloss finishes are usually based on emulsion resins, the most popular type being pure acrylic. The products normally have a PVC range of 14–20 per cent. Recent technical improvements have been in the areas of titanium dioxide manufacture, dispersing agent compatibility and the flow of emulsion and thickeners. These have combined to achieve products with higher initial gloss and better flow.

It is very important to choose an emulsion with suitable hardness. At PVC's of 14–20 per cent there is little pigment reinforcement of the dry thermoplastic film and unacceptable dirt pick-up can result if the polymer is too soft. This pick-up may not simply be dirt collected on the surface but rather engrained foreign matter which cannot be easily removed even with scrubbing.

As discussed earlier, the thermoplastic nature of emulsion polymers makes them well suited to maintain flexibility on exposure and so follow the movement of substrates. Caution is advised in repainting old structures with gloss emulsion paints, however. Situations where thick layers of old, embrittled alkyd or oil-based finishes have been used are not a suitable substrate for gloss latex paints. The gloss emulsion paint will adhere strongly to the topcoat and will then expand and contract with the substrate and ambient conditions, whereas the old thick, brittle layers cannot follow the movement of their new topcoat: total film delamination back down to the substrate will occur, and complete stripping and repainting will be necessary.

The formulation for a typical gloss emulsion paint is given in formula 35.26.

35.8.6 Exterior Opaque Timber Finishes

With the resurgence of timber as a feature material for the exterior of buildings, exterior opaque timber finishes (or 'solid stains', as they are sometimes referred to) have become very popular.

The most common resins used in these formulations are the pure acrylic emulsions, although vinyl acetate–dibutyl maleate and vinyl acetate–acrylic emulsions perform quite well. Generally the acrylics outperform the other types for grain crack resistance and adhesion around the wood crack. These advantages may only be apparent on more difficult timbers and after longer exposure periods. Oil or alkyd modification is sometimes used to upgrade adhesion and water resistance.

The opaque finishes are available in a limited range of traditional wood colours such as Maple, Redwood and Mission Brown. These colours utilise oxide pigments such as iron oxide red and yellow. Recently white has also become available. White finishes can show brown tannin stains on certain timbers, and so these formulae must contain special emulsions or pigments, as discussed above.

The formulation of a typical exterior opaque timber finish is given in formula 35.27.

Method of manufacture. Disperse the first section in a clean porcelain ball mill or bead mill to 25–40 μm. Empty mill and wash out. Add let-down slowly with efficient stirring. Adjust pH and viscosity.

Characteristics

Density	1.227 kg/L
Stormer viscosity	70–75 KU
Mass solids	36.2%
Volume solids	30.4%
PVC	14.6%

Further representative formulas for emulsion-based paints will be found in Chapter 21 (Volume 1).

FORMULA 35.27
Water-based exterior opaque timber finish

	Kilograms	*Litres*
Yellow iron oxide	30.0	7.3
Red iron oxide	120.0	26.1
Black iron oxide	50.0	10.9
Ammonia solution	1.0	1.1
Non-ionic wetting agent	3.0	2.9
Calcium carbonate	60.0	22.2
Anionic dispersant	5.0	3.8
Preservative/fungicide	0.7	0.3
Antifoam	1.5	1.6
Water	150.0	150.0
Water	45.0	45.0
Suitable acrylic emulsion (46% NV)	530.0	500.0
Propylene glycol	30.0	28.9
Coalescent	15.0	15.8
Antifoam	1.5	1.6
Cellulosic thickener solution (3.5% NV)	90.0	88.2
Ammonia solution	1.0	1.1
Water/cellulosic thickener solution	93.5	93.1
	1227.2	1000.0

36 HEAVY-DUTY PROTECTIVE COATINGS

36.1 INTRODUCTION

'*Heavy-duty protective coatings*' have been defined as those coatings used to preserve industrial equipment, pipelines, plant and buildings from deterioration by exposure to water, chemicals or weather. They are also generally referred to as 'industrial maintenance coatings'.

The main purpose of these coatings is the protection of iron and steel, but concrete and other structural and construction materials also require protection from the environment. The term '*maintenance coating*' also includes paints used on ships and marine structures (both on- and offshore), mining and mineral processing, oil and gas production and refining, the nuclear industry, in fuel processing and production and power-generation reactors.

Although generally considered to be functional coatings with the emphasis on substrate corrosion control, there is a growing interest in the decorative aspect of heavy-duty protective coatings. Environmental colours are being specified so that equipment may harmonise with the surroundings, and attempts to provide an attractive and interesting work environment are taking place. A recent example in the United States is the use of architectural colours on pulp and paper plants.

In the past, industrial protective coatings have made little impact on the general public. Because of the lack of decorative properties, little notice is taken of the importance to the Australian economy that these coatings have.

Corrosion of plant and equipment costs the Australian economy not thousands but millions of dollars each year. One large mining company in Northern Australia spent $750 000 in 1979 on maintenance painting alone.

Industrial protective coatings are very specialised and are usually designed to protect a substrate from corrosion by a multi-layer system comprised of a primer, intermediate coat and topcoat. There are some exceptions, and a single zinc silicate coating of 75 μm may outperform a typical organic coating system.

Several other aspects are of extreme importance:

(a) the design of the structure,
(b) the surface preparation,
(c) the coating types, and
(d) the method of application.

36.2 THE DESIGN OF THE STRUCTURE

The key to the effectiveness and life of a coated structure under corrosive conditions is its design and construction. This statement was made in 1959, and yet structures are still designed today without considering the effects of corrosion.

In structures and plant where metals are used, the design engineer, the architect and the chemical engineer must all consider the possibility of corrosion. In fact corrosion should be anticipated, its magnitude clearly defined and its potential cost to an organisation estimated. Only then can one aim for the ultimate: a minimum cost, corrosion-free structure.

Whilst working towards this goal, there are commonsense design features to be taken into consideration: most organic (and inorganic) thin film coatings cannot be applied successfully over sharp corners, edges or narrow cracks. Because of the surface tension and shrinkage effects that occur in most coatings during the drying or curing process, these areas are generally thinner. This drawing-back from a sharp edge effect varies in degree with coating type, film composition and application conditions.

Figures 36.1 to 36.4 show examples of typical design problems and solutions (taken from Australian Standard AS 2312–1980 '*Guide to the Protection of Iron and Steel against Exterior Atmospheric Corrosion*').

36.3 SURFACE PREPARATION

36.3.1 Introduction

With surface preparation it is the 'consequence of failure' that should be kept in perspective.

There is a vast difference in the type of surface preparation that is adequate for a coating system on a child's bike to that specified for a bridge or the inside of a liquid storage tank. The 'consequence of failure' may be $10 for the paint system on the bike or $10 000 a day for reblasting and repainting a bridge simply because insufficient attention was paid to surface preparation.

Designers, engineers and architects tend to place the emphasis on the type of coating, which can be a difficult choice, and pay little attention to the surface preparation. It is very easy to include in a specification 'abrasive blast clean to class 2½' and leave it at that. Whilst this degree of surface preparation may be adequate for a conventional primer on new steel, it is completely inadequate for a zinc silicate primer on deeply pitted steel in a marine environment.

Surface preparation is discussed in greater detail in chapters 48 and 49. The following comments indicate some of the major problems associated with industrial maintenance painting.

36.3.2 Mill Scale, Rust and Contamination

The major problem in preparing steel for the subsequent application of surface coating is to obtain a satisfactory degree of cleanliness by the removal of mill scale, rust and associated contaminants.

Although mill scale when new appears to be tightly adherent and harmless, it can cause accelerated galvanic corrosion. Mill scale is the natural product of the hot rolling of steel sheet and sections during their production, and is a complex mixture of ferric oxide (Fe_2O_3) and ferrous oxide (FeO) in the form of loosely connected layers. This multi-layered material is slightly cathodic to steel and a potential difference of 0.2 to 0.3 volts can exist between it and the underlying steel surface. In the presence of an electrolye solution, the 'corrosion circuit' is completed and the steel will corrode and protect the mill scale. Mill scale is also likely to be detached by thermal expansion and contraction, and coatings applied over it may fail prematurely by apparent loss of adhesion. It is therefore essential, in order to ensure the protection of steel as well as to obtain the maximum coating life, to completely remove all mill scale before coatings are applied.

Rust is the other major problem in preparing steel. In most cases it is a moist and spongy mass containing soluble and insoluble salts and hydroxides resulting from the corrosion reactions. This structure, containing ions, will readily allow a flow of electrons, which encourages the formation of further and larger 'corrosion cells'. The hydroxides formed during the reaction will cause the undercutting of paint films, particularly those sensitive to saponification such as oil-based alkyd paints.

Another form of contamination is soluble salts deposited during or immediately after blasting. In an industrial environment sulfide salts may rapidly be deposited on the freshly blasted surface, and in a marine environment, chloride salts may occur. Some specifying authorities recommend wet abrasive blast cleaning to overcome this problem.

If soluble salts remain on the substrate and the subsequent protective coating is subjected to immersed exposure conditions, osmotic blistering will occur.

36.3.3 Abrasive Blast Cleaning

Abrasive blast cleaning of steel can completely remove rust and scale, resulting in a surface of uniform light grey colour, with no shadows or rust spots. However, varying degrees of surface cleanliness can be produced by this method.

Many years ago there was considerable difference of opinion regarding the exact degree of surface preparation required when blast cleaning was specified. In order to provide an agreed standard throughout the industry, the National Association of Corrosion Engineers and the Steel Structures Painting Council in the U.S.A. established committees to define various degrees of blast cleaning and establish suitable standards. These standards were published by the Steel Structures Painting Council (SSPC) and are now widely recognised and referred to as authoritative definitions and specifications for various degrees of blast cleaning and other methods of preparing steel surfaces. Sample panels enclosed in transparent plastic are now available, illustrating the various degrees of blast cleaning.

Standards using photographs as references were prepared by the Swedish Standardisation Commission and are widely used in Europe. The Swedish photographic standards correlate well with the SSPC Specifications, and in the latest edition of the latter there are cross-references to the photographs.

In recent years the Standards Association of Australia has produced various standards for surface preparation as AS 1627, Parts 0 to 10. Part 4 refers to abrasive blast cleaning. An excellent pictorial guide to the appearance of the steel surfaces prior to and consequent on the varying classes of preparation is set out in AS 1627 Part 9.

As well as freedom from residual rust or mill scale, the depth of the blast or roughness of the surface may also be important with some coating systems. The rougher the surface, the greater the physical adhesion afforded to the coating.

36.3.3.1 Wet Abrasive Blast Cleaning

Conventional abrasive blasting procedures use dry abrasive and dry air. In wet blasting, water is injected at the nozzle, or is mixed with sand in the pot. In this process very little dust is produced, but it results in a surface covered with mud which must be washed off with water. Phosphoric acid is frequently added to the wash water to reduce the rate of rusting. Other chemical inhibitors (such as chromates, phosphates or borates) may be added to the water used for blasting and washing. Wet abrasive cleaning may be the only suitable method in some plants where dust would be a problem.

36.3.3.2 Types of Blasting Material

The paint chemist should be aware of the types of abrasive available (as listed in AS 1627 Part 4) and their respective properties.

Sand is cheap and effective, but banned in most states because of silicosis hazards. Copper slag is effective, but should not be used in immersion conditions because a slight deposit of copper would remain on the steel, leading to corrosion cells. Metallic abrasives such as steel or iron shot (and grit or wire) are widely used.

Problem	Typical application	Solution
Backs of double angles cannot be painted and maintained 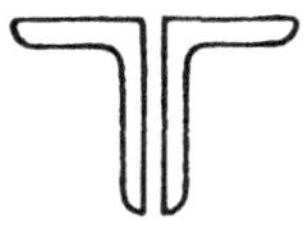	Double angle truss; bearing member	Design as single angle truss, or use T-section 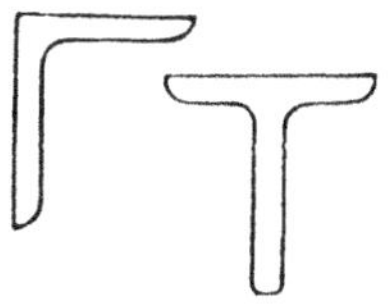
Lower end of angle could collect dirt and moisture 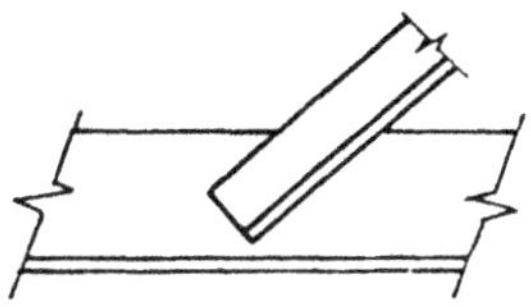	Web member of truss	Invert angle 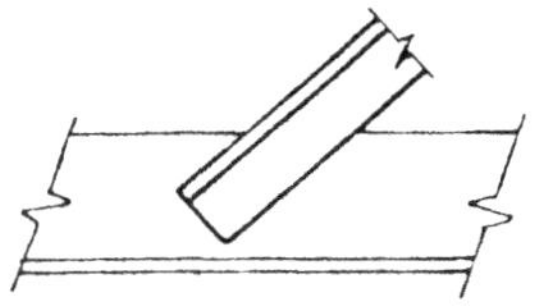
Channels or angles may collect rubbish and moisture 	Sheeting girt; tie or strut member	Invert section 
Sharp corners cause concern about possible loss of protection due to paint draw-back 	Structural members generally	Consider use of circular hollow sections or rectangular hollow sections 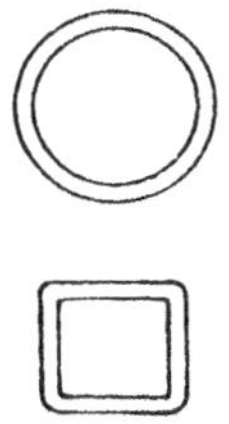

Figure 36.1
Typical Design Problems and Solution

Problem	Typical application	Solution
Gussets cause pockets for dirt and moisture 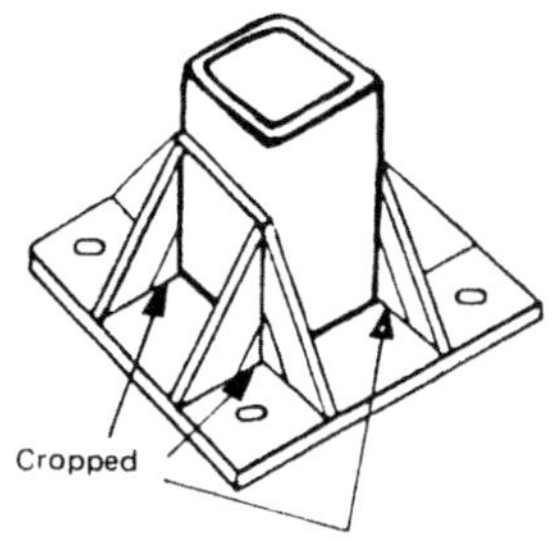	Column baseplate	Design without gussets 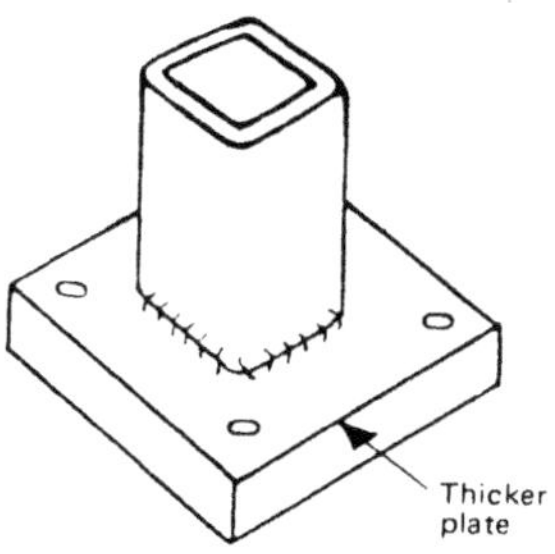
Corrosion point where steel member enters concrete 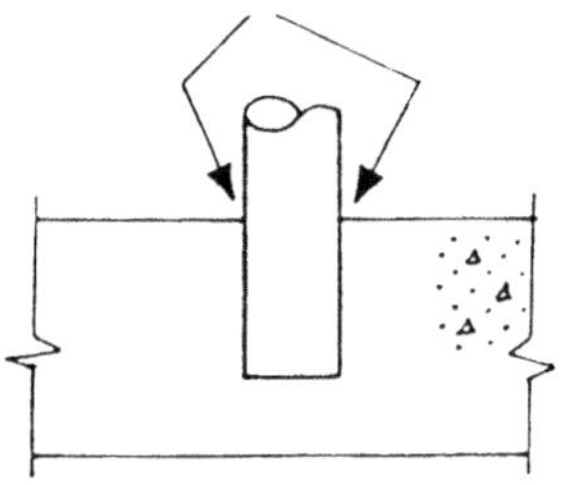	Stanchion for handrail or light structure	Provide mastic sealant coving or sheath 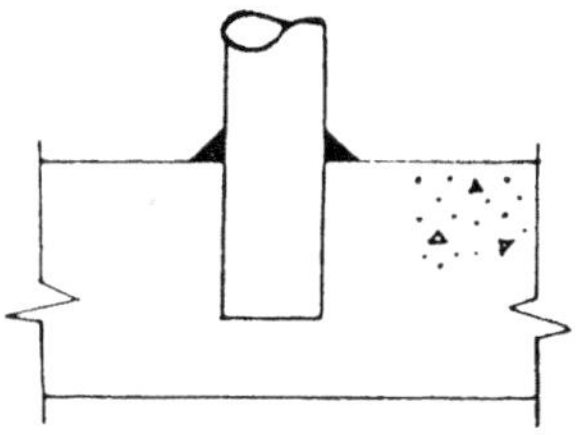
Intermediate web stiffener on plate girder may cause pocket for dirt and moisture 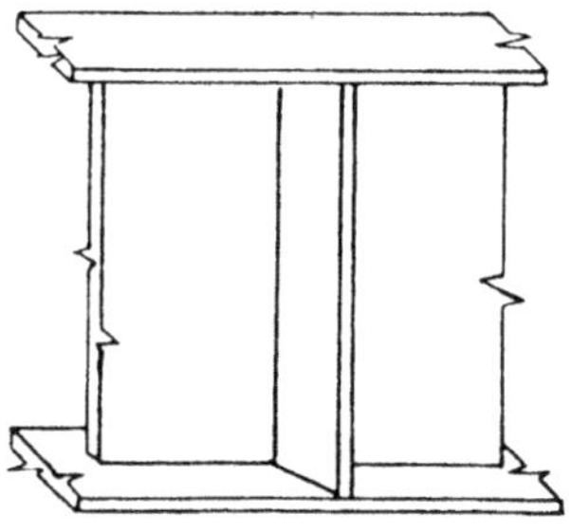	Bridge girder; support beam for machinery or equipment	Stop stiffener short of bottom flange (better and more economic design detail) 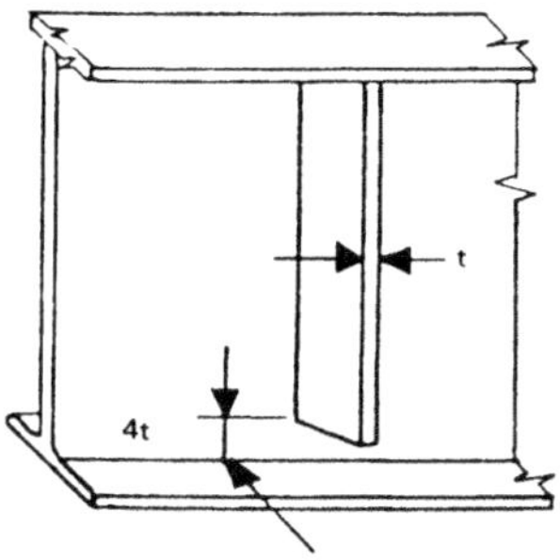

Figure 36.2
Typical Design Problems and Solutions

Problem	Typical application	Solution
Plates in loose contact may encourage capillary ingress of moisture 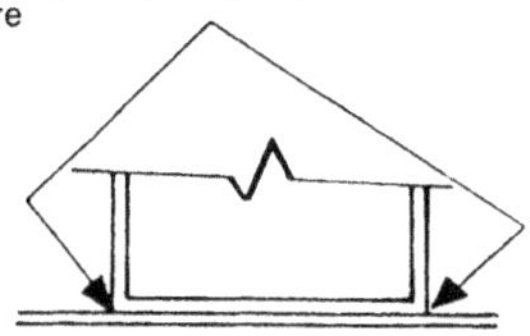	Service tank sitting on steel structure	Locate tank, hoppers on legs 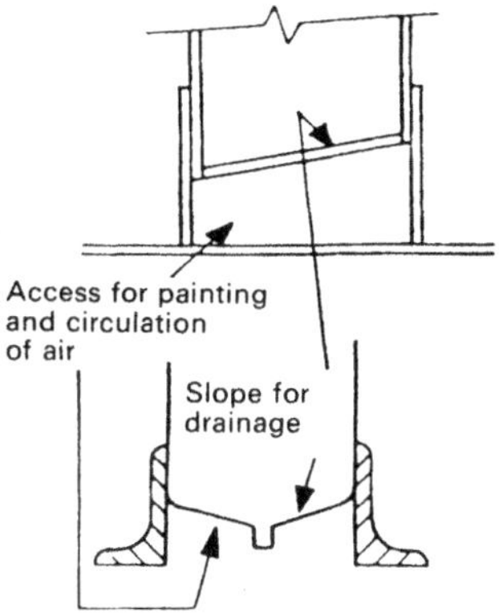
Limited access for maintenance painting 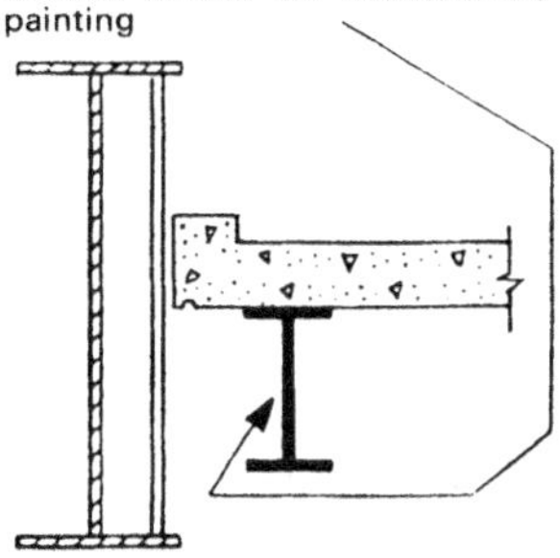	Bridge decking	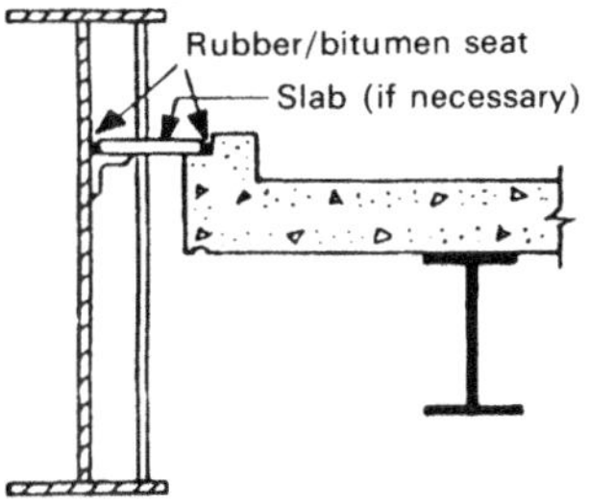
Bolts are difficult to paint properly in the field	Field bolted joints and connections	Use hot-dip galvanized or stainless steel bolts
Strengthening gussets prevent adequate drainage during processing 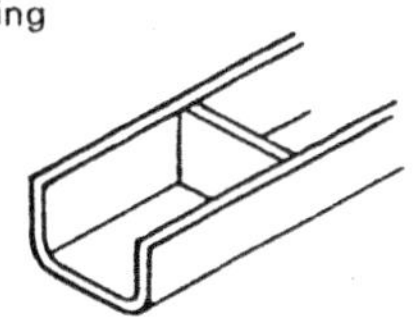	Galvanized members	

Figure 36.3
Typical Design Problems and Solutions

Problem	Typical application	Solution
Lapped joint creates ledge 	Shell of tank, bin or hopper	Arrange joint so that ledge is not on the weather side 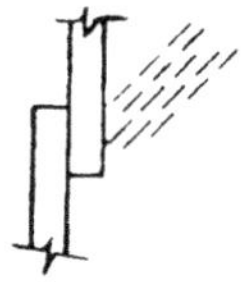
Insides of hollow sections not accessible for painting 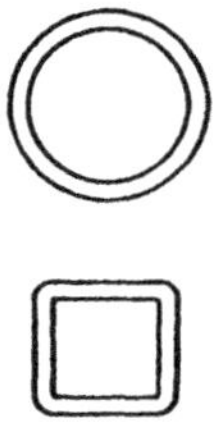	Tubular structures	Prevent entry of air by sealing with suitable welded end plates. These can often be incorporated into the design, e.g. column base and cap plates
Pipe creates corrosion pocket when lying close to flat surface 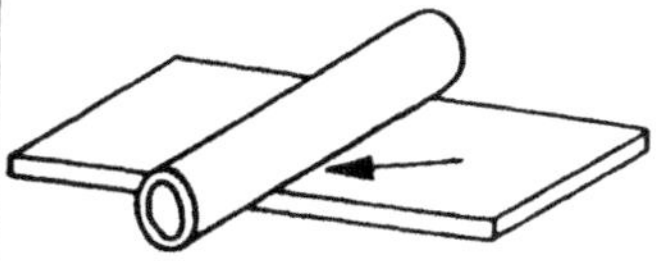	Service pipes running along structural members, tank plating, etc	Keep pipe away from surface 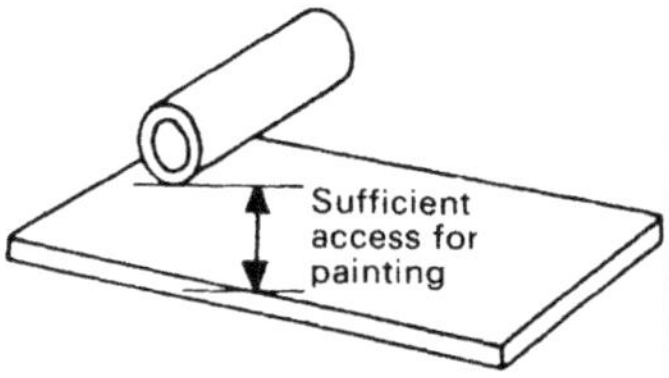 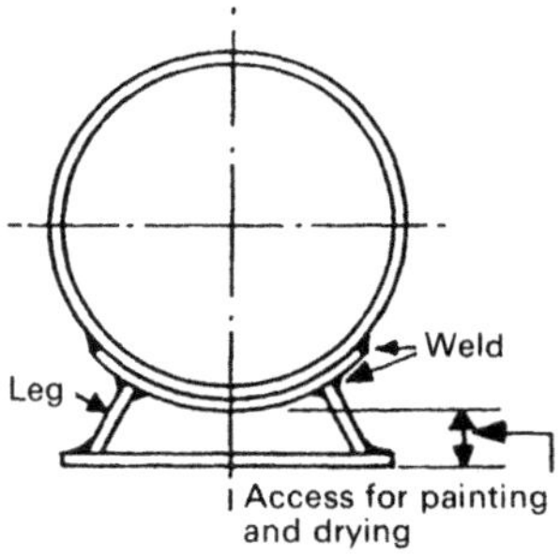

Figure 36.4
Typical Design Problems and Solutions

36.3.4 Other Methods

Although abrasive blast cleaning is the most common process, power tool cleaning and acid pickling are both used, the latter particularly for small shop-fabricated items.

The various methods of surface preparation, in order of decreasing effectiveness, are:

Abrasive blast cleaning	(AS 1627 Part 4)
Pickling	(AS 1627 Part 5)
Flame cleaning	(AS 1627 Part 3)
Power tool cleaning	(AS 1627 Part 2)
Hand cleaning	(AS 1627 Part 7)

36.3.5 Conclusions

As previously indicated, blast cleaning is the best method for preparing steel surfaces. There are circumstances, however, where this procedure is not practical or economical, and a substitute method is desired. In such cases, the most practical method will depend upon the exposure, the type of finish coat to be used and the condition of the existing surface. Each case must be analysed individually.

TABLE 36.1
Preparation methods

	Exposure conditions		
Preparation method	*Tank lining wet*	*Alternate wet and dry*	*Dry, fumes*
White blast Class 3	Ideal	Ideal	Ideal
Blast Class 2½	Compromise	Satisfactory	Satisfactory
Commercial blast	Future problems	Compromise	Compromise
Brush-off blast	Future problems	Future problems	Compromise
Pickling	Compromise	Compromise	Compromise
Power-tool cleaning	Future problems	Future problems	Compromise
Hand cleaning	Future problems	Future problems	Depends on quality of cleaning

36.4 COATING TYPES

36.4.1 Introduction

Unfortunately there is no universal coating or coating system that will satisfy the demands of every environment or set of application conditions.

In industrial protective painting a system approach, using properly selected primer, intermediate and finish coats, usually provides the best answer to the requirements of the problem. Although each coat may be based on the same or similar resin type (an alkyd, for example) there has been a trend in recent years towards mixed 'resin' systems, where each component is formulated on a binder which contributes specific performance and economic benefits to the system (such as a zinc silicate primer, a non-saponifiable intermediate or tie coat, and a decorative and protective topcoat).

This does not imply that all heavy-duty protective systems are necessarily three-coat systems. For example, in the 'deferred topcoat' approach, a single coat of zinc silicate may be used on its own for many months or years, then topcoated with minimum cost to further extend the life of the system. There are obvious economic advantages in this approach.

36.4.2 Primers

Primers can be divided into several classes, each having very distinct and useful properties:

Inhibitive;
Non-inhibitive or non-reactive;
Organic zinc; and
Inorganic zinc.

Irrespective of the type, a primer should meet most if not all of the following requirements:

(a) good adhesion to the surface to be protected, after it has been cleaned or prepared according to specification;
(b) a satisfactory bonding surface for subsequent coats;
(c) the ability to stifle or retard the spread of corrosion from discontinuities such as pinholes, 'holidays' in the film or breaks in the coating system caused by mechanical damage;
(d) sufficient chemical and weather resistance by itself to protect the substrate for a reasonable period (at least the time anticipated before application of subsequent coats in the system); this also implies no adhesion problems when overcoated;
(e) under certain conditions, notably internal coatings for storage tanks, chemical or water resistance equivalent to the remainder of the system.

36.4.2.1 Inhibitive Primers

Inhibitive primers are generally those primers which function by interrupting the reaction at either the anode or the cathode in the corrosion cells that form on ferrous substrates. They are called 'inhibitive' because of the type of pigments used in their formulation.

The oldest, and still considered an excellent pigment, is red lead, Pb_3O_4. Red lead pigment based primers are relatively tolerant of surface preparation and contamination, and give good performance in most environments. The toxicity of lead pigments, and consequent legislation against their use in many countries, prompted the search for alternative inhibitive pigments.

Several metal chromate pigments are excellent inhibitors. The group called 'zinc chromate' and the neutral chromates of strontium and barium fall into this category. 'Zinc chromate' does not exist as such, and the most commonly used anti-corrosive chromate pigments are zinc potassium chromate and zinc tetroxychromate.

The 'zinc chromate' type of pigment functions by releasing the corrosion inhibitive chromate ion when in contact with aqueous permeants. Studies have shown that the particle size of the chromate pigment is most important: decreasing the particle size results in increased protection. Other studies have shown that exposure to chlorine-containing media reduces and may even nullify the protective ability of the pigment. Zinc chromates have a degree of water solubility and are not recommended for primers subjected to immersion conditions. The water-soluble salts lead to osmotic blistering (as indicated under surface preparation).

Chromate type pigments are considered toxic and there is a trend towards alternative inhibitive pigments such as zinc phosphate. It is difficult to explain how zinc phosphate functions: it is virtually insoluble and unreactive but in urethane and chlorinated rubber paints, zinc phosphate is extremely effective.

Other inhibitive pigments include:

Calcium borosilicate	Basic lead silico-chromate
Barium metaborate	Zinc ferrite
Zinc molybdate	Calcium ferrite
Chromium fluoride	'K' pigment.

36.4.2.2 Non-inhibitive Primers

Non-inhibitive primers are usually based on non-reactive water-insoluble pigments, and can be considered as barrier coatings. A typical use for this type of primer is in ship-bottom systems and water tank linings. Such 'pigments' include talc, silica, mica, micaceous iron oxide and aluminium flake.

36.4.2.3 Organic Zinc Primers

Organic zinc primers usually contain a high loading of zinc dust with organic film formers such as epoxy, phenolics or chlorinated rubber. Organic zinc primers have a number of advantages over other primer types:

(a) They have no critical application requirements, and are commonly applied to blast-cleaned steel at 50–75 μm dry film thickness, without problems of cracking or flaking. Being organic and solvent thinned, application by brush, conventional spray or airless spray is usually successful over a wide range of temperatures.
(b) With sufficient zinc dust they offer a degree of galvanic protection, although some zinc ions no doubt contribute by anodic passivation mechanisms.
(c) Many types do not rely on humidity, oxygen or temperature to dry properly.

They do have some disadvantages, particularly when compared with inorganic zinc rich types:

(a) On outdoor exposure they weather in a similar manner to conventional organic-based coatings and frequently have a tendency to chalk, crack and check. For this reason they are generally used in 'systems' rather than being left untopcoated.
(b) They may blister in areas of high humidity (condensation conditions) or if fully immersed in water.
(c) Single-pack products usually have poor solvent resistance.
(d) Their service temperature maximum is usually 150° C.

36.4.2.4 Inorganic Zinc Primers

Inorganic zinc primers are usually referred to as 'zinc silicates'; within this group there are quite significant differences between type and performance. The properties exhibited by inorganic zinc coatings can vary widely depending on how they are formulated and the silicate binder used in the formulation. AS 2105 refers to these types of coatings.

(a) *Alkali metal silicates (water-borne)*
In the manufacture of these coatings, the sodium, potassium, lithium and quaternary ammonium silicates have been used. Alkali silicates are relatively simple chemicals and are available in a range of alkali metal oxide to silica ratios. As well as the contribution of the alkali metal itself, this ratio has a very substantial effect on the curing properties.
   (i) *Baked coatings*. These were the first type of inorganic zinc coatings developed and are based on low $Na_2O : SiO_2$ ratio sodium silicate; this type of silicate was the only one originally available. These coatings were first devloped in Australia. Self-curing tendencies are virtually nil, and heat must be applied to the coating to convert it to a water-insoluble condition. This type of coating is characterised by its extreme hardness and its suitability for application over an acid descaled surface. Baked coatings still have limited use today; their main role is the traditional one of protecting above-ground pipe lines.
   (ii) *Post-cured coatings*. With the advent of higher-ratio sodium silicates (the so-called 'neutral' grade), some degree of self-cure could be obtained, but not reliably. However, it was found that coatings of this type could be converted to a water-insoluble condition by the

post application of an 'acid' wash. This development led to the use of inorganic zinc coatings on large field structures. While post-cured coatings have largely been superseded by self-cured types, they are still in use. Long-term case histories have established them as successful coatings for specific purposes.

(iii) *Alkali silicate self-cured zinc coatings.* With further advances in silicate technology, even higher ratio silicates have become available. Of the cheaper types (potassium and sodium), potassium is preferred; reliable self-curing coatings are available based on high-ratio potassium silicates. If even higher ratios are required and instability is to be avoided, it is necessary to use lithium silicate. This type is also preferred for use in food areas. Excellent curing rates are achieved with some lithium silicates, but at the present time the higher cost tends to restrict use.

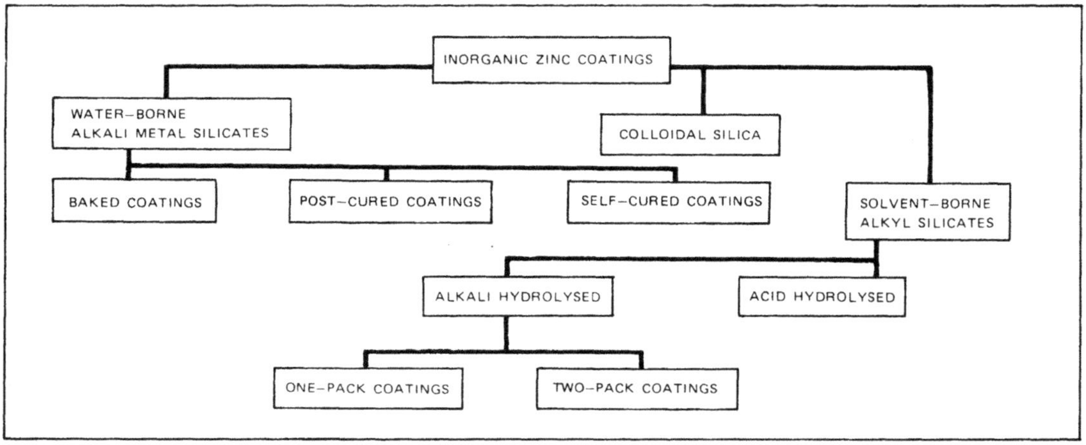

FIGURE 36.5
Classification of inorganic zinc coatings

(b) *Colloidal silica*
Colloidal silica is not normally film forming, but when blended with other reactive materials a coating can be formulated successfully. Unfortunately curing rates are not very reliable due to considerable variation under different weather conditions. When fully cured, this type of coating becomes very hard.

(c) *Alkyl silicates*
Usually one of the commercial forms of ethyl silicate in organic solvent solution is employed. The pH of the silicate is adjusted so that hydrolysis occurs (with the loss of alcohol) when the applied coating is exposed to humid air.

(i) *Acid hydrolysed coatings.* When the pH is below seven, water can be added to the formulation and as hydrolysis has already started, very rapid curing can be achieved under most conditions. However, the period that the partially hydrolysed silicate remains stable is limited, and the product thus has a finite shelf life. The liquid component of these coatings will gel in a period of 6 to 12 months.

(ii) *Alkali hydrolysed coatings.* When the pH is above seven, even a small amount of water will cause the silicate to gel. Steps must therefore be taken to exclude all water at the manufacturing stage, and from application equipment. If water is excluded, the liquid component will remain stable indefinitely.

Two-pack coatings. These were the first types of coatings developed by this method and are the most widely used of the genre.

One-pack coatings. As no reaction occurs in the can, provided water is excluded, it becomes possible to package the powder and liquid together. However, steps must be taken to avoid gassing and settling problems. Several one-pack, alkali hydrolysed, ethyl silicate zinc coatings have recently become available commercially in Australia.

36.4.2.5 The Curing Mechanism of Zinc Silicate Coatings

A simple distinction is that the water-borne alkali silicate coatings lose water during the initial curing stage, whereas the solvent-borne alkyl silicate coatings absorb water, with subsequent release of ethyl alcohol, initially.

As discussed, the principal raw materials used for the vehicle of inorganic zinc coatings are sodium silicate, potassium silicate, lithium silicate, colloidal silica solution and ethyl silicate. Even with all these different basic starting materials, quite similar ultimate reactions occur within the coating (and on the steel surface).

The initial reaction involves the concentration of the silicate/zinc dust film by evaporation of most of the solvent. This provides the initial drying and primary deposition of the coating.

The second reaction is the insolubilization of the silicate matrix by reaction with zinc ions from the surface of the zinc particles and the ferrous ions from the sandblasted steel surface. The vehicle (water soluble or organic silicate) hydrolyses to a form of silicic acid, with which the zinc reacts to form a silica–oxygen–zinc polymer. This is insoluble and forms a strong matrix with zinc powder. At the zinc–silicate steel interface it is believed that an iron silicate–zinc silicate complex is formed. The reaction forms a complex zinc–silicate cement around the zinc particles and an iron–zinc–silicate at the interface of the coating and the steel. This cement matrix is very inert to water, sea water, weather conditions, solvents and many chemicals, including milk acids and salts.

A third reaction taking place over a long period of time (months, or in some cases years) is the combination of the zinc silicate (which can be alkaline) vehicle and carbon dioxide in the air and moisture on the surface. The degree of alkalinity is slowly reduced, and additional silicic acid in the silica polymer allows reaction with any zinc ions still present. The reaction proceeds slowly through the coating and results in an extremely dense and metal-like coating. This curing is very important, since it increases the effectiveness and durability of the inorganic coating with age. Note the contrast with the ageing characteristics of organic coatings.

36.4.2.6 Summary of Advantages of Inorganic Zinc Primers

The advantages of inorganic zinc primers are:

(a) excellent general chemical resistance, compared with metallic zinc;
(b) unaffected by weathering—sunlight, rain, dew, UV, wide ranges in temperature, bacteria, fungus;
(c) provide excellent cathodic protection because of the good metal-to-metal contact of the zinc particles;
(d) as the inorganic binder reacts with the steel substrate chemically, adhesion is excellent;
(e) as a result of this bond, the inorganic silicate coatings form a base or permanent primer which does not undercut or allow underfilm corrosion;
(f) strong film with excellent resistance to wear;
(g) suitable for use in friction grip connection (all other coatings lubricate);
(h) unaffected by gamma rays and neutrons;
(i) heat resistant, even up to 540° C (which is above melting point of zinc);
(j) unaffected by most organic solvents, petrol and other petroleum products (however, they

must not be used for jet fuel as zinc salts can block filters in aircraft fuel lines);
(k) may be welded;
(l) do not shrink on drying or curing (as most organic coatings do) and wet the steel surface well and thus follow surface profile.

36.4.3 Intermediate or Tie-Coats

Many industrial maintenance coating systems are composed of multiple coats of paint. For example, an organic zinc-rich primer, a synthetic resin based tie-coat, a high-build chlorinated rubber build coat, and finally a decorative topcoat of chlorinated rubber may be specified. The reasons for the inclusion in a specification of intermediate and/or tie-coats are:

(a) *Adhesion.* Many high-performance topcoats have marginal adhesion to some types of primers, and a tie-coat between the coatings is essential. For example, unmodified two-component polyurethanes exhibit poor adhesion to inorganic zinc silicate primers. An epoxy-based tie-coat is typically used to provide long-term intercoat adhesion.
(b) *Topcoat problems during application.* Many topcoats 'bubble' during application over zinc rich primers, due mainly to the porosity of the primer. Whilst inorganic zinc coatings are very prone to this, organic zinc primers can exhibit the problem also under certain conditions. Fast-drying topcoats such as vinyls have a high tendency to bubble, and hot windy weather conditions during application accelerate the process. Allowing the zinc primer to weather can reduce the tendency towards bubbling, as can the technique of 'mist coating', the application of an initial thin, dilute film of the topcoat. The most effective method, however, is by the application of a specially designed tie-coat.
(c) *Film thickness and chemical resistance.* Intermediate coatings are included to improve the overall substrate protection offered by a paint system. An example is in the wine industry, where storage tanks are coated with a high-build epoxy for film build and chemical resistance, and topcoated with a thin film of polyurethane for chemical resistance and taintfree performance.

36.4.4 Topcoats

These are the paint surfaces that are exposed to the environment. In multicoat systems such as vinyl water-tank coatings (VR-3 types), the generic type is the same but the composition of each coat may be different. With other systems the topcoat is in fact the primer as well. For example, coal tar epoxy coatings, usually applied directly to blast-cleaned steel to 400 μm dry film thickness, are commonly used in marine environments.

36.4.4.1 Epoxy Coatings

These are based on thermosetting resins and are usually two-component coatings.

The epoxy resins cross-link with an amine-containing chemical. Simple amines are volatile and hazardous, and are often pre-reacted with a stoichiometric equivalent quantity of epoxy resin to form an epoxy-amine adduct.

Polyamide resins are also widely used to cross-link epoxies and are in some respects superior to amines.

Low molecular weight epoxies and polyamide resins are used to form solventless coatings, for use where solvents are hazardous or undesirable.

Water-reducible epoxies are now available in several forms, and are being more widely used as solvent costs increase and environmental legislation becomes stricter.

The main properties of epoxy systems are:

(a) high chemical resistance under a wide range of severe corrosive conditions; they have excellent resistance to alkaline conditions;

(b) good adhesion to a wide range of substrates;
(c) low shrinkage during curing;
(d) good physical properties such as toughness, flexibility and abrasion resistance;
(e) although there are limitations, epoxy resin based coatings are generally superior to thermoplastic coatings for elevated temperature performance.

36.4.4.2 Single-pack Epoxy

Epoxy esters. Prepared by reacting epoxy resins with fatty acids. They provide coatings which air-dry with the assistance of driers; have a high degree of adhesion and flexibility and chemical resistance superior to alkyds.

Thermoplastic systems. Harden purely by solvent evaporation; they are very high molecular weight epoxies dissolved in suitable solvents. Mainly used as binders for zinc.

36.4.4.3 Coal Tar Epoxy Coatings

Two-pack epoxy paints can be modified with suitable grades of tar to produce tar epoxy coatings.

The addition of comparatively low-cost tar to an epoxy system results, generally, in an improvement in water resistance but some loss in chemical and solvent resistance.

They are formulated in a similar manner to normal epoxy coatings and can be applied by most methods, especially airless spray.

36.4.4.4 Vinyl Coatings

Classed as thermoplastic coatings, vinyl resin based coatings form homogeneous and impervious films with excellent toughness and chemical resistance. The film excludes oxygen, water and ions from the surface over which it is applied.

They are generally based on vinyl chloride–vinyl acetate copolymers, which are inert to almost all inorganic acids, alkalies, salts, water, aliphatic hydrocarbons, alcohols and similar materials.

Vinyl coatings are lacquers since the film is formed solely by evaporation of solvents.

They are usually touch dry shortly after application. Water resistance is excellent, and in the U.S.A. some water reservoirs have been coated for twenty years without failure. Proper formulation is critical with these coatings: tests by the U.S. Bureau of Reclamation have shown that some vinyl coatings will remain in perfect condition for long periods while others will fail prematurely.

Vinyl coatings have excellent adhesion to a wide variety of substrates, are tough and flexible, and will withstand tremendous abuse.

These coatings have suffered over the years due to application difficulties and were only available in thin films, thus requiring multiple coats to build a reasonable DFT of, say, 125 μm. These problems have been overcome, however, and 'high build' resins are available which, together with airless spray equipment, produce thick films in one pass. Although these recent developments have reduced the problems, vinyl coatings will not perform adequately unless correctly formulated and carefully applied.

36.4.4.5 Chlorosulfonated Polyethylene

Chlorosulfonated polyethylene coatings, such as Hypalon, resist strong oxidisation from, say, ozone, oxidising acids and hypochlorites. They maintain flexibility and extensibility for an indefinite period, and have good resistance to heat and moisture. Chemical resistance and other properties depend greatly upon the type of curing agent used, and their use, because of application difficulties, has been mainly in special conditions such as fan runners in corrosive atmospheres.

36.4.4.6 Acrylic Coatings

Generally claimed as thermoplastic lacquer coatings, acrylics are characterised by:
(a) transparent water-white films;
(b) good resistance to discolouration from high temperature and UV radiation;
(c) resistance to water, alkali, chemicals;
(d) good electrical properties;
(e) good outdoor durability.

36.4.4.7 Polyurethane Coatings

These are two-pack products consisting of hydroxyl containing resin and an isocyanate component. The exterior durability of the system based on an aliphatic isocyanate is excellent, and is often used in conjunction with epoxy primers to provide lasting decorative properties.

Polyurethanes are notorious for difficulty of overcoating, but acrylic-based improved products are now available. The so-called 'urethane-modified oils' are not used in the field of heavy-duty protective coating.

Polyurethanes are characterised by:
(a) wide range of flexibility and toughness;
(b) high abrasion and impact resistance;
(c) high resistance to chemicals;
(d) excellent electrical properties;
(e) low temperature cure;
(f) excellent weathering characteristics;
(g) high gloss and gloss retention.

36.4.4.8 Chlorinated Rubber Coatings

Chlorinated rubber coatings exhibit good chemical and water resistance, but as a non-convertible coating drying by solvent evaporation they exhibit poor resistance to solvents, animal and vegetable fats and fatty acids.

Adhesion on alkaline substrates, such as concrete, brickwork and asbestos cement is good, and no special priming to prevent saponification is necessary. They are used extensively, particularly when pigmented with micaceous iron oxide, on steel structures in industrial plant or marine environments to resist chemical attack.

Until recently thin-film multicoat application was necessary, but now high build (150–175 μm) versions are available (which do not sag on application).

36.4.4.9 'Bituminous Paints'

These have been used extensively and are probably the oldest type in use. Based on either natural asphalt or processed from crude oil or coal tar, these coatings give a high degree of protection to various substrates when applied either hot or in solvent solution form.

36.4.4.10 Silicones

Silicones will resist temperatures up to 300° C continuously and up to 550° C for short periods. They are used for stacks, furnaces and similar applications. Coatings for 550° C are available in aluminium and black. Silicones offer only moderate chemical resistance.

TABLE 36.2
Lacquer types

	Petroleum hydrocarbon	*Coal tar*	*Chlorinated rubber*	*Chlorinated rubber, oil or alkyd modified*	*Vinyl alkyd*	*Vinyl chloride copolymers*	*Polyvinyl chloride*	*Vinyl–acrylic*	*Acrylic*
Cure type	Lacquer	Lacquer	Lacquer	Drying oil	Lacquer	Lacquer	Lacquer	Lacquer	Lacquer
Effect of sunlight	Minor	Checking	Slow surface chalk	Slow surface chalk	Slow surface chalk	Slow surface chalk	Slow surface chalk	Very slow surface chalk	Prolonged chalk resistance
Wet or humid environments	Excellent	Excellent	Excellent	Fair to good, will yellow	Good, slight yellowing	Excellent	Excellent	Very good	Good
Industrial atmosphere contaminants:									
(a) Acid	Excellent	Excellent	Excellent	Fair to good	Good	Excellent	Outstanding	Very good	Good
(b) Alkali	Excellent	Excellent	Excellent	Fair	Fair	Excellent	Excellent	Very good	Good
(c) Oxidising	Good	Limited	Excellent	Fair	Fair	Excellent	Outstanding	Good	Good
(d) Solvents	Poor	Poor	Limited	Limited	Limited	Limited	Good	Limited	Limited
Spillage & splash of industrial compounds:									
(a) Acids	Good	Good	Very good	Not recommended	Not recommended	Very good	Excellent	Fair	Poor to fair
(b) Alkalis	Good	Good	Very good	Not recommended	Not recommended	Very good	Excellent	Fair	Poor to fair
(c) Oxidising agents	Fair	Not recommended	Good	Not recommended	Not recommended	Good	Excellent	Fair	Poor to fair
(d) Solvents	Not recommended	Not recommended	Not recommended	Not recommended	Not recommended	Not recommended	Limited	Not recommended	Not recommended
Water immersion	Ship bottoms	Submerged and buried pipelines	Ship hulls	Not recommended	Not recommended	As tank lining	As tank lining	Not recommended	Not recommended
Tank linings	Not used	Potable water	Swimming pool	Not used	Not used	Potable water & deionised water, glycols & alcohols, fuels	Caustic solution, fatty acids, alcohols, glycols, animal & vegetable oils	Not used	Not used
Physical properties:									
(a) Abrasion resistance	Poor	Poor	Good	Fair	Fair	Good	Excellent	Good	Good
(b) Heat stability	Softens	Softens	Limited	Fair	Fair	Limited	Limited	Limited	Limited
(c) Hardness	Soft	Poor	Good	Fair-good	Fair	Good	Good	Good	Good
(d) Gloss	None	None	SG to matt	Wide range	Limited range	SG to matt	Gloss to matt	Semi-gloss	Gloss, semi-gloss
(e) Colour	Black & aluminium	Black	Wide range	Full range	Full range	Full range	Most colours	Full range	Full range
Additional notes	Water resistance & ease of repair result in wide use as ship bottom systems		Traditionally low solids & low film build. New high build types becoming available		Useful as primer or topcoat for more chemically resistant vinyl copolymers	Available in high and low build formulas. High build more popular in maintenance applications	Low build. Economic limit to tank lining & other severe exposures	Wide use as finish coat with vinyl systems	Use as finish coat for epoxy systems for improved weathering

TABLE 36.3
Co-reacting and condensation types

	Coal tar epoxy amine or polyamide cure	*Epoxy amine cure*	*Epoxy, polyamide cure*	*Epoxy phenolic*	*Phenolic*	*Urethane moisture cure*	*Urethane, two-pack*
Cure type	Co-reacting	Co-reacting	Co-reacting	Condensation	Condensation	Co-reacting	Co-reacting
Effect of sunlight	Surface chalking	Yellowing & surface chalking	Yellowing & surface chalking	Tank lining	Tank lining	See notes	See notes
Wet or humid environments	Excellent	Very good—may yellow	Very good—may yellow	Tank lining	Tank lining	Very good	Very good
Industrial atmosphere contaminants:							
(a) Acid	Excellent	Good	Good			Good	Very good
(b) Alkali	Excellent	Excellent	Excellent	Tank lining	Tank lining	Good	Excellent
(c) Oxidizing	Limited	Limited	Limited			Poor	Limited
(d) Solvents	Limited	Excellent	Excellent			Excellent	Excellent
Spillage & splash of industrial compounds:				*As tank lining:*	*As tank lining:*		
(a) Acids	Good	Fair	Poor-fair	Some fatty acids	Fair	Fair	Good
(b) Alkalis	Good	Excellent	Excellent	Very good	Not recommended	Not recommended	Not recommended
(c) Oxidising agents	Not recommended	Not recommended	Not recommended	Not recommended	Not recommended	Not recommended	Not recommended
(d) Solvents	Not recommended	Excellent	Very good	Very good	Outstanding	Good	Excellent
Water immersion	Marine pilings, sewage basins	Tanks, sewage basins	Ship hulls	As tank lining	As tank lining	Not recommended	Ship hulls
Tank linings	Marine ballast tanks	Marine cargo/ballast tanks; fuel storage tanks	Marine cargo/ballast tanks; fuel storage tanks	Wide range solvents & foods; caustic cartage	Outstanding for solvents, aqueous solutions & food products	Not used	Marine cargo/ballast tanks; fuel storage tanks
Physical properties:							
(a) Abrasion resistance	Limited	Good	Good		Good	Excellent	Outstanding
(b) Heat stability	Excellent	Good	Good	Outstanding	Excellent	Good	Good
(c) Hardness	Very hard	Very hard	Hard	Very hard	Excellent	Excellent	Excellent
(d) Gloss	None	Wide range	Wide range	Excellent	Excellent	Range	Range
(e) Colour	Black, red	Full range	Full range	Limited	Clear or dark shades	Limited range	Full range
Additional notes	Timing between applications critical in obtaining inter-coat adhesion	Cure rate & pot life affected by temperature (true of all epoxies). Appearance & weathering problems solved by use of acrylic finish coat		Usually hi-bake coating. Lo-bake type has decreased toughness & flexibility	Hi-bake coating. Lining should not be cleaned with alkali solution	Have ability of curing at low temperatures. Urethanes based on aromatic diisocyanates discolour & chalk rapidly in sunlight. Those based on aliphatic diisocyanates have prolonged resistance to yellowing & surface chalking. Care must be taken to avoid moisture contact before cure	

TABLE 36.4
Zinc coatings

	Inorganic zinc, post-cured	*Inorganic zinc, self-cure, water-based*	*Inorganic zinc, self-cure, water-based, ammonium*	*Inorganic zinc, self-cure, solvent-based*	*Organic zinc, one-pack*	*Organic zinc, two-pack*	*Modified inorganic zinc primer*	*Special prime and tie-coats*
Cure type	Inorganic	Inorganic	Inorganic	Hydrolyzable organic silicate	Lacquer	Co-reacting	Co-reacting	Lacquer
Effect of sunlight	Unaffected	Unaffected	Unaffected	Unaffected	Surface chalking	Surface chalking	Very slow chalking	Variable
Wet or humid environments	Outstanding	Outstanding	Outstanding	Outstanding	Very good	Very good	Very good	Usually good
Industrial atmosphere contaminants:								
(a) Acid	Requires topcoat	Requires topcoat	Requires topcoat	Requires topcoat	Requires topcoat	Requires topcoat	Requires topcoat	Requires topcoat
(b) Alkali	Requires topcoat	Requires topcoat	Requires topcoat	Requires topcoat	Requires topcoat	Requires topcoat	Requires topcoat	Requires topcoat
(c) Oxidising	Requires topcoat	Requires topcoat	Requires topcoat	Requires topcoat	Requires topcoat	Requires topcoat	Requires topcoat	Requires topcoat
(d) Solvents	Outstanding	Outstanding	Outstanding	Outstanding	Limited	Excellent	Very good	Limited
Spillage & splash of industrial compounds:								
(a) Acids	Not recommended	Not recommended	Not recommended	Not recommended	Not recommended	Not recommended	Not recommended	See manufacturer's literature
(b) Alkalis	Not recommended	Not recommended	Not recommended	Not recommended	Not recommended	Not recommended	Not recommended	
(c) Oxidising	Not recommended	Not recommended	Not recommended	Not recommended	Not recommended	Not recommended	Not recommended	
(d) Solvents	Outstanding	Outstanding	Outstanding	Outstanding	Limited	Very good	Good	
Water immersion	With suitable topcoat system on ship hulls and other marine structures				Not recommended	With epoxy top-coats on marine structures	See manufacturer's literature	See manufacturer's literature
Tank linings	Marine cargo/ballast tanks. Fuel storage including floating roof tanks				Not used	With suitable topcoat in marine cargo/ballast tanks	See manufacturer's literature	See manufacturer's literature
Physical properties:								
(a) Abrasion resistance	Outstanding	Outstanding	Outstanding	Outstanding	Good	Good	Excellent	Usually good
(b) Heat stability	Outstanding	Outstanding	Outstanding	Outstanding	Good	Good	Good	Usually good
(c) Hardness	Outstanding	Outstanding	Outstanding	Outstanding	Good	Good	Excellent	Usually good
(d) Gloss	None	None	None	None	Flat	Flat	None	Flat
(e) Colour	Grey or tints of grey	Grey or tints of grey	Grey or tints of grey	Grey or tints of grey	Grey or tints of grey	Grey or tints of grey	Grey or tints of grey	Usually red oxide or dark colour
Additional notes	As primers for organic systems, provide greatly extended service life. Special application technique or use of tie-coat may be required to avoid solvent bubbling in organic topcoat. Alkyds always require tie-coat.				Can be used to touch up inorganic primers compatible with topcoats	Choice of topcoat is critical	In some applications, satisfactory over mechanically cleaned surfaces. Excellent for touch up of inorganic zincs	Used to ensure bond of system to substrate

36.4.4.11 Unsaturated Polyesters

These resins are thinned with a reactive diluent such as styrene, and cured by free radical reactions through double bonds in the resin using peroxide catalysts. When formulated with glass flake or other suitable lamellar pigment, they produce virtually 100 per cent solid impervious films, of Barcol hardness 80 per cent, and a single coat of up to 5 mm thickness can be obtained.

36.4.5 Conclusion

Tables 36.2 to 36.6 summarise the many types of systems used in heavy-duty protective coatings.

36.5 APPLICATION METHODS FOR HEAVY-DUTY PROTECTIVE COATINGS

36.5.1 Introduction

Applying protective coatings is a skill learned by practice. No one becomes a painter by reading a book. On the other hand, many painters could improve the quality of their work and avoid costly mistakes by reading the instructions of the coatings and equipment manufacturers.

36.5.2 Effect of Weather

When to paint. The ideal time for painting is when the weather is warm and dry, with little wind. Obviously, many coating projects cannot be delayed until such conditions prevail. The following obstacles to good coating applications can be created by inclement weather.

Damp weather. Under conditions of high humidity, condensation of moisture is likely to occur on surfaces. Condensation on the substrate interferes with bonding of the coating. Condensation on the surface of a freshly applied coating may affect the curing process.

Extremely dry weather. Very low humidities can be a problem with water-based products. Rapid 'flash off' of the water may result in film cracking. It can also cause poor curing rates for solvent-based coatings and alkali zinc silicate primers.

Low temperatures. At low temperatures the film thickness of high build or 'thixotropic' coatings becomes more difficult to achieve. Curing reactions slow down or stop for many materials. Water-based products may freeze. Solvents evaporate more slowly. Furthermore, when the relative humidity is more than 70 per cent, condensation is probable.

High temperatures. Although heat has many beneficial effects in the application of coatings, it often increases overspray (dry fallout), trapped air or solvent bubbles and, in the case of zinc silicate coatings, increased incidence of film cracking. It also reduces the pot life of catalysed materials.

Strong winds. Wind is a nuisance, particularly in spray painting. The material as it leaves the spray gun can be deflected from the target. Solvent tends to flash off, creating excessive 'dry spray' at edges of the spray pattern. Lap marks become more evident. Dirt and other debris may become embedded in the wet film.

Condensation. Condensation becomes a problem when relative humidities exceed 70 per cent. The solution is to avoid coating surfaces that are at temperatures below the ambient (surrounding) air. Unfortunately, on large-scale projects, primers are often applied late in the work day and sometimes at night. Abrasive blasting is a slow process, while applying a primer by spray goes very rapidly. Because of this wide difference in work rates, the contractor may take 6 hours of an 8-hour work day to prepare steel for 1 to $1\frac{1}{2}$ hours of primer application. The contractor should establish that there is no likelihood of condensation during the night becoming a problem. Table

36.7 illustrates the relationship of air temperature, metal temperature and relative humidity to condensation. In tank-lining work, condensation can be avoided by circulating warm dry air in the structure during the coating operation.

TABLE 36.5
Inorganic zinc primers compared to traditional types

Primer type/ requirement	*Based on alkyd or oil binders*	*Based on synthetic binders (excluding alkyds/oils)**	*Inorganic zinc system*
Bonding to surface	Usually have the ability to wet and bond to most surfaces and are somewhat tolerant of substandard surface preparation	Adhesive properties are major consideration of formulation. Not quite as tolerant of substandard surface preparation as oil types	Outstanding adhesion to properly cleaned and roughened surfaces
Adhesion of topcoats	Satisfactory for oil types. Usually unsatisfactory for vinyls, epoxies and other synthetic polymers. Are softened and lose integrity by attack from solvent systems of these topcoats	Formulated for a specific range of topcoats	Fits into wide range of systems. 'Tie-coat' may be required. Specific recommendation should be obtained for immersion systems
Corrosion suppression	Limited alkali produced at cathode & corrosion battery attacks film (saponification) and causes bond failure. Results in spread of under-film corrosion	Usually formulated with good resistance to alkali undercut and contain chromate pigment for a degree of corrosion inhibition	Outstanding in ability to resist bond failure and underfilm corrosion. Anodic property of metallic zinc protects minor film discontinuities
Protection as single coat	Limited by severity of exposure	Limited by severity of exposure	With very few exceptions will protect without topcoat
Chemical resistance	Typical of alkyds	Usually of lower order of resistance than that of topcoat	Not resistant to strong acids and alkalies. Has outstanding solvent resistance

* Sometimes referred to in the trade as 'mixed resin types', these primers may be based on non-convertible resins such as polystyrene or hydrocarbon types, or on similar binders used in topcoats, such as epoxies or chlorinated rubber.

36.5.3 When Weather is a Problem

36.5.3.1 Thinning

Many of the application problems (and drying or curing problems) created by weather conditions can be reduced by lowering the viscosity of the material by the addition of the proper thinner. However, the limits shown in the application instructions should not be exceeded without checking with the manufacturer. The thinner is simply a mixture of solvents compatible with the resins in the coating.

Thinning can provide these benefits:

(a) improved flow and uniformity in application of the material;
(b) reduced overspray, lap marks, bubble entrapment, and film 'mud cracking' caused by rapid solvent flash-off.

TABLE 36.6
Oil types

	Phenolic modified oil	*Epoxy ester*	*Alkyd*	*Uralkyd*
Cure type	Drying Oil	Drying Oil	Drying Oil	Oil Type
Effect of sunlight	Yellowing, surface chalking	Rapid surface chalking	Slow surface chalking	Yellowing & surface chalking
Wet or humid environments	Fair to good, will yellow	Fair to good, will yellow	Poor to good, will yellow	Fair
Industrial atmosphere contaminants:				
(a) Acid	Poor to fair	Fair	Fair to good	Poor to fair
(b) Alkali	Poor	Fair	Poor	Fair
(c) Oxidising	Poor	Poor	Fair	Poor
(d) Solvents	Limited	Limited	Limited	Good
Spillage & splash of industrial compounds:				
(a) Acids	Not recommended	Not recommended	Not recommended	Not recommended
(b) Alkalis	Not recommended	Not recommended	Not recommended	Not recommended
(c) Oxidising	Not recommended	Not recommended	Not recommended	Not recommended
(d) Solvents	Not recommended	Not recommended	Not recommended	Not recommended
Water immersion	Not recommended	Not recommended	Not recommended	Not recommended
Tank linings	Not used	Not used	Not used	Not used
Physical properties:				
(a) Abrasion resistance	Poor	Fair	Fair	Excellent
(b) Heat stability	Fair	Fair	Fair	Good
(c) Hardness	Fair	Good	Fair	Excellent
(d) Gloss	Good	Wide range	Wide range	High gloss
(e) Colour	Limited	Full range	Full range	Limited range

36.5.3.2 Heating

With some materials, heating or warming has an effect similar to thinning. Heating devices are available for use with spray equipment.

36.5.3.3 Increasing Number of Coats

Because thinning reduces the volume solids of a coating, adequate film build may become difficult to obtain. In some situations, reducing the thickness per coat and increasing the number of coats will result in a better job. This is often true in both cold and extremely hot weather. Thinner films permit easier escape of solvent under both conditions. Bubbles and pinholes in hot weather and extremely slow hardening rates of thick films in cold weather are the result of unsuitable solvent release.

36.5.4 Other Environmental Problems

When coating work is being carried out in a plant that is operating, there is always the danger of 'fall out' of dusts or condensed fumes on the substrate or between coats. Where the condition is excessive, major coating work may have to be delayed until there is a plant shutdown. In some instances, 'tenting' of the structure with thin polyethylene or polyvinyl chloride film can be used to prevent this type of contamination.

Coating work should be planned and scheduled to prevent interference from other crafts people such as welders, fitters and electricians.

36.5.5 Methods of Application

There are various methods of applying protective coatings ranging from dipping to electrocoating, but field painting is usually accomplished by brush, roller or spray. Little can be said about the first two which will improve a painter's technique.

36.5.5.1 Brush

Although oils and many alkyds apply readily by brush, vinyls, chlorinated rubbers, epoxies, urethanes and inorganics do not usually lend themselves to this type of application, except for touch-up. These materials become tacky very rapidly and resist the flow-out typical of house paint. However, for some systems, brushing of the primer coat is recommended to ensure that it has been worked into the substrate surface. When this is necessary, the primer should be liberally thinned to allow material flow. In general, brushing produces an uneven film thickness allowing possible premature failure.

36.5.5.2 Roller

Rolling works well with some coatings and is impossible with others. A few materials dry on the roller and do not resolvate on dipping. In a short time the roller becomes too heavy to hold. Some show very poor flow properties. Although vinyls and chlorinated rubbers 'roll' quite well, application by this method results in pick-up of the undercoat. For example, if a white vinyl is applied over a red vinyl undercoat, the finish will be pink or rose-coloured rather than white.

36.5.5.3 Spray

The largest volume of maintenance coatings are applied by spray equipment, which is not only faster than alternative methods, but the resulting films are more uniform in thickness.

The objective in spray painting is to create a mist of atomized (finely dispersed) particles which will cling to the target in a uniform pattern and then flow into a continuous, even film.

In industrial and marine painting, two types of equipment are used:
(a) conventional (air-pressure feed), and
(b) airless (hydraulic pressure).
A third, air-suction feed, is occasionally employed for minor touch-up work.

Air-pressure (conventional) spray. The spray gun, which is the primary component in a spray system, brings the air and paint together. This is accomplished in such a way that the fluid is broken into a spray which can then be directed at the surface to be coated. There are two adjustments in most spray guns: one that regulates the amount of fluid which passes through the gun when the trigger is pulled; and one that controls the amount of air passing through the gun, thus determining the width of the fan. In the external-mix spray gun (which is the most widely used type) the air atomises the fluid stream outside the gun after being filtered through a specially designed air cap. The number, position and size of the holes in the air cap determine the manner in which the air stream is broken up, which in turn governs the break-up of the fluid stream.

The fluid leaves the gun through a small hole in the fluid tip. A needle that is operated by the trigger controls the flow of materials through this tip. As with air caps, fluid tips are manufactured in different sizes to accommodate various materials, the diameter of the orifice in the tip being the differentiating factor. Various coatings require different types of air caps and fluid tips in order to be properly atomised.

It is usually wise to follow the coating manufacturer's recommendation as to the proper spray gun, air cap, and fluid tip for the application of a specific material.

Hydraulic (airless) spray. With this equipment, there is no air used for atomisation. The spray pattern is formed simply by forcing the material under high pressure through a very small orifice in the spray gun. As the material leaves the orifice, it expands and is broken up into fine droplets.

The outstanding advantage of this type of equipment is the reduction of overspray. Smoother applications can be made, especially to corners and crevices, and material loss to the wind is negligible. Total job savings with hydraulic spray because of (a) higher speed, (b) more coverage per litre, and (c) eliminating air consumption, are said to range from 30 to 70 per cent depending on type of material sprayed and other factors.

Adaptability to various types of material is obtained by providing a series of interchangeable fluid tips for the gun: each has a fixed orifice and fan. Since the size and shape of the orifice determines the break-up to the material, the width of the fan and delivery rate, it is important that the proper tip is selected for spraying a particular coating. The only other adjustment is the pressure applied to the fluid. An air-driven pump with a ratio of approximately 28 : 1 or more is commonly used.

36.5.6 Other Equipment

Spray applications which take place in enclosed spaces or which involve tank linings may require additional equipment. In confined spaces, the worker should wear an air mask. Forced air, supplied to the mask through a 6.5 mm hose and discharged in front of the eyes, provides clear visibility and freedom from fumes.

It is also essential in such areas that fans or blowers be provided to remove solvent fumes. If fumes are allowed to accumulate, they cause acute discomfort to the operators and create a serious fire hazard.

For safe application, the concentration of fumes must be kept below 0.5 per cent by volume of air. To best accomplish this, the blower should extract air from inside and exhaust externally. Blowing into a tank normally results in agitating rather than removing the vapours. Since the solvents used in most maintenance coatings are heavier than air, it is necessary to attach a conduit or tube to the blower in order to with draw the vapours from the bottom of the tank.

When used for continuous immersion, many tank lining materials must be force-dried. Small coated units can be dried in an oven with circulating air at 60–90° C. For large tanks, a heater is required to circulate warm air through the tank. One of the most practical systems uses a gas- or butane-fired heater with a blower to force air through the heater and into the tank. *Explosion-proof and spark-proof equipment* should always be provided for coating operations in enclosed areas. This includes motors, junction boxes, lights and any other equipment that could generate a spark or flame.

TABLE 36.7
Relative humidity (percentages) above which moisture will condense on metal surfaces not insulated

Metal surface temp. (°C)	*Surrounding air temperature* (°C) 4	7	10	13	16	18	21	24	27	29	32	35	38	41	43	46	49
2	60	33	11														
4		69	39	20	8												
7			69	45	27	14											
10				71	49	32	20	11									
13					73	53	38	26	17	9							
16						75	56	41	30	21	14	9					
18							78	59	45	34	25	18	13				
21								79	61	48	37	29	22	16	13		
24									80	64	50	40	32	25	20	15	
27										81	66	53	43	35	29	22	16
29											81	68	55	46	37	30	25
32												82	69	58	49	40	32
35													83	70	58	50	40
38														84	70	61	50
41															85	71	61
43																85	72
46																	86

37 ANTI-FOULING PAINTS

37.1 INTRODUCTION

The problem of protecting the bottoms of ships from marine growth and organisms is as old as the history of shipping itself.

There are three main ways in which fouling growth is detrimental to the interest of ship owners and to the owners of small wet-moored craft.

(a) Organisms settling and growing upon the hull destroy its smooth, streamlined contour, which marine architects have gone to a great deal of trouble to design. After leaving dry dock, a ship's underwater surface progressively becomes rougher, with the result that fuel consumption increases to maintain a given speed. It has been estimated that relatively small amounts of fouling can increase fuel costs by as much as 20 per cent. Subsequent dry docking, while providing substantial improvement, does not return the hull to its original smoothness unless all coatings and corrosion products are removed by sand blasting.
(b) Loss of speed means a ship takes longer in travelling between ports, leading to reduced earnings.
(c) There is considerable loss of time and money in taking a ship out of service for dry docking purposes, to remove any fouling growth and apply new anti-fouling paint. Some shipping companies undertake measurement of hull roughness at each two-yearly docking to determine at what point the cost of blasting would be economically justified.

As well as the effect of fouling growth, in the case of wooden vessels, such creatures as the ship worm (*Teredo*) and the gribble (*Limnoria*) acturally bore into the wood of the hull; if allowed to proceed unchecked, they may virtually destroy it in a comparatively short time under favourable conditions.

The underwater hull of a ship is therefore unique and different to any other exposure. Above water in reality is no different, with the exception of a continuously salt-laden atmosphere, to any other atmospheric exposure.

37.2 FOULING ORGANISMS

The settlement of fouling occurs almost exclusively when a ship is either at rest in port or moving very slowly (less than 5 km/h) in an estuary or coastal area. The main organisms involved are shown in table 37.1.

TABLE 37.1
Principal marine organisms involved in fouling

Acorn barnacles:	Anemones:	Molluscs (cont.)
Balanus improvisus	*Metridium* spp.	*Anomia* spp.
Balanus eburneus	*Sagartia* spp.	*Teredo navalis*
Balanus amphitrite		
Balanus tintinnabulum	Corals:	Annelids:
Balanus spp.	*Astrangia* spp.	*Hydroides hexagonis*
Balanus crenatus		*Hydroides norvegica*
Balanus psittacus	Bryozoa:	*Hydroides* spp.
Chelonibia patula	*Membranipora lacroixii*	*Nereis pelagica*
Balanus perforatus	*Membranipora* spp.	*Nereis* spp.
Balanus tulipiformis	*Bowerbankia caudata*	
Chthamalus spp.	*Alcyonidium mytili*	Tunicates:
	Alcyonidium gelatinosum	*Molgula manhattensis*
Goose barnacles:	*Membranipora monostachys*	*Molgula arenata*
Lepas anserifera	*Bugula turrita*	*Botryllus schlosseri*
Conchoderma auritum	*Lepralia pertusa*	*Ascidiella virginea*
Conchoderma virgatum	*Bugula avicularia*	*Diplosoma gelatinosa*
Lepas hillii	*Bugula neritina*	
Lepas anatifera	*Bugula turbinata*	Protoza:
Poecilasma crassa	*Watersipora culcullata*	*Vorticellids*
	Callopora lineata	*Folliculina* spp.
Hydroids:	*Callopora* spp.	
Tubularia spp.	*Alcyonidium* spp.	Algae:
Campanularia spp.	*Membranipora savartii*	*Enteromorpha intestinalis*
Laomedea spp.	*Electra pilosa*	*Enteromorpha* spp.
Clytia spp.	*Schizoporella unicornis*	*Ulva lactuca*
Tubularia crocea	*Scrupocellaria reptans*	*Cladophora* spp.
Campanularia amphora		*Ulothrix flacca*
Eudendrium ramosum	Molluscs:	*Polysiphonia nigrescens*
Laomedea geniculata	*Anomia ephippium*	*Ectocarpus confervoides*
Laomedea sargassi	*Mytilus edulis*	*Ulva* spp.
Campanularia portium	*Ostrea elongata*	*Vaucheria* spp.
Campanularia vorticellata	*Ostrea* spp.	*Stigeoclonium* spp.
Bougainvillia carolinensis	*Mytilus pictus*	*Chaetomorpha fibrosa*
Perigonimus jonsii	*Nudibranchs*	*Acrochaetium* spp.
Podocoryne spp.	*Ostrea parasitica*	*Syphonates* spp.
Plumularidae spp.	*Anomia fidenas*	*Oscillatoria* spp.

The surface waters of the oceans contain immense numbers of microscopic animal and plant life comprising what is known to marine biologists as plankton. The concentration may be so high at times for it to be visible in daytime to the naked eye, and at night the luminous nature of many organisms creates 'phosphorescence'.

Plankton population, at least on the surface, is comparatively sparse in mid-ocean and even then is dependent on currents, temperatures and the presence of nutrients.

Closer to shore, rivers and harbours serve to concentrate the supply of foodstuffs, including phosphates, nitrates and carbonates, causing a denser plankton population. The types present as well as numbers may be further affected by the nature of effluents which may vary from toxic to favouring growth. For example, the presence of phenols restricts many species while others benefit from a food supply of the kind present in sugar-mill waste. As far as fouling organisms are concerned, high salinity levels favour some types while brackish conditions are more suitable to others.

Vessels at rest or moving only slowly run the risk of settlement on their hulls of the larval forms of animal plankton (zooplankton) and spores of phytoplankton. In the absence of natural or artificial constraint, the initial stages of growth of settled organisms is usually rapid, representing a substantial fuel cost by reduction in speed and a maintenance cost in frequency of removal.

The main types of fouling growth in Sydney Harbour are listed in table 37.2.

TABLE 37.2
Principal fouling organisms in Sydney Harbour

Vegetable		*Animal*	
Green algae:	*Enteromorpha* sp.	Acorn barnacles:	*Balanus amphitrite*
Brown algae:	*Ectocarpus*	Bryozoa:	*Watersipora cucullata*
			Bugula neritina
			Conopeum reticulum
		Annelids (tubeworm):	*Hydroides norvegica*
			Spirobis sp.

37.2.1 Algae

Seaweeds multiply and spread by means of extremely minute spores which are liberated in the water and attach themselves to fresh surfaces within seconds and commence to grow in their new situation. With *Enteromorpha*, for example, the reproductive potential is such that a 100 mm frond with a breadth of 10 mm could produce 40 000 000 spores. When this figure is considered in relation to the very large areas of weed growing under natural conditions, then the potential for fouling is tremendous.

37.2.2 Shell Fouling

The two most important of the shell fouling organisms are tubeworms and barnacles. Of the latter, there are several species ranging in size when fully growth from 5 to 60 mm in diameter. In their adult form they are encased in hard calcareous shells firmly attached by an extremely strong adhesive, such that removal by a hand scraper, for example, leaves behind the bottom of the shell which remains as a point of weakness when painted over. The height of the barnacles greatly affects the roughness of the ship's bottom.

Hydroids, Bryozoa and Tunicates comprise the other major fouling organisms commonly encountered on ship's bottoms. These are animals but some Hydroids (e.g. *Tubularia*) and Bryozoa (e.g. *Bugula*) form soft branching bush or tree-like growths easily mistaken for seaweed.

37.3 ANTI-FOULING PAINTS

As the name implies, anti-fouling paints are coatings applied to ships' bottoms which will restrict the attachment and growth of fouling organisms. They function by releasing biocides or toxins which are poisonous to the marine organisms whilst they are in their embryonic state. Once attached and growth is confirmed, most forms, particularly algae, are almost impossible to control.

Anti-fouling paints are similar to most surface coatings, consisting of pigment(s) dispersed in a resin, with additives that are toxic to marine organisms.

37.3.1 Leaching Rate

Anti-fouling paints function correctly only if the rate of release of toxin is controlled at a steady rate. The rate at which toxin is allowed into the laminar layer of water surrounding the ship's bottom is called the 'leaching rate'. Controlling the leaching rate is one of the most difficult problems facing the paint technologist. It must be neither too fast, which would result in overkill and wastage and short effectiveness, nor too slow, when the toxins are locked in the film and the laminar layer does not contain the necessary lethal concentration. Figure 37.1 shows the performance of a satisfactory anti-fouling paint which provides an adequate leaching rate for a prolonged time.

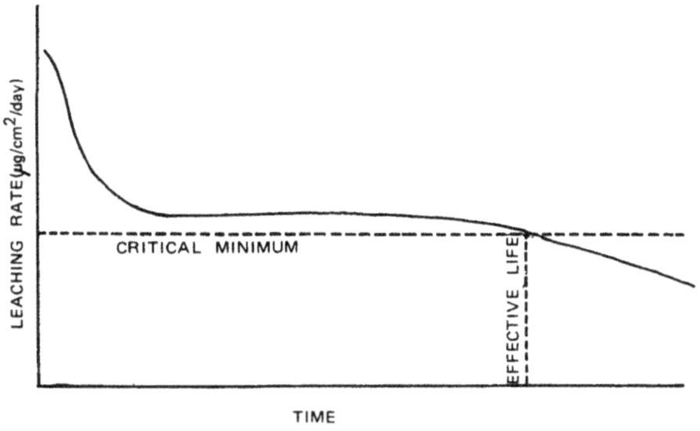

FIGURE 37.1
Leaching rate with time (1)

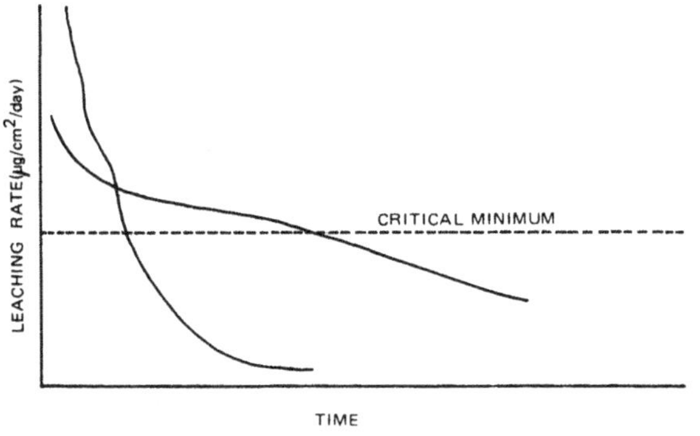

FIGURE 37.2
Leaching rate with time (2)

Figure 37.2 shows two anti-fouling paints with rapid exhaustion of toxins and therefore short effective life.

Of the many thousands of available biocides, only a few have been found satisfactory in anti-fouling paints. Some biocides are too soluble in sea water, most have little effect on marine organisms, some are too toxic and some, such as arsenic and mercury compounds, are today environmentally unacceptable. Those that are used include:

Cuprous oxide and cuprous thiocyanate

Organo tin compounds—tributyl tin oxide and tributyl tin fluoride
Pure organic compunds such as DDT (dichloro diphenyl trichloro ethane) and TMT (tetramethyl thiuram disulfide)

Cuprous oxide is probably the most widely used, either alone or in combination with other toxicants. Anti-foulings used on small aluminium yachts and boats must not contain any cuprous oxide because the galvanic action causes corrosion.

The extremely low solubilities of cupric oxide and cupric hydroxide are the reason they are ineffective, while cuprous chloride is too soluble and its leaching rate cannot be controlled (see table 37.3).

TABLE 37.3
Solubilities of copper compounds

Compound	*Moles/litre*	*g Cu/mL*
Cuprous chloride	5.1×10^{-2}	3100
Cuprous oxide	8.6×10^{-5}	5.4
Cupric oxide	1.4×10^{-8}	0.0009
Basic cupric carbonate	8.0×10^{-6}	0.5
Cupric hydroxide	2.1×10^{-7}	0.013

Experimental results have shown that a steady release of copper at 20 $\mu g/cm^2/day$ will eliminate virtually all fouling. The required rate of release for organotin compounds is about one-fifth of this rate.

37.3.2 Types of Anti-fouling Paints

37.3.2.1 Soluble Matrix

The salinity and pH of sea water varies very little, and the anti-fouling systems in this group depend on these facts. They are based on the natural resins (mainly rosin), drying oils and cuprous oxide either alone or boosted with other toxicants. Rosin is principally abietic acid, which is insoluble in acidic or neutral aqueous solutions, but soluble in alkaline solutions. At pH 8.1, that of sea water, it dissolves slowly so that the toxicant is released.

The inadequacies of this type of anti-fouling arise from the need to make the film permeable to sea water allowing ionic diffusion of the copper ions to occur. These conventional anti-fouling coatings are as a result brittle, water sensitive and transient. An additional defect is that they do not disappear from the surface completely, but remain to break down, delaminate and generally roughen the surface to which they are applied.

$$\underset{\text{rosin}}{RCOOH} + OH^- \rightarrow \underset{\text{resinate ion}}{RCOO^-} + H_2O$$

In sea water the rate of solution of rosin is not constant and gradually decreases due to the formation of the insoluble calcium and magnesium soaps at the surface. Also the deposition of basic cupric carbonate and cupric oxychloride combine to restrict the release of copper or other toxicants, and the system fails to maintain the critical leaching rate.

Such anti-fouling coatings should be immersed within about 12 hours of application, because air exposure causes oxidative changes detrimental to the leaching performance of the paint. In addition, before recoating, soluble matrix anti-foulings should be sealed with undercoat to ensure proper efficiency of the new anti-fouling.

37.3.2.2 Continuous Contact Anti-foulings

The best that conventional technology has been able to achieve is to toughen the anti-fouling matrix with high molecular weight polymers such as vinyl and acrylic types and chlorinated rubbers. The reduced water permeability is offset by raising the volume of toxicant to a high level. Biocide is again liberated at the surface by a process of contact diffusion, but at a rate that becomes progressively slower with time until eventually the situation is reached when it is impossible for toxicant to be released from the depths of the film. However, these anti-foulings can be formulated to give 2 to 3 years' fouling free life. These anti-foulings are not affected by up to 2 months' air exposure.

37.3.2.3 The Lifetime of Two-coat Applications

The lifetime of either the soluble matrix or continuous contact anti-foulings is determined by the period over which leaching is maintained above the critical level required to control all fouling forms—for example, toxicant/unit area plotted against time. The general slope is logarithmic, so a two-coat application does not double the anti-fouling lifetime but only increases it by a factor of log 2—that is, 30 per cent greater life. However, two coats ensure a more even application.

37.4 SEASONAL AND TEMPERATURE EFFECT

In tropical and warm climates the reproduction and growth of most marine organisms is continuous throughout the year.

In colder climates, fouling growth is distinctly seasonal in occurrence. It follows therefore that it is important for ship owners to consider area and season in relation to the type of anti-fouling used.

37.5 LATEST DEVELOPMENTS

The eventual failure of soluble matrix and contact anti-foulings is primarily due to the 'blinding' of the paint surface by degradation products hindering diffusion of toxicity. Self-polishing-type anti-foulings are a new development and have biocide groups (e.g. organo tin groups) grafted to a polymer backbone, which itself is designed for a controlled degree of solubility in sea water. The gradual erosion of the surface is caused by wave action and movement of the vessel through the water. The life of the anti-fouling is in direct proportion to its thickness. By controlling the polishing rate, effective lifetimes from 2 to 3 years, depending on the activity of the ship, can be obtained. At the next routine dry docking, new anti-fouling is applied up to the original thickness. A bonus is that the bottom becomes progressively smoother in service, in marked contrast to the increasing roughness of the soluble matrix or contact anti-fouling films in service. In fact, with polishing systems a ship increases speed at constant power or, alternatively, uses less fuel to maintain constant speed. Both types of performance gain have been demonstrated in commercial ship operation.

There have been many novel ideas tried in the past: such as pumping tributyl tin oxide in kerosene around the hull (the 'Tioxin System'), but this would be regarded as unnecessary pollution today and anyway would be too expensive; pumping chlorinated water around the hull; or likewise pentachlorophenol. The advantage claimed was that they would be used only when needed—while the ship is in port. Each failed because of problems in providing an adequate flow and general engineering difficulties.

Another current approach is the 'non-stick' Teflon concept which, if successful, could lead to the end of conventional anti-fouling technology since no biocide or leaching would be involved.

Periodic removal of minor fouling may be all that is necessary to maintain a clean hull. However, the real practicalities of repair and maintenance as well as problems with the initial application remain to be overcome.

37.6 FUTURE ANTI-FOULING TECHNOLOGY

The past two decades have seen a remarkable transformation in the efficiency, methods and thinking of the shipping industry. The size of vessel has increased dramatically, loading and unloading times have been much shortened, and automation has made possible a reduction in crews. This is the era of supertankers, container ships and bulk-ore carriers. These changes have resulted in a massive increase in the efficiency necessary to keep pace with increasing operating costs around the world.

The growth of marine transport has resulted in many changes, among them:

(a) turn-round times reduced from 10–14 days to 24–48 hours,
(b) dry docking intervals increased from 12 to 30 months.

These changes have altered the type and distribution of foulings. When turn-round was 10–14 days, fouling was mainly animal. With a turn-round of 24–48 hours there is insufficient time for settlement and adhesion of animals; algal fouling—which settles in seconds—is now the predominant type, especially in the waterline to bilge keel sector, the area favoured by light, aeration and nutrients. Thus future work will need to develop anti-fouling systems that are optimum for these changing factors.

Example of the economics of speed loss. Consider the loss of an average of 1.5 km/h on a 25 km/h vessel (due to the drag of fouling growth) on a vessel running at 82 per cent activity (300 days sailing) per annum.

1.5 km/h in 25, i.e.

$$\frac{3}{2} \times \frac{100}{25} = 6\% \text{ loss}$$

6% of 300 days = 18 days lost

The earning capacity of a large carrier or tanker is about \$30 000 per day. Therefore each knot of lost speed is worth 18 × \$30 000 = \$540 000/year/vessel.

It is obvious that the avoidance of fouling and resultant hull roughness can result in very large savings to ship owners.

38 INDUSTRIAL COATINGS

38.1 INTRODUCTION

As the name suggests, industrial coatings are used by industry, often in a factory production line situation, to decorate and protect a wide range of manufactured articles.

The era of industrial coatings began with mass production in Europe and America at the end of the First World War, when the cost of labour rose rapidly and coincided with the demand for cheap mass-produced articles. The first major step was in the motor car industry, when low-viscosity cellulose nitrate (*nitrocellulose*) became commercially available and was used to produce lacquers suitable for this new industry. Production time for painting vehicles was cut from weeks to hours. The furniture industry and then others were quick to follow suit, and the use of industrial coatings had begun.

In the next two decades, technology took over from craft skills, and with the development of the petrochemical industry, nitrocellulose was followed by phenolics, alkyd resins, urea formaldehyde resins, chlorinated rubber, vinyls, epoxies, silicones, and so on. At the same time a wide range of chemical solvents was becoming available, and advances were being made in the production of titanium dioxide and in the development of organic pigments.

Similarly, technology has kept pace with the changing environmental scene, and the escalating cost of petrochemical solvents of the past two decades. We now find our industry producing, in ever-increasing quantities, a number of 'environmentally acceptable' coatings. These include coatings based on non-photochemically reactive solvents—the so-called 'Rule 66' solvents, coatings using water as a solvent or a carrier, 'high-solids' coatings, powder coatings, and radiation-curable coatings.

It is interesting to look at some of the milestones of the industrial coatings industry, and these are listed in table 38.1.

In this chapter, an attempt will be made to show how the groups of raw materials, detailed in earlier chapters of this book, are combined to produce the wide range of coatings used by industry today.

38.2 CUSTOMER'S SPECIFICATION

Industrial coatings are largely tailored to suit the individual requirements of the user. Unfortunately (and historically) the user has dictated the conditions of use, and the formulator has had to formulate to suit these conditions. However, there has been a trend in recent years, particularly

TABLE 38.1
Milestones in industrial coatings

1915–20	Phenolic varnish resins
	Conventional spray gun application
1921–25	Low-viscosity nitrocellulose
	Alkyd resins
	Anatase titanium dioxide
	Maleic treated gum rosin
1926–30	Oil-soluble phenolics
1930–35	Urea formaldehyde–alkyd blends
	Chlorinated rubber
	Molybdate orange pigments
	Vinyl chloride copolymers
	Phosphating and metal pretreatment
1936–40	Phthalocyanine pigments
	Polyurethane resins
	Melamine formaldehyde–alkyd blends
	Wash primers
1941–45	Styrenated, acrylated oils
	Rutile titanium dioxide
	Silicone resins
	UV absorbers
1946–50	Epoxy resins
1951–55	Light-fast organic yellows, oranges, reds, maroons, violets
	Unsaturated polyesters
	Polyamide cured epoxies
	Electrostatic spray
	Powder coatings
1956–60	Acrylic automotive lacquers
	Thermosetting acrylic enamels
	Airless spray
	Urethane oils
	Aqueous industrial primers
	Sand mill–attritor
	Siliconised resins
	Chloride process titanium dioxide
	Coal tar–epoxy
1961–65	Coil coatings–twenty-year durability
	Electrodeposition
	Curtain coating
	Fluorocarbon resins
1966–70	Radiation curable coatings
	Non-aqueous dispersions
	UV curable coatings
	High solids coatings
1971–82	Aqueous industrial enamels
	Electron beam curing
	Cationic electrodeposition
	Rule 66 reformulations
	Mercury and lead replacements

Extracted from a table in 'A Half-Century of Creative Coatings Science' by H L Gerhart in J K Craver and R W Tess eds *Applied Polymer Science* American Chemical Society (Organic Coatings & Plastics Chemistry Division) Washington DC 1975 with some of the newer types of coatings, for customers to modify their techniques to suit the coatings, rather than vice versa.

A specification is usually received by a paint manufacturer when he negotiates with an industrial customer to supply coatings. Each customer is likely to have a specific set of application techniques and performance requirements, and these must always be kept in mind when formulating. The specification may be verbal and brief, such as 'a cheap, quick-drying paint for reconditioned drums' or may be quite detailed in application, composition or performance requirements, as for example in the coatings required for painting a commercial or military aircraft.

Some of the properties that might be covered by the specification include:

(a) substrate type and condition—facilities for pretreatment;
(b) application methods and conditions;
(c) drying time required or baking conditions;
(d) wet paint properties—skinning properties, viscosity, package stability, non-volatile content, degree of dispersion;
(e) decorative requirements—colour, gloss, hiding properties, surface texture;
(f) protective requirements—adhesion, hardness, flexibility, mar resistance, scratch resistance, abrasion resistance, impact resistance, heat resistance, water resistance, alkali-detergent resistance, chemical resistance, solvent or petrol resistance, stain resistance, corrosion resistance, resistance to yellowing, exterior durability;
(g) special requirements—water thinned, lead-free, acceptable odour, non-bleeding, non-skid properties, turps thinned, sanding properties, recoating properties, resistance to fuels and lubricants, non-flammable, environmentally acceptable;
(h) the composition of the coatings required.

Not usually stated in the specification but of some significance is the cost. It must always be remembered that cost is very important in the area of mass production, and the formulator should attempt to provide a coating that meets the specification at the minimum cost.

38.3 THE SUBSTRATE AND SURFACE PREPARATION

The substrates most commonly coated with industrial coatings are iron and steel, but also include other metals such as aluminium and its alloys, zinc-coated steel in various forms, tinplate, diecast alloys (based on zinc, aluminium or magnesium), plated steels (cadmium, chromium or nickel-plated), magnesium and its alloys, brass, bronze, copper and lead. Non-metallic substrates include timber and timber products (hardboard and particle board), concrete, asbestos cement, glass, ceramics, fabric, paper, leather, rubber, bitumen and a wide range of different plastic materials. Sometimes the substrate may have been previously painted, but this is not common in industrial applications. One such example is in traffic marking paints where the substrate may be old traffic markings as well as various road construction surfaces.

When formulating a coating, consideration must always be given to the particular properties of the substrate on which it is to be used.

The substrate and surface preparation are covered in some detail in chapter 48. However, as surface preparation is often an essential part of the industrial coating process, a summary is given here of the common processes related to the surface preparation of metals, since metals are the substrate for most of the products produced by industry. Most metals require degreasing and sometimes the removal of an oxide surface layer.

38.3.1 Iron and Steel

Cleaning—solvent cleaning, manual swabbing or bath—trichlorethylene vapour tank—alkaline cleaning—emulsion cleaning.

De-rusting—wire brushing—grit blasting—flame cleaning—acid pickling.

Pretreatment—phosphating, iron zinc or manganese phosphate—electroplating.

38.3.2 Aluminium and its Alloys

Cleaning—solvent cleaning—trichlorethylene vapour—emulsion cleaning.

Deoxidising and pretreatment—mild alkali or mild acid chemical etching followed by chromate conversion coating.

38.3.3 Zinc-coated Steels

Cleaning—solvent cleaning—trichlorethylene vapour emulsion cleaning.

Pretreatment—phosphating, hot or cold chromate conversion.

38.3.4 Tinplate

Usually painted before fabrication and no pretreatment necessary.

38.3.5 Copper and its Alloys

Not commonly painted—clear coatings require perfectly clean surfaces—for pigmented coatings, a mild grit blast is preferred.

38.4 METHODS OF APPLICATION

The methods of application of industrial coatings are treated in detail in chapter 52. They are summarised here, and some comment is made on the way the different methods have a bearing on the types of coatings used and how the different coatings are modified for the different application techniques. The various spray, dip and rollercoating techniques are of major significance to the industries using industrial coatings.

38.4.1 Brush and Hand Rollercoat

Air-dry coatings only; low capital expense; minimum losses in application; labour intensive; seldom used in industrial application but still widely used in industrial maintenance coatings.

38.4.2 Conventional Spray

Suitable for most types of coatings; relatively low capital expense; requires large supply of clean compressed air for atomization; some overspray losses; ideal for repetitive painting on a production line; can be mounted for automatic operation.

38.4.3 Hot Spray

Similar to conventional spray but paint is brought to atomisation viscosity by heating rather than with added solvent; increased solids at application viscosity; higher builds; major use in nitrocellulose lacquers for furniture industry but becoming more widespread because of less pollution of atmosphere.

38.4.4 Airless Spray

Modification of conventional spray; hydraulic pressure instead of air used to atomise paint; less 'bounceback'; less overspray loss; suitable for inside containers such as cans and drums; applicable to cold or hot spray; faster application than conventional spray; manual or automatic.

38.4.5 Electrostatic Application

Air-assisted or airless; atomised paint is electrically charged; ware to be painted is earthed and paint is attracted to ware; complete 'wrap-around' when painted from one side; minimum over-spray losses; up to 40 per cent less paint used; most benefit on wire work or pipe construction; applicable to most types of coatings, air dry or bake; some modification of electrical resistance of paint usually necessary; controlled with solvent polarity or additives; automatic or hand-held operation.

38.4.6 Dipping

Economical method; minimum losses; usually (but not always) applicable to baking enamels; useful for coating inaccessible surfaces; requires tank agitation for uniformity; good tank stability necessary; high opacity required; additives to control settling, foaming, 'floating' and skinning usually necessary; not suitable for two-pack systems.

38.4.7 Squeegee Application

A modification of dipping; applicable to uniform articles such as broom handles, pencils, wire; usually nitrocellulose lacquers; excess lacquer wiped off by passing article through a rubber end-gland on the tank.

38.4.8 Vacuum Impregnation

Closed dip tank; ware loaded into tank and vacuum applied, then impregnant admitted; deep penetration; applicable to electrical components and impregnating timber with preservatives.

38.4.9 Flow Coating

Variation of dipping; paint is poured or hosed on to article; suitable for large articles where dip tank size prohibitive; similar paint requirements to dipping.

38.4.10 Curtain Coating

Suitable for flat panels; carried on a conveyor through a curtain of paint; continuous curtain maintained by adjustable aperture below a pouring tank; similar paint requirements to dipping but films usually thicker and flash-off times shorter, needing solvent modification to prevent solvent boil; film thickness controlled by conveyor speed, paint viscosity and aperture size.

38.4.11 Tumbling

Suitable for small pieces which would otherwise present handling problems; air-dry or bake; highly decorative finishes not available.

38.4.12 Rollercoating

Applicable to flat panels; paint transferred to surface from an application roller; efficient and economical; no coating losses; usually heat-cured coatings; high boiling solvents to allow flow and levelling and prevent solvent boil; applicable to plywood, hardboards, asbestos cement, tinplate, steel sheets; main use in tinplate for small containers and steel for drums; flexible coatings required to allow post-forming.

38.4.13 Coil Coating

Variation on rollercoating using continuous strip instead of panels; usually aluminium or zinc-coated steels; usually coated both sides; good flow and levelling required; must be flexible for post-forming; most applications are for long-life coatings such as cladding, requiring excellent hardness, adhesion, abrasion resistance and exterior durability.

38.4.14 Electrodeposition

Water-thinned paints; large tanks; similar to electroplating; article to be painted made either anode or cathode; paint insolubilises on application of current; allows painting in inaccessible areas and closed sections without solvent refluxing; no drain off sharp edges; main application is automotive primers (see chapter 39); some use in other industrial areas.

38.5 AIR DRYING COATINGS

Air-drying coatings are not widely used on mass-produced articles because of the large amount of space required for 'drying' the painted articles.

However, they are used in a number of areas where the application of heat is neither practical nor desirable. The larger articles such as fork lifts, tractors, agricultural machinery and earth-moving equipment, where the cost of baking facilities for the through-put involved would be prohibitive, is one area. Another is on substrates subject to distortion by heat, such as timber and certain plastics, or on pre-assembled articles with heat-sensitive components.

In general, the drying time of air-dry coatings can be reduced by the application of infra-red or force drying up to 80° C, without detriment to film properties. Above this temperature, yellowing and film embrittlement is likely to occur with some coatings.

38.5.1 Bituminous Coatings

This group of coatings is based on three main raw materials, asphalt, coal tar and gilsonite. Asphalt is a component of crude oil, coal tar is the residue of coal distillation, and gilsonite is a naturally occurring hard bitumen.

Simple bituminous coatings are generally deposited in solution in hydrocarbon solvents and dry by solvent evaporation. They have traditionally been cheap and have excellent water resistance, but have the disadvantages of poor solvent resistance, thermoplasticity, poor gloss retention on exposure, and restricted colour range (black, brown or pigmented with leafing aluminium paste).

A degree of improvement in the properties of bituminous coatings can be achieved by 'cooking' the asphalt or bitumen with drying oils, rosin and other gums. This introduces a degree of cross-linking in the dried film and hence improved properties.

Some typical uses for bituminous coatings are on underground storage tanks, pipes, foundry castings, rail and truck chassis, car radiators, piers and pontoons, and for waterproofing concrete.

38.5.2 Oleoresinous Varnishes

Oleoresinous varnishes are formed when drying oils are heated with various resins. Driers are usually added during the cooking process. The oils contribute flexibility and durability; the resins upgrade dry time, hardness and gloss.

The common oils used are linseed, bodied linseed, dehydrated castor, and wood oil; and the resins include rosin and its esters, copals, dammar, phenolics and coumarone-indene.

As a class of coating, the oleoresinous varnishes have largely been superseded by the more modern coating resins, but are still used in some industrial applications. They are described in detail in volume 1, chapter 4.

Coumarone varnishes remain the basis for leafing aluminium paints, which find industrial application for painting gas cylinders, wire articles, colliery equipment, tow-bars and trailer frames, structural steel, reservoirs and tanks. Phenolic varnishes are used in electrical insulating varnishes.

Another application for the oleoresinous varnishes is in the metal-container industry where they are used for lining food cans. Unpigmented they are gold coloured and used for fruit and vegetable cans. When pigmented with zinc oxide they are used for linings for foodstuffs containing sulfur, such as corn.

38.5.3 Lacquers

'*Lacquer*' is the name generally applied to the group of coatings that 'dry' by simple evaporation of the carrier solvent. No significant conversion takes place in the drying process and the dried film remains soluble in the solvents from which it was deposited. The films also remain thermoplastic and will soften with heat. Because no conversion takes place, the drying time depends entirely on the rate of solvent release from the film, and this can be widely varied by choice of solvents.

In general the film formers suitable for use in lacquers are linear polymers and high in molecular weight, giving good solvent release. They are usually hard and tough, but must be externally plasticised or modified to give suitable film properties such as adhesion, flexibility, gloss or exterior durability.

Lacquers are generally low in solids, particularly at application viscosities, and with the changes in recent years in solvent costs and availability, are becoming less popular. Because of low solids they are high in cost in terms of film thickness applied, and are wasteful of solvents. However, their very quick air-drying properties make them ideal for use on surfaces that distort with heat, such as timber and some plastics.

The types used in industrial coatings are the cellulosics, the acrylics and the vinyls. Chapter 40 covers lacquers in detail.

38.5.3.1 Nitrocellulose

The lower-viscosity grades of nitrocellulose (cellulose nitrate) are preferred for lacquers to give reasonable solids contents. Industrial lacquers are modified with ester gum or maleic rosin ester at 50–100 per cent of the nitrocellulose content and are suitably plasticised, usually with phthalate esters.

Non-drying alkyds are used at levels in excess of 100 per cent of the nitrocellulose content, to produce lacquers with good durability and polishing properties for use in the automotive and automotive refinish industries.

Nitrocellulose lacquers are widely used unpigmented as finishes for timber furniture. Hot spray application is ideal.

Pigmented lacquers are widely used for timber articles, some typical items being pencils and broom handles (applied by a squeegee process) and paint brush handles (applied by a dip process).

38.5.3.2 Ethyl Cellulose

Ethyl cellulose is soluble in cheaper solvents than nitrocellulose and is a useful lacquer raw material. Being an ether it is more resistant to alkali attack, less readily degraded by ultraviolet and more resistant to heat than nitrocellulose. It is compatible with a wide range of resins and plasticisers including rosin esters, some alkyds, vegetable oils, phthalate plasticisers and nitrocellulose.

The main uses are in pigmented coatings for timber and as an 'extender' for silicone resins in heat-resisting formulations, where as well as lowering the cost, it provides some viscosity or 'body' and makes the coating handleable before 'in-service' application of heat to cure the silicone resin. Such an application is on smoke stacks and on mufflers for cars.

38.5.3.3 Cellulose Acetate

Cellulose acetate has only a limited application in the coatings area. It has poor solubility, requiring expensive solvents, and poor compatibility with resins and plasticisers. It also has high water sensitivity, but it does have excellent resistance to yellowing and has been used in speciality coatings requiring these properties.

38.5.3.4 Cellulose Acetate Butyrate

Cellulose acetate butyrate is a mixed ester of cellulose, and a number of grades are available with varying ratios of acetate : butyrate, and in a range of molecular weights and hence viscosity. Cellulose acetate butyrate is much more soluble than cellulose acetate and has better flexibility and water resistance.

It requires plasticiser levels up to 25 per cent, has very good exterior durability and has been used in clear and pigmented lacquers for metals, timber, fabrics and plastics.

A wide area of usage is in conjunction with thermoplastic acrylics in automotive refinish lacquers, where the cellulose acetate butyrate improves the sprayability and recoating properties.

38.5.3.5 Acrylics

Thermoplastic acrylic resins are characterised by two predominant properties: they are virtually water white in colour; and they are almost unaffected by exposure to ultraviolet light.

They are very hard and brittle materials and need to be plasticised for increased flexibility and adhesion, the most common plasticiser being butyl benzyl phthalate at 25–40 per cent on resin solids. Modification of properties is also obtained with cellulose acetate butyrate. The acrylics generally show poor adhesion to metals, and careful choice of primer is necessary to ensure adequate performance.

The major area of use of pigmented acrylic lacquers is in the automotive and automotive refinish areas (see chapter 39), but they also find use on aluminium boats.

In clear finishes they are used as protective coatings for plated surfaces such as refrigerator shelving, and in long-life coatings for aluminium window frames.

38.5.3.6 Vinyls

Vinyl solution resins, containing predominantly polyvinyl chloride, when suitably plasticised and stabilised are used in industrial coatings, where they exhibit the characteristics of extreme chemical inertness, good water and weather resistance and good toughness and flexibility.

They are extensively used for industrial maintenance in such areas as water-storage dams, tanks and pipelines, and for marine applications, but because of their low solids up to six coats may be necessary for adequate protection.

Because of their good chemical resistance properties and lack of taste and odour, they are used in coatings for lining cans for food, beer and beverage. They also find use as coatings for wire, textiles, paper and aluminium foil. They suffer the disadvantage of requiring expensive solvents for solution.

Another type of vinyl is polyvinyl butyral. This resin has excellent adhesion to metals and is the basis of the so-called 'wash primers' or 'etch primers', where it is used in chemical combination with zinc tetroxy chromate and phosphoric acid and applied at very low film thicknesses (5–10 μm). The wash primers are usually marketed as a two-pack system and have excellent adhesion to most metals but minimal protective properties. They are sometimes called metal conditioners.

Polyvinyl butyral is soluble in alcohols and can be used alone or in combination with nitrocellulose for lacquers for polystyrene plastics. It is also coloured with spirit soluble dyes and used in the decoration of aluminium foils.

38.5.4 Alkyd Resins

Alkyd resins (or 'alkyds') are the reaction product of polybasic acids with polyhydric alcohols, modified with monobasic fatty acids, the fatty acids usually being provided by vegetable oils. As a group the alkyds are the most important of the coatings resins. In 1973 in the U.S.A. they accounted for 34 per cent of all resins used in coatings, ahead of 20.5 per cent for vinyls and 18 per cent for acrylics. No other group of resins exceeded 5 per cent. Alkyd resin applications are reviewed in detail in chapter 7.

In the case of air-drying alkyd resins, drying or semi-drying oils such as linseed, fish, dehydrated castor, soya bean, sunflower, safflower, wood oil and tall oil are used for oil modification. The type of oil used affects the properties. Non-yellowing oils give non-yellowing alkyds and the degree of conjugation of the oil affects the dry time. The drying time and hardness can also be varied by modifying with rosin or rosin esters, and an improvement in water and chemical resistance is obtained by modification with phenolic resins. The drying time is also affected by the 'oil length', which refers to the fatty acid content of the alkyd. Short oil alkyds contain 20–45 per cent fatty acid. Medium oil alkyds contain 45–60 per cent fatty acid, and long oil alkyds contain 60–80 per cent fatty acid. The general effect of varying oil length can be summarised as follows:

Increasing oil length gives increasing flexibility of the film.
Increasing oil length gives increasing solubility in weaker aliphatic solvents.
Decreasing oil length gives increasing viscosity of solution.
Hardness of the film reaches a peak about medium oil length.

38.5.4.1 Short Oil Alkyd Resins

Pure short oil alkyds resins are seldom used in air-dry finishes. They exhibit quick solvent release but poor through-dry and tend to remain thermoplastic. They are usually modified with rosin and phenolic resins to give adequate dry time, hardness and gloss. These modified alkyds do not have good exterior durability, colour and flexibility.

Some typical uses for coatings on short oil alkyds include primers for fabricated steel, reconditioned drums, small tools and agricultural implements, and hardboard primers.

38.5.4.2 Medium Oil Alkyd Resins

These find the widest use in industrial coatings. Exterior durability and film toughness are better than for short oil alkyds. They may also be modified with rosin phenolics.

Some typical uses include coatings for tractors, agricultural machinery, earthmoving equipment, colliery equipment, fork lifts, factory presses, ducting, structural steel, aluminium boats and steel railings. They are also pigmented with anti-corrosive pigments to make primers for steel.

38.5.4.3 Long Oil Alkyd Resins

Long oil alkyds are generally too slow drying to find use in industrial coatings. They are used in industrial maintenance coatings where harsh conditions of exposure are not encountered, as well as their very wide use in architectural coatings.

38.5.5 Styrenated Alkyd Resins

Alkyd resins can be modified with styrene, vinyl toluene and acrylic monomers by a number of different techniques. Those modified with styrene are by far the most widely used.

Varying the proportion of styrene will vary the properties of the resin. When compared with unmodified alkyds, the styrene copolymers offer very quick tack-free times, but often suffer from recoat problems and durability is usually worsened.

Paints formulated on styrenated alkyds have excellent solvent release and approach the speed of dry of nitrocellulose lacquers. They are often used in similar areas and have the advantages of much higher solids at application viscosity using much cheaper solvents.

The fast tack-free time of styrenated alkyds makes them ideally suitable for producing hammer finishes with a minimum of pattern sag on vertical surfaces. Hammer finishes are made by pigmenting a suitable resin with non-leafing aluminium paste, and the addition of a silicone 'hammer-aid' produces the hammered or textured effect of beaten metal. Hammer finishes are often used on metal articles to hide imperfections on the surface, such as welding spots.

Typical examples of the use of styrenated alkyds in pigmented finishes include coatings for small tools and equipment, implements and cabinets.

Hammer finishes are commonly used on die-cast items as well as motors, pumps, industrial machinery, cabinets, photographic accessories and laboratory equipment.

38.5.6 Urethane Oils and Urethane Alkyds

The one-component urethane coatings, urethane oils and urethane alkyds, are formed by reacting drying oils or alkyds with isocyanates. In the case of alkyds the reaction with isocyanate may be with the excess hydroxyl in the alkyd, or part of the polybasic acid may be replaced during the cooking process by an isocyanate. Chapter 11 gives more detailed information.

These coatings dry by oxidation of the fatty acid portion and compared to alkyds they offer faster dry times, much better water and alkali resistance and outstanding abrasion resistance. However, they are usually of poor colour, and whilst they have good film integrity on exterior exposure they usually have poor colour and gloss retention.

Typical uses for urethane oils include clear furniture finishes in a range of gloss levels, and flooring finishes.

They can be pigmented to produce anti-corrosive primers for steel and have been used on exterior garden furniture where full conditions of exposure are not required. Fast-drying road-marking paints are another area of interest.

The urethane alkyds are used in similar areas to conventional alkyds, taking into account the advantages and disadvantages listed above.

38.5.7 Water-soluble Alkyds

The technology of water-soluble alkyds has been available to the industry for decades but it has only been in the past few years that a renewed interest has been shown in these resins. A number of reasons have contributed to this, the main ones being the increasing emphasis by governments on air pollution regulations, the increasing cost and decreasing availability of hydrocarbon solvents, and an increasing safety awareness of flammability. (Refer to chapter 22 for more detail.)

Water-soluble alkyds are produced by cooking the alkyd with an excess of carboxyl groups to give a final acid value of 50 or more. These groups can then be neutralised with a volatile amine—usually ammonia or triethylamine—the salt being soluble in water or water/polar solvent mixtures. The most common solvents used for this purpose are alcohols and glycol ethers.

The usual pigment-grinding techniques are generally suitable for water-soluble alkyds. They may be pigmented by grinding the pigments in a solvent solution of the alkyd, then neutralising and solubilising. Alternatively the alkyd may be preneutralised and dissolved in water and the pigments ground in this solution. In this case suitable antifoams may be necessary. A proportion of anti-corrosive pigment such as strontium chromate is usually included in the coating if it is to be used on steel.

Paints made from these alkyds dry by oxidation in the same manner as conventional alkyds. Water-dispersible metallic driers are used to promote oxidation.

Air-dry paints based on water-soluble alkyds have drying times similar to those of similar oil length conventional alkyds but suffer the disadvantage of retaining a degree of water sensitivity in the dried film. For this reason they are usually used on articles where there is no exterior durability required. Such articles include small steel components, tools, trolleys, etc., and timber articles.

The speed of dry and water resistance of water-soluble alkyds can be markedly improved by incorporating a proportion of acrylic latex into the system. Such paints are then suitable for exterior exposure on suitably primed steel, and have been used on agricultural and earthmoving equipment.

38.5.8 Chlorinated Rubber

Chlorinated rubber paints almost belong in the 'lacquer' category, drying primarily by simple solvent evaporation. However, they are not usually referred to as lacquers.

Chlorinated rubber is soluble in aromatic hydrocarbons, esters and ketones. It is normally plasticised with chlorinated paraffins, phthalate esters, long oil alkyd or drying oils. Stabilisers such as epichlorhydrin or expoxidised oil are usually required.

Paints based on chlorinated rubber are used to resist corrosive conditions. They have excellent water, alkali and acid resistance but poorer resistance to solvent, fats, oils and greases.

Chlorinated rubber has a tendency to cob-web when applied by conventional spray, and solvent adjustments may be necessary for successful spray application.

Apart from their widespread use in heavy-duty industrial maintenance coatings, they are used for pre-coated asbestos cement panels, for swimming pool paints, traffic paints and paints for tennis and other playing courts.

38.5.9 Epoxy Resins

Chapter 12 covers epoxy resins in detail. The main usage of epoxy resins is in the catalysed systems or 'baking epoxies'. However, there are two groups of epoxies used in the air-drying industrial field.

38.5.9.1 Epoxy Esters

These are made by reacting epoxy resins with drying oils, and they dry by oxidation of the fatty acid component in a similar manner to alkyds. They do not compare with catalysed epoxies for chemical resistance but are better than alkyds and find use where mild corrosion conditions are encountered. They are subject to early chalking on exterior exposure. Adhesion to difficult metallic substrates is often superior to alternative resins.

38.5.9.2 High Molecular Weight

The high molecular weight epoxies have good solvent release and high film strengths, and dry in a similar way to lacquers. They are poor pigment wetters and not suitable for finishing coats, but have very good chemical resistance. They are limited in resistance to strong solvents.

These find most use in pigmented corrosion-resistant primers, in particular zinc-rich primers.

38.5.10 Latex Paints

The use of polymer emulsions in architectural coatings is now widespread and ever increasing, and in recent years has become common in industrial maintenance coatings for steel. Some of the technology developed in these areas has been adapted for use in industrial coatings. Acrylics, styrene acrylics, polyvinyl acetate and vinyl acetate copolymers are used.

Latex paints are normally suitable for use on absorbent surfaces, where quick dry times can be obtained by loss of water to the substrate, and subsequent coalescence of the film. They generally have good alkali resistance and can be used on alkaline surfaces such as concrete and asbestos cement.

Typical examples of the use of latex paints in industrial applications include primer and finish coats for prefinished asbestos cement panels—usually applied by automatic spray. Other examples are coatings for concrete or bitumen tennis and other playing courts, and line-marking and runway-marking paints. Chapters 16–21 (volume 1) cover emulsion resin technology in depth.

38.6 AIR-DRY COATINGS—CATALYSED

This group of coatings finds widespread use in the industrial maintenance and marine area, where difficult conditions of exposure require coatings with extreme toughness, abrasion resistance, and chemical and corrosion resistance.

Their use is limited in the general industrial area because of their high cost and the requirements of accurate measuring and mixing prior to use. There are also losses associated with the 'pot life', which is the time the mixed paint remains usable after mixing—sometimes less than eight hours and rarely more than twenty-four hours. For this reason they are obviously unsuitable for use in dipping applications.

38.6.1 Polyurethanes

Two-pack polyurethanes exhibit toughness, abrasion resistance, flexibility and good chemical resistance. They also have extremely good resistance to weathering.

The two-pack polyurethanes use the isocyanate reaction, whereby a polymer containing hydroxyl groups is cross-linked with a polyisocyanate. Any resin containing hydroxyl groups can be reacted in this way, and the list includes alkyds, castor oil, urea and melamine formaldehyde resins, epoxy resins, acrylic resins, cellulose derivatives, etc. In practice, the most popular resins used are the polyethers and the saturated polyesters.

Isocyanates are extremely toxic, and are commonly used as an adduct or prepolymer with a polyol to minimise the quantities of free isocyanate released to the atmosphere.

The isocyanate prepolymers fall into two classes, aromatic and aliphatic. The most important aromatic isocyanates are based on tolylene diisocyanate (TDI) and 4,4'-diphenylmethane diisocyanate (MDI); the major aliphatic isocyanate is based on 1,6-hexamethylene diisocyanate (HDI).

The aromatic isocyanates are quicker curing than the aliphatics, but suffer the disadvantage of yellowing on exposure to sunlight. The aliphatic isocyanates are non-yellowing, and their films have excellent exterior durability.

Catalysts such as tertiary amines and tin salts can be used to speed up the curing reaction of aliphatic isocyanates to equal the aromatics. The use of secondary catalysts also allows curing to take place at temperatures as low as 0° C.

In formulating two-pack urethanes, care must be taken to avoid the use of solvents which preferentially react with isocyanates. Water must be excluded from the system. Alcohols and glycol ethers are unsuitable, and some solvents with a degree of water miscibility can be purchased as 'urethane grade', implying low water content.

Thickeners and flow agents are normally used and it is common to use a water scavenger in the system.

Polyurethanes normally require chemical pretreatment of the substrate, or a suitable primer, such as two-pack etch primer or two-pack epoxy primer.

They find wide usage in industrial maintenance finishes for their good chemical resistance. They are also widely used in the general industrial area, where they give long service on aluminium window frames and metal and timber furniture. Their use as a finish coat over epoxy primers is of major importance for painting commercial and military aircraft, marine craft, trucks, buses and similar products. Chapter 11 (volume 1) describes urethane resin technology in more detail.

38.6.2 Epoxy Finishes

The so-called 'cold-cure' epoxy finishes are formed by cross-linking the terminal epoxide groups of epoxy resins with either amines, amine adducts or polyamides. Because epoxy resins are expensive, their use is limited to areas where their good performance justifies the high cost.

The lower molecular weight epoxies are normally used for two-pack coatings and can be pigmented in the normal manner.

Epoxies cured with simple amines are subject to surface exudation, which can be minimised by allowing an induction period of some hours after mixing, before use.

Amine adducts, or polyamines, are made by pre-reacting epoxy resins with an excess of amine. This offers several advantages—no unpleasant odour, more manageable mixing ratios, faster drying times and freedom from exudation.

Polyamide resins are made by reacting dimer acids with an excess of polyamine, which leaves unreacted amino groups on the end of chains. Polyamide resins increase the flexibility of the film but alkali resistance is slightly reduced. Cure is generally slower, and pot life is longer. Catalysts such as tertiary amino phenols are often used to hasten cure when polyamides are used.

The polyamides are good pigment 'wetters' and can be used for pigment dispersion, usually being better for this purpose than the epoxies.

Solventless finishes can be made using the lower molecular weight or liquid epoxies, cross-linked with low-viscosity polyamines or polyamides. The resultant mixture requires no solvent for application and the entire paint converts to a 100 per cent solids finish. High levels of a thixotrope are usually necessary to give hold up on vertical surfaces at film thicknesses up to 250 μm.

A variation of the solventless finish is coal-tar epoxy. Coal-tar pitch is a material with excellent water resistance but poorer solvent and chemical resistance. It can be used to 'extend' the more expensive epoxy resins to produce cheaper coatings with a good combination of properties. They are not necessarily solventless, but are usually more than 80 per cent solids and thixotropes are normally used to allow high build coatings in the 200–250 μm range.

Suitable liquid epoxy resins can be emulsified in water and used as an aqueous two-pack finish. Curing agents are liquid amino–amido resins, which can also be readily emulsified. These types of finish were initially intended for non-metallic substrates but can now be suitably pigmented for application to steel. As with non-aqueous coatings, coat tar can be used to lower the cost, and it improves the water resistance of the coating. In formulating aqueous epoxies, surfactants, buffers, anti-foams and flow promoters are normally required. Their pot life is usually shorter than for non aqueous epoxies.

Despite the excellent properties of the epoxies, they have a major disadvantage in the decorative area. Glossy finishes are prone to early chalking on exterior exposure and are also subject to yellowing with age. They are used where decoration is of secondary importance to protection.

Solventless epoxies are commonly used as linings for storage tanks and oil tankers. Sand blasting is a prerequisite for ultimate corrosion protection, and minimum film thicknesses of 150 μm are required.

Coal-tar epoxies are mainly used in the industrial maintenance and marine areas, and for coating underground tanks and pipelines, both internally and externally.

In the more conventional paints, the epoxies are pigmented with anti-corrosive pigments and used as primers, especially for asbestos cement panels and over a variety of metals used in the transport and aerospace industries. In the latter applications they are usually finished with two-pack polyurethane coatings.

38.7 HEAT-CURED COATINGS

Heat-cured coatings are commonly referred to as stoving or baking enamels and form by far the largest volume group of paints used by industry. Because an article may be painted, allowed to 'flash-off', then cross-linked with heat to produce a completely cured coating having achieved all its film properties within an hour of painting, they are ideally suited to the conveyorised production lines of most industries.

Stoving schedules may vary over a very wide range in several types of ovens, and the industrial coatings chemist must be able to tailor the coatings to suit these conditions. Stoving schedules vary from 30 minutes at 120° C to a few seconds at 250° C.

Coatings required to bake at temperatures below 120° C are usually referred to as low-bake enamels and are commonly air drying coatings which dry with some heat assistance. Low temperatures are sometimes used on materials that are subject to heat distortion such as timber and some plastics.

38.7.1 Alkyd Finishes

Alkyds based on semi-drying or non-drying oils can be cross-linked with amino resins to produce a major class of heat-cured coatings.

Because of their relatively low cost, good adhesion, hardness and abrasion resistance and their low energy requirements for curing, this group of coatings has long been the 'work-horse' of the industrial coatings manufacturer and for many years has dominated the market for metal coatings applied in factories. We have already seen that in 1973 in the U.S.A., alkyds accounted for 34 per cent of all resins used in coatings.

The alkyds are usually of short oil length and based on tall oil, soya, sunflower, castor, dehydrated castor or coconut oil. They are cross-linked with urea formaldehyde, melamine formaldehyde, triazine formaldehyde or benzoguanamine formaldehyde resins.

Where good colour and colour retention are required, it is desirable to use an alkyd based on non-drying oils such as castor or coconut. However, alkyds based on semi-drying oils, in particular tall-oil, probably have the best all-round combination of properties and form the widest usage.

The commonly used amino resins are melamine formaldehyde (MF) and urea formaldehyde (UF). Alkyd: amino ratios vary from 90 : 10 to 70 : 30 (on solids) for MF and from 70 : 30 to 50 : 50 for UF.

In general MF's are higher in cost than UF's but can be used at lower levels, give quicker cure responses and have much better exterior durability than UF's.

Acid catalysts can be used to lower baking temperatures. Para toluene sulfonic acid, sometimes blocked with amine, is commonly used for this purpose. Where colour will allow, metallic driers such as manganese, iron or cobalt octoates can be used to assist the cure.

Alkyd amino baking enamels can be produced in a wide range of colours and gloss levels. They are also suitable for by most methods of application.

38.7.1.1 Spray Techniques

All types of spraying techniques including electrostatic applications are used to apply alkyd-based enamels, mostly to steel or aluminium substrates. Some typical articles painted in this way include metal shelving and supports, filing cabinets, lockers, drums, steel furniture, display stands and cabinets, ducting, water heater jackets, hospital equipment, fluorescent light fittings and small household appliances.

38.7.1.2 Dipping, Flow Coating, Curtain Coating

Each of these methods is used for alkyd–amino coatings. Dipping and flow coating are used for such articles as metal shelves, toys, metal clips and fittings, tools, tubular steel articles and small motor parts including wheels. Curtain coating is used to apply mirror-backing paint, where a coating is applied to protect the silvered surface of mirrors from mechanical damage and corrosion.

38.7.1.3 Roller Coating, Coil Coating

Alkyds can be formulated with suitable flexibility to allow post-forming of metals coated in the flat state. Roller coating is widely used to coat sheets of steel for pails and drums, pans and trays and metal signs. Alkyds are used as the colour coat and the clear overprint varnish for the exterior of tinplate cans and containers.

In coil applications, they are normally used for their lower cost for coating aluminium for venetian blinds, drapery tracks, canopies and awnings.

38.7.2 Thermosetting Acrylics

Thermoset acrylic finishes are a class of coatings with an excellent combination of properties; the past decades have seen them replace alkyd–amino finishes in many areas. Two such areas are in automotive finishes and in the appliance or white goods industry. Chapter 15 gives a more detailed background.

In general, when compared with alkyd–amino finishes, the acrylics as a group:

(a) are harder and have better scratch and abrasion resistance;
(b) have better colour, colour retention on overbake and colour retention on exterior exposure;
(c) have better water and chemical resistance—particularly alkali-detergent resistance, of interest in the appliance industry;
(d) have better gloss retention and resistance to chalking on exterior exposure;
(e) have superior corrosion resistance;
(f) have superior stain resistance;
(g) are more expensive;
(h) require better metal pretreatment;
(i) require higher curing temperatures.

The properties of thermoset acrylics may be widely altered by varying the type and amount of the different monomers used. As well as methacrylate and acrylate esters, styrene is usually used in the acrylic copolymer. As well as lowering the cost, styrene imparts a high degree of alkali-detergent resistance for washing machine finishes. Minimum amounts of styrene are used if exterior durability is required.

There are two common groups of thermoset acrylics—the acrylamide copolymers and the hydroxy copolymers.

The acrylamide copolymers contain acrylamide in the polymer, and in their manufacture are reacted with formaldehyde, then butylated. This produces copolymers capable of internal cross-linking or they may be externally cross-linked with melamine formaldehyde, alkyds or epoxy resins. This group requires baking schedules in the order of thirty minutes at 160° C or twenty minutes at 170° C.

The hydroxy copolymers contain a proportion of hydroxyethyl or hydroxypropyl acrylate or methacrylate, which leaves excess hydroxy groups on the polymer chain. These groups can then be reacted with melamine formaldehyde in a similar way to alkyds. Lower temperatures are required than with acrylamide acrylics but curing is more critical. Both undercuring and over-curing contribute to embrittlement of the film. Typical baking schedules are thirty minutes at 130° C or twenty minutes at 140° C.

Apart from their almost universal use in the automotive and white goods industry, thermoset acrylics are now used by industry where alkyds were previously used, including small household appliances, kitchen and laundry cabinets. In clear coatings they are used to protect silver-plated articles, polished brass and jewellery.

In a roller coating application, acrylics are used for their good colour retention as a white basecoat on tinplate for printed cans and containers.

In the coil coating-industry, they are used on aluminium for residential siding, awnings and canopies, and exterior caravan sheathing.

38.7.3 Vinyls

38.7.3.1 Thermoplastic Vinyls

Solutions of the vinyl chloride–vinyl acetate–vinylidene chloride resins discussed under air-dry lacquers can be plasticised and heat stabilised for use in baking applications, to make them suitable for use in production line situations. When heated at 170–180° C, near the fusion temperature of the resins, lacquers with poor air-dry adhesion may be found to have excellent adhesion to metal surfaces.

The vinyls have a high degree of flexibility and are free from taste and odour and are thus ideally suited for use as can linings. One of the largest uses is as the internal 'topcoat' for beverage containers. They are also used as the internal coating on closures.

In coil coating applications they have the disadvantage of low solids and high cost per given film thickness, but have good flexibility and single-coat adhesion. They are used on the underside of awnings and deckings and for rainwater gutters.

38.7.3.2 Thermosetting Vinyls

Vinyl resins which contain carboxyl, hydroxy or epoxy groups along the chain are commercially available. These reactive groups can be cross-linked with such resins as epoxies or amino formaldehyde resins at baking temperatures in excess of 120° C.

Whilst the additional modifier resin may detract from the taste and odour properties, the advantage of resistance to softening under heat is gained.

In the container industry the thermosetting vinyls are useful for food containers requiring heat sterilisation and for exterior coating of closures. Coil coated strip is used for fabrication of siding and awnings, and refrigerator and freezer inner liners.

38.7.3.3 Organosols and Plastisols

Vinyl coatings with very high film thicknesses can be applied using a vinyl dispersion technique, rather than by depositing the film from solution. The resins are high molecular weight polymers of vinyl chloride and are dispersed in either a plasticiser (plastisol) or plasticiser plus volatile diluent (organosol). The dispersion resins are virtually insoluble except at elevated temperatures, although there is generally some slight swelling of the resin particles, which is useful in minimising settling.

During the curing cycle, under the effect of heat, the solubility of the resin in the plasticiser increases until fusion occurs; the resin/plasticiser solution flows out, and on cooling forms a tough protective coating. Temperatures required are generally in excess of 180° C, although the lower molecular weight resins and some copolymers can yield coatings with fusion temperatures as low as 150° C.

Viscosity control is achieved by the choice and level of plasticiser, and by the control of particle size in the resin.

Plastisols are used to give tough flexible films at up to 5 mm thickness in one application. They are used in the container industry for closures. In the coil industry they are mainly used where a thick coating or an embossed finish is desired.

Where harder less flexible films are required, it is customary to use organosols. Organosols are used for spray application to suitably primed steel for furniture, plating racks, tool handles, etc.

38.7.4 Epoxies

38.7.4.1 Epoxy Esters

Epoxy esters are formed by reaction of epoxy resins with fatty acids and can be cured with amino resins in a similar way to alkyds. They require similar or slightly higher curing temperatures, and are used where higher chemical resistance is required. They are not suitable for exterior durability and are commonly used in primers, particularly for appliances and motor cars.

38.7.4.2 Solution Epoxies

Solid epoxy resins of higher molecular weights (above 1400) can be cross-linked with urea or phenol formaldehyde resins at temperatures of 180–205° C for twenty minutes. The coatings thus formed have extreme chemical resistance as well as being hard, tough and flexible.

The epoxy-phenolics are among the most chemically resistant known, and the group forms the basis of coatings for beverage and food cans, and drum and tank linings. The epoxy : phenolic ratios are in the range 80 : 20 to 70 : 30, and strong solvent systems are required to maintain solution. Acid catalysts are sometimes used to lower baking schedules. The films have poor colour due to the phenolic resin, but have excellent adhesion, flexibility and impact resistance.

The epoxy-amino resins have better colour, but not quite the same degree of chemical resistance. They cure at lower temperatures than epoxy-phenolics, and are used as primers for coil coating, appliance primers, clear coatings for brass hardware and jewellery, can linings, and coatings for hospital furniture.

38.7.5 Polyesters

Polyesters are the reaction product of polybasic acids with polyhydric alcohols. This same general definition also applies to alkyds, which are further modified with monobasic fatty acids and thus are 'oil-modified polyesters'. The polyesters discussed here are not oil modified and are sometimes referred to as 'non-oil alkyds', 'oil-free alkyds', 'hydroxylated polyesters' or 'linear polyesters'. They are also called 'saturated polyesters' to distinguish them from the unsaturated polyesters which are used with glass fibre in the reinforced plastics industry.

The polyols normally used for oil-free polyesters include trimethylol propane, trimethylol ethane, pentaerythritol and 1,6-hexanediol. The acids include isophthalic and adipic in appliance and automotive topcoats. Where extreme flexibility is required as in coil coatings, azelaic acid and the long chain dimerised fatty acids are also used.

These polyesters have an excess of hydroxyl groups on the chain and are cross-linked by reacting with hexamethoxymethyl melamine, which is a monomeric melamine cross-linking agent. Conventional butylated melamine and urea formaldehyde resins have a tendency to undergo self-condensation at high temperatures and are unsatisfactory for curing this type of polyester.

The acid values of these polyesters are not sufficient to effect cure with hexamethoxymethyl melamine, and acid catalysts are always necessary.

Polyesters have excellent gloss and colour retention, and chemical and stain resistance. They have been used by spraying for appliance and automotive applications, but their biggest usage is in the coil industry, mainly for coating aluminium for use in industrial siding, awnings, transportable sheds and caravan sheathing.

38.7.6 Silicone Polyesters

Silicone polyesters are an important group of resins for use in the coil coating industry, but have limited general use because of their high cost.

They are formed by taking a polyester of the type discussed in the previous section and reacting some of the excess hydroxyls with an organo-siloxane intermediate. The resultant copolymers have greatly improved weathering resistance, the durability varying with the level of silicone modification. A typical silicone polyester might be based on trimethylol propane and isophthalic acid, with 50 per cent silicone incorporation.

Like the polyesters, the silicone polyesters are cross-linked with hexamethoxymethyl melamine in the presence of acid catalyst. They require higher curing temperatures than polyesters, a minimum metal temperature of 232° C being required for efficient curing. If curing is not complete, durability properties will not be achieved.

They are used over suitable primers (usually epoxies) on aluminium and galvanised steel residential siding and building panels, and aluminium roofing tiles and shingles.

38.7.7 Silicone Acrylics

Silicone acrylics are formed by reacting terminal hydroxyl groups on a thermosetting acrylic resin with an organo-siloxane intermediate, in a similar way to silicone polyesters. The proportion of silicone modification is not usually as high as with silicone polyesters. Alternatively, silicone resins have been cold-blended with thermosetting acrylic resins to give a marked upgrading of the durability performance of the acrylic.

The silicone acrylics are used for coil coating, usually in the same areas as the silicone polyesters: over a suitable primer on aluminium and steel residential siding and building panels, and aluminium roofing tiles and shingles.

38.7.8 Fluorocarbons

The fluorocarbons, like the silicone-modified resins, are a group of resins of interest in coil coatings. They are very high cost resins, but they have such excellent resistance to weathering that coatings with an expected minimum life of twenty years can be produced.

The two resins in this group are polyvinyl fluoride and polyvinylidene fluoride. Both polymers have outstanding chemical resistance and resistance to thermal degradation, as well as the excellent resistance to weathering mentioned above.

They are virtually insoluble in any common solvents at ambient temperatures and are used as a dispersion, in a similar manner to the organosols. They require minimum metal temperatures of 232° C for fusion. Like plastisols and organosols they are suitable for embossing.

The fluorocarbon-based coatings are used over suitable primers on coil coated strip where the ultimate in durability is required. This includes aluminium and galvanised steel residential sidings, and aluminium roofing tiles and shingles.

38.7.9 Polyimides

The polyimides are made by the polycondensation of an aromatic dianhydride and an aromatic amine.

Polyimide coatings have exceptional heat resistance, retaining usable properties after exposures to 300° C for months, 400° C for hours and even 500° C for some minutes. Because of this unique property, they find use in coloured coatings for cook-ware, and obviously must be pigmented with inorganic pigments capable of withstanding the required operating temperatures.

38.7.10 Aqueous Systems

Water-reducible systems are one group of coatings developed to cope with problems of pollution, energy shortages and flammability.

The aqueous systems are better suited to heat curing than air drying. Air-dry systems have relatively long drying times related to the low evaporation rate of water (except where the water is absorbed by the substrate), and the air-dried films tend to retain a degree of water sensitivity. Both of these disadvantages are overcome by using heat energy to remove the water and cross-link the coating.

38.7.10.1 Water-soluble Alkyds

As discussed under air-dry coatings, alkyds with acid values of 50 or more can be neutralised with a volatile amine and dissolved in water or water/solvent mixtures. As with solvent-borne alkyds, these can be cross-linked with amino resins, which will react with either hydroxyl or carboxyl groups.

The favoured cross-linking agent for these systems is hexamethoxymethyl melamine which is both water tolerant and stable in water systems.

Similar pigment dispersion techniques are used as with air-drying water-soluble alkyds; but in addition, pigments can be dispersed in the melamine.

The water-soluble alkyds generally find application in similar areas to the solvent-borne alkyd–amino coatings. They are commonly applied by conventional spray or dipping, but as experience increases, it is expected that they will prove suitable for use by other techniques.

Some typical items coated with water-soluble alkyd baking systems include car and truck wheels, tools, wire articles, trolleys and refrigerator evaporators.

38.7.10.2 Aqueous Emulsions and Dispersions

The technology of aqueous emulsions and dispersions began with architectural paints, the polymers for which are thermoplastic and of limited interest in industrial coatings.

Thermosetting polymers are prepared in a similar way to solvent-borne polymers by building reactive groups into the polymer chain, then curing by self-condensation or cross-linking with amino resins.

Acrylics are probably the largest group in commercial use, but polymers based on vinyl acetate and styrene/acrylic can be made thermosetting in a similar way. Binders using low molecular weight polybutadiene are used for primers for their good corrosion resistance on non-phosphated steel substrates.

As with water-soluble alkyds, the preferred external cross-linking agent for emulsions and dispersions is hexamethoxymethyl melamine.

Emulsions generally have poor rheology and associated problems of skinning and blistering ('solvent' boil). Long flash-off times before baking are required.

Water-reducible polymers can be made to behave more like solvent systems by including an excess of acid groups and 'dissolving' in volatile amine/water mixtures (in a similar way to water-soluble alkyds). A disadvantage of this approach is that 'solution' viscosities are high and the coating has low solids at application viscosity, or low molecular weight polymers with poorer performance properties must be used.

A compromise between emulsion and water-soluble polymers can be obtained with solubilisable dispersions. These are prepared using emulsion techniques but they become water soluble under certain pH conditions during application. They offer a favourable balance of properties, with good durability, moderate solids at application viscosity, few application problems, and good pigment dispersion properties and available gloss levels. Viscosity is increased by the addition of co-solvent.

The main area of usage of aqueous emulsions and dispersions is in coil coatings where they have good application properties; the coatings have excellent flexibility and forming properties, and good exterior durability.

38.8 RADIATION-CURED COATING

Radiation curing is carried out using ultraviolet light or electron beams, although at this stage, commercial applications are primarily ultraviolet curing.

Radiation curing is one of the environmentally acceptable methods and offers a number of advantages. It is expected to be one of the techniques which will grow rapidly in years to come.

The coatings are 100 per cent reactive and no evaporation of solvent takes place, thus avoiding pollution problems.

Faster curing rates are available than with other types of coatings, thus giving faster production rates and less 'oven' space. For example a UV-curing oven is only 10–20 per cent as long as a conventional infra-red oven.

The low curing temperatures allow heat-sensitive substrates to be coated; and in the case of printing inks, lower grades of paper can be used.

Radiation-cured coatings are covered in more detail in chapter 43.

38.8.1 Ultraviolet Radiation Curing

Ultraviolet curing uses UV radiation to generate reactive intermediates (free radicals) which initiate addition polymerisation. This takes place without the necessity for any heat application, nor is any significant amount of heat generated in the polymerisation.

The three components necessary for this type of coating are a polymerisable resin, a photo initiator, and a source of UV radiation. A number of factors affect the efficiency of the system, in particular pigmentation which tends to absorb the UV radiation, and film thickness of the coating where incomplete penetration of the UV occurs in thick coatings.

The major industries using UV curing are wood finishing, where clear coatings are applied, and the printing industry where thin films of pigmented coatings are required.

Compared with infra-red cured coatings, the UV curing process requires much lower energy.

The major class of coatings used for wood finishing by this method are clear coatings based on unsaturated polyesters, dissolved in a vinyl monomer such as styrene, vinyl toluene or one of the acrylic monomers. In printing inks, vinyl prepolymers dissolved in vinyl monomers are used.

Other resin systems that are suitable for ultraviolet curing are certain types of acrylics, urethanes and alkyds. These systems are at different stages of development and each system has its own advantages. In all cases a photo-initiator is necessary.

38.8.2 Electron Beam Curing

Electron curing is similar to ultraviolet curing except that electrons are used to form the free radicals which initiate the vinyl polymerisation. No photo-initiator is necessary.

This method can be used to cure all of the types of coatings discussed above under ultraviolet curing, but it has not been commercially developed to the same degree, mainly because of the high cost of equipment and the necessity for shielding the operators from the X-rays produced. In recent years, new cathodes have been developed which operate at lower energies and reduce the amount of shielding required.

The major advantage of electron beam radiation is that curing of pigmented coatings is much better than with ultraviolet.

Both methods have the disadvantage of requiring direct exposure of the coating to the radiation source, and this can only be conveniently arranged for sensibly flat articles.

39 AUTOMOTIVE COATINGS

39.1 DEVELOPMENTS IN AUTOMOTIVE FINISHES

Surface coatings for the mass-production automotive industry are undergoing very significant evolutionary changes because of the influence of *energy conservation*, *environmental controls* and *consumerism*. Thus, in a rather conservative industry where the costs and stakes are very high, there is real incentive to accept and invest in new products and processes that would make the painting of a motor car more economical and effective.

At the current state of knowledge, the industry cannot adequately satisfy the requirements of energy, the environment and consumerism, not only because these requirements are often in conflict, but because the technology, as such, is not yet available.

39.1.1 Environment Protection

Automotive coatings, because of their large usage volumes and basically inefficient methods of application, contribute very significantly to overall environmental pollution. It has been estimated that to coat twenty vehicles in a mass-production automotive plant with 50 μm of finish, utilising a material at 35–40 per cent non-volatile material on application, through conventional spray equipment of 40–45 per cent efficiency, the pollution emitted to the atmosphere is in the vicinity of 200L of solvent and 65 kg of solid contaminant in the form of overspray, either emitted to the atmosphere or collected as sludge in the spray booth. Considering that the average output of a motor plant can vary from 150 to 400 vehicles per day, the total pollution is a most significant factor.

To counter such a situation, industrialised countries are issuing a number of regulations significantly affecting both motor vehicle manufacturers and paint manufacturers. A typical and very explicit recent regulation is the Californian Air Resources Board model rule that limits solvent emission on a mass per coating volume basis (minus water) in automotive assembly painting plants, as follows:

> By December 1980, the solvent content of topcoats for light and medium duty vehicles should be reduced to 520 g/L and that of primer coats to 350 g/L. After December 1984 the limit will be reduced to 275 g/L for both primer and topcoat application. An additional provision proposes to limit the assembly plant emission to 4.2 kg of organic compounds per vehicle produced based on a 24 hour average.

Current estimates indicate that automobile plants using thermosetting and thermoplastic surface coating systems emit up to 8 and 20 kg per motor vehicle, respectively. These figures do not include the effluent evolved during the 'curing' (stoving) of the surface coatings after the solvent has been expelled, nor the solid contaminant emitted to the atmosphere as overspray.

To achieve conformity to such regulations, both the motor and paint industries are actively pursuing various approaches such as:

(a) incineration of volatile effluents combined with carbon absorption techniques;
(b) higher application non-volatile content formulae based on solvents;
(c) high-efficiency application equipment, such as high-voltage electrostatic systems;
(d) water-based compositions low in organic solvent content; and
(e) powder coating.

TABLE 39.1
Surface coating and contamination compositions of four systems for automotive finishing
(based on a car body of 55.65 m² and applying 25 μm of dry paint)

	Examples			
Data	*1*	*2*	*3*	*4*
Weight of dry paint	931 g	931 g	931 g	931 g
Paint: non-volatile	35%	60%	35%	60%
solvents	65%	40%	65%	40%
Applied material	2665 g	1548 g	2665 g	1548 g
non-volatile	931 g	931 g	931 g	931 g
solvents	1734 g	617 g	1734 g	617 g
Application efficiency	43%	43%	85%	85%
Material sprayed	6197 g	3614 g	3132 g	1821 g
non-volatile	2170 g	2170 g	1096 g	1094 g
solvents	4027 g	1444 g	2036 g	727 g
Material lost on overspray	3532 g	2066 g	468 g	282 g
non-volatile	1239 g	1239 g	163 g	163 g
solvents	2293 g	827 g	305 g	119 g
Total contamination				
solid material	1239 g	1239 g	163 g	163 g
solvents	4027 g	1444 g	2038 g	726 g

Table 39.1 illustrates the efficiency of application of four automotive finishes with respect to overall pollution.

39.1.2 Energy Conservation

Most authorities predict that conservation of energy will progressively become a major issue, possibly more so than pollution, and decisively influence many manufacturing operations.

Over the past 25–30 years, the surface coatings used on automobiles have undergone considerable evolution in overall performance and aesthetic appearance. By comparison, the painting process of the motor car has remained substantially unchanged in the method and sequence of operation. Specifically, it still involves three or four applications with at least three stoving/curing stages, all of which are high in energy demand, high in initial cost and relatively inefficient. Consider the following actual situation:

The paint on a motor car represents 0.4–0.8% of the total mass and about 2–5% of the overall material cost. The application and processing of such a quantity of paint (10–20 kg), in a typical automotive assembly plant of 25–30 cars/hour output, involves:

(a) 20–25% of the total available floor space, representing up to 40% of the total assembly plant costs;
(b) 20–25% of the total work force;
(c) 40–50% of the total energy available;
(d) 80–90% of the pollution generated by the plant; and
(e) the use of high stoving/curing temperatures in the region of 140–180° C.

The resolution of these factors in the near future will involve:

(a) simplified automotive coating systems;
(b) more efficient application equipment and techniques; and
(c) lower energy curing mechanisms.

39.1.3 Consumerism

The main concern by consumers for motor vehicles, apart from mechanical aspects, is about corrosion in its many forms. In general, the motor industry considers the topcoats currently in use as adequate for protection and decoration and will not be willing to invest in any further improvements or changes unless forced by pollution and/or energy conservation aspects. However, the industry is not satisfied with the overall corrosion protection, either cosmetic or due to perforation, in the first 3–5 years of motor vehicle life. Corrosion in a car, even of cosmetic nature, degrades its appearance, reduces resale value, implies poor design and workmanship, and reflects adversely on the important issue of safety.

The performance criterion is to aim for no cosmetic corrosion on external surfaces for 3–5 years and no perforation for 5–10 years—preferably the longer period, in each case.

The motor industry has realised that the corrosion of a motor vehicle is due to three main parameters—material of construction, vehicle design, and performance of automotive coatings—and that the most effective resolution is by:

(a) the use of clean, cold-rolled steel, achieved by an electrolytic cleaning operation at the steel mills or motor plants for improved corrosion performance;
(b) the use of improved zinc phosphate pretreatments tailored to the particular priming operation;
(c) attention to the cleanliness of the car body shell before it is presented to the paint shop to reduce carry-over of debris to subsequent operations;
(d) judicious use of zinc, anti-corrosive oils and barrier coatings;
(e) improved corrosion and saponification-resistant primers, electrodeposited or conventionally dipped; and
(f) consideration of the design of the motor vehicle to provide good access of the coating to enclosed sections, large drain holes in door and floor panels, good inner structure ventilation, adhesive bonded flanges, hood, door, deck lids, and plastic spacers for car body mouldings and trims.

39.2 MATERIALS USED IN MOTOR VEHICLE CONSTRUCTION

Steel remains the main construction material for the motor car, on the basis of its cost, availability, strength and formability and the huge investment of the motor industry in tooling and related fabrication processes, including its recycling capabilities. Thus, of the total mass (1000–1200 kg) of an average saloon vehicle, the average distribution of the mass of components is approximately

50 per cent for motor and chassis, 40 per cent for body and floor assembly and 10 per cent for parts made from plastics, elastomers, surface coatings and sealing compounds. The distribution of the materials employed is shown in table 39.2.

TABLE 39.2
Percentage by mass of typical automotive materials

Material	*Percentage by mass*
Ferrous metal	82.5
Non-ferrous metal	6.0
Plastic	4.5
Glass	2.5
Rubber	3.0
Paints & sealing compounds	1.0
Soundproofing	0.5
	100.0

Over the past few years, considerable advances have been made in steel technology as used by the motor industry, as follows:

(a) the use of higher tensile strengths to reduce mass by using thinner gauges;
(b) improved processing techniques during annealing to avoid formation of carbon deposits on the surface which tend to cause poor phosphating and paint adhesion, resulting in inferior corrosion performance and paint appearance; and
(c) the use of one-sided zinc-coated steel applied electrolytically ('electrogalv'), hot dip (galvanising) or as organometallic coatings ('Zincrometal') in order to protect vulnerable interior areas on a car body such as rocker panels, fenders and door inners, extending to a two-sided version for more critical areas such as floor pans and side sills.

The trends that will affect material usage can be grouped into four main categories:

(a) changes in car design due to the requirements for increasing fuel economy, reduced exhaust emission, safety legislation, improved reliability and serviceability;
(b) changes in the method of manufacturing aimed at economies in energy usage and increased productivity;
(c) requirements to increase the recyclability of materials; and
(d) availability of new alloys with improved properties leading to economies in fabrication, cost of raw materials and running costs.

Thus, in the early 1980s, the average family car includes up to 100 kg of aluminium and an equal quantity of plastics of various types for exterior and interior parts. With the objective being to reduce running costs, a kerb mass reduction of 50 kg can reduce fuel usage by 1.5–2 per cent. Of the total energy consumed in making and using a motor vehicle, 10 per cent is consumed in producing the raw materials, 5 per cent by the various manufacturing processes, and 85 per cent as fuel during the life of the car.

39.3 AUTOMOTIVE SURFACE COATING SYSTEMS

The function of an automotive coating system is to protect and decorate, under a variety of adverse and challenging conditions, for long periods of time.

Vehicle design, the use of dissimilar metals in assembly operation, warranty clauses required by the manufacturers, combined with energy and economic pressures, present a real challenge to the automotive surface coatings technologist.

The systems developed for such applications comprising surface preparation, priming, filling and topcoating, are totally integrated to supplement each other whilst performing their individual specific functions.

39.3.1 Surface Preparation

39.3.1.1 Inorganic Conversion Coatings

(See also chapters 41, 49, 51.) This stage involves the chemical modification of the metal surface prior to the application of an organic protective coating. In automotive finishing, this is usually carried out by the application of a *zinc phosphate conversion coating* in five to eight stages. The principal three operations are *cleaning*, *phosphating* and *rinsing*. When properly performed, each operation contributes significantly to good corrosion performance, adhesion and cleanliness of the complete finish.

As with most processes, effective zinc phosphating relies on the condition of the car body shell ('unit') before it reaches this important stage. The unit must be pre-cleaned in the body shop and be free from rust, temper discoloration, chemical etching, metal particles, welding smudges and excess sealing compounds from pre-priming operations, such as vinyl sealing and zinc-rich priming.

Commercial 'phosphate' systems used by the automotive industry consist of metal phosphates dissolved in carefully balanced solutions of phosphoric acid so as to remain in solution. When a reactive metal is immersed in the phosphating solution, some 'pickling' takes place and the acid concentration is reduced at the liquid–metal interface; iron is dissolved, hydrogen is evolved and a phosphate coating is precipitated. The fact that these coatings are formed in place at the metal surface, incorporating metal ions dissolved from the surface, leads to their description of *conversion coatings* which are integrally bonded to the metal. In this respect, these phosphate coatings differ from electrodeposited coatings which are superimposed on the metal; typical zinc phosphate coatings used in automotive surface coatings range in film weight between 100 and 350 g/m^2 and are of amorphous or crystalline nature depending on the particular process.

Conversion phosphate coatings are electrically inert and, by covering and insulating the anodic and cathodic areas characteristic of steel surfaces, they prevent electrochemical corrosion under subsequent coatings.

The greatest challenge to the technology of conversion coatings occurred with the introduction of electrodeposited anodic and cathodic primers. Zinc phosphate systems, which have given reliable performance over many years under conventional solvent-type dip primers, proved deficient and unreliable under the new electrodeposited primers. Typical failures were:

(a) rapid under-paint film failures as corrosion blisters ('scab') up to several centimetres in diameter originating from minor paint film damage and leading to loss of paint adhesion within a very short period of time;
(b) removal of the painting system, carrying some phosphate film with it and exposing the chemically active substrate, due to sudden bends and impacts resulting from collision.

One of the proposed mechanisms for this type of failure is that, during electrodeposition, the very high electrical field strengths that are developed rupture many of the metal–phosphate bonds, weakening the adhesion of the phosphate layer to the steel surface.

A new set of requirements for phosphate conversion coating was therefore developed regarding film thicknesses, density, crystallinity, uniformity, chemical composition and final chemical rinses.

39.3.1.2 Methods of Application

Conversion phosphate coatings used by the automotive mass-production industry are invariably spray-applied for performance and production reasons. However, there is an increasing trend towards spray-dip phosphate processes where the lower part of the car body (floor panel, box sections) are dipped for more efficient cleaning and phosphating whilst the rest of the vehicle is sprayed.

A typical six-stage zinc phosphate process utilises the following steps:

(a) Pre-clean stage: 60–80 s contact time with alkaline solution at 60–100° C.
(b) Rinse stage: 30–50 s rinse under recirculated water at 40–60° C.
(c) Rinse stage: 30–40 s rinse, with water plus additives/activators for zinc phosphate conditioning, at 40–50° C.
(d) Zinc phosphate stage: 60–90 s contact time with zinc phosphate systems at 40–60° C.
(e) Rinse stage: 20–30 s rinse under recirculated water at 40–50° C.
(f) Acidified rinse: 20–30 s acidified rinse at 60–70° C.

An installation for a conventional 20 cars/hour plant will be up to 50 m long, occupying 5–10 per cent of the total paintshop area.

39.3.1.3 Comparative Advantages of Iron and Zinc Phosphate Coatings

As pre-paint treatments, both the iron phosphate process and the zinc phosphate process have inherent advantages which determine their selection for a particular application, often based on a balance between cost and overall quality.

A clear economic advantage of the iron phosphate method is that it usually requires fewer processing stages, because cleaning and phosphating can generally be accomplished in one step; the precleaning and rinsing stage associated with other processes is not required. Another economic advantage of the iron phosphating process is that the special acid-proof equipment construction frequently associated with other processes is not required. Thus, the initial capital investment for the iron phosphate process is usually considerably lower.

On the other hand, zinc phosphate coatings, on account of their physical structure, chemical bonding to the metal substrate and film thickness, impart better adhesion and corrosion performance to organic coatings. For these reasons, manufacturers of products with long life expectancy in corrosive environments—as is the case with the automotive industry—predominantly use zinc phosphate coatings despite their higher cost per vehicle.

39.4 PRIMING

Despite advances in the metallurgy of steel, conversion coatings and surfacers (fillers), priming still remains a very fundamental stage in the automotive surface coating system.

The function of the primer is to form a protective coating on the metal, resistant to corrosion and chipping, to which the subsequent system can adhere. Such primers as a rule need not hide metal defects to the extent expected from body surfacers; but they must incorporate a high order of film integrity, because they provide the sole protection for the vulnerable internal sections and undersides of a car body.

Until the late 1960s and early 1970s the priming systems were of the solvent-borne dipping type with occasional water-borne compositions. A major breakthrough was achieved with the advent of electrodeposited primers, initially of the anodic type and subsequently of the cathodic type. Their wide and unqualified acceptance by the motor industry is due to improved performance, economies in operation, pollution reduction and greater degree of automation.

39.4.1 Solvent-borne Dip Primers

Products of this type have a typical composition shown in table 39.3. The application is invariably carried out by dipping the car body into the primer, operating at relatively low viscosities, such as 14–16 s No. 4 Ford cup and, after a suitable draining period, curing in the region of 150 ± 20° C for 20–30 minutes. Under normal dipping conditions, a film thickness gradient of 5–20 μm (dry) is achieved consistent with flow, appearance and performance over the length of a car door.

TABLE 39.3
Typical solvent-borne dip primer composition

Characteristic	*Composition*
Vehicle	Based on medium oil length alkyd or epoxy ester resins
Oil type	Semi-drying types
Pigmentation	Mainly extenders typified by barytes and/or carbonates, with small percentages of carbon black or iron oxide for colour and coverage, and zinc chromate or lead silico-chromate as anti-corrosive agents in small percentages (2–5%)
Solvents	Medium to low aromatic solvents such as white spirit and mineral turpentine
Pigment volume	Normally low, typically 20–30%
Non-volatile material by volume	35–40% approximately

A dipping operation's success is highly dependent on the composition of the primer, its viscosity, solvent type, rheological properties, the shape of the article and rate and angle of withdrawal. Although such primers can be formulated with an excellent balance of corrosion and mechanical properties, they invariably exhibit two handicaps:

(a) Poor aesthetic appearance, unless the manufacturer limits the dipping operation to a practical shallow dip (3000–5000 mm). 'Washing off' of the outside critical areas is commonly a time-consuming and labour-intensive operation.
(b) Solvent refluxing of enclosed sections (inner doors, box section, etc.) during drying and stoving removes the primer substantially from these critical areas.

Alternative supplementary processes are occasionally employed, such as flow-coating and roto-dipping, for overcoming these limitations.

39.4.2 Electrodeposition Primers

Electropriming of motor vehicles began early in 1960, some time after the principle of electrodeposition of paint had been established in various other areas. Since then there has been widespread adoption of the process throughout the automotive industry, and now almost all mass-produced cars are primed in this manner.

The main advantages claimed for this process are:

(a) uniform film thickness (in the range of 15–30 μm) on all surfaces, including areas normally considered inaccessible and difficult to paint; refluxing is absent;
(b) excellent adhesion and corrosion resistance;
(c) the deposited film contains less than 5 per cent water and thus requires no 'flash-off' before curing;
(d) greatly reduced fire hazard and air pollution, because the main solvent is water;
(e) fully automated and mechanised processes utilising up to 95 per cent of the coating materials;
(f) low maintenance and man-power costs.

The only real disadvantage of the process is the high installation cost, limiting its use to the treatment of a very large number of similar articles.

Electrodeposition (EPD) is a process analogous to electroplating; an anode and cathode are immersed in an aqueous medium through which a current is passed.

39.4.2.1 Anodic EPD

In *anodic EPD* the article is made the anode (positive) with the negatively charged paint solids being attracted to the article and deposited more or less uniformly over the surface(s) in contact with the paint, to the desired film thickness.

During the process of film deposition, the film thickness increases very rapidly at first, then more slowly; the electrical resistance of the film increases sharply towards the end of the deposition. The applied current consequently diminishes with the increased resistance; insufficient protons are therefore produced at the anode to reduce the pH and cause resin precipitation (see later) and therefore deposition virtually ceases. In practice, the areas of the anode (article) closest to the cathode are coated first and, as the film build increases and builds up resistance, the more remote areas are then coated.

Typical operating characteristics of an electrocoat installation are shown in table 39.4.

The most significant part of the EPD process is the composition of the paint and, in this area, most of the progress has been made during the past decade.

Essentially electropaints consist of two components: the anodic system comprises an acidic component, which is the film former, and an alkali component. The acidic system on its own is completely insoluble in water. However, by including an alkaline material it can be completely solubilised. During the electrodeposition process the acidic and alkaline components are separated. The former then forms the adherent paint film on the article, and the latter is liberated. The alkaline component is therefore essential to solubilise the paint initially, but forms no part of the paint film.

TABLE 39.4
Typical operating conditions of an anodic electrocoat bath

Operating parameter	*Typical figures*
Non-volatile content	10–12% by mass
Operating temperature	25–30° C
Applied voltage	150–250 V
Coating time	2–3 min
Distance between article and cathode	30–50 cm
pH	7.8–8.4 at 30° C
Specific conductivity	3200–6500 microsiemens
Coulomb yield	minimum 10 mg/C
Stoving	150–180° C (20–30 min)

It follows that as more and more articles are painted, more and more liberated alkali builds up in the paint bath. Unless this is prevented, the paint will finally become so alkaline that it will no longer be possible to deposit the acidic component, because this dissolves as rapidly as it is formed in the concentrated alkali surrounding it.

Therefore, it is essential to devise some means of preventing this build-up of alkali—that is, some means of pH control. There is also a need, of course, to replace the acidic component of the paint bath which is removed on each painted article.

The two principal methods used for paint control are the *base-deficient system* and the *membrane system*. The former method involves the addition of acidic or base-deficient paint which in turn reacts with, and is solubilised by, the excess of base present in the dip tank. In the latter method, the alkaline component is removed from the paint by passing it through a semi-permeable membrane. There are many automotive plants around the world using either method of control successfully.

Since the introduction of electrodeposition to the automotive industry, numerous anodic polymer families have been developed to achieve optimum coating characteristics combined with the desired corrosion resistance. Such anodic polymers are maleinised oils, polyesters, epoxy esters, graft polymers, maleinised polybutadiene; combined with specific anti-corrosive pigmentation, they achieve the optimum balance of performance properties, particularly over well-phosphated steel surfaces.

39.4.2.2 Cathodic EPD

Further significant improvements in anti-corrosion (particularly over poorly phosphated and/or bare steel surfaces), improved mechanical properties such as chip resistance, and higher resistance to saponification in contact with dissimilar metals, were achieved with the advent of cathodic electrodeposition. Two of the most important inherent characteristics of cathodic polymers responsible for improved performance are:

(a) The cathodic resin migrates to the cathode in the same direction as metallic ions.
(b) On electrodeposition, the cathodic resin is deposited as an alkaline-type polymer which tends to be an inherent corrosion inhibitor.

The necessary characteristics of a cathodic primer were achieved by specific polymer development of non-oxidising epoxy-type resins solubilised at low pH (3–7) and the selection of pigments compatible with, and stable in, this polymer system.

Operating parameters for a cathodic EPD System are shown in table 39.5.

TABLE 39.5
Typical operating conditions for a cathodic electrocoat bath

Operating parameter	*Typical figures*
Non-volatile content	20–22% by mass
pH	6.0–7.0
Conductivity	1100–1600 μS
Tank temperature	16–28° C
Voltage	250–350 V
Deposited thickness	15–22 μm
Curing temperature	30 min @ 175–180° C
Effective throwing power	25–30 cm

The basic equipment for cathodic electrodeposition is similar to that of anodic systems. Some specific differences characteristic of the process are:

(a) The part to be coated must be the cathode.
(b) Mild steel tanks must be lined with a resistant lining to prevent attack from the acid environment characteristic of cathodic paint systems.
(c) Special quality stainless steel electrodes are used to resist attack by the acid solubiliser that migrates to the anode.
(d) Membrane control is desirable to prevent low molecular weight acids produced in the EPD process from accumulating in the paint bath.
(e) Ultrafiltration is an important part of cathodic systems to provide nearly 100 per cent utilisation of paint, minimise waste treatment and remove any water-soluble contaminants carried into the tank.

Because of the improved corrosion performance—unmatched currently by any other organic-type primer—cathodic electrodeposition has found wide acceptance throughout the motor industry and is displacing the anodic types.

39.4.3 Ancillary Primers

During the assembly of a motor vehicle and after the complete painting operation, several ancillary primers are used to enhance corrosion protection in specific areas where the design and assembly operations prevent adequate protection by the main automotive painting system. Typical of the range of products used are zinc-rich primers and anti-corrosive waxes.

Automotive-type zinc-rich primers are of the 'weld-through' quality. They are electrically conductive and, therefore, parts coated in this way must be welded together and subsequently coated during electrodeposition priming. They are formulated using special grades of zinc dust and low acid number chemical resistant polymers, such as the epoxy resins, the zinc forming 90–95 per cent of the formula. They are used in the early stages of motor vehicle assembly in areas such as hem flanges in hoods, doors, deck lids, box sections and wheel fenders.

Anti-corrosive waxes are normally formulated on natural oils (typically fish oils) or synthetic polymeric waxes, such as polyethylene, modified with thickeners and anti-corrosive organic compositions. They are normally applied at the final stages of the painting operation to provide additional sealing of the internal areas of motor vehicles and thus enhance corrosion protection due to their inertness and low water permeability.

39.5 SURFACER/FILLERS

The function of the *surfacer/filler* in the automotive coating system is three-fold:

(a) to provide a foundation for the subsequent topcoat;
(b) to supplement the mechanical and corrosion performance properties of the overall system; and
(c) to permit rectification of metal and primer imperfections mainly for aesthetic reasons.

These properties are normally achieved at an average dry film thickness of 20–30 μm in a curing cycle of 20–30 minutes at 150–170° C, depending on the composition.

Specific characteristic properties of surfacer/fillers are good filling, easy sanding, good flow, compatible in performance with the primer and topcoats, good humidity and corrosion performance and good overall adhesion.

Currently the predominant type of surfacer/filler is solvent based, although water-borne compositions will make inroads slowly as developments in their technology make them more adaptable to existing car plant facilities and processes.

39.5.1 Solvent-based Surfacer/Fillers

Products of this type are usually grey, buff or red in colour. Their pigmentation consists of a combination of hiding pigments such as carbon black, titanium dioxide, iron oxides and extenders such as barytes, clays and talcs. Anti-corrosive pigments containing chromates or lead are not commonly used because of toxicity and allergy problems probably caused by the dust created during the sanding and rectification operations.

Pigmentation, although simple in concept, is most important in surfacer/fillers, and properties such as particle size and shape, contaminants, and type and amount of water-soluble chemicals often determine the performance of the product.

However, over the past few years, the most significant change in the formulation of these products has been in the vehicle system, where the traditional epoxy-ester types are gradually replaced by oil-free polyesters to improve weathering and, particularly, resistance to chalking. This requirement received importance with the widespread introduction of metallic topcoats: the transparency and low film thickness of these coatings permits sunlight to reach the surfacer/filler, causing a type of deterioration that weakens topcoat adhesion, resulting in massive film failure, referred to as 'sheeting' or delamination. Incorrect pigment selection such as the inclusion of lithopone will aggravate this type of failure.

For a given pigmentation and vehicle system, the level of pigmentation varies considerably, depending on such factors as gloss, filling, sanding properties and whether the topcoat is of a thermosetting or thermoplastic type. Generally, in thermosetting systems the pigment content of surfacer/fillers is relatively low (pigment : binder ratio 1–2 : 1) whilst in thermoplastic coatings it is higher (typically 3–4 : 1) for specific adhesion and performance requirements.

The trend in surfacer/fillers is towards lower pigment contents, higher application solids and improved flow. The flow is particularly important when applied with high-voltage electrostatic equipment such as the Ransburg No. 2 Bell type.

A representative formula for a thermosetting surfacer/filler is shown in formula 39.1.

39.5.2 Water-borne Dip Primers/Spray Surfacers

Products of this type are usually based on high acid number vehicles of the same chemical structure as are used in their solvent-thinned counterparts and pigmented on similar lines. The high number of acid groups, normally obtained through incorporation of maleic anhydride, is reacted with amines to impart water solubility. Water-miscible solvents such as ethylene glycol monobutyl ether are normally used up to 20–30 per cent to facilitate solubilisation, and to improve flow and wetting of metal substrates.

Cross-linking is effected or assisted through water-soluble melamine resins of the hexamethoxymethyl melamine types and/or metallic driers.

Apart from reduced solvent pollution and fire hazard properties, water-borne dipping primers exhibit the same limitations in dipping characteristics and performance as their solvent based counterparts. Their application is therefore limited to small output (5–10 vehicles/hour) automotive plants and generally have been superseded by the electrodeposition process.

Water-borne surfacers have also found limited application in mass-production automotive plants mainly in the U.S.A. under pressure from the environmental authorities. Their problem areas are:

(a) limited application latitude in conventional spray-booth facilities, although not to the extent of water-based topcoats;
(b) they are not easily adaptable to electrostatic application facilities currently in use in most automotive plants. However, they exhibit inherently very good flow, holdout and balance of performance properties.

FORMULA 39.1
Formula for a light-grey non-sanding surfacer/filler (thermosetting type)

	Parts by mass
Carbon black	0.1
Titanium dioxide (rutile)	3.8
China clay	3.4
Barytes	23.9
Tall oil epoxy ester (50% NV)*	42.0
Melamine formaldehyde (60% NV)	0.5
Urea formaldehyde (60% NV)	3.5
Aromatic solvent	22.2
Flow control agent	0.2
Curing catalyst	0.4
	100.0
Constants	
Pigment : binder ratio	135 : 100
Non-volatile material	55% by mass
Application viscosity	21–23 s Ford cup @ 20° C
Thinner reduction	20–30% by volume
Application method	DeVilbiss JGA502
	Binks No. 19
	Ransburg No. 2 Bell type
Fluid rate/stage	conventional spray
	300–400 mL/min
Bake	30 min @ 150–160° C
Film thickness	25–30 μm

* Usually 60 : 40 : oil : epoxy resin.

39.5.3 Low Volatile Organic Content Surfacer/Fillers

Future trends in primer surfacers are towards high solids, *low volatile organic content* (VOC) primers.

Water-based primers have found limited use as a means of reducing the VOC because their use presents problems with application and with equipment requirements. Application problems, due to the low volatility of water as a solvent and corrosion problems associated with the use of water-based paints, necessitate stainless steel plant and equipment.

Because of these problems, high solids primer surfacers, based upon organic solvents, are proving to be more widely acceptable. This is because they do not involve any significant changes in plant and equipment or in application technique. Oil-free polyester resins are the most common vehicles used for high solids primers.

39.5.4 New Developments

39.5.4.1 Powder Primers

These represent the ultimate goal in emission control, having no volatile organic content. Powder primers have excellent corrosion resistance and durability, but there are problems with film build control when applied by conventional means. They are not very widely used but could be in the future.

39.5.4.2 Electrophoretic Powder Coatings

These are powder coatings applied electrophoretically in a dip tank. The system is used in conjunction with a conventional electrocoat tank in the so-called 'reverse electrocoat process'. In this process, a coat of primer surfacer is applied to the exterior surfaces—the electrophoretic powder primer—followed by a coat of conventional electrocoat primer to the interior surfaces in a second tank. The films are co-baked. This process has the benefit of allowing the electrocoat primer to be formulated for maximum corrosion resistance and throwing power without the incumbrance of any exterior film requirements to compromise these properties. In addition, the advantages of the use of a powder primer on exterior surfaces are realised.

This process is capital intensive and does lend itself very readily to automation.

TABLE 39.6
Comparison of thermosetting and thermoplastic topcoat systems

Properties	*Thermosetting enamels*	*Thermoplastic lacquers*
Composition	Acrylic/melamine formaldehyde	Acrylic/external plasticizer; cellulosic modifer
Molecular weight	10 000–25 000	70 000–120 000
Curing mechanism	Cross-linking	Solvent evaporation
Curing temperature	125–150° C	140–160° C
Non-volatile content by mass	50–55% solid colour 40–45% metallic colour	40–45% solid colour 30–35% metallic colour
Reducing thinner for application	Aromatic type (xylene/toluene)	Blend of aromatic/ ketonic/ester
Reduction to application viscosity 20 ± 2 s Ford 4 cup	10–25% by volume	80–100% by volume
Application solids/mass	35–45%	15–20%
No. of coats to achieve 50 μm dry film thickness	2–3	3–4
Fluid flow	500–800 mL/min	900–1200 mL/min
Application properties and appearance:		
Solid colour	Good	Good
Metallic colour	Fair to good	Excellent
Polishing capabilities	Fair	Excellent
Repairability	Fair panel repair	Good, can be spot-repaired
Durability	Similar	Similar
Solvent resistance	Very good	Good to fair
Chemical resistance	Similar	Similar
General performance properties	Similar	Similar

39.6 AUTOMOTIVE TOPCOATS

If the present is an indication of the future, we should expect significant developments in automotive topcoats in application, composition and curing under the influence of consumerism, energy conservation and environmental pressures. Significant evolutionary advances have already been made in the two major technologies that serve the automotive industry, namely thermosetting enamels and thermoplastic acrylic lacquer topcoats.

The estimated world production of cars for 1980 was 30 000 000 units. A quarter of these were manufactured by General Motors, using acrylic lacquers for both solid and metallic colours. For the remainder, the manufacturers—represented by Ford and Chrysler in the U.S.A. and all the Japanese and most of the European manufacturers—used thermosetting enamels. Although thermosetting acrylic technology is used invariably for metallic colours, alkyd enamels still predominate, outside the U.S.A., for solid colour formulae.

The basic differences between thermosetting and thermoplastic topcoats are shown in table 39.6.

39.6.1 Thermosetting Topcoats

39.6.1.1 Alkyd Types

The basic technology for *alkyd*-type automotive coatings based on oil-modified polyesters has remained relatively unchanged since the early 1960s in comparison to their acrylic counterparts. Their continued use by many European and Japanese motor manufacturers is based on advantages claimed in good appearance, cost and adaptability in their operations.

A typical alkyd automotive finish is based on a non-drying alkyd resin cross-linked with a 25–35 per cent butylated melamine-formaldehyde resin and cured at 120–130° C for 20–30 minutes. Minor developments in alkyd technology over the past few years, in order to improve exterior durability, include:

(a) substitution of the dibasic acid, employed in the alkyd resin component, from *ortho*-phthalic to *iso*-phthalic acid;
(b) use of trimethylolethane/propane in place of glycerol/glycol;
(c) use of selected fatty acids (myristic, pelargonic) as partial or total replacement for the basic coconut oil fatty acid component.

Available information suggests that alkyd-type finishes exhibit good durability in solid pastel colours, adequate in dark solid colours (reds, blues) and unacceptable in metallic finishes under Australian and U.S.A. field conditions.

Some European alkyd formulations incorporate a nitrocellulose modification to reduce the curing temperature to 80–100° C and improve application and processing properties, such as repair work. Invariably such modifications adversely affect durability, and as such they have found limited application in mass-production applications.

39.6.1.2 Conventional Acrylic Types

Acrylic topcoats are based on acrylic copolymers cross-linked with butylated or isobutylated melamine-formaldehyde in the ratio of 60 : 40 to 80 : 20, depending on the desired balance of properties such as durability, flexibility, intercoat adhesion and chemical resistance.

Typical acrylic copolymers comprise:
(a) styrene, between 20 and 40%;
(b) an acrylic copolymer (40–60%), based on a blend of hard (methyl methacrylate, butyl methacrylate) and soft (butyl acrylate, 2-ethylhexyl acrylate) monomers, forms the backbone of the acrylic resin;
(c) a hydroxyl monomer (10–30%), such as hydroxypropyl methacrylate or hydroxyethyl acrylate, to provide the cross-linking sites with melamine components.

The effects of these components are shown further in table 39.7.

Considering that the molecular weight of such an acrylic copolymer is relatively low (10 000–25 000), its performance properties and durability requirements for an automotive finish largely depend on the degree and mode of cross-linking during stoving, normally in the region of 130–140° C for 20–30 minutes. Representative formulae for acrylic copolymers are shown in formulas 39.2 and 39.3.

TABLE 39.7
Effect of monomer composition on acrylic copolymer properties

Monomer	*Effect on polymer properties*
Methyl methacrylate Styrene Vinyl toluene Acrylonitrile	Hardness
Butyl acrylate Ethyl acrylate 2-ethylhexyl methacrylate	Flexibility
Acrylamide Butoxymethyl acrylamide Hydroxyalkyl acrylates Glycidyl acrylate	Cross-linking sites
Acrylic acid Methacrylic acid Maleic anhydride	Curing acceleration

FORMULA 39.2
Formula for a typical acrylic copolymer, suitable for metallic finish systems

	Parts by mass
Styrene	12.0
Methyl methacrylate	8.0
Butyl acrylate	12.0
Butyl methacrylate	9.0
Hydroxypropyl methacrylate	8.0
Acrylic acid	1.0
Xylene	50.0
	100.0

Constants

Non-volatile material	50% by mass
Viscosity (Gardner-Holdt)	X-Y

Typical monomers used in the design of thermosetting acrylic resins and their specific contribution are shown in table 39.7.

Table 39.8 shows the effect of varying the styrene and 'soft' acrylate levels.

Acrylic resin technology is described further in chapter 15, polyesters in chapter 8, and alkyd resins in chapters 5, 6 and 7. A typical thermosetting acrylic topcoat formulation is given in formula 39.4.

39.6.1.3 Non-aqueous acrylic dispersons (NAD)

These compositions are based on dispersions of acrylic copolymers, similar in composition to those described in section 39.6.1.2, usually in aliphatic solvents, and sterically stabilised either by a low molecular weight aliphatic polymer, or by an alkylated melamine-formaldehyde stabiliser. A typical formula for a polymer of this type is given in formula 39.5.

The manufacturing procedure involves three basic stages:

(a) a conventional *seed stage* of methyl methacrylate/stabiliser/initiators/aliphatic solvents;
(b) a *feed stage*, where the rest of the monomers are added to complete the polymerisation; and
(c) addition of aromatic/alcohol solvent/ester blends to swell the particles thus formed and to reduce the viscosity of the continuous phase.

TABLE 39.8
Effect of styrene and 'soft' acrylate levels

Effect of increased styrene levels	*Effect of increased acrylate levels*
Cost reduces	Flexibility increases
Hardness increases	
Chemical resistance increases	
Solubility in solvent increases	Durability increases
Compatibility with amino resin decreases	

The development of these finishes was fostered by the advantages of less-expensive, less-polluting solvents and ease of application in mass-production automotive lines. With respect to the latter property, NAD finishes made possible the application of topcoats in two coats instead of the normal three, particularly in metallic colours, without sacrificing appearance, due to their higher application solids, higher spraying efficiency and rheological characteristics.

FORMULA 39.3
Formula for a typical acrylic copolymer, suitable for solid colour finishes

	Parts by mass
Styrene	21.0
Ethyl acrylate	5.0
2-ethylhexyl acrylate	13.0
Hydroxypropyl methacrylate	10.0
Acrylic acid	1.0
Xylene	50.0
	100.0
Constants	
Non-volatile material	50% by mass
Viscosity (Gardner-Holdt)	X-Y

The basic formulating principles of NAD enamels, including pigmentation, cross-linking, application and overall performance properties, are similar to those of conventional thermosetting acrylic topcoats described in section 39.6.1.2.

39.6.1.4 Basecoat/Clearcoat Systems

Clear on Basecoat technology originated in Germany some time ago and has developed up to and through the 1970s to the products now in use. It is widely used by European and Japanese automotive manufacturers in metallic colours to impart characteristic high gloss, depth and chemical resistance.

The process involves the application of a highly pigmented basecoat to a film thickness of 15–20 μm to provide the coverage and colour, air-flashed for 2–3 minutes, followed by a clear coat to an average film thickness of 30–40 μm. The composite system is baked for 30 minutes at 130° C,

FORMULA 39.4
Formula for a typical thermosetting acrylic topcoat

	Parts by mass	*Instructions*
Rutile titanium dioxide	17.5	Premix and disperse to less than 5 μm
Butylated melamine-formaldehyde (60% NV)	4.2	
Xylene	3.0	
Butylated melamine-formaldehyde (60% NV)	3.0	Premix and add. Continue dispersion 30–60 min
Acrylic copolymer (50% NV)	3.0	
Toluene	2.0	
Xylene	3.0	Washout
Butylated melamine-formaldehyde (60 NV)	10.8	Make-up
Acrylic copolymer*	45.0	
High boiling aromatic solvent†	3.0	
Flow control additive	0.2	
Flow control solvents (aromatic/esters/ethers)	5.3	
	100.0	
Constants		
Non-volatile material	50 ± 1% by mass	
Viscosity	30–34 s Ford 4	
Pigment : binder ratio	0.5 : 1	
Acrylic : MF ratio	70 : 30	
Gloss	78–80%/20° head	
Hardness	36–40 Sward units	
Application thinner	xylene/toluene/high boiling aromatic blend	
Application viscosity	17–20 s Ford 4	
Typical gun set-up	JGA502 Cup 78/Tip and needle FF	
Application conditions	Atomising pressure 400–500 kPa	
Fluid flow	700–800 mL/min	

*i.e. Formula 39.3 type.
†e.g. *Solvesso*. 150.

similar to conventional acrylic enamel topcoats. The current technology involves two types of basecoat:

(a) high solids thermosetting acrylic melamine-formaldehyde systems; and
(b) low solids polyester systems cross-linked with melamine-formaldehyde.

All-acrylic thermosetting clearcoats are employed for optimum durability. Both basecoats and clearcoats are suitably modified for application control and performance.

A typical basecoat/clearcoat application in the main automotive colour spray booth can be represented as follows:

First station. Basecoat to all car body interiors invariably by manual application.
Second station. Basecoat to all exterior surfaces normally by automatic application.
Third station. Clearcoat to all interiors as at first station.
Fourth station. Clearcoat to all exterior surfaces as at second station.
Fifth station. Clearcoat manual application to all exterior surfaces not covered by the automatic process.
Sixth station. Clearcoat application to all exterior surfaces by an automatic process.

The airflash time between the exterior coats of basecoat and clearcoat (second and fourth stations) can vary between 3 and 6 minutes and that between exterior clearcoats up to 3 minutes.

FORMULA 39.5
Formula for a typical NAD polymer composition

	Parts by mass
Methyl methacrylate	8.0
Stabiliser for dispersion	8.0
Styrene	10.0
Butyl methacrylate	9.0
Hydroxypropyl methacrylate	12.0
Methacrylic acid	1.0
2-ethylhexyl acrylate	7.0
Aliphatic solvent	35.0
Aromatic solvent	5.0
Alcohols	5.0
	100.0

39.6.1.5 Water-based Thermosetting Topcoats

The advent of water-based automotive topcoats has been decisively influenced by pollution regulations, particularly on the west coast of the U.S.A., where they are currently in use in General Motors and Ford plants. Additional pressure from the increased cost of solvents and their uncertain availability has also prompted the use of water-borne systems. However, their limited application in the motor industry is due to two main reasons:

(a) In the current stage of development, expensive modification to the existing spray and bake facilities is required to give acceptable film appearance. This basically involves controlled humidity and temperature in spray booths and subsequent split baking schedules with elaborate precautions to prevent marring of the film by ambient dust and dirt. Consequently, they are high in energy and cost requirements by comparison to other potential technologies, such as high solids solution acrylics or urethanes.
(b) There have been difficulties in producing the full automotive colour range, particularly in metallics, to meet the established aesthetic standards.

Water-thinned enamels currently in use are based on acrylic polymers similar to those used in the solvent-thinned types, except that the percentage of hydroxyl containing monomer is lower and that of the acid (carboxyl) monomer considerably higher. The carboxyl group is reacted with solubilising amines (triethylamine and dimethylethanolamine are examples) to obtain water solubility or dispersibility. The cross-linking resin is normally of the hexamethoxymethyl melamine type.

Two basic types of polymers have been developed, used either individually or in blends, to achieve a optimum balance of application and performance properties: water-soluble polymers, and water-dispersed polymers.

Water-soluble polymers. Usually in the form of amine salts, these polymers form clear true solutions in water containing 20–30% of a water-miscible organic solvent blend such as ethylene glycol monobutyl ether/butanol. Their rheological and performance characteristics depend on (a) type of amine; (b) type of solution copolymer; (c) non-volatile content; and (d) type of organic solvent.

The preparation of a typical copolymer is shown in formula 39.6. The equipment required includes an efficient stirrer, condenser and monomer addition facilities.

Copolymers such as these in surface coatings have the disadvantages of high water retention during drying, with a tendency to sag and/or solvent 'boil'; high amine content imparts odour, weakens humidity performance and often adversely affects stability.

FORMULA 39.6
Formula a water-soluble acrylic polymer for automotive topcoats

	Parts by mass
Isopropanol	12.2
Ethylene glycol monobutyl ether	3.5
Heat to reflux and add the following over 2 hours	
Methyl methacrylate	9.0
Butyl acrylate	22.0
Hydroxypropyl methacrylate	6.2
Acrylic acid	1.0
Catalyst	0.2
Reflux for 3–4 hours until desired molecular weight is achieved, then add	
Ethylene glycol monobutyl ether	10.0
Remove low-boiling distillate. Cool to 100–110° C and add the following. Maintain stirring.	
Dimethylaminoethanol	1.3
Distilled water	34.6
	100.0

Water-dispersed polymers. These are prepared using similar monomer compositions to the water-soluble polymers, but with much reduced organic solvent levels and with significantly higher molecular weights. The latter property contributes greatly to improved mechanical properties, durability, and the achievement of high application solids with less tendency to solvent boil.

The physical state of the dispersion system allows faster water evaporation during application and curing, but causes greater difficulties with flow, gloss and pigmentation. For this reason, most of the practical formulations are based on blends of water-soluble and water-dispersed polymers.

In these water-based compositions, the pigment is normally dispersed in a melamine resin and extended with a water-soluble acrylic polymer. A typical formulation is given in formula 39.7.

Further acrylic polymer additions are made to give the required pigment : binder ratio.

FORMULA 39.7

Formula for a typical acrylic dispersion topcoat

	Parts by mass
Titanium dioxide	64.5
Deionised water	21.5
Dimethylaminoethanol	0.3
Anti-foam additive	As required
Ethylene glycol monobutyl ether	4.3
Melamine resin	2.9
Water-soluble acrylic copolymer (50% NV)	6.5
Mix for 20–30 min and disperse to less than 5 μm in a sand mill. Maintain pH at 7.8–8.5	100.0

Application characteristics of water-based finishes. The evaporation characteristics of water-borne paints are dominated by the fact that water has a single boiling point, high latent heat of evaporation and is very dependent on the ambient relative humidity. In addition, water differs from most other solvents in that it readily forms azeotropes with many other solvents that can alter its volatility. These characteristics reduce the application and flexibility of such finishes in conventional automotive facilities designed for solvent-based systems, unless substantial investment is made to control temperature, humidity and air-flow movement in the spray booth and pre-bake stages.

39.6.1.6 High Solids Automotive Topcoats

The development of high solids automotive topcoats stems from the EPA (Environment Protection Agency) requirement to reduce total solvent emission to the atmosphere, whether photochemically active or not.

A composition of 40 per cent non-volatile content by volume (NVV) at the time of application is estimated to contain 520 g/L of solvent. Similarly, a composition of 61 per cent NVV contains 350 g/L of solvent. To meet the required maximum level of 275 g/L, as expected, a composition of 70 per cent NVV is necessary. Considering that most of the automotive coatings currently in use can barely meet the 40–50 per cent NVV requirement, any demands to exceed these limits impose severe technical demands on surface coating technology.

The present method of achieving such high-solids solvent-borne coatings is by the use of low molecular weight polymers (oligomers) of alkyd or polyester type cross-linked with amino or isocyanate resins.

Although such compositions can be adapted to existing automotive facilities with small additional capital investment and can meet high transfer efficiency requirements by means of high-speed electrostatic equipment, their acceptance by the industry is limited, for the following reasons:

(a) Low molecular weight systems are generally sensitive to the condition of the substrate. Any contamination often shows as poor wetting and 'telegraphs' through the coating, which does not bridge surface profiles as well as a high molecular weight coating.
(b) The need for higher cure temperatures and effective cross-linking to achieve performance properties specified by the automotive industry. Cross-linking agents such as isocyanates can be expensive and toxic.
(c) Uncertain exterior durability particularly in metallic colours.
(d) Film thickness is difficult to control, except on high-speed electrostatic disc applications.

39.6.2 Conventional Thermoplastic Acrylic Topcoats

Thermoplastic automotive topcoats currently based on acrylic resins are almost exclusively used by General Motors in the U.S.A. and by most of the overseas General Motors plants.

The formulation of these finishes, referred to as *lacquers*, is a much more rigorous exercise than that of thermosetting enamels because their overall performance depends not only on the composition of the lacquer but also on the undercoating system.

Thus, whilst in the progressive development of acrylic enamels from the earlier alkyd enamels the undercoating system remained basically the same, the advent of acrylic lacquers necessitated the development of specific undercoats for such critical properties as adhesion, solvent craze on recoating and cracking in the field.

39.6.2.1 Basic Formulating Principles

Acrylic lacquers consist essentially of four main constituents:

Acrylic polymer
Cellulosic modifier
Plasticiser
Solvent

Their types and proportions determine the performance and processing properties of lacquers in any given automotive plant.

Acrylic polymer. This is the basic ingredient in acrylic lacquers and constitutes 50 per cent or more of the vehicle system. The acrylic resin is usually produced by free radical polymerisation of methyl methacrylate with or without small percentages (5–30 per cent) of other acrylates or methacrylates, in solution, to an average molecular weight of 100 000. Significantly higher molecular weights improve durability and mechanical properties, but reduce application solids and smoothness of the film and affect processing in general; on the other hand, lower molecular weights adversely affect durability and overall performance.

Cellulosic modifier. This is invariably cellulose acetate butyrate and normally is used in the range of 15–25 per cent on vehicle solids to improve the application properties, particularly of metallics, without sacrificing performance and appearance.

Plasticiser. This is used, at 15–25 per cent depending on the composition of the acrylic polymer, and can be of the monomeric type such as butyl benzyl phthalate, or polymeric oil, or an oil-free polyester type. It contributes to the gloss on reflow, gloss retention on exposure, and cold crack performance and solvent craze on rectification.

Solvents. The high molecular weight of the acrylic polymer and diversity of ingredients demands a solvent blend to impart high solvency, compatibility and acceptable application properties without adversely affecting durability on fully-cured and low-baked repair films. A typical solvent balance is shown in formula 39.8.

The formula for a typical metallic acrylic lacquer is shown in formula 39.9.

FORMULA 39.8

Formula for a typical solvent blend for thermoplastic acrylic topcoats

Solvent type	*Examples*	*Parts by mass*
Aromatic	Toluene, xylene	40
Ketones	Acetone, methyl ethyl ketone	30
Glycol ethers and esters	Ethylene glycol monoethyl ether acetate	30

FORMULA 39.9
Formula for a typical metallic thermoplastic acrylic lacquer topcoat

	Parts by mass
Aluminium pigment paste (60% NV)	1.6
Acrylic resin (90 : 10 methyl methacrylate : butyl acrylate—40% NV in toluene)	43.0
Cellulose acetate butyrate	4.4
Plasticiser (oil-free polyester)	8.2
Acetone	10.0
Methyl ethyl ketone	12.0
Xylene	3.0
Ethylene glycol monoethyl ether acetate	17.8
	100.0

39.6.2.2 Processing Parameters

Acrylic lacquers are normally reduced to low application viscosities (16–18 s 4 Ford) with a blended thinner from 80–100 per cent reduction by volume and applied at relatively high fluid flow (800–1200 mL/min).

To take advantage of the reflow properties of acrylic lacquers, the film is first baked at 80–100° C for 10–15 minutes to achieve a working film hardness at room temperature (3–4 Knoops). Film imperfections are wet sanded (P1000–P12000 paper), and subsequently baked at 150–160° C for 30 minutes. The film reflows, eliminating the sanding marks, and producing high gloss and smoothness.

39.6.3 Non-aqueous Thermoplastic Dispersion (NAD) Lacquers

NAD lacquers, more commonly referred to as LDL (Lucite Dispersion Lacquers), were originated, and are currently supplied, by Du Pont to a limited number of U.S.A. automotive plants on the basis of their higher application solids. Their package non-volatile contents are approximately the same as regular lacquers, but they need only 50–60 per cent reduction with solvent to give satisfactory spraying viscosity. Thus the non-volatile contents by volume on application can be increased from the average 13–15 per cent to 21 per cent with potential to reach 27 per cent.

Advantages in reduced pollution, lower costs through the reduction of solvent usage, use of a large percentage of aliphatic hydrocarbon solvents, and reduction of the number of applied coats have been confirmed.

The formulation of the basic NAD thermoplastic polymer and of the lacquer formulation itself is difficult, as the balance of partial solubility is not easy to achieve along with required package and operating stability. The coalescence of the film requires the use of slow-evaporation active solvents with good solubility for the basic NAD polymer, and consequently relatively high baking temperatures in the region of 155–170° C.

The molecular weight of NAD polymers is significantly higher than solution types, and this results in the improvement of certain properties such as resistance to checking, cracking and improved curing characteristics.

39.6.4 Powder Coatings

From the viewpoint of air pollution control, powder coatings have the considerable advantage that they have almost 100 per cent non-volatile content. Because they are based on high molecular weight polymers, their film performance is of a very high order particularly for chemical resistance, impact resistance, hardness, and durability in solid colours. Application efficiency is high, and any overspray can be reclaimed and re-used. As there is no solvent to flash-off, extended air flash times before bake are not necessary.

Despite the relatively high curing temperatures (20 minutes at 170–180° C), they require a lower overall energy input. Spray booths and ovens do not need 'make-up' air to remove solvent vapours, resulting in considerable cost savings.

Full exploitation of these coatings by the automotive industry is hindered by four main factors:

(a) high capital investment in the application and curing equipment which is dependent on the number of colours to be applied;
(b) film thicknesses in the region of 40–50 μm cannot reliably be controlled and thus material costs are unfavourable by comparison with wet coating systems;
(c) metallic colours suffer in appearance and durability; and
(d) the cost of equipment for the manufacture of powder topcoats and surfacers is considerable.

There are three basic types of powder coatings under consideration by the automotive industry:

Epoxy types: for primers and surfacers where adhesion, mechanical properties and costs are the main requirements.
Polyester types: for exterior durability and colour retention.
Urethane types: a version of the polyester type using a different curing mechanism for optimum exterior durability.

Powder coatings are described in chapter 42.

39.6.5 Coatings for Plastic Parts

A number of special surface coatings are currently used to coat plastic components used by the motor industry. The great variety of plastics used—such as SMC (sheet moulding compound), ABS, nylon, flexible urethane, and polypropylene—differ greatly in surface characterisation, flexibility, rigidity and solvent sensitivity.

Surface coatings are used on plastics for protection and decoration and the type is carefully chosen for overall compatibility with the intrinsic characteristic properties of the plastic.

To meet the requirements of the motor industry, a primer is invariably used to provide adhesion and seal the plastic from any adverse effect of the topcoat solvents. Such primers are based on chlorinated polyolefins or urethanes applied at very low film thicknesses. For rigid plastics, low temperature curing (70–100° C) thermoplastic and thermosetting compositions can be used, whilst for flexible components such as bumper-bars, the composition is normally urethane-based.

40 LACQUERS

A *lacquer* is usually defined as a solution of organic film forming material(s) in organic solvent(s) from which the solvent evaporates, during and after application to a substrate, and in which the drying mechanism is solely via evaporation. Films of true lacquers stay permanently soluble in similar solvents to those used as a 'carrier' for the film-forming components. The binder in these *non-convertible* coatings remains thermoplastic, in contrast to *convertible* coatings which thermoset due to chemical reaction.

Lacquers include vehicles based on resins dissolved in water but as these do not have a significant application except as temporary coatings they are not described further here.

The solvent mixture present in a lacquer has the function of enabling the solid components to be applied in a homogeneous film to the substrate, but usually it has a significant effect on the properties of the dry film. Solvent requirements vary widely with the type of resins used in the lacquer, and will be considered in this chapter when the main film formers or resins are described.

The use of the Solubility Parameter concept is recommended in optimising solvent mixtures in lacquers. Refer also to chapter 24 in volume 1.

The properties of lacquers vary with the main type of film-forming resin used, and their main advantage is rapid drying speed. They are made for application to a wide variety of substrates, from rigid to very flexible. In general, they are applied at relatively low volume solids; another disadvantage is that the degree of toughness and abrasion resistance which can be achieved is inferior to that obtained in a film in which cross-linking or other chemical reactions occur during and/or after application of the film.

By judicious choice of vehicle components, lacquers with excellent weathering properties are now available, the best example being lacquers based on polymethyl methacrylate resins.

40.1 TYPES OF LACQUERS

The main types of lacquers, listed under the main film former on which they are based, and which may be modified with resins, plasticisers, etc., are:

40.1.1 Cellulose Derivatives

Cellulose nitrate (NC)
Cellulose acetate
Cellulose acetate butyrate (CAB)
Cellulose acetate propionate (CAP)
Ethyl cellulose (EC)

Ethyl hydroxyethyl cellulose (EHEC)

40.1.2 Acrylics

Polyalkyl methacrylates and acrylates and copolymers
Polyalkyl acrylates–styrene copolymers

40.1.3 Vinyl Resins

Polyvinyl chloride (PVC)
Vinyl chloride–vinyl acetate copolymers
Polyvinyl acetate (PVA)
Vinyl chloride–vinylidene chloride copolymers
Vinyl chloride–vinyl iso-butyl ether copolymers
Vinyl chloride–acrylic ester copolymers
Polyvinyl acetate–polyvinyl alcohol resins
Vinyl acetate–ethylene copolymers
Vinylidene chloride–acrylonitrile copolymers
Polyvinyl butyral (PVB)
Polystyrene
Styrene–butadiene copolymers
Vinyl toluene–butadiene copolymers
Vinyl toluene–acrylic copolymers

40.1.4 Rubber Derivatives

Chlorinated rubber
Cyclised (isomerised) rubber

40.1.5 Polyurethane Elastomers

40.1.6 Chlorinated Polyolefins

40.1.7 Natural Resins

Shellac
Copal
Other gums

40.1.8 Miscellaneous

Coumarone-indene
Hydrocarbon resins
Polyamide resins
Pitches and tars
Phenol-formaldehyde resins
Ketone resins

40.1.9 Typical Modifying Resins

Non-oxidizing oil modified alkyds
Oil-free polyesters
Chlorinated terphenyl resins
Ester gum and modified ester gums
Phenol and maleic modified ester gums
Zinc and calcium resinates

Toluene sulfonamide formaldehyde
Terpene resins
Chlorinated paraffin
Dewaxed damar
Vinyl chloride–acetate copolymers
Acrylic resins
Ketone aldehyde resins

40.2 CELLULOSE DERIVATIVES

40.2.1 Cellulose Nitrate (NC)

Cellulose nitrate—also known as nitrocellulose, nitrocotton, collodion and pyroxylin—is the most widely used film former for the manufacture of lacquers. NC lacquers are the fastest drying types and will be dealt with more comprehensively than other types.

40.2.1.1 History of NC Lacquers

Nitrocellulose was discovered by Schönbein in 1846, with the consequent advent of smokeless powders, and nitrocellulose lacquers were first made commercially by Parkes, who in 1855 was granted a patent covering the use of a liquid nitrate composition. Early progress was slow, because of the limited range of solvents then available. Ether and alcohol solvent mixtures were used, but the very rapid evaporation of these solvents cooled the lacquer film below the dew point of the surrounding air. This resulted in moisture being deposited on the surface, giving a milky white appearance (known as 'blushing') to the coating.

In 1881 Stevens patented the use of amyl acetate as a solvent for nitrocellulose, and laid the foundation of the modern nitrocellulose lacquer industry. Amyl acetate, being a much slower evaporating solvent, deposited a smooth uniform film, free from 'blushing'. At that time, however, the nitrocellulose available gave only solutions of high viscosity and low solids content, making an excessive number of applications necessary in order to obtain an adequate thickness of coating.

It was not until just after the First World War that low-viscosity nitrocellulose became available. This enabled the preparation of lacquers having a viscosity low enough to permit application by a spray gun, whilst having a relatively high concentration of solid, film-forming material. At this time, the automotive industry created a demand for quick-drying finishes suitable for assembly-line manufacturing methods.

40.2.1.2 Manufacture of Nitrocellulose

Cellulose, derived mainly from cotton linters and wood pulp, is the main raw material for nitrocellulose manufacture (cotton linters are the short fibres remaining on the cotton seed, after the long fibres have been removed for textile purposes).

Cellulose fibres are freed of impurities by boiling in caustic solution and, after washing, are treated with a mixture of nitric and sulfuric acids in agitated stainless steel reactors. After the desired level of nitration (which controls the solubility characteristics) is obtained, the spent acid is removed by centrifuging and 'drowning' rapidly in water. The cellulose nitrate is then heated in water until the correct viscosity characteristics are obtained. The slurry of nitrocellulose and water goes to centrifuges or presses, where most of the water is removed. The remaining water is removed by pumping ethyl (or butyl) alcohol through the nitrocellulose, which is packed and sold in the alcohol-wet state; the average alcohol content is 35 per cent mass, for safe transport and handling.

Nitrocellulose is made in various grades, and solutions at the same solids content vary from very low to very high viscosity. High-viscosity nitrocellulose gives the most flexible films and is used mainly for non-rigid surfaces such as leather and textiles. Low-viscosity grades are used to prepare lacquers of high solids content at low viscosity, where the maximum thickness of film per application is required.

Nitrocellulose is manufactured in Australia by ICI in two main types: '*H*' high nitrogen content (11.7–12.2 per cent) and '*L*' low nitrogen content (10.7–11.2 per cent). The former are conventional ester-soluble grades, and the latter so-called 'alcohol soluble'. The high nitrogen version is the main type used for NC lacquers. The low nitrogen types, which are more thermoplastic, are used for special purposes such as paper coatings and lacquers for aluminium foil. They are also used in flexographic and gravure printing inks, which are mostly low-viscosity lacquers.

TABLE 40.1
Effect of viscosity grade of nitrocellulose on lacquer viscosity and solids (Nitrocellulose : resin : DBP—40 : 40 : 20)

Nitrocellulose grade	*Viscosity at 12.5% solids (5% nitrocellulose) cP*	*Solids at 100 cP viscosity %*
DHX3–5	5	33
DHX8–13	7	28.5
DHX30–50	20	22
DHL25–45	80	13.5
DHL120–170	270	10
FHH9–20	55 000	3

40.2.1.3 Viscosity

NC is available in a wide range of viscosity types. A low-viscosity grade has a lower molecular weight and low tensile strength, whereas the high-viscosity type has higher molecular weight, high tensile strength and greater flexibility.

The effect of the grade on lacquer viscosity and solids in a typical lacquer containing 40 per cent NC on total resin solids is shown in table 40.1.

The lower-viscosity types are the most commonly used grades, because they enable the application of relatively high solids content coatings. *DHX–13* NC is the most popular type.

40.2.1.4 Solvents

The high nitrogen content grade is normally dissolved in a suitable mixture of:

Active solvents
Latent solvents
Diluent solvents

The evaporation rate of the components must be such that the remaining mixture in the evaporating film always remains a solvent for the NC.

The low nitrogen 'alcohol-soluble' grades still require addition of active solvent to the alcohol in order to obtain complete solubility. The amount varies between 5 and 30 per cent depending on the solids content required and the NC viscosity grade used.

Active solvents. These solvents are true solvents for NC, and the main types used commercially are:

Ester solvents:	butyl, ethyl and ethylene glycol mono-ethyl ether acetates
Ketone solvents:	acetone, methyl iso-butyl and methyl ethyl ketones
Glycol ethers:	mono-methyl, mono-ethyl and mono-butyl ethers of ethylene glycol

Latent solvents. Latent solvents for NC become solvents in the presence of true solvents and, although non-solvents by themselves, often produce a stronger solvent mixture than the true solvents alone. They are usually alcohols, and may be used to reduce the cost of the solvent mixture. Alcohols may improve the solvent power of an active solvent: a 50 : 50 mixture of ethyl alcohol and butyl acetate has a toluene dilution ratio of 3.0 compared to 2.7 for the ester alone. (The toluene dilution ratio is the amount of toluene that a solution of NC in the solvent under test will tolerate before precipitation of the resin.)

TABLE 40.2
Properties of common active, latent and diluent solvents for lacquers

	Relative evaporation rate	*Toluene dilution ratio*	*Flash point TOC °C*	*Solubility parameter*
	n-butyl acetate = 1			
Active				
Acetone	7.7	4.6	−19	10.0
Methyl ethyl ketone	4.6	4.3	−3	9.3
Methyl acetate 99%	4.2	3.1	−1	9.1
Methyl iso-butyl ketone	1.6	3.5	20	8.4
n-butyl acetate	1.0	2.7	29	8.5
Ethylene glycol mono-ethyl ether	0.2	4.9	48	9.9
Ethylene glycol mono-ethyl ether acetate	0.2	2.5	59	8.7
Ethyl glycol mono-butyl ether	0.1	3.4	70	8.9
Latent				
Ethyl alcohol 95%	1.7	—	15	12.7
Sec. butyl alcohol	0.9	—	27	11.1
n-Butyl alcohol	0.5	—	42	11.4
Diluent				
Toluene	1.9	—	7	8.9
Xylene	0.6	—	28	8.8
Aliphatic–aromatic hydrocarbons				
Shell *X3B	2.2	—	−1	8.2
Shell *X55	4.4	—	below −30	7.4

* Abel closed cup flash points.

Diluent solvents. Diluents for NC are usually aliphatic and/or aromatic hydrocarbons, the latter giving much higher dilution ratios than the former. It is particularly important that the mixture of diluents has sufficient solvent power to keep the resinous components in solution during the drying of the film. Short oil coconut oil modified alkyds are very sensitive to aliphatic hydrocarbon diluents, for example.

The speed of evaporation of the solvent balance should be such that sufficient active solvent remains in the film during the drying process. Diluents which have an appreciably faster evaporation rate than the active solvents are used Normal butyl acetate is a typical active NC lacquer solvent, and the speed of evaporation of solvents and diluents is often given in relation to n-butyl acetate as 1. A figure greater than 1 indicates a faster evaporation rate than butyl acetate.

Table 40.2 presents some key solvents and properties. Additional information is given in chapter 24, 'Solvents', in volume 1.

Retarders. If 'blushing' is encountered only occasionally, a normal solvent balance can be used in the lacquer and thinner, and a small addition of a strong slower solvent called a 'Retarder' can be added when atmospheric conditions are liable to cause 'blushing'. Ethylene glycol mono-butyl ether is particularly useful.

Solvent strength and solids content. A balance must be struck between strength of solvent mixture and solids content. Too weak a solvent mixture will produce a higher viscosity, with consequent need for additional solvent to reduce to the application viscosity; if it is too strong, difficulty may be encountered because of solvent attack on previously applied coats of primer, sealer or finish.

40.2.1.5 Plasticisers

Nitrocellulose films are too brittle for most applications and require plasticisers to impart the necessary flexibility. NC is compatible with a wide range of plasticisers (such as vegetable oils, monomeric and polymeric esters) which may be divided into two main classes:

Solvent plasticisers. This is the preferred class of plasticisers, because higher tensile strengths are obtained at similar degrees of flexibility (than with non-solvent plasticisers). Dibutyl and dioctyl phthalates are common examples of this type. Dioctyl phthalate (DOP) is preferred in cases where permanent flexibility is required. Dibutyl phthalate is much more volatile, although it is a better solvent for NC than DOP.

Non-solvent plasticisers. These are often of use in very flexible lacquers, typically for leather, because of the much softer finish they produce. Raw and blown castor oil are typical examples of this class of plasticiser. It is desirable to always have some proportion of solvent plasticiser present to obtain satisfactory stability of the mixture. Alone, the oil plasticisers are prone to 'spew' or migrate to the surface of the lacquer after drying. A typical lacquer for leather finishing contains 40 per cent DHL 120–170 NC, 56 per cent blown castor oil and 4 per cent dibutyl phthalate.

40.2.1.6 Resins

These are necessary modifiers for many types of NC lacquers in order to produce satisfactory gloss and specific adhesion. Typical resins are:

Coconut and castor oil modified alkyds
Ester gum and maleic modified ester gum
Toluene sulfonamide formaldehyde resins
Dewaxed damar
Vinyl chloride–acetate copolymers
Acrylic resins
Cyclic ketone resins
Sugar-based resins
Styrene–maleic anhydride resins

Coconut and castor oil modified alkyds. These resins are in common use in lacquers requiring good durability, polishing properties and colour retention. A typical example is an automotive refinishing lacquer containing approximately 1 part coconut oil modified alkyd solids (36 per cent oil length) to 1 part *DHX8–13* NC solids and 0.4 parts dibutyl phthalate.

Pure and maleic modified ester gum. These 'hard resins' are in wide use in glossy hard lacquers. Large proportions result in brittle films, but this can be overcome by plasticising with dibutyl phthalate and a non-oxidising alkyd. They are used to produce high solids type, economical industrial lacquers because of their fairly low cost, but are not used for exterior use because of poor weathering resistance. As a general rule, the higher melting point grades give harder films and faster drying because of better solvent release properties.

The vehicle solids in a typical general-purpose NC industrial lacquer suitable for use on wood, metal or hardboard, is:

38% Dry NC
52% Ester gum (capillary tube MP approx. 65° C)
5% Castor oil
5% Dibutyl phthalate

If the ester gum is replaced by a maleic modified ester gum, a faster drying lacquer which has a harder surface is obtained.

Toluene sulfonamide formaldehyde resins. This type of resin is used with NC for special-purpose lacquers, particularly where moisture resistance is required. A notable example is their use in nail polishes which typically contain 30 per cent NC solid, 50 per cent resin and 20 per cent dibutyl phthalate. Monsanto *Santolite MHP* is typical of this type of resin.

Dewaxed damar. This is the only 'natural' resin which has been used in significant quantities in NC lacquers. As obtained, damar contains approximately 10 per cent of damar wax (beta-resene), which must first be removed by a dewaxing process; this is carried out by dissolving the damar in toluene and adding ethyl alcohol to precipitate the wax. Damar is particularly useful in clear nitrocellulose furniture lacquers. It imparts good flexibility, gloss, rubbing and 'pull-over' characteristics, and it is also useful in exterior lacquers as it produces quite durable films with NC. However, its use has declined in recent years because the dewaxing process is labour intensive, and a long settling time is necessary to remove the wax.

Vinyl chloride–acetate copolymers. Special vinyl chloride–vinyl acetate copolymers, which contain a much higher vinyl acetate content than normal, are compatible with NC, and have advantages of water-white colour, good exterior exposure characteristics and perspiration resistance. Union Carbide *Bakelite VYNC* is an example of this type of resin.

Acrylic resins. These water-white resins are often used with nitrocellulose for special applications. They impart good gloss and gloss retention, and are sometimes used where a degree of chemical resistance is required (e.g. perspiration-resistant finishes such as racquet lacquers), 'stop-off' lacquers for plating baths, etc. They also give improved colour and yellowing resistance, but some yellowing is always encountered with NC lacquers. Further information on these resins is given in the section of this chapter dealing with acrylic lacquers.

Cyclic ketone resins. These resins, which are produced by a condensation reaction between cyclohexanone (usually) and formaldehyde, are particularly useful for yellowing resistant and water and oil resistant NC lacquers. They are neutral resins and may be used in special instances where acidic resins cannot—for instance, luminous lacquer vehicles, bronzing media and bases for aluminium pastes. They confer excellent adhesion and high gloss; their very pale colour gives almost water-white furniture lacquers.

Sugar-based resins:

Sucrose acetate isobutyrate (SAIB). This mixed ester of sucrose is manufactured by the controlled esterification of natural sugar with acetic and isobutyric anhydrides. It is a water-white, very viscous (approximately 1000 pa. s) liquid resin and is soluble in a wide range of solvents, including aliphatic and aromatic hydrocarbons. It has a very good viscosity–solids relationship, which makes it useful as a plasticising resin in place of non-oxidising alkyds in NC wood lacquers. Here a higher solids content is required without using lower-viscosity grades of NC, which affect the cold check resistance of the finish. In such an application the water-white colour of SAIB, together with its colour stability on exposure to heat and light, may also be used to advantage. This resin type is marketed by Eastman as *SAIB*.

Sucrose benzoate. This water-white ester of sucrose is a hard non-crystalline resin; it is an excellent modifying resin for NC, and promotes high gloss, clarity and hardness without brittleness. It has good resistance to water and alcohol and discoloration by UV light; it is particularly useful in clear NC lacquers for use on pale-coloured timber in furniture finishing, as very good cold check resistance properties result. A typical example of this type of resin is Velsicol Chemical Corporation's product, which has a melting point (Ball and Ring) of 98° C.

A typical clear wood lacquer which will withstand 18 cold check cycles may be made from a resin solids mixture of:

40% *DHX30–50* NC
36% sucrose benzoate
12% dibutyl phthalate
12% castor oil

Styrene–maleic anhydride resins. SMA resins are water-white (colourless) hard resins which are soluble in alcohols, esters and ketones. They have good solvent release properties, good water resistance and very good resistance to yellowing by UV light. A typical resin of this type is BASF *Suprapal AP20* which has a DIN softening point of 140–160° C. The high acid value (approximately 150) limits use in coloured lacquers, but they are particularly useful in NC clear lacquers for use on wood.

Other resins. These are many other resins that are suitable for use with nitrocellulose, but one of the most important properties which must be considered is compatibility with NC.

Whilst oxidising alkyd resins (and many other resins that dry by chemical reaction) are often used with nitrocellulose, such coatings fall outside the definition of lacquers, as they dry partly by chemical reactions.

40.2.1.7 Pigments for NC Lacquers

Most of the pigments suitable for use in surface coatings, as discussed separately in volume 1, are suitable for use in NC lacquers.

Where second coats of another coloured lacquer are to be applied over the lacquer, the pigment should be insoluble in lacquer solvents or a 'bleed' of the soluble colour may discolour the second coat. Many organic red pigments are particularly prone to this defect.

40.2.1.8 Compatibility Tests

Because the concept of compatibility or mutual solubility of lacquer components is very important, methods of determination of compatibility of resins and plasticisers with nitrocellulose, as an example that should be followed when formulating any type of lacquer, will be covered here.

As in many cases the solvent mixture has an effect on the compatibility of a resin with NC (or other film former used) the specific solvent mixture, which has been selected for the particular use, should be used in the compatibility test. Concentration of main film former, resins and plasticisers also plays a part in compatibility; where possible, the tests should be carried out using separate solutions of NC and resin such that when they are mixed the non-volatile vehicle concentration in the mixture approximates that to be used in the lacquer.

However, for convenience in preliminary tests, only one solution each of resin and NC needs to be made, the concentrations being made as high as possible whilst maintaining a workable viscosity. In this way various proportions of resin solution and NC solution may be easily mixed. A number of mixtures should be made over a somewhat wider range of ratios of resin to NC than those likely to be used, and films on glass prepared of each. This may be conveniently done using a 150 μm wet film draw-down gauge. If a clear solution is not obtained on mixing the two solutions, it is, of course, not necessary to cast out films. After the films are thoroughly dry they should be closely inspected for any sign of turbidity, the presence of which is an indication of incompatibility. Slight incompatibility is sometimes observed only when the glass panel is held at a particular angle to the observer's eye so that the observer receives refracted light from the film.

Minor incompatibility does not necessarily prevent use of the particular resin, as satisfactory properties may often be achieved by either adjustment of solvent balance or addition of small amounts of another mutually compatible resin or plasticiser, which can act as a 'coupler' to induce compatibility between the resins and the NC.

In the case of spraying lacquers, films of the final formulation, less any pigment that may be present, should be applied by spray gun to glass panels and, after drying, the film checked for compatibility; the different solvent balance in the film applied in this way may affect the compatibility of the components.

40.2.1.9 Manufacture of NC Lacquers

Clear lacquers. Clear NC lacquers may be conveniently made in most cases by dissolving the NC and resins in the solvent mixture in fairly high speed mixers. However, unless whole drums of NC can be used in the batch, a modification of this procedure is usually desirable because of the comparatively poor homogeneity of the mixture of nitrocellulose and alcohol sold as Industrial Nitrocellulose. The alcohol in this mixture separates towards the bottom of the NC drum and even if the drums are continually inverted before use (which is often impracticable) there is little certainty that the portion of NC weighed contains the stated percentage of nitrocellulose solids. For this reason, bases or stock solutions of nitrocellulose are often made. As the manufacturer usually guarantees that a whole drum of NC contains a specified amount of 'dry' NC, provided whole drums are used in the bases, these can be manufactured with a specified nitrocellulose content.

Pigmented lacquers. These are made usually by addition of a pigment dispersion to a clear lacquer prepared as above. The pigment dispersions may be:

(a) A dispersion of the pigment(s) in the clear lacquer (or NC base or resin solution or plasticiser) prepared in a ball mill, pebble mill, sand mill or bead mill.

(b) Separate dispersions of the individual pigments in nitrocellulose plus plasticiser made on a differential two-roll mill or Banbury mixer. The dispersions are usually made into 'dry' chips which are dissolved into the lacquer on a high speed mixer resulting in dispersion of the pigments in the lacquer. This method is very convenient in that the chip may be stored safely for long periods and a dispersion in any of the main types of lacquers made very quickly, simply by solution of the chip. It also has the advantage of producing lacquers having a slightly higher gloss than those pigmented by other methods. The main disadvantages involve cost and sporadic lapses in product quality.

Some pigments are available in micronised form, in a limited range, and these may often be dispersed by a simple high speed stirring operation, but usually special care must be taken to formulate a satisfactory mixing base. A high shear cavitator is very suitable for this work.

Dyed lacquers. These transparent coatings are prepared by adding a concentrated solution of dyestuff in suitable solvents to a clear lacquer. Special dyes are available having suitable solubilities in the types of solvents used in lacquers.

In general, dyed lacquers do not have good durability, as the dyes are transparent to UV radiation and thus do not protect the lacquer against degradation. Most dyes have poor lightfastness and are fairly fugitive in sunlight.

40.2.1.10 Application of NC Lacquers

NC lacquers may be formulated for application by most of the conventional methods—cold spraying, hot spraying, dipping, squeegee coating and electrostatic application.

Cold spraying. This is still the most widely used method of lacquer application. The lacquer is thinned to a viscosity of approximately 20 s at 25° C determined by Ford 4 cup (approximately 40 cP) and atomised by jets of compressed air at the spray gun nozzle. It may be drawn from a syphon feed gun by a venturi effect at the nozzle or fed to the nozzle by means of a pressure or gravity feed tank.

An interesting concept, which was developed some years ago, was that of 'High Build' (or High–Low) solvent balance: a high proportion of very low boiling solvents (which are almost all lost during passage of the lacquer from spray nozzle to article) and sufficient high boiling solvent left in the film to obtain satisfactory flow and resistance to blushing (the rapid loss of considerable quantities of low boiling solvent causes considerable cooling of the lacquer). Formulas for typical normal and 'High Build' thinners are given in formulas 40.1 and 40.2.

FORMULAS 40.1 AND 40.2

Formulas for normal and High Build NC lacquer thinners

	Normal Parts by mass	*High Build* Parts by mass
Acetone	—	30.0
Ethyl acetate	5.0	—
Butyl acetate	15.0	—
Alcohols	20.0	10.0
Toluene	60.0	—
Ethylene glycol mono-ethyl ether acetate	—	15.0
Ethylene glycol mono-butyl ether	—	5.0
Heptane	—	40.0
	100.0	100.0

Hot spraying. Instead of using thinners to reduce the viscosity sufficiently to enable application by spray gun, the effect of heat may be used. Thus a lacquer having a viscosity of 600 cP at 25° C may be lowered to approximately 90 cP by heating to 70° C.

For hot spray application the solvent balance should be varied somewhat from conventional cold spray formulations. The formula for a typical hot spray lacquer solvent is given in formula 40.3.

Various types of hot spray equipment suitable for the application of lacquer are available. Basically, they consist of a heat exchanger, spray gun and suitable pumps and heated hoses. This type of application is used in the furniture industry for application of clear nitrocellulose lacquers. The main advantages of this process are:

(a) application of lacquers having a higher solids content is possible without resort to the lower-strength very low viscosity grades of nitrocellulose; this often enables the number of coats to be halved with consequent savings in application man-hours and solvent costs;

(b) less reduction in volume of the film during drying, resulting in less grain exposure and a smoother surface;

(c) reduced sagging, enabling 'full' coats having good flow properties to be applied;

(d) blushing seldom encountered, due to the lacquer reaching the article at approximately room temperature instead of considerably below it.

FORMULA 40.3
Formula for a typical hot spray lacquer solvent

	Parts by mass
Butyl acetate	37.5
n-Butyl alcohol	12.5
Ethyl alcohol	10.0
Xylene	40.0
	100.0

Dipping

(a) *Normal method.* In the conventional dipping process the lacquer is adjusted to approx 40 s Ford 4 cup at 25° C; the article is dipped and withdrawn at a fairly fast rate of approximately 100 mm per second. Only relatively thin coats can be applied by this process, and lacquers must be formulated as high in solids as possible.

(b) *Slow withdrawal method.* In this process a very high viscosity lacquer (approximately 10 000 cP) is used, and after immersion the article is withdrawn at a slow rate of 20–60 mm per minute. Because of a surface tension effect the thickness of coat is controlled by the withdrawal rate: the film thickness is reduced proportionally with the withdrawal rate. Adequate film builds may be obtained, in one dip, without sagging or running off of the coating using this method.

(c) *Squeegee coating.* Certain types of articles of even cross-section may be effectively coated using this process. The lacquer and process used are similar to those used in the slow withdrawal dip process. The withdrawal rate is rapid, however, and the surplus lacquer is removed by passing the article through an aperture, which is the same shape as the cross-section of the article, using a flexible rubber washer called a squeegee. The thickness of coat depends upon the speed of passage through the squeegee, the viscosity and solids of the lacquer and the size of the aperture in the washer. Usually the size of the aperture is smaller than, or equal to, the size of the uncoated article. The thickness of the coating obtained is due to expansion of the size of the hole due to the force exerted in removing the surplus lacquer. The process is used widely for such articles as broom handles, pencils and cables.

Electrostatic application and 'detearing'

(a) *Spray gun type.* This is an adaptation of the standard spraying process, where an electrostatic potential of about 100 kV is maintained between conventional or airless spray guns and the article being sprayed (which is usually earthed). Due to the opposite polarity the mist of sprayed coating is attracted to the article without any substantial loss by overspray. The process is claimed to result in savings of 40 per cent by eliminating much of the loss of spray mist to atmosphere or spray-booth exhaust.

(b) *Disc and bell methods.* The basic process is similar to the spray gun method, but atomisation of the lacquer is effected by pumping it to the centre of a spinning disc or bell. The lacquer passes to the edge of the bell or disc by centrifugal force and is atomised into a mist as it is thrown from the edge.

(c) *'Detearing'*. Use is made of electrostatic attraction in this process to remove the blob of lacquer which forms at the drain point of a dipped article. Before the lacquer is completely dry the article is passed over a grid maintained at a potential of about 40 kV relative to the article, which is usually on an earthed conveyor. The drop of surplus paint is removed by electrostatic attraction, resulting in a much more satisfactory appearance.

Solvent composition. Careful attention must be given to the formulation of the solvents in lacquers that are to be applied by electrostatic equipment; the solvent balance of normal lacquers is often too polar, because of the presence of alcohols and ketones, resulting in the lacquer being too high in conductivity for optimum application properties. The various suppliers of electrostatic equipment give conductivity guidelines for their various processes and are able to supply equipment to measure the conductivity of coating formulations. Application equipment is discussed in greater detail in chapter 52.

40.2.2. Cellulose Acetate

This film former, which is made by reaction of acetic acid and cellulose, is available in several lacquer grades varying in viscosity between low (2–4 seconds) and high (60–80 seconds). The acetic acid content of 54–56 per cent is somewhat higher than the plastic-grade cellulose acetate.

Cellulose acetate has a similar tensile strength–viscosity grade relationship to that of nitrocellulose, but is not quite as extensible as nitrocellulose. The main advantages for lacquer manufacture are:

Water-white colour
Resistance to light and heat
Low flammability
High electrical resistance

The main disadvantages are a very limited range of suitable solvents, and the expensive mixtures necessary due to poor tolerance for diluents. Very poor compatibility with resins and plasticisers and lower water resistance than NC are also limitations.

40.2.2.1 Solvents

Single solvents for cellulose acetate are acetone and methyl acetate, whilst good solvent mixtures include methanol and MEK (1 : 9), and ethanol and methylene chloride (1 : 9.) Ethyl lactate and diacetone alcohol are higher boiling solvents suitable for use with cellulose acetate but they give high viscosity solutions due to poorer solvency.

40.2.2.2 Resins

Toluene sulfonamide formaldehyde resins and certain alkyds and phenolics are the main types suitable for use with cellulose acetate.

40.2.2.3 Uses

Cellulose acetate is useful for preparation of aircraft dopes, cable lacquers and heat-resisting coatings (for such articles as electric light bulbs).

40.2.3 Cellulose Acetate Butyrate (CAB)

With this mixed ester, many of the disadvantages inherent in the acetic acid ester have been overcome and some of the advantages retained. CAB has a similar degree of water resistance to NC. A greater degree of solubility and compatibility with resins and plasticisers than cellulose acetate is offered, but its compatibility with resins is much more limited than NC and its tensile strength is lower than that of cellulose acetate.

CAB is available in a range of viscosities between 0.01 second and 40 seconds, with butyryl contents between 53 and 17 per cent and acetyl contents between 2 and 29.5 per cent. The most widely used types for lacquer manufacture have 37 per cent butyryl and 13 per cent acetyl contents ($\frac{1}{2}$ second and 20 second grades are typical) which gives the highest solubility and compatibility with resins and plasticisers.

40.2.3.1 Solvents

Good solvents for the most widely used grades of CAB include acetone, methyl ethyl ketone, ethyl acetate and butyl acetate. Up to 4 parts of diluents such as xylene and toluene may be added to 1 part of these solvents.

40.2.3.2 Resins

Resins having good compatibility with CAB include acrylics, chlorinated diphenyl, some non-oxidising alkyds, silicones, epoxies, toluene sulfonamide formaldehyde resins and rosin. However, the range of resins compatible with CAB is much more restricted than in the case of NC.

40.2.3.3 Uses

CAB is particularly useful for manufacture of clear lacquers for wood, aluminium and brass, and for aeroplane dopes and cable lacquers. It is also used to modify the properties of automotive finishes.

40.2.4 Cellulose Acetate Propionate (CAP)

This mixed ester, which contains about 2.5 per cent acetyl and 40–46 per cent propionyl, has similar advantages to CAB over cellulose acetate, but is not quite as soluble as CAB. However, it has a greater tautening effect than CAB and for that reason is a key raw material for aircraft fabric dopes. It is not widely used in other lacquers but one important area is in lacquers for grease-barrier applications, such as coatings for fibreboard containers for greases, oils or oil-containing foods. In such cases a lacquer is made without modification by other resins or plasticisers.

40.2.5 Ethyl Cellulose

Ethyl cellulose is made by reacting alkaline cellulose with ethyl chloride under pressure, and is available in a range of ethoxyl contents from 44.5 to 49 per cent. The type containing 47.5–49 per cent ethoxyl is the most widely used in lacquers and is available in a range of viscosities from 4 to 300 centipoise. The tensile strengths of ethyl cellulose films are somewhat lower than those of nitrocellulose, but flexibility and elongation are considerably higher. As with NC and other cellulose polymers, the high-viscosity grades have the highest flexibility, elongation and tensile strength.

EC has better solubility characteristics than NC and it is soluble in much cheaper solvents; it has greater resistance to burning than NC.

40.2.5.1 Solvents

EC is soluble in a wide range of solvents, including aromatic hydrocarbons. Mixtures of alcohols and aromatic hydrocarbons are particularly good solvents for ethyl cellulose, tolerating considerable quantities of aliphatic hydrocarbons. EC is also soluble in most of the solvents suitable for NC.

40.2.5.2 Resins

EC is compatible with a very wide range of resins, but in general is not as compatible as NC (except that it is compatible with phenol formaldehyde resins).

40.2.5.3 Uses

EC may be used for the manufacture of many of the types of lacquer that are normally made from NC. The major advantages of EC over NC are flexibility and resistance to cracking and cold checking, and the lower cost of the solvents needed.

However, the adhesion of EC lacquers in general is not as good as that of NC lacquers, and special care in formulation must be taken to ensure satisfactory adhesion by careful choice of modifying resin.

40.2.6 Ethyl Hydroxyethyl Cellulose (EHEC)

This mixed ether has recently found popularity for speciality lacquers. Its particular virtue is solubility in hydrocarbons containing about 30 per cent aromatics. It is also soluble in aliphatic hydrocarbons containing 10 per cent alcohol.

EHEC is compatible with a similar range of resins to ethyl cellulose.

It is useful for lacquers where low-solvency solvents are necessary, either for cost reasons or if the surface on which the lacquer is to be applied is sensitive to 'stronger' solvents. EHEC enjoys considerable use in screen printing inks and is occasionally employed to rectify low-viscosity batches of architectural enamels.

40.3 ACRYLICS

40.3.1 Polyalkyl Methacrylates, Acrylates and Copolymers

The main thermoplastic acrylic resins used in acrylic lacquers are polymethyl methacrylate and copolymers with ethyl, n-propyl, n-butyl methacrylates and/or methyl or ethyl acrylates which are produced by bulk polymerisation or copolymerisation in solvents. The polyalkyl methacrylate resins are much harder than the corresponding acrylates; the flexibility of both homologous series increases, whilst their hardness decreases as the chain length of the alkyl group increases. Acrylic lacquers for use on rigid surfaces are usually formulated with polymethyl methacrylate homopolymers, or copolymer resins in which the major monomer is methyl methacrylate. Acrylic resins are reviewed in greater detail in chapter 15.

Acrylic lacquers are not as fast drying as NC lacquers, but they are water-white in colour and resistant to yellowing on exposure to heat and UV radiation.

The molecular weight of a polymethyl methacrylate resin for use in sprayable exterior lacquers must be controlled carefully for the optimum balance of toughness, solubility in solvents and sprayability without 'cobwebbing' which can occur if the polymer contains too high a proportion of high molecular weight material.

Typical optimum weight average molecular weights for, respectively, bulk and solution copolymerized methyl methacrylate (with 2–5 per cent methacrylic acid) are 55 000 and 105 000, in order to obtain a desirable balance of application and film properties and viscosity.

40.3.1.1 Formulation of Acrylic Lacquers

Acrylic resins are often used as modifiers for other film formers such as NC. Acrylic lacquers are not usually modified with other resins, with the exception of small amounts of resins such as CAB which can be used to improve flow, petrol resistance, application and recoating properties, and to increase viscosity. When colour retention is not important, nitrocellulose can be used similarly subject to the solvent being suitable. Commercial acrylic resins are available which are internally plasticised by appropriate flexibilising co-monomers to suit a wide range of substrates from rigid to very flexible—that is metal lacquers to rubber lacquers. Where maximum durability is required (such as automotive lacquers) an essentially polymethyl methacrylate resin is chosen; benzyl butyl phthalate has been described as the 'preferred' plasticiser for such applications.

The durability of acrylic automotive lacquers is so much better than NC lacquers that they have almost completely superseded them.

40.3.1.2 Solvents

Acrylic resins are generally soluble in aromatic hydrocarbons such as toluene and xylene, acetate esters of ethyl and butyl alcohols, ethylene glycol monoethyl ether esters, and ketones such as methyl ethyl ketone and methyl iso-butyl ketone. Some grades are soluble in aromatic-aliphatic hydrocarbon mixtures such as mineral turpentine.

The spraying properties and resistance to crazing of polymethyl methacrylate lacquers by UV light and weather are greatly influenced by the choice of solvent. For applications where maximum durability is required, special attention should be given to solvent composition and film integrity.

A typical formula for a solvent composition suitable for acrylic lacquer applications is shown in formula 40.4.

FORMULA 40.4
Typical solvent composition for an acrylic lacquer

Solvent	*Parts by mass*
Toluene	56.5
Acetone	21.0
Ethyl glycol monoethyl ether acetate	13.2
Methyl ethyl ketone	8.0
n-Butanol	1.3
	100.0

40.4 VINYL RESINS

40.4.1 Polyvinyl Chloride (PVC)

As a pure homopolymer, PVC is rarely used in lacquers due to poor solubility in common solvents.

40.4.2 Vinyl Chloride–Vinyl Acetate Copolymers

A range of such resins is available for use in lacquers. The acetate modification confers reasonable solubility in organic solvents but until recently most of these resins required fairly strong solvent mixtures. Relatively low solids contents of, say, 20 per cent at a lacquer viscosity of 400 cP are achieved in a 1 : 1 blend of MIBK and toluene.

A typical resin suitable for use in lacquers is *Bakelite VYHH* which contains:

14 per cent vinyl acetate
86 per cent vinyl chloride

This copolymer does not have good adhesion characteristics without baking at approximately 175° C, but when 1 per cent vinyl acetate is replaced by maleic anhydride (as in *Bakelite VMCH*), good adhesion is obtained at ambient temperature.

Vinyl chloride–acetate lacquers have excellent resistance to alkalis, mineral acids, alcohols, greases and oils. However, most of these resins have poor compatibility with resins and this limits their use. Where compatibility with non-oxidising alkyds or epoxies is required in order to obtain particular lacquer properties, one of the special copolymers that have good compatibility with such resins should be chosen. A typical resin that has good compatibility contains 6 per cent vinyl alcohol, 91 per cent vinyl chloride and 3 per cent vinyl acetate: this resin allows the lacquer formulator to choose from a wide range of film formers for modification of the properties obtained from the vinyl resin.

40.4.3 Vinyl Chloride–Vinyl iso-Butyl Ether Copolymers

By copolymerising vinyl iso-butyl ether with vinyl chloride, a much more soluble resin than PVC/PVA copolymers is obtained. This is soluble in aromatic hydrocarbons and aromatic–aliphatic hydrocarbon mixtures.

A range of viscosity grades is available: by using the lowest viscosity types, similar lacquer solids contents to those obtained in PVC/PVA copolymers prepared by solution copolymerisation can be achieved in aromatic hydrocarbons; higher solids contents will result from use of higher strength ester and ketone solvents.

These resins are non-hydrolysable, have good water and chemical resistance, and are recommended for use in lacquers for metal, concrete, plastics and asbestos cement. Their properties are similar in many applications to those of chlorinated rubber, but they have the advantages of higher aliphatic hydrocarbon solvent tolerance and greater flexibility. For many applications these resins have sufficient flexibility and adhesion without modification, but where additional flexibility is needed 10–25 per cent of a phthalate plasticiser can be used. Where additional gloss and hardness are required, modification can be made with cyclic ketone–formaldehyde resins, which will not detract markedly from the water and chemical resistance of the lacquer, although some loss of weathering resistance will result.

This type of resin, which is marketed as *Laroflex MP* by BASF, is particularly recommended for use in anti-corrosion and marine primers and finishes, and fire-retardant lacquers.

40.4.4 Vinyl Chloride–Acrylic Ester Copolymers

By copolymerising vinyl chloride with methyl methacrylate, superior resistance to UV light and heat is obtained in comparison to PVC/PVA copolymers; in many cases, additions of stabilisers are not needed. These copolymers also exhibit wider compatibility with other resins than the PVC/PVA types, and have excellent water resistance and weathering properties. Their good adhesion properties make them useful for non-yellowing lacquers for metal.

Solubility and other properties of these resins are somewhat similar to those of the PVC/PVA type. Rhône-Poulenc *Rhodapas ACVX* is typical of this class of polymer, which is usually modified with 20–25 per cent monomeric plasticiser (such as dioctyl phthalate) in order to produce a lacquer with good adhesion to steel.

40.4.5 Polyvinyl Acetate–Polyvinyl Alcohol Resins

These resins (made by hydrolysis of PVA) are available in several degrees of hydrolysis, but as most are soluble in cold water they have little application in lacquers. Some grades, which are insoluble in cold water and soluble in mixtures of ethyl alcohol and water, can be used effectively in clear lacquers for paper and board, to impart resistance to animal and vegetable oils, greases and many solvents, including petroleum hydrocarbons.

40.4.6 Vinyl Acetate–Ethylene Copolymers

The grades suitable for use in lacquers are based on a monomer mixture containing 40–45 per cent vinyl acetate. They are more accurately described as ethylene–vinyl acetate copolymers (EVA), but the title has been reversed here in order to classify them under the general heading of 'Vinyl Resins', which is the category used by their manufacturers. Unlike other EVA copolymers, which are based on lower vinyl acetate concentrations, these grades of resin (exemplified by Du Pont *Elvax 40* and Bayer *Levapren 450* series) can be formulated into lacquers at a practical solids content. They are soluble in aromatic and chlorinated hydrocarbons, and such solutions will tolerate appreciable dilution with aliphatic hydrocarbon solvents.

Whilst it is possible to use this type of resin as the main binder in lacquers for wood and metal, the major application, because of its high degree of flexibility, is for lacquers for flexible substrates such as aluminium foil, paper and board. It is also very effective in heat seal lacquer coatings for backing board used in blister packaging; modification with small amounts of nitrocellulose to reduce thermoplasticity is usual.

40.4.7 Vinylidene Chloride–Acrylonitrile Copolymers

This type of copolymer (best known as *Saran*) is high in vinylidene chloride content and low in acrylonitrile. It forms lacquer films having very low moisture vapour permeability, and these can be used for coating paper and board for food packaging. It is compatible with a limited range of resins, including styrene-rosin esters and acrylates, and it has very poor solubility, the most suitable solvent being methyl ethyl ketone. This type of resin is usually used without plasticiser, particularly where its excellent moisture barrier properties are required; as lacquers based on it have poor spraying properties, they are usually applied by dipping, roller coater or doctor blade.

40.4.8 Polyvinyl Butyral (PVB)

This resin is obtained by partially butyrating polyvinyl alcohol, and a typical resin contains 17–21 per cent polyvinyl alcohol and 75–80 per cent polyvinyl butyral, with small amounts of polyvinyl acetate. This polymer has excellent hardness and impact strength, and gives colourless clear coatings which are resistant to yellowing by sunlight. A wide range of molecular weight resins are marketed, varying from 30 000 to 150 000.

Solubilities of these polymers vary with composition but, in general, the main solvents used are alcohols. Fairly large amounts of aromatic hydrocarbon diluent may also be used.

Polyvinyl butyral resins have a high degree of flexibility and some grades have excellent adhesion to metal. Adhesion can be further improved by modification with certain phenolic resins, and alcohol-soluble natural gums, with which it is compatible, may be used to increase gloss.

A major usage of polyvinyl butyral lacquers is in coatings for aluminium and for flexible substrates such as aluminium foil and paper. It is widely used in architectural and marine primers.

40.4.9 Polystyrene

A wide range of styrene polymers is available for use as the main resin in lacquers. They have a high degree of chemical resistance and are manufactured in a range of molecular weights, the higher resins having melting points of 125–145° C. They are soluble in aromatic hydrocarbon solvents and will tolerate appreciable dilution with aliphatic hydrocarbons; grades are available that have good compatibility with other resins. Some commercial products are available that are internally plasticised such that additional plasticiser is not required; but when necessary, ester-type plasticisers (which have an aromatic structure, such as dicyclohexyl phthalate) have good compatibility. Polystyrene resins may be used in primers for concrete and are useful in zinc-rich primers.

40.4.10 Styrene–Butadiene Copolymers

These resins are made by copolymerising a small amount of butadiene with styrene; the butadiene increases the toughness of the polymer. They are exemplified by the Goodyear *Pliolite S* range, which have good solubility in aromatic hydrocarbon and aromatic–aliphatic petroleum solvents. Although they have good chemical resistance, resistance to UV light is poor. As they have good moisture resistance, they are particularly recommended for marine paints and for concrete swimming pools.

They are compatible with many resins, but 70 per cent chlorinated paraffin is particularly useful where chemical resistance in the film is needed. They are fairly hard and require plasticising for most applications, and are compatible with a wide range of chemical plasticisers.

40.4.11 Vinyl Toluene–Butadiene Copolymers

These resins, such as Goodyear *Pliolite VTL*, have similar properties to the styrene–butadiene copolymers. They have increased solubility in aliphatic hydrocarbons, being soluble in low aromatic petroleum hydrocarbons. Otherwise, properties and applications for this type are similar to those described for styrene–butadiene resins.

40.4.12 Styrene–Acrylic Copolymers

By copolymerising styrene with an acrylic monomer (such as ethyl acrylate) a resin with improved UV resistance is obtained. A typical resin is Goodyear *Pliolite AC-L*. In general, it requires more plasticiser to achieve the required film flexibility, but it is compatible with a very wide range of plasticisers. It can be used for similar applications to the styrene–butadiene resins and is preferable in applications where good weathering resistance is required. This class of resin has limited compatibility with other resins, and may be less soluble in low aromatic hydrocarbon solvents than its butadiene counterpart. Paving paints are a typical end use.

40.4.13 Vinyl Toluene–Acrylic Copolymers

These have similar properties to the styrene–acrylic copolymers, but are much more soluble in low aromatic petroleum hydrocarbons (having similar solubility to the vinyl toluene–butadiene copolymers). A typical example is Goodyear *Pliolite VTAC-L*, which has particularly good resistance to fats and oils. Properties, apart from solubility, are similar to those of styrene–acrylic copolymers.

40.5 RUBBER AND DERIVATIVES

40.5.1 Chlorinated Rubber (CR)

This resin, produced by chlorination of natural rubber, is used to produce lacquers for special purposes where a high degree of chemical resistance is required. It is soluble in aromatic hydrocarbons and is compatible with a wide range of resins. It has low tensile strength and very low elongation, so must be fairly highly plasticised. Typical applications for CR include chemical resistant lacquers (in which the plasticiser is often a chlorinated paraffin containing about 50 per cent chlorine); where a hard resin modification is required to obtain additional gloss, a 70 per cent chlorinated paraffin resin is the usual recommendation.

CR is available in a range of viscosity grades which vary between 10 cP and 90 cP (the viscosity of a 20 per cent solution in toluene), which allows the formulation of lacquers that have a satisfactory viscosity–solids content relationship. Major uses include primers and finishes for chemical plants and marine applications.

The most commonly used grades of chlorinated rubber are Bayer *Pergut*, Hercules *Parlon* and ICI *Alloprene*.

The solvent mixture used must include a proportion of high boilers in order to achieve suitable application properties and to eliminate 'cobwebbing'.

40.5.2 Cyclised (Isomerised) Rubber

This resin is manufactured from natural rubber (by dissolving it in benzine and reacting with a metallic halide such as stannic chloride or titanium tetrachloride) to form an isomer of rubber, which contains less unsaturation and has better solubility. It gives higher solids content films than natural rubber, but forms films that are hard and brittle (but tough when suitably plasticised). It is compatible with a very limited range of plasticisers and lacquer resins.

This type of resin (Croda *Plastoprene* is a typical example) has excellent chemical resistance, good adhesion properties and heat resistance, and is soluble in aliphatic hydrocarbon solvents. It does not have good exterior durability but has exceptionally low water absorption properties. By plasticising with chlorinated diphenyl, it is possible to formulate lacquers with excellent chemical and water resistance and excellent adhesion to metal.

40.6 POLYURETHANE ELASTOMERS

These film formers are fully reacted urethane elastomers, which are manufactured from polyesters and aromatic or aliphatic isocyanates, the latter type having superior light stability.

The main application for lacquers which are based on PU elastomers is for flexible and semi-flexible substrates such as thermoplastics, rubber, PVC, leather and foamed polystyrene, for which lacquers can be formulated with a high degree of abrasion resistance, excellent adhesion and low temperature flexibility properties. The weather resistance of urethane elastomers is not as good as that of two component urethanes, but satisfactory durability can be obtained for applications such as extruded flexible rubber or plastic weather stripping.

These resins in general require fairly strong solvents, although some commercially available products are soluble in low boiling alcohols. N-methyl–2 pyrrolidone and dimethyl formamide are recommended by some manufacturers as strong solvents to promote additional adhesion to difficult substrates.

A novel application for lacquers based on PU elastomers is for 'in the mold' finishing of polyurethane foamed articles. Here the lacquer is applied to the mold before the foam is injected, which is the reverse of normal procedures in that the article is applied to the lacquer.

Urethane lacquers are usually formulated using PU elastomers only, but where special properties are needed vinyl chloride–acetate copolymers, nitrocellulose or CAB can be used with certain grades.

Plasticisers are not normally required or recommended for use in PU elastomer lacquers, as best results are obtained by choosing a grade of resin that has the required flexibility.

40.7 CHLORINATED POLYOLEFINS

This type of polymer is available in various grades which have a wide range of applications in lacquers. A typical commercial resin, Eastman *Chlorinated Polyolefin 310–6*, is soluble in most solvents except alcohols and aliphatic hydrocarbons. It has similar properties and applications to those described for chlorinated rubber.

Other specialised types of chlorinated polyolefin, which have more restricted solubility, have been successfully used in the manufacture of primers for polypropylene, allowing application of conventional topcoats with satisfactory adhesion to this 'difficult' substrate.

40.8 NATURAL RESINS

40.8.1 Shellac

This resin has the longest history of any currently used lacquer medium, being the basis of French Polish; it is still widely used in spirit lacquers in solution in alcohol. Normal shellac, which is refined from an insect excretion on the bark of trees, contains a waxy component which is not soluble in alcohol. Dewaxed shellac can be purchased as such, or a solution of regular shellac in ethyl alcohol can be allowed to stand until all insolubles have settled and the clear solution decanted.

Shellac is a flexible resin which gives tough films without plasticiser, but small quantities of castor oil can be used for additional flexibility. Shellac-based spirit lacquers are useful in paper lacquer and wood finishing. A minor application is in 'knotting' lacquer for timber before painting.

40.8.2 Copal

The main copal resin used in lacquers is soft Manilla resin, an alcohol-soluble resin which is widely used in spirit lacquers for printed labels; whilst not as tough as shellac, it has sufficient flexibility without the use of plasticisers.

40.9 MISCELLANEOUS SYNTHETIC RESINS

The miscellaneous synthetic resins (listed at the beginning of this chapter) could all be used in a lacquer as the main film former but, because of their generally brittle nature, they do not form strong films in most cases, and should be considered mainly as modifying resins for other film formers. However, they are used for certain speciality lacquers where tensile strength is not important and where low cost is the main consideration.

40.10 GENERAL NOTES ON FORMULATION

Chapter 40 has covered the main aspects of formulation of lacquers based on the various main film forming resins discussed. However, a comprehensive treatment of the formulating principles would occupy a volume in itself because of the wide variety of polymers available and, more particularly, the differences in properties and composition of the ranges of proprietary resins available from different manufacturers.

In the case of all of the resin types described, detailed information is available from their manufacturers in the form of technical publications which give comprehensive details regarding solubility and viscosity–solids relationships in solvents, compatibility with plasticisers and resins, adhesion properties, suitability of pigment types and any special requirements (such as the need for stabilising, where necessary, to heat and/or light). It is strongly recommended that the lacquer formulator obtains as much information as possible on all of the available grades and studies this carefully before attempting to produce lacquers.

Evaluation of the suitability of a particular system will, of course, depend on the end use, but the general principles described under NC lacquers should be followed in order to develop a stable and reproducible product.

41 CONVERSION COATINGS

41.1 INTRODUCTION

A conversion coating may be defined as one formed by a chemical reaction *which converts the surface of a metal substrate into a compound which becomes part of the coating.*

In the case of *oxidising* and *anodising*, the coating is almost entirely composed of a reaction product of the substrate, but in *chromating* and *phosphating* the major part of the coating is derived from reaction products of the processing chemicals. These coatings have many practical uses, and these will be discussed in more detail below but, to generalise, they are used in the main for the following purposes:

(a) decoration;
(b) corrosion resistance; and
(c) provision of a base for a supplementary coating.

From the decorative point of view, they can produce a pleasing colour or gloss and a more even appearance. As corrosion inhibitive coatings they provide a barrier to the ingress of aggressive gases or liquids, or act as carriers for corrosion inhibitors such as chromates, which are absorbed into the matrix of the coating. As a base for supplementary coatings they can act as a carrier for soft coatings such as oils and greases, as an inert barrier between certain metals and paint films between which some reaction may occur, and to provide further anti-corrosive effects by limiting spread of corrosion from damage sites.

In the modern coatings field there is an enormous range of conversion processes available, the processes for colouring metals alone running literally into thousands. Some of these processes provide excellent finishes, but many are quite mediocre as they were designed for specialist applications under conditions which may not be universally reproducible. Some of the more complex formulations which have been published from time to time appear to have been developed empirically, since it is evident that not all of the components perform a useful function in the process. However, it is proposed to discuss only those processes which are of some historical interest, or those which are at present in common use in the industry. Conversion coatings, particularly those based on phosphates and chromates, are also discussed in chapter 49.

41.2 OXIDE COATINGS

The colouring of metals by oxidising has been carried out for centuries, not only to enhance the appearance of an article, but also to improve the corrosion resistance. While the appearance is generally improved by the imparting of a pleasing colour, attractive gloss finishes may also be produced as well as the production of a more even surface. The required colour and gloss can be obtained by choice of process or by varying the conditions of the one process, but brightening or dulling before or after application of the coating may be necessary in some cases to achieve the correct degree of gloss. Attractively coloured, glossy surfaces are usually produced for decorative purposes such as in jewellery, while the more sombre matt surfaces may be used for machine parts or, as in military equipment, to render the objects less conspicuous.

The corrosion resistance of these coatings is usually quite poor, but they have the advantage of good absorptive properties so that they can hold oil or grease on the substrate with less opportunity for it to be wiped off. In the case of supplementary paint films, they can also improve adhesion.

41.2.1 Oxide Processes on Iron

41.2.1.1 Temper Colours

Probably the oldest of the oxide processes is the heating in air of iron at controlled temperatures, between 220 and 360° C for up to 30 minutes, to produce a thin film of magnetite (Fe_3O_4). Under these conditions a range of colours may be produced varying from a straw-yellow through brown, purple and blue to bluish-grey, the colour produced being dependent upon both temperature and time of treatment. The colours are due to light interference, since the coating thickness is of the order of 100 nm. The longer the process continues and the higher the temperature, the thicker will be the oxide layer, and a deeper colour will result. Articles to be coated by this process are sometimes dipped in linseed oil or preferably sulfonated oils, or smeared with a thin layer of tallow before heating, to produce thicker coatings and more uniform colours.

41.2.1.2 Inoxyde Process

The Inoxyde process produces grey coatings of magnetite, up to 1 mm thick, by annealing clean steel parts at about 800 to 900° C for 20 minutes in a weakly oxidising atmosphere. This is frequently followed by annealing in a weakly reducing atmosphere for 20 minutes which improves the adhesion and thickens the coating. A coating consisting of magnetite and haematite is produced in the first or oxidising stage, which is fully converted to magnetite in the second or reducing stage. While this type of coating is relatively resistant to mechanical wear, it will not withstand deformation by drawing or pressure as it is fairly brittle and tends to peel.

41.2.1.3 Fused Salts Process

There is a wide call for the production of a matt black finish on steel articles, and some of the earlier processes were quite complicated or difficult to control. A simplified procedure was ultimately introduced and consists of immersing the articles in a bath of molten sodium nitrate at 300° C for about 20 minutes. An alternative melt is a mixture of sodium and potassium nitrates, which has a lower fusion temperature. This process is much faster, gives more uniform heating and produces a pleasing blue-black colour.

41.2.1.4 Caustic Alkali–Nitrate Process

Amongst the most widely used processes for producing black oxide finishes is one which consists of immersion in a hot solution of sodium hydroxide containing oxidising agents such as sodium nitrate, although nitrites, permanganates, dichromates and chlorates may also be used.

This process goes back to the First World War, and the original version was called the Browning process, used for finishing gun barrels. It required a treatment time of upwards of 72 hours. The modern baths require from 10 to 30 minutes immersion at a temperature of 135 to 150° C, the longer times being used for higher alloys. A typical commercial formulation would be:

Sodium hydroxide	750 g/L
Sodium nitrate	250 g/L

Activators such as cyanides, tartrates and tannates which complex and inactivate dissolved iron and copper are frequently added to minimise the formation of red oxides.

Sometimes, two solutions are used instead of one, but the combined time for the two tanks is the same as for a one-tank process. The first tank is operated at a lower temperature (140° C), the second at a higher temperature (150° C), although both solutions must be at boiling point which is accomplished by varying the concentration of the bath. This system eliminates the formation of a reddish colour at higher temperatures and on certain types of case hardening. Because the bath is operated at boiling point and the boiling point is a function of concentration, the temperature is used as a convenient method of control. Chemical control is difficult because of the high alkalinity of the bath. The widely used Jetal process uses sodium nitrite, replacing the sodium nitrate, and operates at or near 140° C for up to 30 minutes.

The coatings obtained by these processes may be up to 50 μm thick, but dimensional change is rarely more than 1 μm. Sealants such as oil or dewatering fluid are often used, but care should be taken so as not to apply an excess, or the coating will never lose its tacky feel. Dewatering fluids can be used alone, or to prepare the surface for painting.

41.2.2 Oxide Processes on Copper and Copper Alloys

41.2.2.1 Alkali–Process

A bath similar to the alkali–nitrate process for steel produces a black coating on copper which, unlike the amorphous coating on iron, is acicular in structure and gives the surface a velvety appearance. Unfortunately, this beautiful upper layer is easily rubbed off, leaving a smooth black film underneath, whose absorptive properties are quite good and which increases the bond strength of adhesives, lacquers and enamels.

Bright finishes are used for jewellery and the dull finishes for optical components and thermal sensing apparatus, as the emissive and absorptive characteristics approach those of a black body.

41.2.2.2 Ammoniacal Copper Carbonate

A blue-black coating can be produced by immersion in an ammoniacal solution of copper carbonate (200 g/L) at about 80° C for a few minutes. The colour can be converted to a jet black by the addition of sodium carbonate to the solution. The colour may be fixed by a subsequent dip in 2½ per cent sodium hydroxide solution.

As in the case of iron, these coatings have little protective value, hence corrosion protection must be supplied by a supplementary coating of oil, lacquer or enamel.

41.3 CHROMATE COATINGS

41.3.1 Zinc

41.3.1.1 Introduction

Chromate passivation of zinc was introduced many years ago to improve the corrosion resistance of zinc applied to a steel substrate. The original process was patented in the United States in 1936 and is now known as the Cronak process. It was based on a combination of sodium dichromate and chromic acid with some sulfuric acid. Other variations were covered in the original patent, but these are of little importance. While the original composition suggested a range of concentrations, the present-day process, as found in the literature and in specifications, is:

Sodium dichromate dihydrate	200 g/L
Sulfuric acid	7 ml/L

This process produces coatings ranging in colour from faint iridescence to iridescent red, green and yellow when used at room temperatures from about 10 seconds contact time.

During the Second World War formulations were developed that produced heavier coatings and were bronze or olive-drab in colour. These coatings imparted an enhanced corrosion resistance necessary for the protection of military equipment under the conditions of high temperature and humidity encountered in the South Pacific. Since the war, many commercial variations on the original formulation have been developed to give a wide range of coatings for specific purposes, so that now they can be used in the following applications on zinc substrates:

(a) to impart corrosion resistance;
(b) to enhance the appearance;
(c) to give better paint bonding properties.

As expected, no one formulation can produce a coating exhibiting all these properties at their maximum level, so that formulations must be adjusted to emphasise different characteristics depending upon end use.

41.3.1.2 Film Formation

The mechanism of the formation of the film is still not completely understood, but the present knowledge of the process provides a reasonable explanation of what occurs.

When the zinc is immersed in the passivation bath, it is attacked by the acid, bringing about a fall in the acidity near the surface accompanied by the production of hydrogen. Some hexavalent chromium (as dichromate) is reduced to trivalent chromium and some converted to the hexavalent chromate form, which results in the formation of hydrogen ions to replace those lost in the solution of the zinc. The result is the formation of a complex basic chromic chromate gel of somewhat uncertain composition, which bonds itself to the zinc. The following set of equations give a reasonably accurate picture of the overall reaction:

(a) $Zn + 2H^+ \rightarrow Zn^{++} + H_2$
(b) $HCr_2O_7^- + 3H_2 \rightarrow 2Cr(OH)_3 + OH^-$
At this stage, the acidity falls due to reactions (a) and (b), giving rise to:
(c) $HCr_2O_7^- + H_2O \rightarrow 2CrO_4^= + 3H^+$
(d) $2Cr(OH)_3 + CrO_4^= + 2H^+ \rightarrow Cr(OH)_3.Cr(OH)CrO_4 + 2H_2O$

It is believed that a thin gel film is formed with the chromic hydroxide providing an insoluble matrix and the basic chromic chromate absorbed in it as a soluble component. The film thus formed is quite soft and easily damaged, so that care is to be exercised in handling it for at least 24 hours after treatment, after which time the film hardens and becomes relatively abrasion resistant.

41.3.1.3 Coating Characteristics and Properties

With the present range of both formulations, the appearance of the coating can be varied from a highly polished, silvery, clear coating (not unlike chromium plate) through ranges of bright blue or iridescent bright red, yellow, green and blue up to the heavy chromate colours of deep yellow, bronze and olive-drab.

Also available are dyed coatings where, by the use of acidified dye solutions, red, black, blue, orange, green or violet films can be produced. The intensity of these colours can be varied from light pastel to dark shades, depending upon the colour of the chromate film. These coloured films, however, being based on organic dyes, often exhibit poor lightfastness and are usually used only for identification purposes.

The colour of the film in the Cronak process deepens as it gets thicker, passing from iridescent hues of green, yellow, blue and red, thence to a brassy yellow and finally a deep brown. With a very short time of immersion (of 1 second or so) the film is scarcely visible, or faintly coloured, and would be much thinner than normal. The colour of the thinner coatings is simply due to interference of the light transmitted and reflected through the film, while the thicker yellow films show the colour of the hexavalent chromium, and the brown films show the presence also of trivalent chromium. It should be noted that the iridescent and yellow films are firmly adherent, but the brown films are powdery and non-adherent and the appearance of this colour is an indication that the passivation process has been allowed to proceed for too long. Generally, the process should not be allowed to proceed for less than 5 seconds nor more than 20 seconds, although this may vary for some commercial formulations.

The deeper bronze and olive-drab colours are obtained in films of normal thickness by the addition of mild reducing agents such as formic acid which increase the trivalent chromium content of the film. Bleached or colourless films are sometimes produced in commercial applications to enhance the appearance. However, this practice is to be avoided where corrosion resistance is of prime importance as the chromate ion is the major inhibiting factor, and it is the removal of this coloured ion by leaching in hot water that brings about the bleaching effect. The films are said to be self-sealing, in that a damage point is re-passivated by chromate ion leaching out of the film into the damaged area.

A common fault in the industry is to accelerate the drying of the film by air heating or immersion in boiling water. Unfortunately, this has the effect of removing water from the gel matrix leading to a phenomenon called 'mudcracking' with a serious drop in the corrosion resistance.

Passivated films exhibit low electrical resistance, which makes them useful for electrical contacts and radio chasses, an earth being possible without damaging the film. They serve not only as a good base for paints, but also for the bonding of rubber and adhesives. When used for this purpose, the applied film should be allowed to dry out for at least 24 hours, otherwise poor adhesion will result.

41.3.2 Cadmium and Galvanised Zinc

These films may also be applied to hot dipped zinc and cadmium plating, and the foregoing comments on the treatment of zinc plating apply equally to these other substrates. However, a special alkaline pre-treatment should be used for hot dipped zinc. Highly decorative iridescent or black finishes can be produced.

41.3.3 Copper

A somewhat similar film can also be applied to copper and copper alloys using a bath of similar composition. The film is designed to minimise the effect of handling during fabrication of components and as a surface preparation prior to painting. The film gives the surface of the metal an opalescent appearance, but it is sometimes difficult to determine by visual examination whether the process has been applied. However, some brilliant primary colours on copper alloys are produced from thiosulfate solutions such as:

Sodium thiosulfate	240 g/L
Lead acetate	15 g/L

the colour depending upon temperature, up to 50° C, and immersion time. Additions of organic acids, such as oxalic or citric acids, can further expand the range of brilliant coating colours.

41.3.4 Magnesium

Magnesium and its alloys are sometimes protected by a chromate passivated film, which usually gives the surface a dull to lustrous dark brown or black appearance. The film alone confers quite good corrosion protection to the surface, but is also often used as a base for paints. A more durable dense black passivated finish can be obtained by 30 minute dip at boiling point with a solution of:

Sodium dichromate	160 g/L
Calcium fluoride	2.5 g/L
at pH 4.5–5.5	

whereas a matt grey to an iridescent reddish-yellow coating can be obtained with a 2 minute dip in a solution of:

Sodium dichromate	180 g/L
Nitric acid	141 g/L

41.3.5 Aluminium and Aluminium Alloys

41.3.5.1 Alkaline Processes

Chromate passivation processes for aluminium were first developed as early as 1915 by Bauer and Vogel, and a modified process called the MBV (Modified Bauer Vogel) process was developed by Eckert in 1930.

This process and others similar to it such as the E.W. (Eftwerk), the L.W. (Lautawerk), the Pylumin and Alrok processes are all of an alkaline type, being based on sodium carbonate and sodium chromate with some also containing basic chromic carbonate. The operating temperature is usually about 90° C with a time of treatment ranging from 5 to 15 minutes.

The coating is formed by a chemical reaction between the aluminium and the alkali, which produces aluminium hydroxide and hydrogen. The hydrogen reduces hexavalent chromium to trivalent chromium, which forms as a precipitate of hydrated chromic oxide intermixed with hydrated aluminium oxide. These mixed hydrated oxides bond themselves to the metal surface in the form of a gel coating.

The colour of the coating depends upon the type of alloy treated, but is usually an opaque grey to greyish-green. Like any of the chromate processes, it may be used as a base for painting or other treatment, but it is also used alone because it produces a film with a uniform grey colour and fair corrosion resistance. Undyed colours vary from a lustrous light grey to grey-black. The introduction of permanganate can produce brown coatings. However, the use of the alkaline processes has declined in recent years as the acid processes have gained more importance.

41.3.5.2 Acid Processes

The first of the acidic chromate processes was developed in 1945 by the American Chemical Paint Co. (now Amchem Products), and was the well-known Alodine process, also called Alocrom in the U.K. Other processes, such as the Tridur, Alchrome and Iridite, to name a few, have been developed since, but are all more or less equivalent to the original method. At the present time there are basically two types of process being used, the chromate/fluoride and the chromate/fluoride/phosphate.

As in the case of the Cronak process, a basic chromic chromate gel is formed but, in addition, hydrated aluminium oxide is also formed so that the film is a complex mixture of the two (with some absorbed phosphate in the case of the phosphate types).

The chromate/fluoride process provides thin, dense amorphous films, ranging in colour from transparent to iridescent greenish-yellow. They are often used to treat aluminium where the original colour of the surface is to be maintained. The phosphate type process produces somewhat thicker crystalline films ranging in colour from yellowish-green to light brown.

In the aircraft industry in particular, chromate treatments are used as an alternative or even as a replacement for anodising. Chromate films generally give better corrosion protection than anodising and are better suited to paint bonding, yet are far simpler to apply and require much shorter treatment times. On the other hand, they display lower resistance to abrasion and the dyed films are not as attractive as those produced in anodised films, nor can they be produced with any degree of light fastness.

Like other chromate films, they have low electrical resistance, good weldability, self-healing of damaged areas and ability to be cold formed without flaking.

41.4 ANODISING

41.4.1 Introduction

Most conversion coatings are applied as the result of a chemical reaction taking place at the surface. Anodising is an electrochemical process whereby an aluminium object is made the anode in an electrolytic cell containing an appropriate electrolyte, whereupon a thin film of aluminium oxide is caused to grow by the oxidising effect of the anodic current. If the electrolyte is of the type that cannot dissolve the oxide layer, such as boric acid, then the process is self-terminating and a relatively thin, non-porous coating is formed. These films are useful in electrical applications because of their low electrical resistance compared to that of the thicker, porous films. Most electrolytes are strongly acidic and exert a solvent action on the oxide film. However, the film grows at a faster rate than that at which it dissolves, and the result is a film having porous upper layers separated from the metal surface by a thin barrier layer, which is of the order of 0.1 μm thick. The coating grows by migration of ions across this layer, but it is not known whether the aluminium ions migrate to the solution/oxide interface or oxygen ions to the aluminium/oxide interface.

Prolonged action of the electrolyte causes the pores to take on a conical rather than cylindrical cross-section, and this results in an upper limit of thickness being achieved when formation rate equals solution rate. These porous coatings are much thicker than the non-porous types and have many uses in the protective and decorative fields.

41.4.2 Anodising Baths

There are many different electrolytes that can be used in anodising, but the most common in Australia are the chromic acid, sulfuric acid and sulfuric/oxalic acid baths, all of which produce a more or less porous film.

41.4.2.2 Chromic Acid

This was the first anodising bath used and was patented by Bengough and Stuart in 1924 in the U.K. It operates at a concentration of about 3 per cent chromic acid, at a temperature of about 40° C and the voltage being raised in steps from 0 to 50 V over a period of about 1 hour. An alternative ('Accelerated') process operates at 9 per cent concentration at 30° C for 30 minutes at a voltage of 40 V. This is called the N.B.S. Process.

The films produced in these baths are thinner than in the sulfuric acid baths (see below), being of the order of 1–4 μm. Although thinner, they have an equivalent corrosion resistance, some authorities stating that this is due to the presence of absorbed chromate ions which have an inhibitory effect. The chromic process is favoured wherever the electrolyte may be trapped in folds or joints, particularly in aircraft components. However, it is not suitable for alloys containing more than 5 per cent copper. The film is usually grey and opaque and, being fairly ductile, is more suitable for cold worked components than the sulfuric process film; however, it is not as suitable for colouring by dyeing.

41.4.2.3 Sulfuric Acid

The most widely used of all the anodising processes is that using sulfuric acid as the electrolyte. The first of these was patented by Gower and O'Brien in 1937 in the U.K.

This bath generally operates at a concentration of about 15 per cent (although up to 70 per cent is possible), at about 20° C and the potential difference between 12 and 22 V. The films produced are thicker, from 2.5 μm up to 30 μm, more transparent and cleaner looking than those produced in the chromic bath. Because of these characteristics this process is favoured when a dyed finish is required. Other than this, the advantages over the chromic process are cheapness, less process control necessary (in voltage maintenance) and harder, more abrasion resistant films.

41.4.2.4 Hard Anodising

This process employs a solution of about 10 per cent sulfuric acid operating at a low temperature of 4–10° C (cooling coils are necessary) and a potential difference between 10 and 50 V. Sometimes oxalic acid (2 per cent) is used as a bath additive, in which case the temperature is lowered to 0° C and the potential difference raised to about 75 V.

At the lower temperature and higher voltage there is a lowering of solvent action on the film which results in a denser, harder and more abrasion resistant coating of a thickness ranging from 25 μm to 125 μm. These films are useful in such applications as gears, valves, bearings, etc.

41.4.3 Sealing

In both the chromic and sulfuric processes, a sealing stage must be carried out if the film is not to be over-coated. On the other hand, if the film is to be painted it must be left unsealed. The sealing step is usually carried out by immersion in boiling water or dichromate solution for about 20 minutes. This hydrates the oxide at the surface, causing it to swell and seal the pores. The dichromate seal renders the film yellowish in colour, and should not therefore be used when dyeing is to be carried out.

41.4.4 Coloured Anodising

Modern processes allow many pleasing colour effects to be achieved with anodised aluminium. There are a number of processes available, the more important being impregnation colour, integral colour and electrolytic oxide pigmentation.

41.4.4.1 Impregnation Colour

Many dyes can be applied to the film, before sealing, by immersion in a solution of the dye, which is absorbed into the pores, possibly helped by the mordant action of the aluminium hydroxide. Another method consists of precipitating coloured inorganic compounds into the pores, usually by means of a double dip, one for each of the precipitating chemicals; the well-known bronze colour, for example, is obtained by using cobalt acetate and potassium permanganate. A single dip inorganic colouring is also used, one of the best known being the precipitation of iron oxide into the pores from a solution of ferric ammonium oxalate, giving a pleasing gold colour. It is essential to seal the coating after impregnation of the coloured substance to prevent subsequent leaching out. It should be noted that dyed coatings are suitable only for interior applications as their lightfastness is quite poor.

41.4.4.2 Integral Colour

Certain alloying elements such as silicon and chromium cause the anodised coating to assume a dark colour, which is useful in many architectural applications because it is lightfast and durable. Silicon gives a dark grey and chromium a gold colour. The reason for the colour is not well understood although, in the case of silicon alloys, it is thought to be due to the presence of tiny particles of silicon which come out of solid solution as the aluminium is oxidised.

41.4.4.3 Electrolytic Oxide Pigmentation

When certain metal salts are dissolved in the sulfuric acid electrolyte, the resultant film has metallic oxides deposited in it, due to anodic oxidation of the metal. The colour is independent of the alloy composition and the thickness of the coating and is quite durable and lightfast.

41.4.4.4 Subsequent Painting

Anodising makes an excellent base for paint because it is somewhat porous and also has good anti-corrosive properties in itself. Painting should, however, be carried out on unsealed films, otherwise adhesion will be poor. If the film has already been sealed, it should be given a mild alkaline etch before painting. A recommended process, if the aluminium article is to be painted, is to produce a thin film up to about one micrometer (1 μm) in thickness by using either DC or AC anodising.

41.5 PHOSPHATING*

41.5.1 Introduction

The phosphating process is said to go back to the third century AD; it is claimed that the Romans protected iron articles using phosphates. In 1869 a British patent was issued which described a process of protecting steel articles by heating to red heat and plunging into phosphoric acid. However, the father of the modern phosphate process can truly be said to be Thomas Coslett who, in 1906, introduced the Coslettising process which involved the immersing of steel components in a boiling bath of phosphoric acid and ferrous phosphate. The bath operated for about four hours, producing a thin film of ferrous phosphate on the metal. He developed a Zinc Coslettising process in 1909 in which steel articles were immersed in a boiling solution of dilute phosphoric acid and primary zinc phosphate, operating for about two hours. This was the original zinc phosphating process from which the modern baths have been developed.

Richards in 1911 and Allen in 1916 patented the manganese phosphate bath, the latter patent being taken over by the Parker Rust Proof Co. of America, who called it Parkerising. Other types or modifications have been developed since then, so that there now exists a fair range of phosphate treatments for various purposes, the main types being the thin and thick iron phosphates, thin, medium and heavy zinc phosphates, and heavy manganese phosphate.

41.5.2 Coating Formation

The mechanism of the formation of a phosphate film on a metal is basically the same for all types. The bath consists of a primary phosphate dissolved in a dilute solution of phosphoric acid. When a metal component is immersed in the solution, the acid attacks the metal surface, causing a local lowering of the acid concentration at the liquid/metal interface. The soluble primary phosphate decomposes to the sparingly soluble secondary and insoluble tertiary phosphates, which precipitate onto the surface, being molecularly bonded in the process. The decomposition releases hydrogen ions which tend to restore the acid level in that region. This process continues until the surface is completely coated with the phosphate film, whereupon the deposition slows to a stop.

41.5.3 Modern Processes

The earlier processes were quite slow in their action and required to be operated at near boiling temperatures, with the result that faster processes were developed containing accelerators which reduce the operating times to as little as a couple of minutes, as in the case of iron phosphate.

---

* Some aspects of phosphate conversion coating are discussed in chapter 49 (section 3).

The zinc phosphate process usually takes 5 to 10 minutes and the manganese phosphate up to 30 minutes. Operating temperatures have also been reduced so that some may even be used at room temperature, but the usual bath formulations require temperatures at about 50 to 70° C.

The action of accelerators is still not completely understood, but as most of them are oxidising agents such as nitrates, nitrites, perchlorates or nitro-organics, they are thought to have a two-fold effect in that (a) they act as cathodic depolarisers to eliminate gaseous hydrogen formed in the initial reaction and (b) they increase the number of local cathodic areas, since it is at these points that it is thought that precipitation occurs (although there is another school of thought that says precipitation occurs at anodic areas).

In the case of zinc phosphate coatings, a major factor in nucleation of the crystals is the settling of tiny suspended particles from the solution onto the metal surface.

41.5.4 Factors Affecting Coating Formation

The state of the metal surface has a pronounced effect on the kind of coating obtained, particularly in the case of non-accelerated baths. Heavily worked surfaces lead to patchy coatings, while grit blasting tends to give more uniform crystals. Contaminants such as oil, dirt or mill scale lead to thin, non-continuous coatings and therefore must be completely removed before treatment. However, the method of removal is quite important. Pickling in strong mineral acids or alkalis can result in coarse coatings having poor corrosion resistance. Degreasing in trichlorethylene and pickling in mild alkali or phosphoric acid is recommended. However, as phosphoric acid pickling will render the affected area relatively passivated to further treatments, the use of mild organic acid pickling such as citric or oxalic acids should be considered in these cases. If strong acid pickling is necessary, the components should be rinsed thoroughly and preferably pre-dipped in a solution of mild alkaline titanium salt, which has the effect of precipitating myriads of nuclei (probably a complex titanium phosphate) on to the surface which will result in fine even coatings. Nickel and zirconium salts are also used for this purpose, and nickel salts are also used to promote nucleation on some high alloy steels, which are otherwise difficult to phosphate.

A very common factor causing poor coating formation is the chemical imbalance which can be brought about by overworking the bath. Since the solution is fairly delicately poised to precipitate phosphates when the acid balance is affected, it follows that the bath will give poor results if the imbalance temporarily produced becomes permanent. Too much acid results in little if any precipitation and, as a result, thin, non-continuous coatings. Too little acid results in general precipitation throughout the bath, resulting in loose, powdery coatings and excessive bath sludging.

The bath should be controlled by regular analysis to ensure that total acid and free acid and their ratio are kept within the recommended limits. A final brief acid rinse, usually of dilute chromic acid, will further passivate and seal any potential corrosion centres to maximise the coating's anti-corrosion performance in the field, particularly for iron phosphate coatings.

41.5.5 Applications of Phosphating

Phosphate coatings are invariably used as a base for a supplementary coating such as paint, oil, grease or soap. The supplementary coating is necessary because phosphates are fairly porous and have little corrosion resistance on their own. However, their use is often recommended because they have the capacity to increase the service life of the applied coating; in the case of paint, the increase may be up to five times normal life. This is accomplished in a number of ways, but two of the most important factors are the increase in the adhesion of the paint film to the substrate and the prevention of spread of corrosion from damage sites. A phosphate coating can also act as an inert barrier between a metal and a paint with which it may react, such as zinc and oil-based paints. The phosphating process produces no change in the mechanical properties of the base metal, nor does it produce hydrogen embrittlement in high tensile steels, which often occurs in the case of electroplating. Thus, it may safely be used as a preparation of these steels before painting.

The light and medium coatings such as iron and zinc phosphate are the most appropriate for the application of paint. Being fairly thin and fine grained, they permit a smoother, more even finish to be obtained. Paint applied over the heavier, coarser phosphates may easily lose gloss through penetration of the film by the larger crystals. Although iron phosphate gives a much thinner and less corrosion resistant film than zinc phosphate, it is often preferred as a paint base, particularly in an automated system where the phosphate is to be sprayed. Iron phosphate solutions are less prone to sludging, which can lead to frequent blockage of the jets. However, the more recent low sludging zinc phosphate formulations have largely overcome this problem and are now tending to replace the iron phosphates in this application. The heavy iron, zinc and manganese phosphates are used where the supplementary coating is to be grease, oil, wax or soap, as they have a high absorptive capacity, retain the film longer and are more resistant to abrasion than the lighter films. For example, one useful application is in metal dies where the life of the die can be increased up to about three times. Piston rings, high speed cutting tools, etc., are often given a heavy phosphate coating impregnated with oil or grease to reduce both wear and corrosion.

Where abrasion resistance coupled with low friction surface is of prime importance, the impregnating material preferred is a non-phenolic oil that dries to a non-tacky surface. In recent years, molybdenum disulfide has also been used in this application with excellent results.

The following are trademarks:

Alodine	Eftwerk	Lautawerk	Pylumin
Alrok	Alocrom	Tridur	Iridite
Alchrome	Parkerizing		

42 POWDER COATINGS

42.1 INTRODUCTION

This chapter describes the technology and application of powder coatings as used by the paint industry.

At present there are three major types of powder, applied by broadly similar processes:

(a) thermoplastic,
(b) thermosetting, and
(c) vitreous enamel.

Thermoplastic powders are generally regarded as products of the plastics industry and *vitreous enamel powders* as part of the range offered by specialist producers. *Thermosetting powder coatings*, however, have largely been within the province of the paint industry because:

(i) Customers were converted from paint or were regarded as paint users. Customers' existing paint stoving equipment could often be used for curing powder.
(ii) There was a common source of many raw materials, especially pigments, suitable for both conventional and thermosetting powder coatings.
(iii) The technology of thermosetting powder formulation has many similarities to that of paint.

The first thermoset powder coatings were based on a mixture of solid epoxy resins, pigments and a hardener blended by tall milling. Application was by fluidised bed. Progress in the late 1950s was limited by lack of control of particle size, and the fact that the pigment was not incorporated into or wetted by the resin. It was not until the mid 1960s that hotmelt extrusion compounding processes and controlled grinding/classifying techniques, availability of electrostatic guns for application, and development of specialised epoxy resins and hardeners allowed this class of powder coatings to develop rapidly. The next fifteen years saw the introduction of acrylic, urethane and polyester powders and an extension of the opportunities for use into many new fields, especially exterior applications.

42.2 POWDER APPLICATION

42.2.1 Electrostatic Gun Application

The predominant method of application is by use of an electrostatic powder gun, whereby a low-pressure air stream conveys powder to a gun containing electrodes (typically at 50–90 kV) that impart a charge to the powder. The charged particles are directed towards an earthed object to be coated.

The object must have a conductive surface in order to allow a uniform powder film to be built up. Powder is deposited and held to the surface by *electrostatic forces* and subsequent heating melts the powder, which passes through a flow-out stage prior to thermoset cure.

The efficiency of the application system is affected by many variables as shown in table 42.1.

TABLE 42.1
Variables in powder coating application

Air flow rate to gun
Concentration of powder in the air stream
Ease of flow of the fluidised powder
Particle size and distribution of the powder
Condition of the air stream (should be dry and clean)
Design of gun charging system
The electrode voltage and current employed
Resistivity of the particles
Velocity and direction of air within the spraying area
Distance of gun from the object
Conductivity of the object
Efficiency of earthing the object
Shape of the object
Design and materials of construction of the spraying area

The powder can be *fluidised*—that is, suspended in air so that it behaves like a fluid—and then transported by an air stream. The movement of the powder to the gun is capable of fine control within wide limits of air pressure/powder concentration, but it is essential that clean, dry air and free-flowing powder is used. There are many designs of electrodes for charging the powder, and guns are also produced that impart an electrostatic charge by frictional forces alone (*tribo-electric charging*). In some cases, a low voltage is delivered to the gun (avoiding the need for heavy cables) and an internal transformer develops the high voltage required.

The deposition efficiency can be markedly affected by the particle size range and distribution within that range; it is also affected by the resistivity of the powder particles. Some resins have a greater ability than others to accept a charge, and consequently powders vary in their electrostatic properties. This is particularly noticeable when spraying objects with deep recesses. The walls of the recess set up a 'Faraday Cage' effect, which limits the penetration of powder into the recess. In severe cases there may be no penetration at all. However, by control of electrode voltage, good gun design, correct choice of powder and control of powder feed rate/air pressure, these problems can in most cases be overcome.

For many years, spray booths were conventionally designed, but deposition efficiency has been improved recently by a tunnel shape where a powder cloud is directed either with or against the path of travel of a conveyorised object. Non-conductive surfaces (e.g. acrylic sheets) of the tunnel, combined with methods for deflecting or re-ionising the powder, improve efficiency.

In most cases, powder that is not deposited on the object is recovered by an extraction system. Cyclones are frequently used, but tunnel systems are being produced with a moving floor where powder is attracted and returned for use. Recovery rates of 95 per cent are not unusual which makes powder coating economically attractive, as well as avoiding a waste disposal problem or pollution of the environment.

Electrostatic application of powder produces film builds after cure of 35–100 μm. The low end of the range generally requires specially produced fine particle size powders. Above 100 μm there can be a self-limiting effect, whereby the coated object acquires sufficient surface charge to repel approaching powder.

Powders can be applied to hot or cold surfaces. On hot surfaces, often in excess of 200° C, the powder fuses to the surface on contact. The advantages of this method of application are that high build can be achieved, typically up to 400 μm, and the powder applied to surfaces that might otherwise be unsuitable for electrostatic application, such as phenolic mouldings or glass. Pipe coating is an important example of the application of 'hot coating' to obtain high build.

42.2.2 Fluidised bed application

Powder can be applied by fluidised bed techniques, whereby powder is fluidised in a container which typically has a porous diaphragm in its base through which air is admitted. This method has long been used for thermoplastic powders. A heated object is immersed in the fluidised powder which melts on to the surface. More recently, the fluidised bed has been further developed to allow electrostatic charging of the particles; cold objects can be coated in a simple dip process.

42.2.3 Heat Cure Considerations

Unlike paints, powder coatings should be brought to their curing temperature as rapidly as possible. This allows the resin to flow out before the thermosetting cure process takes place. Cure times quoted by powder manufacturers are based on the length of time required at a specified object surface temperature; the time taken to bring the object surface up to temperature is additional and depends on the size of object and oven design. Conventional convection and infra-red ovens are used. There is a significant energy saving with powder compared to wet paint stoving enamels, because there are no solvent fumes to extract from the oven with consequent heat loss. Induction heating is frequently used for pipes before powder coating, and this method is also suitable for small metal objects of regular size.

42.2.4 Pretreatment

Clean, dry surfaces are a prerequiste for powder coating, and a chemical conversion process is a preferred method of surface preparation, especially where anti-corrosive performance is required. Powder coatings do not assimilate traces of oil or other contaminants to the same extext as solvent-based paints, and hence a higher standard of surface cleanliness is required.

42.3 POWDER CHARACTERISTICS AND USES

42.3.1 The Powder Particle

The nature of the grinding process is such that powder particles are produced in a wide range of sizes, varying generally from 10 to 100 μm. For most applications, the ideal size range is 20–40 μm with a small tail on each end. A particle size classifier is used in the manufacturing process to limit or reject larger particles, and it is also possible to classify the lower end of the range to eliminate 'fines'. Over- or undersized particles can be re-processed.

Maintaining particle size within a narrow range is important because there is a loss of efficiency in the electrostatic application of powders outside the optimum range. The size and mass of the particles and optimal charging efficiency in relation to particle surface area are important factors. Considerable research has been done on these aspects, notably by the late Professor Bright of the University of Southampton.

The softening point of resins suitable for powder coatings is generally in the region of 60–100° C. The molecular weight of the resins will largely determine the extent to which resin viscosity drops as the temperature increases and hence affects the flow characteristics of the fused coating in the early stages of heat curing. However, the packing density of the powder particles on the surface of the coated object is also a factor affecting good flow—by reducing trapped air in the powder film—and particle size range is relevant.

42.3.2 Types and Uses of Powder

It is difficult to generalise on resin properties and end uses, because wide variations can occur in the performance of, for example, epoxy resins by altering the type of hardener or by the use of hybrid systems containing a stoichiometric proportion of a carboxy functional polyester. Table 42.2 (below) is therefore intended as a guide and should not be taken as necessarily excluding some resins from certain applications. One resin supplier lists at least thirteen polyester resins for use in powder coatings, and the range of properties is necessarily wider than the simplified table can indicate.

Table 42.2 gives strong pointers to the type of system that might be chosen for a particular end use; this is summarised in table 42.3.

For reasons of lower cost, fast curing rate and some degree of compatibility, epoxy, epoxy/polyester and polyester systems have tended to be the preferred systems amongst manufacturers and users in many parts of the world. Polyurethanes enjoy less popularity currently due to higher cost, slower cure and the volatile content of the blocking agent in the commonly used hardener. Acrylics perform satisfactorily but their lack of compatibility with other systems can lead to severe contamination problems in manufacturers' and users' plants.

TABLE 42.2
Comparative performance of major powder classes

System	*1*	*2*	*3*	*4*	*5*	*6*	*7*
Resin curing agent	*Epoxy DICY*[(a)]	*Epoxy anhydride*[(b)]	*Epoxy imidazole*	*Epoxy polyester*	*Polyester TGIC*[(c)]	*Polyester blocked isocyanate*	*Acrylic poly carboxylic acid*
Property							
Gloss	G–VG	G	M–G	VG	G–VG	G–VG	G–VG
Cure rate	G	G–VG	VG	M–G	M–G	M	M
Hardness	G–VG	VG	VG	G	G	G	G
Chalk resistance	P	P	P	M–G	VG	G–VG	G
Chemical resistance	G	VG	VG	M–G	M–G	M–G	M–G
Adhesion	G	VG	VG	G	G	G	M–G
Colour	M	G	M	G–VG	VG	G–VG	VG
Solvent resistance	G	G	VG	M	M	M	P–M
Flexibility	VG	G	VG	G	G	G	M

Key: P = poor M = moderate G = good VG = very good
(a) Accelerated or substituted dicyandiamide
(b) Acid anhydride or adduct
(c) Triglycidyl isocyanurate

42.3.3 Powder Stability

Fast curing powders with curing temperatures as low as 130° C and capable of curing within 30 to 60 seconds at temperatures of 220 to 240° C may have a limited shelf life, possibly of six months at ambient temperatures. They also present manufacturing problems, as any extended dwell time in the extruder during manufacture may be sufficient to initiate cure. Normally, powders curing above 160° C are stable and most powders have a storage life of several years.

It is, however, important to protect from high humidity or moisture, heat and compression (for example, by poor stacking) as pressure blocking can occur, resulting in caking of the powder.

TABLE 42.3
Suggested powder classes for specified applications

Application	*Suitable classes (refer table 42.2)*
Building products/exterior use	5, 6 or 7
Metal furniture	1, 4
Chemical plant	1, 2, 3
Electrical equipment	1, 2, 3
Domestic appliances	2,7
Pipe coatings	2, 3

42.3.4 Safety Precautions

Powder/air mixtures are capable of providing a dust explosion hazard. Proportions of powder to air that will give a minimum explosive concentration vary according to type of powder and are typically between 10 and 50 g/m^3. It is common, however, to use the lower limit in safety-related calculations.

The energy required to ignite a power/air mixture is 50 to 100 times more than is necessary for the same concentration of solvent vapour in air so the process is intrinsically safer than many paint processes. In addition, the energy developed in an electrostatic gun is generally less than a tenth of that needed to ignite a powder/air mixture at an explosive concentration. However, unless the object being coated is properly earthed, it acts as a capacitor as charged powder is deposited on its surface and is capable of discharging at higher energy levels than will be developed by the gun. It is prudent to employ air flows that keep the powder concentration below the explosive limit and to design equipment (booths, ducting, cyclones) with explosion panels that are immediately released under the pressure of an explosion.

The toxicity of powders varies according to the pigments used and types of hardeners. Acid anhydrides and other hardeners can be hazardous. Powders should not be inhaled, and suitable masks should be worn wherever there is a likelihood of inhalation, especially where toxic pigments are used. Powders should be washed from the skin after contact.

Personal protection equipment should not, however, be considered a substitute for good dust extraction systems and good housekeeping to minimise airborne powder.

42.4 POWDER COATINGS MANUFACTURE

42.4.1 Raw Material Selection

The currently preferred method of powder coating manufacture is to melt-mix a uniform blend of predominantly solid raw materials in an extruder using heat to melt the resin. Mechanical mixing of the melt results in dispersion of the pigment and other additives in the molten resin. The extrudate is then cooled and ground to give the required particle size of the powder.

The equipment used has generally been adapted to powder coating manufacture and as such the process is discontinuous, requiring transfer of the material between the various stages of the manufacturing sequence. This limits the quantities that can be readily handled, and larger batches must be manufactured as a series of identical pre-mixes run through the plant in sequence. Weighing of raw materials must thus be accurately carried out to ensure identical pre-mixes, as tinting or other changes cannot be carried out on the finished powder.

The raw materials used for powder manufacture should be of a high standard of cleanliness. Resins should be free of gel particles and extenders and pigments free of large particles. Any of these contaminants will be evident in the final product as imperfections in the powder film and cannot be separated by sieving.

The resins used should be in flake or granular form and no larger than 1 cm in size. Resins in lump form will have to be reduced by grinding a hammer mill, for example.

42.4.2 Pre-mixing of the Raw Materials

The powder coating raw materials must be thoroughly dry-blended to give a uniform feed to the extruder. This pre-blending may consist of tumbling in a hopper or more extensive mixing using a ribbon or high-speed paddle mixer. Care must be exercised that the heat generated during mixing is not excessive, otherwise the resin may partially soften and cause agglomerates to form.

Acrylic flow control additives or other liquid components can be added at the pre-mix stage, although intensive mixing is required to uniformly distribute the liquid and still retain a free-flowing dry pre-mix. Alternatively these liquid additives can be master-batched with the resin before use and added as a solid.

If colour or other properties of the product must be changed or adjusted, the pre-mix is the last stage at which this can be done. Once the pre-mix is extruded, no change can be made and any subsequent additions to the powder will be as a non-homogeneous dry-blend.

42.4.3 Extrusion of the Pre-mix

In this stage the raw material pre-mix is heated to melt the resinous components, and mechanically agitated to effect dispersion of the pigments and other components to provide a homogenous melt. This is usually achieved in an extruder. Three types of extruder are used: planetary screw, twin screw and reciprocating single screw, with the latter being the most common type. Output of a typical extruder is in the range 200 to 400 kg/h. Heat applied to the extruder barrel melts the resins in the pre-mix, creating a stiff paste. The dispersion results from the mechanical mixing via the screws and internal motion of the melt.

As the total residence time of the material in the extruder barrel is of the order of 30 seconds, the mixing time is extremely short. Pigments that are difficult to disperse, especially organic pigments, do not develop their full colour strength.

The temperature of the extrudate needs to be limited to 100 to 120° C to avoid any premature reaction in the melt, resulting in poor flow-out of the powder. This applies especially to fast cure or low-temperature cure powders which may gel in the extruder unless conditions are carefully controlled. After extrusion the melt is rapidly cooled, again to prevent any pre-reaction occurring.

42.4.4 Cooling the Extrudate

The molten extrudate passes from the extruder to a pair of water-cooled rolls which chill and squeeze the extrudate to a thin band. The extrudate band is then further cooled by either a water- or air-cooled conveyor until it becomes brittle. It is then crushed by either toothed rollers or a hammer mill to 1 to 2 cm granules suitable for further grinding to a fine powder.

42.4.5 Fine Grinding and Classifying

The granulated extrudate is reduced to a fine powder by a pin mill or high-speed hammer mill, which grinds the powder by impact with the grinding elements. A stream of cooled air sweeps the ground powder from the milling chamber, and thus also serves to minimise heat build-up.

The mill is usually equipped with either an internal or external vane-type classifier, which acts to separate any powder particles above the desired size range. The oversize powder is recycled through the mill.

42.4.6 Powder Collection

As the powder is swept from the grinder by a stream of air, it must subsequently be separated from it. This is achieved either with a fabric filter or a cyclone. Cyclones provide a convenient method for collecting the powder, being relatively inexpensive and easy to clean. However, cyclone efficiency is such that recovery of the order of 95 to 98 per cent only can be expected. The residual powder, mostly of small particle size, must then be removed by a final fabric-type filter and is usually rejected as waste.

Fabric filters alone are nearly 100 per cent efficient, and the air stream requires no further filtration. .The recovered powder is continuously removed from the filter element by either mechanical shaking or an air blast. Whilst the fabric filters are more efficient collectors than cyclones, cleaning presents a major drawback. The filter elements must be removed and cleaned, usually by thorough washing, after every batch. Cleaning times for fabric filters are nearly double those for cyclones, and hence the small powder losses from a cyclone are compensated by faster plant colour changes.

42.4.7 Sieving

Most powder manufacturers sieve the powder after grinding to ensure no oversize particles are present.

Either vibratory or cylindrical sieves are used, with mesh sizes commonly in the range 100 to 200 μm. Oversize powder rejected by the sieve is reprocessed though the grinder.

42.4.8 Plant Cleaning

Thorough cleaning of all equipment in a powder manufacturing plant is essential when changing colour or product type. Any trace of a previously manufactured colour, especially if present in the grinding or collection equipment, will result in contamination of the product, visible as a speckled finish. This imposes a higher standard of cleanliness on the powder manufacturer than would be required for many liquid paints, where small amounts of a previously made product may be absorbed by the new production.

Because of the many stages of manufacture required for production of the powder and the complexity of the equipment, cleaning between batches can represent a large percentage of plant time. This makes small batches uneconomical, as cleaning time can, in these cases, be longer than the product processing time.

42.4.9 Alternative Methods of Powder Manufacture

Whilst extrusion melt-mixing is currently universally used for powder manufacture, other methods have been used or proposed.

Ball milling of the pigment, hardener and resin was used in the early stages of powder development, but this does not allow a high gloss powder to be produced. In addition, the powder consists of discrete particles of its components and may result in segregation during spraying.

Z-blade mixers or similar equipment have been proposed to replace the extruder, but the extended time at melt temperature results in a pre-reacted product and poor flow.

Precipitation of powder particles from a solvent-based pre-mix by addition of a non-solvent has been developed to a commercial scale by an American company, and spray drying of a solvent-based paint has been tried on a pilot plant scale.

Both of the last-mentioned methods allow greater flexibility in production. Current paint manufacturing equipment can be used to prepare the pre-mix. The 'powder' can be adjusted for colour and other properties in the solvent-based state before being precipitated and/or dried. This overcomes one of the major drawbacks of the current extruder method, where the powder cannot be modified after the pre-mix has been extruded. Both precipitation and spray drying of liquid pre-mixes point to powder manufacturing methods of the future.

42.5 FORMULATION

42.5.1 Choice of Resin System

For practical purposes the choice can be limited to polyester, hybrid epoxy–polyester and epoxy types. If good exterior durability is required, especially for retention of gloss and resistance to chalking, polyester types must be used.

The polyester type of greatest importance is the carboxyl functional polyester cross-linked during stoving with triglycidyl isocyanurate (TGIC). Generally the acid value of this type of polyester is fairly low (30 to 50 mg KOH/g) and an approximately stoichiometric proportion of TGIC (6 to 10 p.h.r.) is used. This binder system can be considered as a general-purpose type, and it is usually only the price premium of this type that dictates the use of other resin systems.

Hybrid epoxy–polyester powders are used where the good appearance qualities of polyesters, both flow and gloss, are required, but where UV resistance is not required. This binder system is lower in cost than the TGIC cured polyester, and competes more directly with unmodified epoxy systems. The hybrid type also incorporates a polyester component with carboxyl end groups but is higher in functionality and acid value, which is generally in the range of 50 to 100 mg KOH/g.

About 50 per cent of the binder system is a conventional Bisphenol A epoxy resin of epoxide equivalent weight in the range 700 to 900. The chemical curing reaction, as in the polyester type, is ester formation between the epoxide groups and the carboxyl groups. It is possible to produce polyesters designed for hybrid systems of a lower AV, and these are used at higher proportions with epoxies (70 : 30, for example) to give properties intermediate between the normal hybrid type and polyesters. In particular the UV resistance is usefully improved and the tendency for epoxy-rich systems to yellow is reduced.

Epoxy powder coatings pre-date the other two major types in development and still hold a large share of the market. Their earlier price advantage over hybrid types is now largely eroded. The more common type of epoxy powders use amine-type compounds, generally accelerated for cross-linking, or substituted dicyandiamide. Acid anhydrides are also suitable cross-linking agents, but are not used as widely because of their toxicity and moisture sensitivity. Epoxies have the advantages of faster cure rates and better chemical and solvent resistance than polyester-containing powder coatings. With anhydride curing agents and the use of organic tin catalysts, very fast curing rates are achievable. Such powders are most suitable for residual heat curing, where the object is preheated and coated hot with no subsequent stoving. Large-diameter tubes for pipelines are powder coated by this method.

Other resin systems for thermoset powder coatings include those based on acrylic and polyurethane resins. Acrylic resins generally possess epoxide functional groups, pendent to the polymer chain, and cross-linking is achieved with dicarboxylic acids. Acrylic systems have not been a major force in the powder coating market except in Japan and offer limited advantages over polyesters, yet have the disadvantages of being prone to caking on storage and of cratering other powders.

So-called 'polyurethane' powders are hydroxyl functional polyester resins which are cross-linked with isocyanate curing agents. The isocyanate group on the curing agent must be chemically 'blocked' to stop reaction at extrusion and ambient temperatures, and caprolactam is normally used as the blocking agent. The blocking agent is released during the stoving process of the coating, exposing the isocyanate groups which form urethane linkages with the hydroxyl groups in the polyester. The volatile by-product and the slower cure rate of 'polyurethane' powders have limited their acceptance. Their exterior durability is good but no better than the polyester type, but flow and general appearance are usually excellent.

42.5.2 Pigmentation

42.5.2.1 Prime Pigments

Factors of specific importance to powder coatings in the choice of pigmentation are:

(a) *Heat resistance of the pigment.* Powders are stoved at about 200° C for 10 minutes—higher than most wet stoving enamels. Most inorganic pigments have adequate heat resistance, but many organic pigments are unstable at this cure schedule. In all cases, pigments must be evaluated for heat stability before being adopted.

(b) *Chemical resistance of the pigment.* Components of the resin system may affect coloured pigments. The dicyandiamide curing agent in epoxies is a common offender and discolours many azo-pigments and lead chromes.

(c) *Ease of dispersion.* Extrusion compounding is not as efficient a dispersion process as ball milling or sand grinding of wet paint. Easy dispersing grades of pigments should be used wherever possible, especially in the difficult-to-disperse organic pigments such as phthalocyanines.

(d) *Hiding powder.* Because of the low pigment levels used in powder coatings, the hiding power and tint strength of pigments are particularly important. Especially with certain organic pigments, levels above 2 to 3 per cent in a formula result in poor levelling (flow).

42.5.2.2 Extender Pigments

Extender pigments are used to reduce the cost of formulations and/or to reduce gloss level. With full-gloss products (above 90 per cent at 60°), very fine extender pigments are needed and precipitated grades of barium sulfate (blanc fixe) or calcium carbonate are used. The particle size of these grades should not be above 1–2 μm if good gloss is to be maintained. Generally, having satisfied the requirements of colour and hiding powder with prime pigments, the resin is progressively replaced by extender by the formulator until the point where physical properties and levelling deteriorate.

Coarser extenders, up to 30 μm, are used at quite high levels to achieve a satin gloss (about 50 per cent at 60°). Ground calcite is generally used, but barytes can also be employed. Extenders such as unmodified china clay are unsuitable because of high vehicle demand and chemically bound water.

With white finishes or pastel shades, little extender can be used without too great a sacrifice in flow and mechanical properties, due to the high loading of prime pigment.

Extender pigments can be chosen which limit the flow-out of powder particles on stoving, resulting in textured finishes. These can vary from a rounded 'ripple' finish to a sharp sandy texture, according to the choice and level of extender used. For these effects talc is normally employed since its plate-like structure and high vehicle demand inhibit flow. Powdered thermoplastics such as polypropylene are sometimes used to introduce textured effects.

Gloss levels below the normal 'satin' level (50 to 60 per cent at 60°) cannot be achieved by the use of coarse extenders alone; generally changes to the resin/hardener system and the use of additives are required to formulate matt finishes.

It is worth noting that an increase in particle size of the extender beyond 20 to 30 μm does not result in a significant reduction of gloss and usually results in a 'sand-papery' effect, especially at low film build where the extender size is greater than the film thickness.

42.5.2.3 Metallic Pigments

Bronze and aluminium pigments are used in powder coatings for metallic effects. They are used in two ways: extrusion compounded with the other pigments or dry blended with the final powder.

Metallic pigments when extrusion compounded show no tendency to float and leaf as in wet paints and do not result in any degree of colour 'flip'. Additionally the metallic flakes tend to become crumpled in extrusion, thus resulting in loss of metallic glitter. An additional problem is the tarnishing of the surface of the metallic flakes under the influences of temperature and the chemical environment of the resin system. For this reason it is normal practice to use coated, tarnish-resistant grades of metallic pigments.

In order to obtain desired metallic effects it is often necessary to dry-blend small amounts of metallic pigments with the final powder. In this case the metallic pigment behaves very differently on stoving the film; it tends to float to the surface of the coating and is much more efficiently used. The disadvantage with this approach is that the coating is non-uniform and the metallic pigment is vulnerable. Thus mechanical damage on the surface of the article can be very noticeable, and tarnishing of the metallic pigment under service conditions can readily occur.

42.5.3 Additives

Certain additives are essential for satisfactory powder coatings, but many are optional or specific certain end-uses.

42.5.3.1 Wetting Additives ('Flow Aids')

Substrate wetting by the resin systems used in powder coatings is often inadequate by itself, resulting in cratering and de-wetting defects with most practical substrates. This problem is worse if the substrate is slightly oily. Wetting additives, sometimes known as 'flow aids', must be used to achieve good substrate wetting. By far the most common types are the low glass-transition temperature acrylics such as poly (butyl acrylate), poly (2-ethylhexyl acrylate) and various copolymers. These are used at between 0.5 and 1 per cent in the formulation. They are slightly incompatible with powder coating resins and are exuded to the interface where they aid the substrate wetting process. Certain silicone resins can also be used for substrate wetting but are not as efficient as the acrylics.

42.5.3.2 Air Release Agents

The release of interstitial air as the powder fuses is important, and benzoin is often used as an aid in air release during stoving and helps to prevent surface pinholes which result from entrapped air. The mechanism of the action of benzoin is not certain, but it is possible that it acts as a flux providing a very low viscosity coating on powder particles as they fuse. It is normally used at about 0.5 per cent, principally in hybrid and polyester powders.

42.5.3.3 Catalysts

The curing reaction of each of the resin systems can be accelerated by the use of catalysts. Acceleration, however, results in the penalty of poorer flow-out and hence more pronounced 'orange peel' on the finish. Some proprietary curing agents have an appropriate level of catalyst incorporated. Catalyst levels are usually very low, often in the 0.1–0.5 per cent range.

42.5.3.4 Slip Aids

Low molecular weight polypropylene or polyethylene waxes can be used to enhance the surface slip properties of powder coatings. Marring which occurs from the rubbing of painted surfaces together can be much improved by the use of slip aids. Fine PTFE powders and certain silicones are also effective slip aids. Slip aids are effective at 0.5–1 per cent.

42.5.3.5 Gloss Control Agents

Reduction of gloss below the level normally achievable by use of coarse extenders can be made with hydrogenated castor oil or similar waxes at a level of 2 to 4 per cent. The gloss can be reduced in epoxy powder coatings by the use of proprietary curing agents with more than one type of reactive group; *Huls* B55 and B68 curing agents are of this type. Such curing agents result in a very fine 'microwrinkle' on the film surface and reduce the gloss in the process. Introducing two types of curing mechanism into other resin systems has also been successful in producing matt powder coatings. Similarly, physical mixtures of two powders which differ in cure rate can be used to reduce gloss.

42.5.4 Special Effects

Because powders possessing differing levels of wetting aid give rise to surface tension variation in the melt, the addition at a powder rich in wetting aid will cause craters in a mix with normal wetting aid levels. A multiplicity of craters can be formed by such powder mixture to produce a hammer finish pattern.

By the use of pigments which tend to float under the action of crater formation, and metallic Typical starting-point powder coating formulae are shown in formulas 42.1, 42.2 and 42.3.

An interesting effect is produced if metallic pigments are blended externally with a 'hammer-finish' powder. In this case the metallic pigment is thrown to the ridges of the hammer pattern producing a decorative metallic veining on the surface. This is the basis of the so-called 'spangle' finishes.

It should be noted that hammer finishes of the various types must be applied much more heavily than normal powders in order to allow the undulating texture to form and to prevent craters being formed which expose the substrate.

42.5.5 Typical Formulations

Typical starting-point powder coating formulae are shown in formulas 42.1, 42.2 and 42.3.

FORMULA 42.1
Formula for a white gloss epoxy

	Parts by mass
Epoxy resin (EEW = 900)	60.0
Accelerated dicyandiamide	3.5
Acrylic flow aid	1.0
Titanium dioxide	35.5
	100.0

FORMULA 42.2
Formula for a white hybrid epoxy–polyester

	Parts by mass
Polyester resin (AV = 90)	30.0
Epoxy resin (EEW = 900)	30.0
Acrylic flow aid	0.5
Benzoin	0.5
Titanium dioxide	39.0
	100.0

FORMULA 42.3
Formula for a white polyester

	Parts by mass
Polyester resin (AV = 35)	57.0
Triglycidyl isocyanurate	4.3
Acrylic flow aid	0.5
Benzoin	0.5
Titanium dioxide	37.7
	100.0

42.6 TEST METHODS FOR POWDER COATINGS

Test methods for thermoset epoxy powders for pipelines have become well established because of the critical performance requirements. The tests are primarily devoted to the applied coating rather than the powder itself; British Gas and DIN 30671 Specifications are typical. Adhesion, flexibility, impact resistance and cathodic bond failure are important variables measured in these tests. In other end use areas, performance tests appropriate to paint finishes—such as salt spray resistance, hardness, gloss, for example—are applied to powder coatings.

Problems arise, however, in defining pre-delivery standards for powder. In a number of areas there is no general agreement on the test methods that should be employed, although draft standards exist in France (NF T 30–500 to 504), Germany (DIN 55990 parts 1–8) and the U.S.A. (ANSI/ASTM D 3451–76). The following review is therefore intended to emphasise physical properties of powders which have a direct bearing on their application and which are different from those tests normally performed on liquid coatings.

42.6.1 Sampling

A batch of paint can usually be regarded as homogeneous and a wet sample to be representative of the whole. Powder is manufactured in comparatively small unit quantities, typically 200 to 500 kg, although the total batch may be many tonnes. This can allow production variation to occur throughout the batch, and hence a single sample may not represent the whole. Sampling methods need to be agreed between supplier and user.

42.6.2 Particle Size Analysis

In practical terms, vacuum sieving is the only simple and effective technique available, but this is limited to 32 μm particles and above. Whilst the percentage below this size can be measured, it is only possibly by this method to estimate distribution. Microscopic and sedimentation measurements are tedious, whilst the Coulter Counter, although effective across the range, is probably too expensive for general use.

Most analytical methods assume a spherical particle shape, whereas in fact shapes are irregular as a result of the impact grinding process and they behave differently from spherical particles.

42.6.3 Melt Flow

As previously noted, this is a function of the resin viscosity at the curing temperature, the time taken to reach this temperature, air entrapment and the components of the powder formulation. An indication of flow can be assessed from the length of flow of a pellet of powder heated to the cure temperature on an inclined panel.

42.6.4 Reactivity

Gelation time can be determined by placing a small quantity of powder (0.5 g) on a pre-heated hot plate at a specified temperature and noting the time taken for the molten mass to change from a flowing melt to a rubbery gel when stirred with a probe.

Minimum cure time at a given temperature can be established by stoving thin aluminium panels coated with the powder under test and noting the time taken for a film to be produced which does not crack when the substrate is sharply folded. Solvent cure tests are also widely used and consist of assessing the degree of surface softening by MEK or MIBK after contact with the solvent for 30 seconds and immediately wiping off. Although subjective and operator-dependent, solvent cure tests can be very helpful. They should be combined with cross-hatch adhesion and/or impact adhesion tests for cross-reference.

It is most important to note that comparatively small differences in cure rate of powders of the same generic type can give rise to a reduction in gloss when these powders are intermixed. Checks should therefore be made by intermixing batches, whenever formula alterations are made, to ensure that any variation of cure rate has not affected gloss.

42.6.5 Volatile Content

Most powders used have a volatile content of less than 1 per cent, most of which will be absorbed moisture. However, where a blocked hardener is used, as with polyurethane powders, then volatile content may be of the order of 4 per cent. This can conveniently be checked by measuring the loss in mass of powder on a tared dish after curing at the recommended schedule.

42.6.6 Density

If there is a low (1 per cent or less) volatile content, then a theoretical density can be calculated for the powder formulation with reasonably accuracy. This can be measured by a liquid displacement method using kerosene or water/detergent, but results are liable to error because of difficulty in wetting powders. A gas pycnometer provides very much more accurate results.

Density is important in determining powder utilisation. The theoretical consumption, assuming total reclaim of powder, can be calculated:

$$\text{Coverage } (\text{m}^2/\text{kg}) = \frac{1000}{\text{Density} \times \text{coating thickness } (\mu\text{m})}.$$

Cost comparisons between powders should take account of density, as this will affect powder utilisation.

42.6.7 Powder Flow

The dry powder must readily fluidise for ease of application. Test methods involving flow through a funnel or the angle of respose of a cone of powder after pouring will not properly demonstrate this. A test for powder flow in the fluidised state is described by Shell Chemicals in their literature.

42.6.8 Storage Stability

Two types of deterioration can occur on storage. Physical effects such as moisture (humidity), heat and pressure can cause blocking or caking and affect powder flow and fluidisation. Heat and pressure can be simulated by placing a test quantity of powder in a cylindrical vessel with a weighted disc in contact with the powder and storing in an oven at 40 to 50° C for 7 days. Chemical changes may occur, especially where fast curing hardeners and accelerators are used. This may also give rise to flow problems as above and may affect the cure of the powder.

42.7 DISADVANTAGES IN POWDER UTILISATION

Some of the negative factors of powder coating need to be considered and the extent to which these could affect growth of the market:

42.7.1 Powder Mixtures

Powders cannot be intermixed to obtain blends or for tinting purposes. Such mixtures give a 'pepper and salt' appearance. As a consequence, small batches are impracticable.

42.7.2 High Capital Cost of Manufacturing Equipment

The cost of setting up a production unit for 300 tonnes p.a. output for a general range of products was of the order of $A500 000 in 1980. Economic sense suggests that a manufacturer entering the market or increasing capacity should be assured of sufficient sales opportunity to utilise new capacity within a reasonable time. Experience has shown in some countries that such considerations have been ignored, resulting in overcapacity and low profitability.

42.7.3 Colour Change by Applicators

Until recently, colour change has necessitated duplication of application equipment, especially spray booths, ducts and cyclones, to reduce the down-time for cleaning the system. Small colour runs are often performed without reclaim of overspray powder to avoid this problem.

42.7.4 Powder Cost

Coating film thicknesses are normally not less than 50 μm, and the resins used in powders would be classed amongst the more expensive types used by the paint industry.

42.7.5 Capital Cost of Application Equipment

Although the cost of installing a line for applying, reclaiming and stoving powder compares favourably with the cost of a wet paint line, a paint user may nevertheless decide to defer a decision to use powder until he needs to expand or replace existing equipment.

TABLE 42.4
1979 market shares (%) for electrostatic powder coatings*

Category	*Percent*
General metal finishing	43
Pipe	22
Major appliance	13
Electrical equipment	12
Transportation	10
	100

* Moulding and thermoplastic textile powders not included.

42.8 FUTURE PROSPECTS

Although the production of coloured powders has generally made it necessary to manufacture minimum batches of about 400 kg to offset the cleaning downtime in production plants, this has not been a deterrent to the growth of the market for powder coatings. It is significant that a large number of jobbing enamellers have switched much of their work to these systems.

The supply capacity for powder coatings has matched demand despite the capital cost of manufacturing equipment, and this situation is likely to continue. It seems probable, however, that capital costs and the specialised nature of powder coatings will restrict proliferation of manufacturers, although this has not been the situation in the U.K.

Colour change has been made easier as equipment suppliers address themselves to the problem, and further developments in this area can be expected. Already, systems are available that eliminate the need for lengthy ductwork and cyclones. Colour change 'at the gun' has also been vastly simplified with changeover powder hoppers and feed lines.

Where a single powder application replaces two wet coats of paint, the materal cost alone is invariably in favour of powder. For single coats of wet paint, the economic justification for powder can in many cases still be proven through other factors:

(a) reduced energy costs (especially oven and factory heat losses);
(b) ease of adaptability to automatic application;
(c) high levels of reclaim of overspray; and
(d) improved performance and reduced rejects.

Powders are being produced with fine, controlled particle size to allow lower film builds (35 to 40 μm) to be applied, and wider availability of these powders can be expected.

Applicators with existing wet paint equipment have in some instances found that it is more expensive to modify their operations to comply with requirements of environmental protection authorities than to convert to powder systems.

Many of the early predictions about the rate of growth of the powder coating market failed to take account of improvements that have occurred in conventional paint technology, such as the high solids, waterborne and electrophoretic application techniques.

TABLE 42.5
1979 product mixed for electrostatic powder coatings*
tonnes

Class of powder	*Usage*
Epoxy	13 830
Polyester and hybrids	5 440
Acrylic	680
Nylon 11 and vinyl	450
Total electrostatic powder coatings	20 400

* Moulding and thermoplastic textile powders not included.

Tables 42.4–42.6 are excerpts from the results of a 13-month study covering 1000 major metal finishing installations in the U.S.A.

Table 42.4 shows industry segment market shares for electrostatic powders (it does not include powders applied by fluidised bed). In that group using electrostatically applied powder coatings, most (43 per cent) are categorised as general metal finishers; a significant amount of powder is used in pipe coating and rebar (concrete reinforcing bar) applications.

The breakdown of powder consumption (electrostatic) by generic types appears in Table 42.5. Not surprisingly, epoxy systems predominate by a wide margin.

Table 42.6 shows the market share for electrostatic powder coatings of 9 per cent in 1979, with predicted growth to an 11.5 per cent share by 1984.

TABLE 42.6
1979 and 1984 available metal finishing* market shares (%) for electrostatic powder coatings

Application	*1979*	*1984*
Electrical equipment	23	28
Major appliances	16	17
General metal finishing	11	14
Transportation	3	6
Total metal coatings*	9	11.5

*Coil, metal container, marine, maintenance and electrical insulation not included.

43 ULTRAVIOLET-CURED COATINGS

43.1 THE CONCEPT

The cross-linking free-radical curing of organic coatings has long been known, particularly the electron beam process. This process, however, is limited in application because of the high capital investment in plant required. In contrast, ultraviolet curing has a relatively low capital cost and is more widely applicable.

The curing process is based on photo-initiated free radical mechanisms; the background technology was obtained via the electron beam process developments. The process requires the presence of a photo-initiator (with or without a synergist) to absorb ultraviolet radiation from the source into the polymer film in order to initiate the reaction. Free radicals are then rapidly formed to promote curing. Not all compounds capable of producing initiation can be used, nor will any particular unsaturated monomer necessarily copolymerise. A typical initiator is methyl benzoin ether, but hydroperoxides have also been used. Selection of ingredients is still largely empirical and the mechanisms are not fully understood.

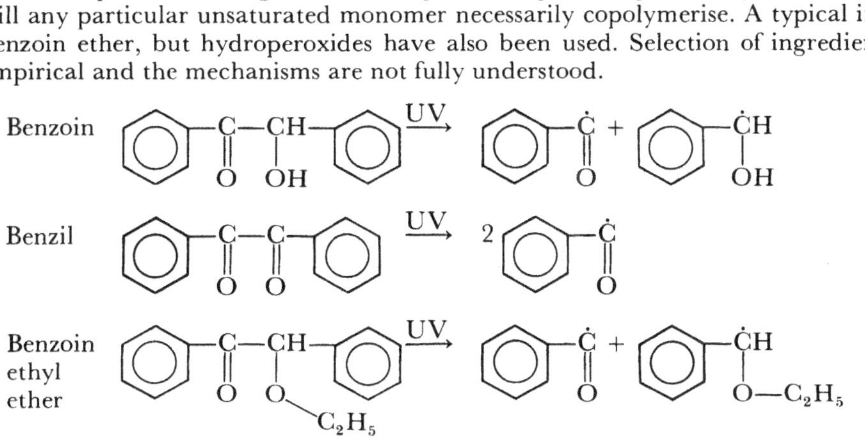

The advantages of ultraviolet curing are:

(a) Rapid cure, as low as one-fifth second in thin films.
(b) Low temperature cure in the film, hence little distortion.
(c) Long pot-life in the uncured polymer package, which may be further lengthened by the addition of inhibitors.
(d) High solids systems without solvents.
(e) High line speeds and compact 'ovens'.
(f) Rapid on/off times, as radiation is quickly available—as little as five minutes warm-up time.

(g) Reduction of air pollution.

Curing rate depends on:
(a) The chemical compound to be cured.
(b) The amount of ultraviolet energy per unit area. The relationship is not linear: doubling the ultraviolet radiation flux reduces cure times to less than half.
(c) The ultraviolet spectrum emitted by the source. Individual UV sensitisers require different emitted wave lengths for correct curing. Medium pressure mercury vapour lamps emitting over a wide range in the UV spectrum (180–400 nm) are suitable for most ultraviolet curing applications.
(d) Thickness and transparency of the film—absorption of the ultraviolet energy decreases logarithmically with film thickness, with 75 μm dry as an operative maximum. A twofold increase in thickness of film increases cure time more than three times.
(e) The initiator or sensitiser used; for example, with benzoin butyl ether and 2-methyl anthraquinone and mixtures, the cure times are as shown in table 43.1.
In the presence of higher viscosity monomers and/or pigment, the cure times are as shown in table 43.2.

TABLE 43.1
Effect of initiator choice

Initiator	*Benzoin butyl ether*	*1 : 1 blend BBE/MAQ*	*2-MAQ*	*Mixed Benzoin ethers*	*Benzophenone*
Cure time (seconds)	0.55	0.09	0.16	0.48	0.59

TABLE 43.2
Effect of pigmentation

Initiator	*Benzoin butyl ether*		*1 : 1 blend Benzoin alkyl ether : 2-MAQ*	
Monomer	*TMPTA*	*PETA*	*TMPTA*	*PETA*
Cure time Clear (seconds)	0.25	0.17	0.19	0.10
Cure time 1 : 1 TiO_2 (seconds)	1.48	1.30	1.18	0.90

(f) The efficiency of the initiator may be increased by the addition of synergists such as amines, as shown in table 43.3.

Current data available on this reaction allow the selection of a very wide range of initiator/synergist combinations, the only limit being that some combinations may give rise to colour problems.

TABLE 43.3
Effect of synergists (Cure times in seconds)

Amine (2% NV)	*Benzil (3%)*	*2-Chloro-Thioxanthone*	*2-MAQ*
Nil	0.22	0.13	0.09
Ethanolamine	0.09	0.06	0.22
Diethanolamine	0.09	0.06	0.13
Triethanolamine	0.09	0.06	0.13
Triethylamine	0.09	0.09	0.09
Dimethyl aminoethanol	0.06	0.06	0.09
Morpholine	0.03	0.03	0.09

The features of ultraviolet curing relative to conventional methods may be summarised as:
(a) Reduced air pollution—system as high as 100% NV.
(b) Saving in plant space because of smaller equipment.
(c) Labour saving—fully automated lines are available.
(d) Rapid cure—seconds rather than minutes, typically 9 to 15 seconds on pineboard, 2 to 5 seconds on tinplate—coupled with immediate handling after the radiation source.
(e) Superior film build, hence higher quality of finish.
(f) No increase in total finishing costs.

43.2 THE EQUIPMENT

A complete ultraviolet-curing unit consists of:
the ultraviolet source;
an irradiator—housing and reflector assembly;
electrical controls;
shielding, cooling and safety equipment.

The source is generally a medium-pressure mercury vapour lamp consisting of a quartz tube, up to 1.8 m long, filled with measured amounts of argon, xenon and mercury. The tube ends are electrodes between which an arc is struck to bring the mercury to the operating temperature, with a typical warm-up period of 2 to 5 minutes. Output is approximately 80 watts/linear centimetre (200 watts/inch), and such a lamp may be expected to operate for 1000 hours at above 85 per cent efficiency, slowly deteriorating with time, on a 240 V input. The lamp also has considerable infra-red output. The latest lamps available take the output to 120 W/cm and installation of aluminium reflectors over the lamps increases the efficiency a further 30 per cent. The newest reflectors are coated with quartz for an even better performance. Hildebrand-Stuttgart have an Impulse Radiation technique yielding higher ultraviolet and lower infra-red radiation, bringing the wave length down to 190 nm. This allows the work to be further away, or permits penetration of thicker or more opaque films at the same distance; the further lowering of the substrate temperature ensures better dimensional stability.

There are generally 3 or 4 lamps per metre of tunnel length. Major peaks in wavelength amount to only 9 per cent of output, and the total output of usable wavelengths is approximately 16 per cent of the electrical energy input. For this reason the photo-initiator must absorb the radiation shower as efficiently as possible.

Longer lamps than substrate dimensions should be installed, because the lamps darken near the electrodes with use, thus decreasing efficiency.

The irradiator is a lamp housing reflector assembly which supports the lamp over the curing surface. The reflector is generally elliptical, with the lamp socket placed at one focus and the work to be cured running through the other focal point. This gives fast curing, good cooling of the lamp seals and sockets, and less heating of the cured surface. The lamp wall reaches 700° C and the air temperature within the oven may rise to 370° C; with the UV radiation this produces ozone in the oven space. Ozone is pollutive and inhibits the polymerisation reaction, hence the oven must be provided with good exhaust draught facilities. Shielding and safety cut-out equipment are necessary to protect the crew from ultraviolet burns.

Single-unit irradiators are used for inks and metal decorative coatings, whereas multi-unit irradiators find use with thicker, slow-moving films such as wood finishes and chipboard sealers. The clearance of the lamp from the substrate varies from 7 to 10 cm; the board temperature reaches only 43° C.

43.3 THE MATERIALS

The materials currently utilised in ultraviolet curing processes are:

(a) Oil-free polyesters, typically based on trimethylol ethane, neopentyl glycol and isophthalic/adipic acids.

(b) Unsaturated acrylic—methacrylic oligomers dissolved in styrene, acrylate or methacrylate esters capable of co-polymerising with the base resin during the reaction. A wide range of products are available, with varying boiling points and viscosities. Vinyl acetate monomer should not be ignored, as it is a convenient viscosity-reducing comonomer. As these solvent monomers copolymerise with the base resin, the total formulation may be regarded as being 100 per cent solids and thus pollution free. Copolymerising high molecular weight monomers may also be used to reduce pollution further—trimethylol propane triacrylate, for example. Many epoxy reactive oligomers are also available.

Most development has occurred with oil-free polyesters as the vehicle, because they do not suffer from air inhibition of the reaction as epoxy-acrylics do; the latter require the presence of a nitrogen blanket in order to cure. In contrast to electron beam curing, waxes may be used in ultraviolet coatings to assist cure, particularly if only low-pressure lamps such as blue daylight fluorescent tubes are available.

Lower molecular weight monomers give harder films, while higher monomers tend to give toughness with better abrasion resistance and superior sanding properties. The higher molecular weight monomers find greater use in inks and can coatings. There is currently a trend towards methacrylates as they are considered less toxic than acrylates.

For the *initiator*, benzoins are favoured for polyesters whereas acetophenones are preferred for acrylics. Some initiators have chlorine in the compound to increase the energy level, but hydrogen chloride is released when disproportionation takes place. Not only is this dangerous, but also the gas appears to promote non-uniform cross-link density.

It was noted earlier that the initiator/synergist combination may cause colour; for example, benzophenone and Michler's ketone produce a green hue:

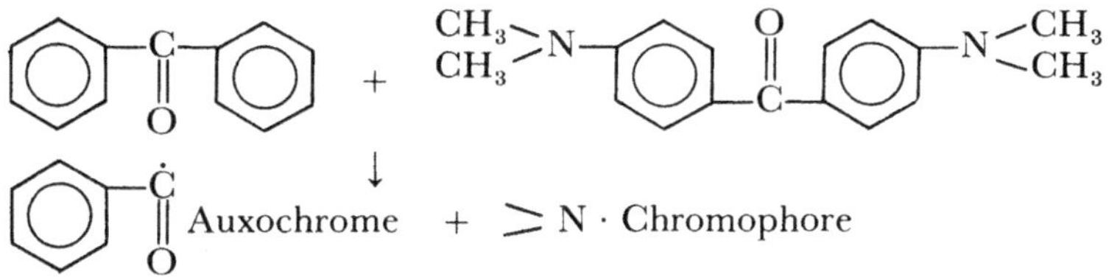

Later derivatives are capable of secondary reactions to produce more free radicals, thus:

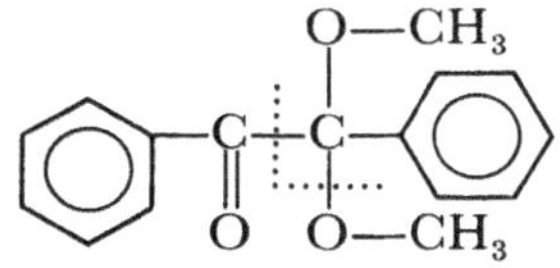

the breakdown being non-toxic, occurring in the range 340–354 nm.

Photo-initiators improve performance up to an optimum level and above this activity falls away; more initiator will not help to cure a thicker or opaque film, but time of exposure and radiation flux density will achieve this. For instance, the presence of titanium dioxide in a 40 μm film requires 6 to 10 times the exposure time, requiring either slower or longer lines and ovens than with, say, semi-transparent coatings. The effectiveness of a photo-initiator may be measured in terms of the range of polymers it will cure and their residual unsaturation.

43.4 APPLICATION AND FORMULATION

At present, the major outlet for ultraviolet coatings is in the furniture field as sealers for chipboard, masonite and plywood. These may be finished in single colours or subjected to a multicoat printing process leading to imitation grained timbers, which may be used in the furniture field, including television cabinets.

A typical system:

1. Fine sand panel and clean.
2. Ultraviolet filler—reverse roller coater.
3. Cure approximately 10 seconds, ultraviolet oven 7.5 m long with 12 lamps.
4. Sand panel (auto sander, vacuum offtake).
5. Apply base coat by direct roll coater.
6. Dry in infra-red oven, 7.5 m long, peak board temperature 55° C.
7. Emboss or gravure print the grain pattern.
8. Ultraviolet topcoat, cure in oven 9 m long with 15 lamps.
9. Offload conveyor.

Obviously, materials that contribute ultraviolet absorbing properties or opacity markedly reduce the efficiency of the process, as loss of ultraviolet transparency within the film rapidly increases cure times. In early development the process was restricted to clear or semi-transparent sealers on surfaces such as chipboard. Today, however, greater understanding of the role of the synergist has extended application to opaque coatings.

Ink films are only 10 per cent of the thickness of paint films, and inks generally aim for transparency and not opacity. The particular colour printed will, however, affect the cure rate—obviously, maximum ultraviolet sorbency would be expected in yellows. The process has allowed a major speed increase and immediate reeling of paper web. Hybrid systems exist, with normal vehicles to which peroxides are added—such inks will set in 1/2 second under ultraviolet cure. The principal disadvantage is that the high degree of cross-linking achieved makes rework of such paper almost impossible.

When formulating coloured products the effect of pigment volume concentration must be taken into account. Low PVC films will allow passage of ultraviolet radiation; high PVC films will give rise to more diffuse reflectance from the top of the film and more scatter within the film, such that a 10 μm film pigmented at 10 per cent PVC will be more effectively cured than a

5 μm film pigmented at 20 per cent PVC—little functional difference would exist in opacity of these two films.

Today many pigments are both surface coated and aftertreated. These processes have a bearing on the cure rates obtained, as shown in table 43.5.

TABLE 43.4
Effect of resin type (Cure times in seconds)

Polymer	*Epoxy-acrylic*		*Urethane-acrylic*		*Polyester*
Monomer	*PETA*	*TMPTA*	*PETA*	*TMPTA*	*PETA*
Initiator					
Benzoin butyl ether	0.48	0.15	0.65	0.90	0.50
BBE + 2-MAQ	0.10	0.06	0.49	0.69	0.16

TABLE 43.5
Effect of pigment grade (Cure time in urethane/acrylic resin)

Grade	*Anatase A*	*Anatase B*	*Untreated rutile*	*Treated rutile*
Time (seconds)	0.28	0.47	0.36	0.47

Dispersion should be as effective as is practicable in order to maximise transparency in the film, but the milling temperature should be controlled below 55° C in order not to induce polymerisation of the reactive monomer. Small additions of quinones are often made to ensure can stability. Ullage space in containers should be the minimum possible, in order to obviate evaporation of monomer (free of stabiliser) and possible polymerisation in the air space.

Control over the flexibility of the final film can be exercised by varying both the polymer and the monomer; suitable chain lengths of the polymer reactants may be selected. The original monomer used was styrene, tending to give slightly brittle films—the same hardness but superior toughness can be obtained by using methyl methacrylate. Similarly, flexibility can be conferred by the use of butyl acrylate or 2-ethylhexyl acrylate. Where maximum toughness and low odour are required, higher molecular weight monomers such as trimethylol propane triacrylate may be used. Where printing tack is needed to effect roll transfer, pentaerythritol triacrylate is available —its viscosity is Gardner $Z_3 - Z_4$ as monomer.

43.5 POLLUTION CONTROL

It was noted earlier that ultraviolet polyesters and acrylics are dissolved in reactive monomers. To date, styrene has been preferred because its reaction is faster in the process and less is lost to the atmosphere; in a correctly run oven, as little as 2 per cent styrene escapes to atmosphere. With the acrylic monomers, butyl acrylate is similar to styrene in loss, but ethyl and methyl acrylates can lose 10 per cent and acrylonitrile 15 per cent. After-burners are not necessary with styrene, but they are with the others. Further, these losses may be completely eliminated in a correctly cured system by the selective use of waxes.

Due to low temperature in the curing film, the volatilisation of the co-monomers is kept to a low level—about 5 per cent of that from a 120 to 140° C bake by thermal processes.

It should also be emphasised that the materials used in ultraviolet curing coatings are in general of higher toxicity than the ingredients in conventional coatings. Speciality monomers (for example hydroxy substituted acrylic esters) should be handled with particular care. Additional ventilation and protective clothing are essential, and efforts should be made to reduce operator contact prior to cure. The fully cured coatings are of course inert and comparable in properties to similar materials applied by alternative processes.

44 PRINTING INKS

44.1 INTRODUCTION

Printing technology, until the early 1960s, was more an art than a science. Until that time the most valuable commodity that an ink technologist could possess was experience. The industry as a general rule was not encouraging scientific input or instrumentation in ink development, but was rather relying on techniques and principles passed from one employee to the next.

Since that time great strides have been made in improving the situation, both overseas and in Australia. All major ink manufacturers in Australia now encourage the employment of qualified people, and the old dependence on experience has given way to instrumentation. This has resulted in recent years in the development of UV-curing inks, microwave-reactive systems and much-increased adoption of scientific principles to the drying and final film properties of the ink itself.

There are many terms used in describing properties and characteristics of printing ink that are peculiar to the industry. To assist the reader to understand these terms, and the manner in which they are used, the following list is included.

Reel or web-fed machine. A printing machine fed by a continuous length, or *web*, of *substrate* such as paper, film, board, etc., originating from a cylinder or reel.

Blocking. This problem occurs on reel-fed machines where the web is rewound after printing, before the ink is thoroughly dry, causing the printed side to stick to the web or film above it in the reel and thus 'block'. It can also occur in sheet-fed printing systems, where the printed surface is brought in contact with another surface while the ink is wet or tacky.

Sheet-fed machine. A printing machine where the substrate to be printed is in the form of sheets. The sheets are transferred to the machine by the aid of suckers which pick the sheets up and move them to the entry position on the machine.

Ink set. An ink after printing is said to be *set* when it will not smear or transfer when lightly rubbed.

Set-off is the transfer of ink from a printed sheet to the back of the next sheet falling upon it in the delivery of the printing press.

Grind of an ink is the particle size to which an ink has been ground, as measured on a *grindometer*.

Duct of a printing press is the trough from which the ink starts the distribution process to the printing plate.

Fountain. A term applicable only to the lithographic process. The trough and roller system used to apply water to the printing plate.

Tack refers to high shear rate viscosity. It is measured on a specially designed instrument called an *inkometer*.

Impression is the force applied to the paper or film by the printing plate or block during printing.

Register. The alignment of several colours to each other during the printing process.

Ink body. The manner with which an ink flows through the printing press ink distribution system; rheology of an ink.

44.2 THE PRINTING PROCESSES

There are five major printing processes. These are: *gravure*; *flexography*; *screen process*; *letterpress*; and *lithography*. The first three are usually grouped together under the heading 'liquid inks', while the latter pair are generally referred to as 'oil inks'. The liquid inks are all based on volatile solvent systems, are therefore hazardous, and dry almost entirely by evaporation; oil inks are almost entirely non-volatile and dry by penetration and oxidation. The application of oil inks requires a complex roller distribution system for reducing the ink thickness to a level where it can be controlled as it is applied to the printed surface. The various processes are discussed individually below.

44.3 GRAVURE AND FLEXOGRAPHIC PRINTING PROCESSES

44.3.1 Gravure Printing

Generally rotary gravure printing presses are reel-fed and are regarded as the largest consumers of ink in the various printing processes. Gravure printing is achieved from an etched image plane surface on a copper cylinder which is frequently chromium plated to give a longer life.

The etched cylinder revolves in the ink duct and picks up a film of ink in the etched cells. Excess ink is removed by a flat metal strip called the *doctor blade*, leaving ink in the etched cells; it is removed from the unetched areas and between the cells.

The web of paper, foil or film is forced into contact with the printing cylinder by an *impression roller*, allowing the ink to flow from the cells to the substrate, thus forming the required image. The intensity of the printed colour may be varied either by changing the depth of the cells or by increasing the area of each cell, but quite often the change of intensity is achieved by a variation in both.

The printed web passes from the printing station through an ink-drying system and is then rewound or sheeted.

Advantages of the Gravure Process

(a) No restriction on solvent selection, because no rubber rollers are in contact with the ink. Rubber rollers swell in the presence of aromatic hydrocarbon solvents.
(b) The printing of almost any substrate is possible, from foil to paper or film. The substrate must be relatively flat to achieve tonal printing—if the surface is rough, some cells will not print and parts of the design will be missing.
(c) Very high printing speeds, often in excess of 300 m/min.

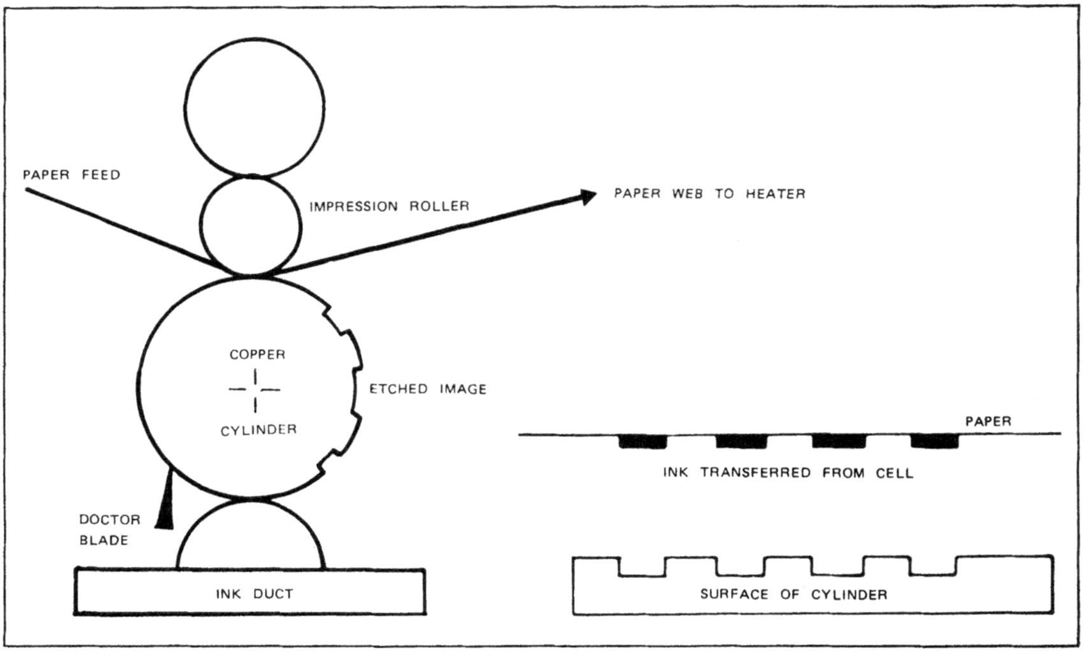

FIGURE 44.1
Cross-section of gravure printing

Disadvantages of the Gravure Process

(a) Very high cost of producing the printing cylinder.
(b) Highly volatile solvent system.
(c) Ink vapours are hazardous and quite often toxic.

44.3.2 Flexographic Printing

The image or impression in flexography is taken from a raised surface and is therefore described as *relief printing*. It is similar in this respect to letterpress printing, which will be discussed in a later section.

The printing surface is called a *stereo* and is made from natural or synthetic rubber or, more recently, from various polymers. These types of materials create problems for ink formulators because of the actions of various solvents upon them. Aromatic hydrocarbons, esters and ketones cause swelling and breakdown when in contact with rollers or stereos made from these materials. Minor quantities of some of these solvents can, however, be tolerated without damage.

The ink is picked up from the ink duct by the duct roller and transferred to the *Anilox roller* (see page 615). The Anilox roller, unlike the duct roller, is not rubber covered but is generally made from engraved or grained steel. The graining allows a given depth or film weight of ink to be transferred to the plate cylinder. The plate cylinder makes contact with the web by means of the impression cylinder, thus transferring the image to the substrate.

There are two basic types of flexographic presses in use:

A stack press, in which the printing stations are 'stacked' on top of each other and each station has its own impression cylinder.

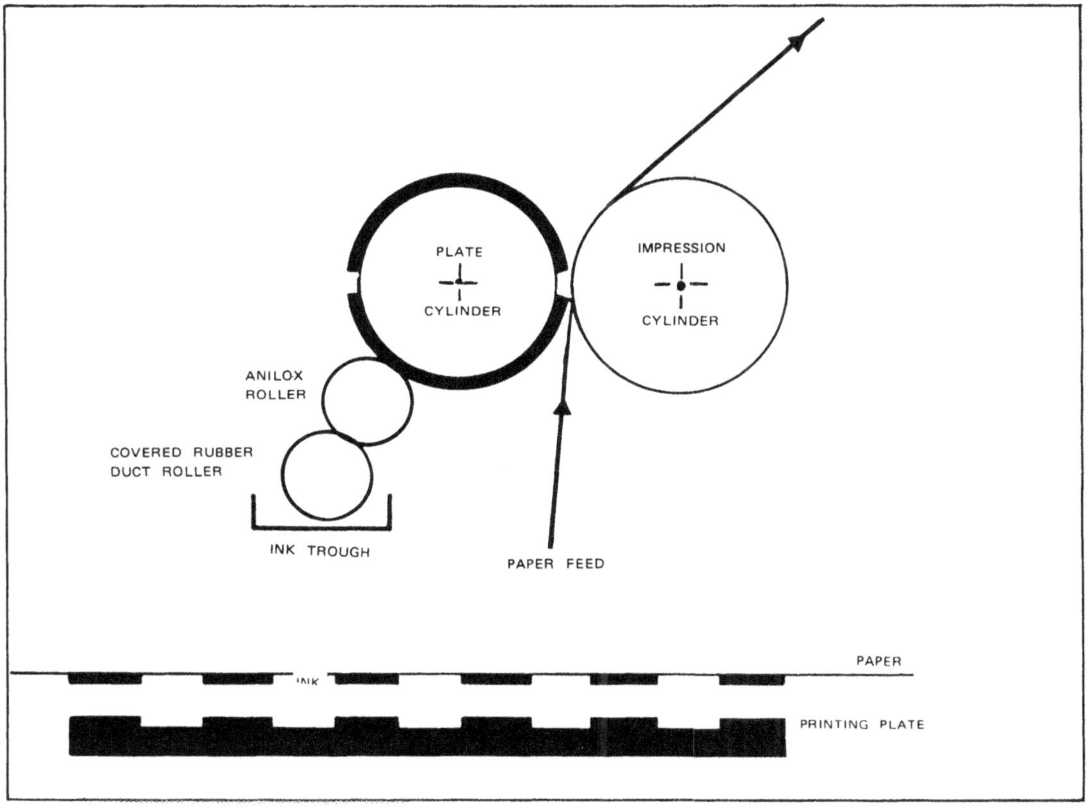

FIGURE 44.2
Cross-section of flexographic printing (relief printing)

A central impression press, in which the web to be printed passes around a large metal drum around which the printing stations are situated; this drum then acts as a common impression cylinder. This type of press is used for extensible films such as polythene where the combination of heat and tension can cause register problems during printing.

Flexographic inks when printed on absorbent papers do not require heating to force-dry the print; the heat generated within the press plus the draught caused by the moving web are sufficient. However, in most cases, and especially with pigmented inks, some form of heating is necessary: this usually takes the form of a steam or electrically heated drum, infra-red lamps or a hot air blast within a tunnel. It is essential with all of these force-drying methods that the vapours be removed quickly and completely to avoid resoftening of the film and subsequent blocking on rewinding of the web.

Advantages of the Flexographic Process

(a) The printing stereo is easy to make and low in price.
(b) It is a very fast printing process with the ability to print a wide range of substrates.
(c) It is suitable for short runs.
(d) The solvent system is low in cost—usually alcohol-based—thus ensuring a price advantage in comparison to inks supplied for most other processes.

Disadvantages of the Flexographic Process

(a) The colour strength and gloss are difficult to achieve because half of the ink is composed of solvent (i.e. low resin solids).
(b) Solvent restriction, because of the use of rubber stereos and rollers.
(c) The inks are all flammable.
(d) The odour is sometimes unpleasant, and ink solvent vapours may be toxic.

44.4 GRAVURE AND FLEXOGRAPHIC INKS

Inks suitable for the gravure and flexographic processes are normally of the *non-convertible* variety—that is, they undergo no chemical change on drying but remain soluble in the original solvent system. A *convertible* coating may be defined as one that undergoes a chemical change during the drying process, forming an oxidised or cross-linked product that is resistant to solvent attack; letterpress and lithographic inks normally fall into the convertible category. (Convertible and non-convertible coatings are discussed more fully in chapters 40 and 41.)

Non-convertible coatings dry by one or more of the following methods:
(a) evaporation of solvent;
(b) absorption into the *stock*;
(c) solvent attack on the film;
(d) heat fusion.

44.4.1 Raw Materials Used in Liquid Inks

44.4.1.1 Pigments

Pigments are used in liquid inks to impart colour and opacity and may be based on insoluble inorganic or organic chemicals or soluble dyestuffs. The selection of a pigment for a particular purpose has to be considered carefully to ensure that it will not decompose in the ink system, contravene health regulations (e.g. lead content in lead chromes), fade too quickly on exposure, or dissolve or discolour upon application.

44.4.1.2 Resins

Resins promote adhesion of the ink to the substrate, flow of the varnish (or vehicle) and provide print gloss. They can be either *film formers or non film formers*.

Film formers are linear polymers which produce continuous self-supporting films. Nitrocellulose and polyamide resins are examples of this type. Inks based on this class of resin generally possess low solids contents at normal working viscosities.

Non film formers are polymers which are not capable of forming a continuous film. Use of these resins leads to inks of high solids content possessing very good gloss but which lack the toughness that results from the film forming group.

Other important properties of resins for ink application include melting point, hardness, solubility, solution viscosity and solvent release.

The various resins used in gravure and flexographic printing processes are described below.

Nitrocellulose. This is used extensively in liquid inks based on alcohol or esters/ketones to promote hard tough continuous films. A plasticiser is normally added to increase the flexibility of the dried film. Resin modifiers are also used to improve the gloss of the ink.

Ethyl cellulose. This material is soluble in alcohol/hydrocarbon mixtures and is more flexible than nitrocellulose. It is used in both flexographic and gravure inks.

Polyamide resins. The long aliphatic linear chains in polyamide resins ensure the great flexibility of these materials. They are soluble only in alcohol/aromatic solvent mixtures and thus have a limited use in flexography. Polyamide resins will print on, and adhere to, a great variety of surfaces including paper, foil, treated polythene, glassine and coated and uncoated cellophane.

Rubber derivatives. There are two main types: *chlorinated* and *cyclised.*

The chlorinated variety is soluble in aromatic hydrocarbons, esters and some ketones and may be diluted with aliphatic hydrocarbons; thus it can be employed only in gravure printing systems and requires some plasticiser to provide satisfactory flexibility.

The cyclised rubber type is soluble in hydrocarbon solvents but insoluble in ketones, esters and alcohols; it is a flexible resin of good gloss and adhesion to certain substrates.

Vinyl resins. PVA and PVC copolymers are mainly used in inks for printing on PVC films and foils and, because they are fairly soft, are generally modified with a harder resin. They are used mainly in the gravure process, with ketones and aromatic hydrocarbons as the solvents.

Acrylic resins. These resins are soluble in nearly all solvent types with the exception of alcohols and aliphatic hydrocarbons. They are used in both gravure and flexographic inks and generally provide good adhesion and grease resistance to the finished ink. Care must be taken with solvent selection to ensure adequate volatility, as retained solvent tends to cause odour problems.

Ethyl hydroxyethyl cellulose. EHEC is the only aliphatic solvent-soluble cellulose resin. It finds some use in gravure inks for printing on paper.

Rosin derivatives. The group includes the maleic and fumaric modified phenolics, together with limed and zinc resinates. They are used to promote gloss and adhesion.

Shellac. Shellac is an alcohol-soluble resin which can be rendered water soluble by reaction with a suitable base, such as monethanolamine. This resin, because of its ability to oxidise and become water insoluble once cast from solution, is used extensively in flexographic printing. Its inherent film-forming properties and wax content also help to produce highly scuff-resistant prints.

44.4.1.3 Solvents

Solvents are used to control consistency and viscosity and are generally classified according to their solvency, chemical composition and evaporation rate. The solvency may be defined in three ways—namely those regarded as true solvents (complete solubility), or latent solvents (partial solubility) or diluents (no solubility).

The resin being dissolved determines into which category the solvent will fall; thus a solvent may fall into any one of the above three groups depending on the resin being dissolved. For example, toluol is a solvent for EHEC but a diluent for nitrocellulose, whereas alcohol is a solvent for shellac, a latent solvent for nitrocellulose, but only a diluent for certain acrylic resins.

The chemical composition of flexographic and gravure solvents include alcohols, glycols, esters, ketones and hydrocarbons, together with smaller groups of chlorinated hydrocarbons and nitroparaffins.

Alcohols: methyl, ethyl, butyl.
Glycol ethers: methyl, ethyl, butyl.

Esters: methyl, ethyl, propyl and butyl acetate.
Hydrocarbons: hexane, heptane, toluol, xylol and many more which fall into special petroleum cuts or fractions.
Ketones: acetone, methyl ethyl ketone, methyl isobutyl ketone.

The evaporation rate of a solvent for use in flexographic and gravure inks is of great importance; such inks dry by evaporation. There are published figures for the rate of evaporation of solvents usually based on a standard such as ethyl or butyl acetate or ether. These standards, however, do not take into account the effect of solvent mixtures on individual rates of evaporation, or the effect of the solvent release from the resin.

It is important to note here that, where an ink contains a mixture of solvents, the slowest must be a true solvent for the resin and not just a diluent, otherwise precipitation of the resin may take place resulting in a loss of gloss and adhesion.

44.4.1.4 Additives

Plasticisers promote flexibility in film-forming resins and improve gloss and adhesion.

Waxes promote slip and thus improve scuff resistance.

Dispersants improve the pigment dispersion and may improve gloss, flow of the ink and stability of the ink in the can.

Gelling agents are used to control ink body and to prevent pigment settlement.

Antifoams are used mainly in water-based flexographic inks to prevent foaming during printing.

44.4.2 Pigment Content of Liquid Inks

To provide colour strength and opacity, an ink must contain a minimum quantity of pigment. This level varies according to the class and individual properties of the selected pigment. The *pigment : binder ratio* is also influenced by the choice of pigment; this ratio is the ratio by weight of pigment to the non-volatile binder (resin) and has a great bearing on the flow and gloss of the printed film.

The following list is intended to show the various chemical types of pigments commonly used in liquid inks, together with a guide to the percentage of each used. (For further detail on inorganic and organic pigments, see chapters 25 and 27.)

Red Pigments (10–15 per cent)
Lake Red C
4B Toners—calcium and barium
Toluidine Red
PMTA Pinks and Magentas

Green and Blue Pigments (10–15 per cent)
Phthalocyanine Green and Blue
PMTA's
Bronze Blue

Yellow Pigments (10–40 per cent)
Primrose, Lemon and Mid-Chrome (30–40 per cent)
Benzidine Yellow (10–15 per cent)

Orange Pigments (10–40 per cent)
Molybate Orange (30–40 per cent)
Benzidine Orange (10–15 per cent)

White Pigments (30–45 per cent)
Rutile and anatase titanium dioxide

Black Pigments (10–15 per cent)
Carbon Black

44.4.3 Gravure Ink Classification

There are several ways of classifying the various types of gravure inks now manufactured, but only two will be discussed here. The first classifies the inks into the two groups according to the end use: either publication gravure printing or commercial gravure printing. The second method, which is the more universal, uses a letter system to denote the type; the letter also gives an indication of the correct ink reducer, the range of compatible groups and the type of resin system employed.

44.4.3.1 'A Type' Inks

This is the most general and widely used class in the publication field and is used in newspapers, magazines, catalogues and preprinted inserts. In general these are the cheapest gravure inks, employing aliphatic hydrocarbons and low aromatic petroleum solvent cuts. Limed resins are principally used as binders, with a little EHEC included to lower the viscosity (reduce 'ink body').

44.4.3.2 'B Type' Inks

These inks are modified 'A types' which use up to 50% aromatic solvent in the solvent system. They use cheap hard resins, although the percentage of EHEC is generally higher. 'B type' inks are designed to replace the 'A type' where clay-coated or highly calendered stocks are to be printed.

44.4.3.3 'C Type' Inks

'C type' inks are based on nitrocellulose, which may be modified in many ways with various resins, plasticisers and waxes. The nitrocellulose gives the film strength; the resin additive promotes gloss and flow with improved printability; the plasticiser increases the flexibility; and the wax adds slip and thus scuff resistance.

The solvent system consists of ketones and esters, although latent solvents such as alcohol, and diluents such as toluol, are used to keep the cost to a minimum.

The 'C type' inks are used to print on paper, board, foil and cellophane.

44.4.3.4 'D Type' Inks

These inks are based on polyamide resins and are sometimes referred to as *cosolvent inks* because of the combination of solvents that are necessary to solubilise the resin. Alcohols and aromatic hydrocarbons are good solvents for the 'D type' systems, ethyl alcohol and toluol being preferred.

This group is very important because they give adhesion to very difficult printing surfaces such as treated polythene, foil and all types of cellophane.

44.4.3.5 'E Type' Inks

These inks are alcohol-reducible and are generally based on alcohol-soluble nitrocellulose. They are similar to 'C type' inks, although they are lower in residual odour and thus are preferred where odour retention might be a problem.

44.4.3.6 'F Type' Inks

Chlorinated rubber is the resin employed in 'F type' inks, with toluol as the major solvent. Other resins may be used, together with a plasticiser, where the conditions demand better gloss.

These inks are used on board, particularly for cigarette packets where they give very hard abrasion-resistant films. They also have the property of excellent alkali resistance and are therefore used for printing soap and laundry products.

44.4.3.7 'W Type' Inks

'W type' inks are water soluble and of limited use in Australia. They may be based on water-soluble EHEC or polyvinyl alcohol and employ small amounts of wax and silicone antifoam. Their greatest use is in printing the background colour for cheques.

44.4.3.8 'X Type' Inks

This is a miscellaneous category covering inks not fitting into any of the previous categories. One of the most popular members of this group is an ink designed for printing on vinyl sheeting and upholstery; it is based on vinyl copolymers and uses a ketone/hydrocarbon solvent system. Representative formulas for gravure inks are provided in formulas 44.1 to 44.9.

44.4.4 Flexographic ink classification

Flexographic inks may be divided into three groups according to their pigmentation: *aniline* or *dye ink*; *pigmented ink*; and *semi-pigmented ink*.

44.4.4.1 Dye Inks

These inks consists of a basic dye which is laked with an organic acid, usually tannic acid, resin and an alcohol solvent. The basic dye alone is water soluble and therefore unsuitable for use in printing inks because of the lack of water or moisture resistance. The laking process provides water insolubility, but permits alcohol and glycol solubility. The resin, frequently shellac, improves printability, prevents ink absorbtion and allows a more economical use of the ink. Glycol is sometimes added to control the drying speed and to improve printability. The formulation for a typical dye ink is given in formula 44.10.

The formulation can be varied to meet various specified requirements, including increased colour strength, more gloss holdout or a change in viscosity. It should be emphasised that these inks are very transparent and thus unsuitable for printing over coloured surfaces or cheap unbleached boards.

Dye-type flexographic inks are mainly used to print wrapping paper, bread wraps and glassine bags, where their poor light resistance is not restrictive.

44.4.4.2 Pigmented Inks

These inks consist of pigments, resins and solvent, together with several additives as discussed previously. The formulation is varied to suit the stock porosity and printing speed. The type and amount of resin is dependent on the porosity and chemical type of the printing surfaces, so that for very porous stocks the resin percentage is raised and a film former is often preferred. Shellac is a very useful resin in such formulations, because it exhibits film-forming properties together with low solution viscosity and good pigment-binding qualities. As the printing surface becomes less porous, it is important to ensure that the ink is not over-resinated as this can prevent penetration of the ink into the substrate, resulting in lack of drying and blocking in the reel. The formulations for two typical pigmented ink are given in formulas 44.11 and 44.12.

44.4.4.3 Semi-pigmented Inks

These inks may be considered as a mixture of dye and pigmented inks, but they are, however, generally confined to paper printing. A white shellac ink tinted with a dye, or a coloured pigment ink tinted with a similarly coloured dye, are two examples of this category. In these cases, the

FORMULAS 44.1 TO 44.9
Formulations for typical gravure inks

	Parts by mass
'A Type'	
1. Publication Black	
Limed rosin	8.0
Gilsonite	30.0
Modified furnace black	8.0
Prussian blue	2.0
Lecithin	0.5
Methyl violet dye	0.5
Heptane	46.0
Solvent[a]	5.0
	100.0
2. Publication Red	
Ba or Ca Lithol red	8.0
China clay	13.0
Limed rosin	30.0
EHEC	1.0
Lecithin	0.5
Ink oil	1.0
Heptane	24.0
Solvent[a]	22.5
	100.0
'B Type'	
Lithol red	6.0
China clay	16.0
Limed rosin	23.0
EHEC	1.5
Toluol	10.0
Solvent[b]	39.0
Wax compound	4.0
Lecithin	0.5
	100.0
'C Type'	
Carbon black	14.0
Nitrocellulose	12.0
Maleic resin	6.0
Plasticiser	3.0
Wax compound	4.0
Normal propyl acetate	18.0
Alcohol	43.0
	100.0

	Parts by mass
'D Type'	
Titanium dioxide	40.0
Polyamide resin	22.0
Wax compound	3.0
Toluol	16.5
Alcohol	17.0
Butanol	1.0
Water	0.5
	100.0
'E Type'	
Titanium dioxide	42.0
Nitrocellulose	9.0
Plasticiser	2.0
Wax compound	4.0
Normal propyl alcohol	8.0
Alcohol	35.0
	100.0
'F Type'	
Phthalocyanine Green	10.0
Chlorinated rubber resin	15.0
Toluol	57.0
Solvent[c]	15.0
Plasticiser	3.0
	100.0
'W Type'	
Acidic dye	5.0
China clay	15.0
Water soluble EHEC	10.0
Water	69.9
Antifoam	0.1
	100.0
'X Type'	
Titanium dioxide	25.0
Phthalocyanine blue	5.0
PVC resin	16.0
Plasticiser	4.0
Methyl isobutyl ketone	30.0
Methyl ethyl ketone	19.5
Wetting agent	0.5
	100.0

[a]Shell × 3B [b]Shell × 2 [c]Shell × 95

pigment allows the ink to be printed on other than just white kraft, while the dye adds colour strength and brilliance. The formulation for a typical semi-pigmented ink is given in formula 44.13.

44.5 SCREEN PROCESS (STENCIL PRINTING)

Screen printing is a method of printing from a stencil marked on a silk or nylon cloth which is supported on a timber frame. The ink is forced through the cloth in the image area by means of a squeegee.

The stencil may be hand-cut in paper or a lacquer film, or prepared by a photographic process: a light-sensitive film is exposed through a positive image and the unexposed areas removed by a solvent wash. The stencil is then adhered to the screen, ready for printing.

FORMULA 44.10
Formulation for a typical dye ink

	Parts by mass
Basic dye	10.0
Tannic acid	15.0
Shellac	10.0
Glycol ether	10.0
Alcohol	55.0
	100.0

FORMULA 44.11
Formulation for a typical water-based pigmented ink

	Parts by mass
Titanium dioxide	35.0
4B Calcium toner	5.0
Shellac	10.0
Monoethanolamine	1.0
Alcohol	3.0
Water	42.0
Wax compound	4.0
	100.0

The silk screen process is largely a hand-fed operation, although automation is taking place, involving a mechanical squeegee action. The most modern machines have an automatic feed with a mechanical squeegee, drying ovens and automatic delivery, giving a printing speed of up to 2000 impressions per hour.

The deposit of ink on the paper or stock by this printing method is considerably greater than for any other process. The quantity of deposit depends on the diameter of the thread used to make the cloth. This is also usually proportional to the number of threads per centimetre; for instance, a 40 threads/cm screen would produce a very heavy deposit, while a 160 threads/cm screen would yield a very fine ink deposit.

FORMULA 44.12
Formulation for a typical alcohol-based pigmented ink

	Parts by mass
Benzidine yellow	10.0
Shellac	22.0
Alcohol	49.0
Glycol ether	15.0
Wax compound	4.0
	100.0

FORMULA 44.13
Formulation for a typical semi-pigmented ink

	Parts by mass
Titanium dioxide	18.0
Basic green dye	3.0
Laking agent (tannic acid)	3.0
Nitrocellulose	18.0
Shellac	5.0
Alcohol	39.0
Glycol ether	10.0
Wax compound	4.0
	100.0

Since the deposit of ink applied is high, the pigment content of *screen inks* is quite low. For organic pigments the content is about 4–8 per cent, while for inorganic types it may be as much as 20 per cent. The pigments used are the same as for flexographic and gravure inks although far greater emphasis is placed on the lightfast varieties. This is because a great deal of screen printing is used for posters which are subject to outdoor light and weather exposure.

These inks generally dry by evaporation, although some oxidising types are manufactured for metal and plastic printing. The viscosity is kept low to allow the ink to easily pass through the screen but, unlike other liquid inks, the body is not 'long flowing' (thixotropic); the ink is deliberately gelled with bentonite clay, or aluminium octoate, in order to promote a sharper print or 'snap off' from the screen. This also prevents the ink from dripping through the screen between each pass of the squeegee.

There are many different types of screen inks:

Poster ink (paper printing)
Lacquer
Enamel or oxidising
PVC
Mylar and plastic

It can be seen from this list that the screen printing system is versatile and can be used to print such items as paper cards, fabric, plastic bottles, wood, glass and even circuit boards. Representative formulations for screen inks are given in formulas 44.14 to 44.16.

FORMULAS 44.14 TO 44.16
Formulations for three typical screen inks

	Parts by mass
Iron blue	4.0
Calcium carbonate	40.0
Resin/Stand oil varnish	47.0
White spirit	8.0
Mixed drier	1.0
	100.0
Phthalocyanine green	6.0
EHEC	6.0
Zinc resinate	25.0
Plasticiser	4.0
Bentonite clay	3.0
Mineral turps	40.0
Xylol	16.0
	100.0
Lemon chrome	20.0
EHEC	8.0
Zinc resinate	15.0
Plasticiser	2.0
Mineral turps	50.0
Bentonite clay	5.0
	100.0

Advantages of Screen Printing

(a) Low cost of printing a small quantity.
(b) Startling colour strength and brilliance.
(c) Unlimited size of printed sheet.

Disadvantages of Screen Printing

(a) Poor process work definition.
(b) Inks are flammable.
(c) Very slow printing process, with high labour cost per printed unit.

Other types follow a similar pattern, the solvent system being changed to suit the type of resin used. In all cases a structure is built into the ink either by the nature of the resin/solvent or by chemical or physical gelation. It can also be achieved by a large percentage of extender pigment such as calcium carbonate; formula 44.14 is a good example of this technique.

44.6 LETTERPRESS AND OFFSET LITHOGRAPHY

44.6.1 The Letterpress Process

The letterpress process is a *relief* process: the impression or image is taken from a raised surface in much the same manner as a rubber stamp is used to transfer ink from a stamp pad.

The printing surface may consist of alloys of antimony and tin, copper and zinc, or rubber, nylon or PVC. The blocks used to print type matter, line or tone work are produced by a negative photographic process which allows an etching of the relief area after exposure. The etch used may be an acid (in the case of metallic plates or blocks) or simply alcohol (in the case of the nylon plate).

In both letterpress and offset lithography, the inks used are of a paste-type consistency, and must therefore be distributed in a fundamentally dissimilar manner to liquid inks. Liquid inks are thin and the printing plate is very near to the drier; paste inks are relatively thick and have to be spread thinly by pressing them through an ink roller train before applying them to the printing plates.

The different letterpress machine types are described below.

44.6.1.1 Platen Press

This press operates on a clam-shell action, where the paper is hand-fed to the machine.

44.6.1.2 Flat-bed Machine

This machine has a reciprocating action in which the type or image is carried on the bed of the press and the paper on the cylinder. The impression or contact takes place as the cylinder and paper pass across the image. There are three types of flat-bed letterpress machines:

(a) *Stop cylinder* (good registration)
(b) *Two revolution* (small cylinder)
(c) *Single revolution* (long cylinder)

44.6.1.3 Reel-fed Web Rotary Press

The rotary machine carries curved printing plates, generally two to each cylinder, so that two prints are taken for each revolution of the cylinder. A second cylinder, called the *impression cylinder*, is used to enable pressure to be applied to the paper web as it passes across the image area. This is a continuous process, where each reel of paper is automatically joined to the next while the machine is printing. This system is employed to print newspapers.

44.6.1.4 Sheet-fed Rotary Press

Used for short-run/medium-run quality printing of such articles as books, magazines and brochures. The printing plate is curved and may be of metal or synthetic base, although recent developments with nylon relief plates suggest that they may well be the plate of the future for sheet-fed rotary presses.

Advantages of Letterpress Printing

(a) Excellent reproduction, expecially of type matter.
(b) Low cost of small-run work, such as business cards.
(c) No toxic vapours.
(d) Almost no fire hazard.

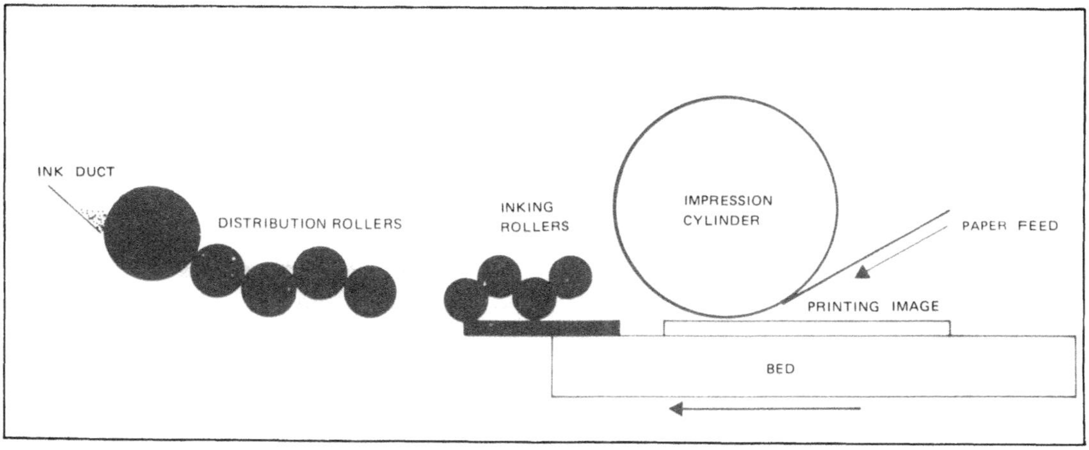

FIGURE 44.3
Cross-section of a stop cylinder letterpress machine at the commencement of the printing stroke. The machine has been inked up and is now ready to take a print

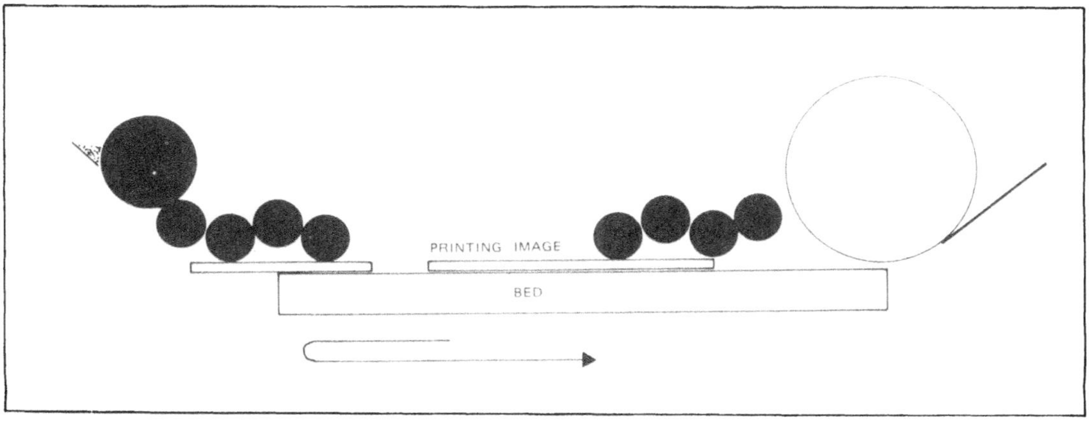

FIGURE 44.4
The stop cylinder machine at the end of the printing stroke. The bed will return to the commencement position via inking rollers where the image will be re-inked ready for the next impression. During this stroke of the bed, the impression cylinder is stationary

Disadvantages of Letterpress Printing

(a) Costly typesetting or platemaking.
(b) Limitation of reciprocating action (flat-bed machines).
(c) Hard appearance of prints because of impression.
(d) Solvent restriction which in turn limits resin selection.

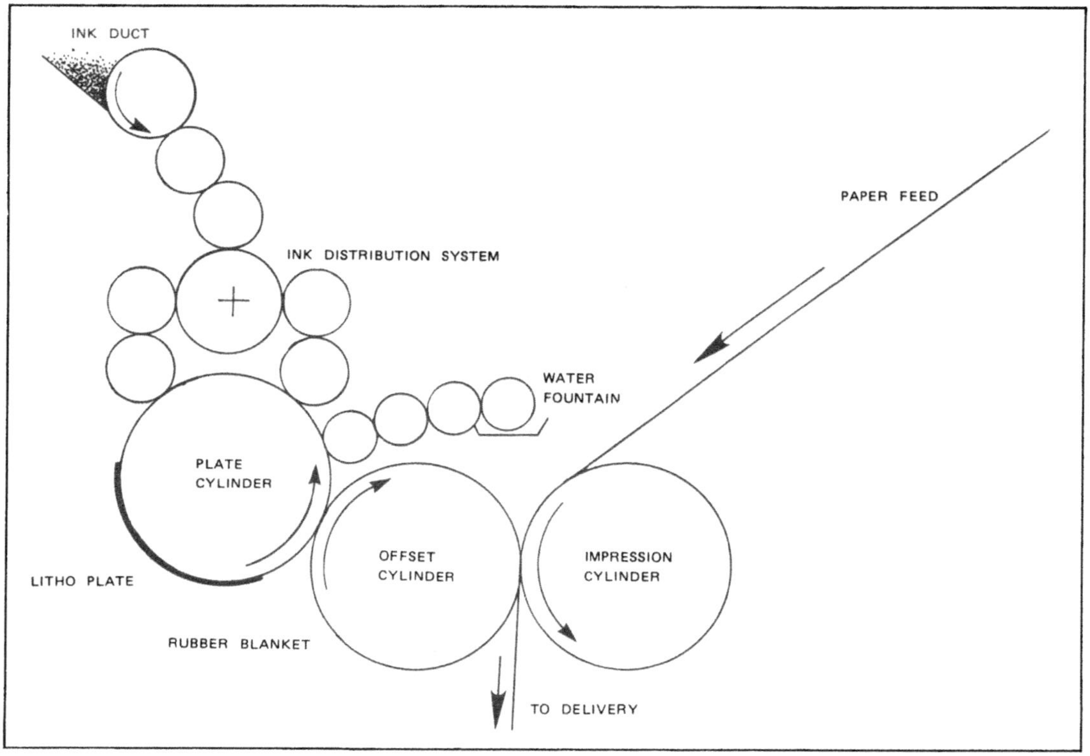

FIGURE 44.5
Cross-section of offset lithography

44.6.2 The Offset Lithographic Process

Offset lithography is a planographic process where the image and non-image are, to all intents and purposes, in same plane. The image area is oil receptive and the non-image area is water receptive so that, providing that the plate is initially wet, the ink when rolled across the plate will only be attracted to the oil-receptive areas. The press consists of:

(a) an inking system to deliver the ink, by means of an ink duct and a system of rollers, to the plate;
(b) a fountain system to deliver the water from a trough by means of a series of cotton-covered rollers to the plate;
(c) a blanket (resilient rubber) cylinder to lift the ink from the plate and transfer it to the paper by exerting pressure on an impression cylinder.

The use of the resilient rubber blanket allows minimum pressure for satisfactory image transfer, thus eliminating the hard appearance of letterpress prints.

44.6.2.1 Types of Press

(a) Sheet-fed single-colour, two-, four-, five- and six-colour machines.
(b) Web-fed blanket-to-blanket or *perfecting* machines.
(c) Web-fed common impression machines.
These machines range in size from small office duplicators to 18 × 36 m giants.

44.6.2.2 Printing Plates

The plates used have to be flexible, easy to handle, durable and of excellent stability. The most common are aluminium, 'Trimetal', nylon and to a lesser extent zinc.

Aluminium plates are available as negative or positive working and may be presensitised or uncoated. These plates are used for short to medium-run length (10 000–100 000 impressions).

'Trimetal' consist of a layer of chromium over copper on a steel backing. The chromium forms the non-printing area, while the copper forms the image area. The steel supplies the strength and flexibility. These plates are expensive to buy and to process and are designed only for long runs of 100 000 to 1 million impressions.

Nylon plates are actually relief plates and are used on offset machines with the fountain disconnected. The process is therefore best described as offset rotary letterpress.

Zinc plates are now almost non-existent, having been replaced largely by the aluminium plate.

Advantages of Offset Lithographic Printing

(a) Speed (all rotary action).
(b) Low cost of plates.
(c) 'Soft' look of printed result (no impression embossed from raised printing plate).
(d) Able to print any stock (tinplate to paper).
(e) Better colour and halftone reproduction than rotary letterpress.
(f) Low-odour inks.
(g) Low fire hazard in storage of inks.

Disadvantages of Offset Lithographic Printing

(a) The ink/water balance is critical.
(b) Double or offset transfer of ink means a very thin film of ink on the printed sheet, resulting in poor gloss.
(c) Water in the process restricts the pigment range to the non-water-bleeding types.
(d) Severe restriction on solvent selection and thus resin choice.

44.7 LETTERPRESS AND OFFSET LITHOGRAPHY INKS

Both of these inks are of the convertible (oxidising) type, with probably only one exception: the letterpress fibreboard type. These inks dry by any one or more of the following techniques:
(a) absorption into stock;
(b) evaporation of solvent;
(c) oxidation;
(d) polymerisation;
(e) precipitation.

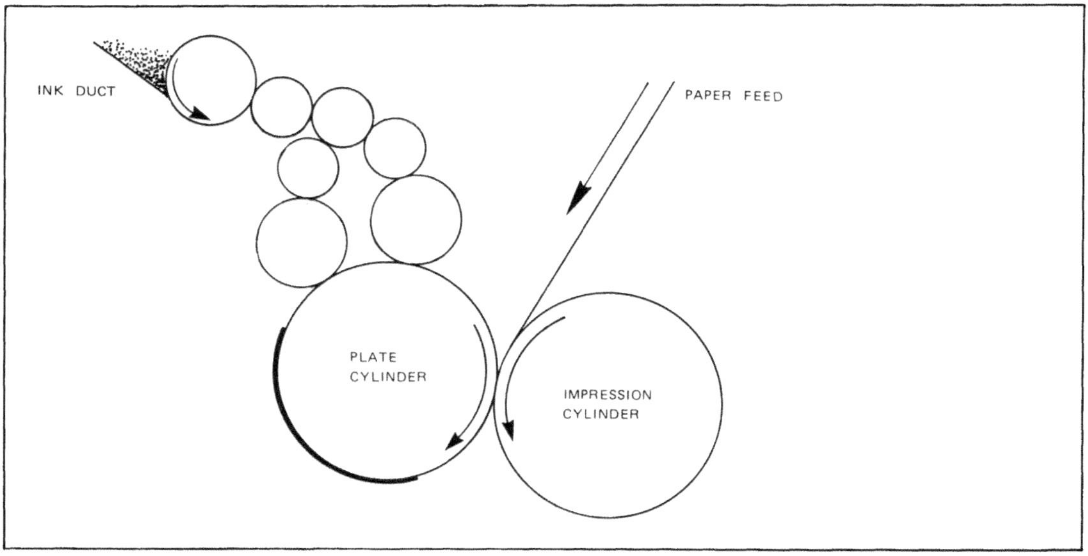

FIGURE 44.6
Cross-section of rotary letterpress printing

44.7.1 Raw Materials Used in Letterpress and Offset Lithography Inks

44.7.1.1 Resins

Resins promote binding, adhesion, gloss and pigment wetting. They can be either film formers or non film formers (as discussed in section 44.4.1.2). Their main properties to be considered are:

(a) melting point;
(b) solubility and solution viscosity;
(c) pigment-wetting properties;
(d) solvent release.

These are very similar to the properties required by flexographic and gravure inks, although solubility is more critical because of the restricted solvent selection available for use on offset and letterpress machines. The restriction is caused by the large number of rubber rollers that are necessary to distribute the ink, and their susceptibility to aromatic, ketone and ester solvents.

The main resins used are described below.

Phenolics

Pure phenolics (reactive and otherwise). Very seldom used because of their poor solubility in the appropriate solvents. The high viscosity of solutions (when possible) leads to low solids and little gloss. They are generally used only in speciality applications, for example, where soap resistance is required.

*Modified phenolics.** Prepared by the reaction of phenol and formaldehyde in the presence of excess highly acidic natural rosin (usually above 10 per cent). The rosin inhibits the formation of insoluble complexes. This gives a resin soluble in vegetable oil and low aromatic solubility. Esterification with glycerine gives even better solubility and much lower acid value. The final result is a resin

* See also chapter 10 (volume 1).

of excellent hardness, gloss, water resistance, durability and reagent fastness, with good drying properties coupled with fast solvent release. The resin is normally cooked into a vegetable oil at 100–240° C, depending on the degree and type of modification; the higher the temperature of the cook, the better the ester interchange between ingredients and the better the diluent stability or resistance to precipitation when diluted with low aromatic ink oils.

Maleics. These are prepared by the reaction of rosin (abietic acid) and maleic anhydride. There are two types:
Type 1. Alcohol soluble, not esterified
Type 2. Oil soluble, esterified with a polyol (e.g. glycerol, pentacrythritol)
Type 1 is used in glycol-based/moisture-setting letterpress inks for carton printing, in conjunction with shellac.
Type 2 is used for general letterpress and offset application; fair-to-good gloss, high solids, good wetting and reasonable solvent release.

Cyclised rubber. Made by treating natural rubber with a catalyst, thus forming a cyclic product. Soluble in esters and vegetable mineral oils of low aromaticity. Insoluble in alcohols and ketones. Compatible with most alkyds, resins, cellulose derivatives. Main advantages: quick solvent release tack-free film, very little gloss under most circumstances, limited by ability to *string* or *fly* on the press (as ink film splits at impression, a spray of ink flies at random).

Rosin (wood, gum, tall oil). These all have low melting point, high acid value, and poor 'consistency'. All are cheap and are important for this reason as a starting point in manufacture of more valuable resins, as already discussed. Their largest use is in cheap publication letterpress, drying by penetration and giving no gloss. Small percentages may be used to promote better pigment wetting but their high acid value restricts their unmodified use in offset inks (emulsification).

Shellac. From the secretion from or an excretion of the lac insect *Laccifer lacca.* Soluble in alcohol, glycols, water, and insoluble in mineral oils and vegetable oils. Consists almost entirely of fatty acid compounds, wax and lac dye. Used in letterpress fibreboard inks, where it gives good scuff resistance and pigment wetting.

Tung Oil is not used as much now as previously due to its high price. It shows good pigment wetting and speed of dry. It must be cooked correctly to give best results. Because of the high degree of water resistance, Tung Oil is used in one form or another in offset inks.

Dehydrated castor oil and safflower oil have very limited use, their main advantages being colour retention, particularly on stoving. They are rarely used unmodified but frequently in the alkyd form. Litho use is restricted because of their ability to be easily emulsified.

Alkyds. The relevant types are the vegetable oil modified alkyds which may be considered similar to oleoresinous varnishes, and are mostly derived from phthalic or isophthalic anhydride. The oils used include linseed, tung, dehydrated castor and safflower selected to impart certain properties: linseed, for example, for pigment wetting and flow; tung for pigment wetting, quick drying; DCO flow, colour retention; safflower colour and reasonable drying. The viscosity may vary from 20–40 Pa.s to over 1000 Pa.s, the application determining the preferred viscosity. General properties are good flexibility, adhesion, hard drying, good gloss, pigment wetting and water resistance.

Petroleum resin. Slow solvent release; likely to be used in the future as the price of the natural resins increases. Low cost incentive for work at this stage.

Zinc–calcium resinate. Very little use in letterpress and offset inks. Low-cost publications only.

Polyamides. Work now under way; may have a future in letterpress and offset systems.

Epoxy resins. Again, work is under way and the future could see a greater use of these resins (described in chapter 12).

44.7.1.2 Solvents

Solvents control consistency, viscosity, tack and to some extent flow and gloss, and may help with adhesion, particularly polar types. Same classification as in flexographic and gravure systems; that is,
(a) solvent power (as discussed previously);
(b) evaporation rate (as discussed previously);
(c) chemical composition.

Alcohols. Limited use in letterpress and offset inks; some use of long carbon chain alcohol solvents.

Glycols. Ethylene, propylene, diethylene.

Esters, aromatics, ketones. No use—these solvents are either too fast or attack rubber rollers, plates and blankets.

Hydrocarbons. Most important group by far; usually low aromatic hydrocarbons (0–8 per cent). Boiling range up to 590° C. Many narrow cut oils are available in this range.

44.7.1.3 Vegetable Oils

Linseed oil is probably the most important. It has good flow, pigment wetting, drying and water resistance. Used in nearly all letterpress and offset inks, from 5 to 20 per cent, as the raw alkali refined or bodied/polymerised oil. It is further used in alkyd manufacture.

44.7.2 General Formulating Techniques for Letterpress and Offset Lithography Inks

A general formulation for these works is shown in formula 44.17.

44.7.2.1 Pigment Content

Pigment content is determined by two factors:
(a) colour strength required;
(b) ability of varnish system to carry pigment.
The second factor is the limiting function because colour strength has to be sacrificed if the varnish system is unable to wet and carry the pigment percentage required. The following is a list of pigments commonly used in letterpress and offset inks. The higher figures quoted relate to the offset process only.

FORMULA 44.17
The general formulation for letterpress and lithography inks

	Per cent by mass
Pigments	10–30
Varnish (resin, vegetable oil, mineral oil)	30–70
Wax compounds	4– 8
Driers	1– 4
Solvent or diluent	1–10
Additives	1– 2

Red pigments (10–30 per cent)
Lake Red C
Permanent Red 2B (barium, calcium and manganese)
Permanent Red 4B (max. of 20 per cent only)
PMTA Magenta and Rose

Green pigments (10–25 per cent)
Phthalocyanine green (max. of 20 per cent only)
PMTA green

Blue pigments (10–40 per cent)
Phthalocyanine blue (max. of 18 per cent only)
Bronze blue
PMTA blue

Violet pigments (10–20 per cent)
PMTA violet

Yellow pigments (10–50 per cent)
Primrose, Lemon, Mid and Scarlet Chrome (max. of 50 per cent)
Benzidine Yellow (max. of 22 per cent)
Benzidine Orange (max. of 18 per cent)

White pigments (30–50 per cent)
Rutile and anatase titanium dioxide

Black pigments (10–18 per cent)
Carbon black—mainly furnace type

44.7.2.2 Varnish System

This is probably the most important component of any ink formulation. It determines or influences the following properties of the finished ink:
(a) gloss;
(b) setting and drying speed;
(c) adhesion;
(d) scuff resistance;
(e) water resistance (offset process);
(f) chemical resistance;
(g) ease of pigment dispersion;
(h) flow;
(i) limiting pigment percentage, (i.e. maximum percentage);
(j) stability.
The main varnish types are described below.

Resin varnish (120–200 Pa.s). This is a cook of a rosin modified phenolic or maleic resin with linseed or tung oil or in many cases with both oils. This is carried out by heating the resin and tung oil to 200–240° C and holding for a clear pill. The time of holding varies depending on the particular resin used and the properties desired in the final vehicle. In general the longer the cook the slower the drying rate and the higher the acid value of the varnish.

When cooking varnishes with tung oil, care must be taken (particularly at temperatures of 200° C or higher) to ensure that gelation does not occur. This can result, if not controlled, in the formation of a sponge-like mass of very limited use. For control of this exothermic reaction most resin varnishes of the tung oil type are chilled with linseed oil upon reaching the clear pill stage. This quickly lowers the temperature and thus ensures a satisfactory result.

Example:

Modified phenolic	30–35 per cent by mass
Tung oil	20–25 per cent by mass
Heat to 220° C and hold for clear pill. Add	
Linseed oil	40–45 per cent by mass

In many cases a mineral oil percentage of 10–20 per cent may be used instead of the entire proportion of linseed oil.

These varnishes give a good gloss and scuff resistance coupled with high water resistance. The setting speed is slow although the final dry oxidised film may be achieved at a rapid rate. They are good all-round vehicles for letterpress and offset, and often one varnish is used to make ink for both processes.

Quickset varnish. This is a mineral oil solution of a resin especially chosen for its limited solubility in the respective oil. It contains a small percentage of vegetable oil to enable stability to be maintained.

These are manufactured, for best results, by the so-called cold technique. This is the principle of force dissolving the resin at about 140° C by rapid high shear mixing. In most cases all ingredients are mixed together although sometimes the vegetable oil is left out until solution is effected. The resin chosen is in most cases of the modified phenolic variety, although pure phenolics in some proportion are used.

Example:

Modified phenolic	40–50 per cent by mass
Linseed oil	5–15 per cent by mass
Mineral oil (BP 270° C)	35–55 per cent by mass

Quickset varnishes are valuable in that they set very quickly to a tack-free film. The process is one of precipitation where a rapid increase in viscosity results from the loss of the mineral oil component to the paper. This leaves a very viscous resin/pigment, vegetable oil combination on the surface of the paper, to undergo further drying by oxidation in the same way as the previously discussed varnish.

Speed of tack-free time, together with moderate gloss and scuff resistance are the main advantages. This varnish is widely used in both letterpress and offset inks for sheet-fed work.

Heatset varnish. These varnishes are almost exclusively used in the offset process. They are designed to dry mainly by evaporation and are printed only on web-fed machines. In composition they resemble the quickset type, except that the distillate or mineral oil is lower in boiling range. They do not precipitate but dry by forced evaporation of the solvent as the printed web passes through an oven or over a steam-heated drum.

Vegetable oil is kept to a minimum to ensure immediate and complete drying. The main use is as a plasticiser for the resulting film, and to improve the pigment wetting powers of the system.

Example:

Maleic resin	40–50 per cent by mass
Mineral oil (BP 210–245° C)	40–50 per cent by mass
Linseed oil	5–10 per cent by mass

Metal decorating lithographic varnish. This varnish consists of a simple cook (180° C) of a maleic resin and a suitable alkyd. The alkyd may be a linseed or safflower type of long oil length. The viscosity is of the order of 1000 Pa.s and in many cases this varnish is mixed with a high-viscosity alkyd before manufacture. The main properties desired are effective drying at 150° C in less than 10 minutes, good water resistance and a minimum of yellowing.

Example:

Maleic resin	35–40 per cent by mass
Isophthalic alkyd (20 Pa.s)	60–65 per cent by mass

Bodied oils and alkyds. Included in this category are the heat bodied linseed oils (known as litho varnishes), tung stand oils and long oil alkyds. These materials are seldom used as a major component in an ink formulation but rather to supplement one of the varnishes previously discussed. This is a more recent development.

The use of litho varnishes is diminishing in favour of the superior isophthalic alkyds and urethane oils, and only remains because of their low price.

Alkyds, in particular, have many advantages. They promote good flow, pigment wetting, ease of grinding, hard drying and high water resistance. They are found in most letterpress and offset inks at levels between 5 and 25 per cent.

High-viscosity safflower and DCO alkyds are used in stoving enamels or metal-decorating lithographic inks. In this case the percentage of the vehicle (i.e. alkyd) may be as high as 100 per cent. This is possible because the alkyd is stoved and thus force dried. It would not be possible on most papers, as the drying and tack-free times would be excessive.

44.7.2.3 Wax Compounds

These are designed to give the printed ink film slip, thus protecting the print from rubbing. The usual method of manufacture is to dissolve a wax by heating in a suitable varnish, and grinding while still molten. Waxes include the polyethylene, Polymekon, microcrystalline and polytetrafluoroethylene (PTFE) types. The percentage of solid wax added to a finished ink rarely exceeds 3 per cent.

44.7.2.4 Driers

The driers used include nearly all of the more common types including cobalt, manganese lead, calcium, zirconium and zinc, as various organic acid salts. These may be of the linoleate, napthenate, octoate, tallate or resinate variety. The acid used generally has little to do with the effectiveness of the metal. The degree of dispersion of the metal, however, is extremely important, and a small percentage of drier properly dispersed through an ink is very much more effective than a large percentage inadequately dispersed.

Letterpress and offset inks generally are formulated with cobalt and manganese salts as the chief metal driers. These driers may range from equal parts of each to a 1 : 3 ratio of cobalt to manganese. The total percentage will usually lie between 1 and 3 per cent. The other metals mentioned earlier are only used in speciality applications such as non-yellowing whites (certain driers cause white ink films to turn yellow over time); as wetting aids to pigment dispersion; as aids to the normal Co/Mn combination.

Until recently the drier content was carefully matched to the drying oil content and type of pigment. With the introduction of modern synthetic inks, the drying oil content has been substantially reduced and is now at such a low level (10 per cent) that this is no longer necessary.

The pigment type is still important to drying speed, however, and must be considered when calculating drier content. Iron blue and some benzidine yellows promote oxidation, while carbon black and PMTA's retard drying. The drier content and type should be varied in accordance with this factor.

44.7.2.5 Solvent or Diluent

These are added to the ink formula as a means of controlling the final viscosity or tack. This is necessary to allow for the various viscosity gradients that occur with different pigments. For example, a 20 per cent dispersion of Phthalocyanine Blue in a 40 Pa.s alkyd may produce a viscosity of 140 Pa.s, whereas a 20 per cent dispersion of a 4B calcium toner in the same alkyd may give a viscosity of 240 Pa.s. The diluent or solvent could be added to the 4B ink in the example to reduce its viscosity to 140 Pa.s with very little change to the colour strength. The percentage addition necessary to make this adjustment may be only 2–3 per cent of oil.

The oil added to any ink is usually of the same type as that used in the varnish system of that ink, and whether a solvent or a diluent is added generally depends on the particular formulator concerned. It may depend on circumstances known only to the ink chemist concerned.

Letterpress and offset inks contain the following 'reducers':

Mineral oils (low aromatic; BP 210–310° C)
Glycols (ethylene, propylene, etc.)—letterpress only
Low-viscosity vegetable oils (tung, linseed)

44.7.2.6 Additives

These consist of wetting agents, stabilising compounds, flow promoters, and emulsifiers, and are only added in quantities for special applications.

44.7.3 Examples of Letterpress Inks

44.7.3.1 Penetration Drying Types (such as 'news inks')

These are used to print from rotary letterpress machines running at more than 480 m/min. The output from one machine may be as high as 65 000 daily newspapers per hour. The paper must be very absorbent to soak up the ink. The viscosity is about 12–30 Pa.s and ink is soft and long flowing. See formula 44.18.

For use on a platen-type press, the carbon black is increased to about 22 per cent and the body is thus shortened. The viscosity is increased to approximately 100 Pa.s.

FORMULA 44.18
Formulation for a letterpress ink, penetrating type

	Per cent by mass
Pelletised carbon	12–15
Induline blue toner	1– 3
Asphaltum complexes	1– 3
Mineral oil (5 Pa.s)	60–70
Mineral oil (0.5 Pa.s)	10–15

44.7.3.2 Penetration/Oxidation Types

These are used to print better-quality papers where some degree of finish is expected. An ink drying by penetration only would smear easily; thus oxidation to a hard film is used to aid the drying properties. These inks are more expensive than the preceding type as they contain more toner and a percentage of drying vegetable oil. See formula 44.19.

44.7.3.3 Oxidation Types

These are used to print papers and stocks of poor absorbency where penetration types are far too slow. They are also used to print quality coated paper where gloss is of prime importance.

FORMULA 44.19
Formulation for a letterpress ink, penetration/oxidation type

	Per cent by mass
Carbon black	16–20
Reflex blue paste	5–10
Linseed stand oil	60–65
Mixed drier	3– 5

These inks contain not only vegetable oil but also hard resin. This is added to improve the gloss and hardness of final print. See formulas 44.20 and 44.21.

These inks are generally run on cylinder machines and are manufactured at a viscosity of about 120–140 Pa.s or a 'tack' of '7–9 units'. The flow is a compromise between the thin news inks and the short platen types discussed earlier.

FORMULAS 44.20 AND 44.21
Formulations for two letterpress inks, oxidation types

	Per cent by mass
Uncoated black	
Carbon black	18
Bronze blue	8
Modified phenolic varnish	54
Thin 'litho' varnish	18
Cobalt naphthenate (8%)	1
Manganese linoleate (6%)	1
Coated yellow (Process Yellow)	
Benzidine yellow	12
Modified phenolic varnish	64
Linseed alkyd (40 Pa.s)	15
Wax compound	4
Cobalt naphthenate 8%	1
Manganese naphthenate 6%	1
Mineral oil	3

44.7.3.4 Precipitation Types

Moisture-set (cylinder presses). These are used to print food wrappers, cartons and ice cream tubs. The stock is absorbent and contains a high level of moisture. This moisture causes a precipitation of the maleic adduct from the diethylene glycol solution, thus forming a solid. Drying takes place in about 10–15 minutes. See formula 44.22.

Oil-based (cylinder presses). Used to print quality process work on good coated papers. They are a combination of drying and setting types based on oxidation, and penetration. These inks contain a high percentage of a quick setting varnish as discussed earlier and a low percentage of oxidisable vegetable oil. The resin precipitates from solution as the solvent system penetrates the stock. See formula 44.23.

FORMULA 44.22
Formulation for a letterpress ink, moisture-set precipitation type

	Per cent by mass
Barium 2B Red	18
Maleic/glycol varnish	76
Diethylene glycol	6

FORMULA 44.23
Formulation for a letterpress ink, oil-based precipitation type

	Per cent by mass
Phthalocyanine blue	10
Quickest varnish	65
Linseed alkyd	10
Wax compound	6
Cobalt naphthenate	1
Manganese linoleate	2
Mineral oil (270° C)	6

44.7.4 Example of Lithographic Inks

When formulating an offset ink, it must always be remembered that the raw materials chosen should be resistant to water. Pigments must not bleed in mildly acidic water, and emulsification of the varnish system must not occur even under high shear.

The pigment strength, too, is most important. It must be remembered that this is a double transfer process where the ink is transferred from the plate to the blanket and then to the paper. This means that there is an extra split of the ink film weight as compared to letterpress. Thus offset ink must be 70–100 per cent stronger than a similar letterpress ink. Because of this, some pigments are difficult to use in offset systems as their colour strength or tint strength is inadequate. Such pigments are:

Ultramarine blue
Lake Red C
Iron blue

The first of these is seldom used. The other two are used in offset inks, but extra care must be taken with the formulation to ensure that the high pigment loading necessary does not adversely affect the varnish system. The major reason that they are used at all is their low cost and, in the case of Lake Red C, its clean bright colour.

44.7.4.1 Penetration Drying Types

Used mainly on web-fed machines for the printing of publication work (newspapers, magazines). The ink is similar to the letterpress news ink, with two exceptions: it does not contain news ink dye toner, and the oils are chosen more carefully to avoid emulsification.

These inks, like all other offset inks, are higher in viscosity than their letterpress counterparts. This is necessary to help promote water resistance and sharp definition in the printed job. Their viscosity is typically 50–100 Pa.s. See formula 44.24.

44.7.4.2 Penetration/Oxidation Types

These inks are the most common ink types for sheet-fed presses. They are printed on paper and board where gloss and scuff resistance are important factors. The setting rate is slow and several hours are usually necessary before the printed matter is tack free and able to be handled. Their viscosity is typically 160–260 Pa.s. See formula 44.25.

44.7.4.3 Oxidation Types

Used on less porous stocks than the former type (e.g. foil, plastic) but, because of the great advancements in penetration/oxidation inks, are of decreasing importance. Printed sheet-fed oxidation inks require great care in stacking the sheets to avoid set-off and sticking. Their viscosity is typically 160–260 Pa.s. Their formulation is as for penetration/oxidation types but replacing the quickset varnish with resin varnish.

FORMULA 44.24
Formulation for a lithographic ink, penetration drying type

	Per cent by mass
Carbon black	18–25
Asphaltum complexes	4–8
Bodied linseed oil	10–20
Mineral oil (5 Pa.s)	40–65
Reflex toner	2–4

FORMULA 44.25
Formulation for a lithographic ink, penetration/oxidation type

	Per cent by mass
Rubine 4B (Calcium)	16–20
Resin varnish	30–50
Quickset varnish	20–30
Wax compound	4–10
Driers (Co/Mn type)	2–3
Mineral oil (270° C)	4–8

44.7.4.4 Precipitation Types

For obvious reasons there is no moisture-set type, but there is an oil-based variety. The offset ink follows closely the letterpress precipitation type but with the usual increase in colour strength. This ink is fast becoming the most important sheet-fed lithographic ink. There are several reasons for this:

(a) fast setting speed;
(b) good working properties (high water resistance);
(c) moderate to good gloss;
(d) moderate to good scuff resistance;
(e) all-round suitability (general-purpose ink);
(f) and because of the above, fast turn-around of production runs.

They are used only on sheet fed machines. Their viscosity is typically 160–260 Pa.s. See formula 44.26.

44.7.4.5 Heatset Type

This ink is used only on reel-fed or web offset machines where as many as ten different colours may be printed by the one machine. The web is seldom rewound but is generally cut and folded as the web leaves the machine, all in a continuous operation.

The ink must be dry and scuff resistant immediately it leaves the drying system of the press. This may be a steam-heated drum over which the web passes, or a gas oven. The ink therefore is formulated to dry almost entirely by evaporation in a similar manner to flexographic and gravure inks.

Used to print coated and uncoated papers at speeds of up to 270 m/min. Their viscosity is typically 60–120 Pa.s. See formula 44.27.

FORMULA 44.26
Formulation for a lithographic ink, precipitation type

	Per cent by mass
Phthalocyanine blue	14–18
Quickset varnish	45–70
Linseed alkyd	8–18
Wax compound	4–8
Driers (Co/Mn type)	1–3
Mineral oil (270° C)	4–8

FORMULA 44.27
Formulation for a lithographic ink, heatset type

	Per cent by mass
Benzidine yellow	10–12
Heatset varnish	70–89
Wax compound	6–14

This type of formulation usually requires a two-part grind. This means that some of the varnish is left out of the initial weigh-up of the batch of ink, and the ink ground as a concentrate. The remainder of the varnish is added when the desired grind reading is achieved, and the ink again passed over the mill. The varnish retained for the post grind with addition may be as much as 50 per cent of the total. The advantages of this system are:

(a) small volume of ink to grind, thus cutting down on mill time;
(b) increased gloss due to addition of varnish after pigment has been dispersed;
(c) better pigment dispersion owing to higher viscosity of mill base.

It might be added here that many inks are manufactured this way when gloss is of prime importance included; in this category are all of the offset inks other than the penetration drying types, and similar letterpress categories.

44.7.4.6 Metal Decorating

These inks are basically of an oxidation type designed for printing by offset lithography sheet metal, tin plate, aluminium or blackplate. They differ from the oxidation inks discussed earlier in that they are of much higher viscosity and lower drier content and are designed to be dried in an oven at elevated temperatures. The temperature of the oven can be in the range 140–200° C. The pigments used in these inks must be carefully selected to be suitable for the litho process and, secondly, stable to heat. Some Benzidine yellows, Lake Red, PMTA's, etc., are unsatisfactory because of their instability at high temperatures.

These inks are used to print soft drink cans, biscuit tins, food tins, 5 to 25 litre drums, cosmetic and aerosol cans. Typical viscosities are 400–800 Pa.s. See formulas 44.28 and 44.29.

FORMULAS 44.28 AND 44.29
Formulations for two metal-decorating offset inks

	Per cent by mass		*Per cent by mass*
Example 1		*Example 2*	
Phthalocyanine green	18–24	Calcium 2B	15–30
Metal decorating varnish	60–70	Metal decorating varnish	40–60
Wax compound	0–4	Safflower alkyd	10–20
Manganese linoleate (6% Mn)	1–2	Wax compound	0–2
Mineral oil (245° C)	0–10	Manganese lineolate (6% Mn)	1–2
		Mineral oil (245° C)	0–8

The second example is included to show that, when high pigment loadings are encountered, an alkyd resin is necessary to impart a degree of flow to the ink. It also aids in ink transfer and smooth printing on the printed sheet.

45 COMPONENTS AND TECHNICAL ASPECTS OF A PAINT-TINTING SYSTEM

45.1 INTRODUCTION

What is a tinting system and why is it such an important part of a company's product range in the decorative market?

A tinting system is a means of providing colour, by combining a range of universal tinters with a range of tinting bases by means of tinting machines. Its importance lies in the fact that tinting systems supply the bulk of the Australian decorative market with the colours the market has come to expect. The very livelihood of manufacturers in the Australian paint market depends on their ability to use their tinting system to its maximum advantage, giving a wide choice of colour and at the same time minimising stock items for distributors and dealers.

The definition of a tinting system as described above mentions both universal tinters and tinting bases. Universal tinters can be incorporated into emulsion- and alkyd-based paints. They are basically coloured pigments dispersed in a glycol/surfactant blend and require certain basic properties:

(a) good acceptance in emulsion- and alkyd-based paints;
(b) a certain degree of fluidity to enable their use in tinting machines;
(c) a minimal effect on the basic properties of the paints into which they are incorporated;
(d) no appreciable tendency to dry out in the tinting machines;
(e) no appreciable tendency to settle out on storage.

Tinting bases are the base paints and as such contain the resins, solvents, driers and other additives that determine the characteristics of the paint. Tinting bases usually contain the bulk of the prime pigments which impart the covering properties to the dry, applied paint film.

45.2 MANUFACTURE OF TINTERS

The manufacture of tinters, once a suitable formulation has been established, is not a particularly difficult task. However, there are one or two points to consider because tinters are, by nature, dispersions of pigments at close to maximum pigment volume concentration. Firstly, concerning equipment, there are three main milling machines that are in common use in the production of tinters. These are:

(a) the family of mills typified by the triple-roll mill;

(b) the sand mill; and
(c) the cavitation mill (and others in this family).

A fourth category, that of the speed bead mill, is really a combination of the sand mill and the cavitation mill.

The triple-roll mill is the traditional mill for the manufacture of tinters, mainly because it lends itself to the handling of highly viscous pastes. This mill requires the pugging of the ingredients, prior to passing through the mill. A number of passes is usually required to achieve a satisfactory degree of dispersion.

The sand mill has gained in popularity in recent years, as it is amenable to continuous production. Typically two heavy-duty tanks positioned side by side, with a portable sand mill between them, is the usual configuration. The tinter is mixed in one tank, and then pumped through the sand mill into the other tank. Alternatively, a single tank can be used with the material recirculating after passing through the sand mill.

These two mills are, in general, suitable for a full range of tinters. The cavitation mill is suitable for only a limited number of tinters, such as yellow ochre, white and possibly red oxide. However, linked with a sand mill, the cavitation mill can be extended to cover all tinters. This configuration would use the cavitation mill in place of the first tank in the previous discussion on the sand mill. The material, having spent, say, thirty minutes in the disperser, would be discharged through the sand mill into a reducing tank.

Other mills that can be used (but are not in common use) might include steel ball mills, attritor mills and sigma-blade mixers. The particular configuration that a producer will use depends primarily on the equipment already in use. It is then the task of the surface coatings technologist to adapt the formulation to enable its use on available equipment.

A potential problem is that the fineness of dispersion achieved by a given mill configuration may not result in full colour development of the pigment used. Pigments of which this is especially true include organic oranges and violets. With these pigments the fineness of dispersion may meet specification, yet the pigment may not have reached its true strength or tone. Such pigments require special care in their conversion from dry powder to liquid tinter. If poor development results on milling, it is usually a sign that the formulation is sub-optimal, and re-evaluation of the amount or type of surfactant may be required.

The temperature achieved in the mill during the dispersion process can also cause problems. Two examples are excessive thickening of the tinter because of chemical changes in either the pigment or the surfactants used, or loss of strength and alteration of tone, particularly with Hansa Yellow. Mills are usually water-cooled, but rarely is this sufficient to maintain a suitable temperature in Australian conditions. Some overseas companies have overcome this by installing refrigeration plants to cool the cooling water, before pumping it through the water-jackets of their mills. Other ways of controlling heat build up are: reduction of the dispersion time, provided that the pigment is still developed; monitoring the temperature; and switching off the mill to allow time to cool down. This is unlikely to be popular with production personnel, but may well be preferred to the production of unusable material.

Once the batch has been tested and adjusted to the satisfaction of the quality-control officer, it is filled through a de-aerator, usually a thin film, rotating drum under vacuum, into the nominated containers. A check should be made to ensure that the level of aeration of the filled material does not exceed a nominated level, otherwise a further inaccuracy will be built into the tinting system, as it is usual to measure by volume. A typical maximum aeration level is 3 per cent by volume.

45.3 STANDARDISATION OF TINTERS

A range of tinters would of course vary from company to company, in terms of the number of tinters, their colours, their tones and their strengths. A typical system would include some or all of the following:

Yellow (two types)	Orange
Red (two types)	Yellow ochre
Blue	Red oxide
Green	Chromium oxide green
Black	Umber
White	Brown
Violet	

Where two types are indicated, it means that there are available a lower-cost, relatively low light-fast version, and a higher-cost high light-fast version.

No matter how many tinters, or what shades are chosen, all will require careful standarisation before they can be used to produce colours that will be reproducible. Because tinters are designed to be used with machines that deliver on a volume basis, it is obvious that they must be standarised by volume, not by mass. Consequently, the first requirement in standarising a tinter is to accurately determine its density. Most tinters aerate during the manufacture, and have to be de-aerated before filling. If, as is the usual case, de-aeration occurs immediately before filling, and after adjustments have been made to the batch, the material on which strength testing is done will contain between 5 and 10 per cent air. To obtain a correct measure of the density and to measure the amount of air in the batch, a pressure density cup—for example, the ICI Pressurised Weight per Gallon Cup—should be used. This gives what is known as the 'true' density. The nominal density is taken by using a conventional density cup, and the two results used to calculate the percentage of air. The true density is always used for strength adjustment.

Standardisation then proceeds by adding a known volume of tinter, weighed accurately on a balance, into a known mass of base paint (standard light tint base), and this is compared against the standard for that tinter, prepared in the same manner. The comparison can be visual, and/or instrumental. If the colour of the prepared paint is deeper than its standard, then the batch of tinter is too strong and additions of either glycol or extender base are required. The choice between glycol or extender base is made on the basis of the required consistency or viscosity of the tinter. This process is repeated until a satisfactory result is achieved.

45.4 STANDARDISATION OF THE BASES

As with the range of tinters, a particular company will select from a range of bases those that best meet that company's requirements, both in terms of total colour offer, and also in terms of stockholding required at the retail level. Bases can be divided for standardisation purposes into two groups: single-pigment bases, and multiple-pigment bases.

Single-pigment bases commonly in use include:

Base	*Pigment*
Light tint	0.30 kg/L titanium dioxide
Mid tint	0.20 kg/L titanium dioxide
Deep tint	0.10 kg/L titanium dioxide
Neutral	No prime pigment
Red	Organic red
Blue	Phthalocyanine blue

Brown	Iron oxide brown
Ochre	Iron oxide yellow
Red oxide	Iron oxide red

Multiple-pigment bases, generally a coloured pigment with titanium dioxide, include those listed below. It is apparent that there is some overlap with single bases, a company choosing only one of the alternatives.

Base	*Titanium Dioxide Level*	*Other Pigment*
Blue	0.10 kg/L	Phthalocyanine blue
Brown	0.10 kg/L	Iron oxide brown
Ochre	0.10 kg/L	Iron oxide yellow
Yellow	0.10 kg/L	Organic yellow

The reason for having these multiple-pigment bases is that they provide a means of achieving colours that will be more opaque than the corresponding single-base colours. Generally, however, they are not as useful as their single-pigment counterparts as they cannot produce clean bright colours, always having a milkiness; and they are also much more difficult to standardise. For these reasons they are becoming less popular with paint manufacturers.

To standardise a single-pigment base, a known volume of the batch under test is weighed accurately and added to a set mass of tinter, either black or blue. The same procedure is repeated for the standard and the two samples compared to determine if adjustments are required.

To standardise a multiple-pigment base, the balance of pigments is checked to ascertain if correct. This is done by comparing the untinted base with its standard. If the balance of, for example, blue and white pigments is not correct, the batch will show up as either too blue or too white. Factory shaders are then required to restore the balance before any testing of strength can be done. Failure to approach the testing in this manner will inevitably lead to a situation where the person doing the testing does not know whether the batch is at the correct strength or not, because the tone of the prepared test colour will not match the tone of the standard test colour.

Testing of white bases is usually done with either a blue or black tinter, but the coloured bases require careful selection of an appropriate tinter. It is possible to use black or blue for all coloured bases other than blue bases, where experience shows that red is a better choice.

45.5 TESTING OF COMPATIBILITY OF TINTERS AND BASES

Because tinters are designed to be accepted by both alkyd- and emulsion-based paints and because the requirements of these two systems are quite different, it should be recognised that any particular formulation for a tinter is bound to be a compromise. The art of formulation lies in knowing firstly how to adjust the properties of a tinter to change the balance of properties, and secondly where to set that balance. There are two main tests that can assist a surface coatings technologist in determining whether the balance of properties of any system of tinters and bases is technically sound.

The first of these is an acceptance test. This is performed by adding a known quantity of tinter to a known quantity of the base, stirring for a set time (say, 2 minutes) and then preparing a panel about 100 × 150 mm, brushing with normal shear along the length of the panel. The brush is then wiped to remove excess paint, and one end of the panel is over-brushed vigorously. The panel is then spotted with a drop of the paint, and allowed to dry in a vertical position; this allows the drop of paint to run down the face of the panel. By comparing the three areas of the panel, it is possible to rate that combination of tinter and base for compatibility. Where all three areas of the panel are the same colour, the compatibility is excellent. Where there is a marked difference in colour between the three areas, this is an indication that, depending largely on the severity of the colour difference, there may be problems with this combination of tinter and base.

The second test uses the same paint as the first. A panel is brushed out within 5 minutes of the tinter being added to the paint. A second brush-out is prepared after a period of 24 hours, and the two are compared. An increase in the depth of the colour indicates that the tinter takes time to wet up in that paint. Similarly, a decrease in the depth of colour, particularly in a white base, could mean that the base itself has wet-up, due perhaps to the increased surfactant level from the tinter addition. It could also mean that the coloured pigment has flocculated. In either case, there is an indication that the tinter and base are not fully compatible.

The results of these tests should be compared with similar tests performed on a commercially available system, so as to build up the background knowledge necessary to be able to use these tests to evaluate a new system.

45.6 THE EFFECTS OF TINTERS ON PAINT

Tinters contain large concentrations of both glycols and surfactants. Both of these can have an adverse effect on the properties of the paints into which the tinters are incorporated. It is necessary to be aware of these effects, and to test for them, in order to be able to exercise control over the basic properties of a tinting system. Particular problem areas commonly experienced include the following:

(a) retardation of the drying times of alkyd paints;
(b) increased water sensitivity of the paint film;
(c) change in the rheology of the paint; and
(d) change in the effect of weathering on the paint film.

45.6.1 Drying

Retardation of the drying of alkyd paints can be a serious technical and commercial problem. After the basic acceptance tests, this test is the most important in evaluating a tinting system. If the level of retardation exceeds four hours at 10° C, 50 per cent relative humidity, then there may well be cause for concern. So far it has not proved possible to eliminate all retardation of drying. For this reason (as well as for other reasons), there has been a tendency for levels of tinter addition to paints to be kept below about 8 per cent by volume, with some companies restricting to about 6 per cent.

45.6.2 Water Sensitivity

The addition of extra surfactant can have a marked effect on the dry paint film. This is most often experienced with latex paints, but some instances have occured with alkyd paints. The practical effect of this is to reduce the wet-adhesion of the film, increase the incidence of water-spotting (the gloss increase and/or colour change that occurs on some fresh paintwork after a shower), and reduce the strength of the film. Water sensitivity can be improved by reducing the surfactant content of the base paint, or by the use of surfactants that are more water resistant.

45.6.3 Change in Rheology

Different types of tinters can have different effects on the rheology of the tinted paint. It should not be assumed, therefore, that adjusting the rheology of an untinted base paint will be sufficient to achieve good application properties of the tinted paint. There is little that can be done to correct this, except to test the particular product with a normal level of tinter added, and make any corrections that are necessary to provide a compromise between the application properties of the paint at different tinter levels.

45.6.4 Weathering

In general the weathering properties of tinted colours will be equal to those of ready-mixed colours. Possible exceptions are pale tints using the low light-fastness tinters mentioned earlier in this chapter. However, certain combinations of tinters will be inherently better than others, and this should be taken into account in colour designing and development of mixing instructions for the colours. For example, the addition of a white tinter to a red base may adversely affect the weathering properties of that base.

It can be seen that a tinting system is a very complex affair, with a number of compromises. It is therefore important for adequate testing to be carried out before implementing any new system or making any changes to an existing system. The exposure results in table 45.1 indicate typical performance.

TABLE 45.1
Exposure results for five ready-mix paints and five tinted paints

Length of Expoure: 12 months *System:* 1 coat primer, 1 coat undercoat, 2 coats finish
Situation: 45°N Queensland

Colour	*Tint*	*Ready-mix*	*Chalk rating**	*General appearance*	*Gloss washed**
Blue Water		√	9	7	30
BS 4800—18E53	√		9	8	26
Charcoal		√	7	7	10
BS 4800—OOA13	√		10	8	9
Mid Green		√	8	9	7
BS 4800—16D45	√		8	9	10
Bright Red		√	8	10	7
Bright Red	√		7	9	8
Brown		√	8	9	19
BS 4800—08B29	√		10	8	8

* Chalk ratings: 0 (worst) to 10 (best)
Gloss: 60°, per cent

45.7 TINTING MACHINES AND OTHER EQUIPMENT REQUIRED

Tinting machines form the third aspect of a tinting system, and they are just as important as the other two, for they form the link between tinters and bases. Tinting machines can be divided into three groups: manual, semi-automatic, and automatic.

Manual machines, such as the Australian produced Strazdins, contain 10–12 canisters, fitted with a dispensing pump and either a manual or electric stirring mechanism. The pump consists of a cylinder, usually of stainless steel but sometimes of plastic, and a plunger. Attached to the plunger is a graduated scale of notches. The scale is divided into fluid ounces (U.S.) (designated 1Y, 2Y, etc.) and notches. There are two common graduations in use in Australia, the 64 notch = 1Y, and the 48 notch = 1Y. (1Y = 1 U.S. fluid ounce = 29.57 mL). The 48 notch machines often have the capability of dispensing a half notch, bringing their smallest dispensing unit down to $\frac{1}{96}$ of a Y. Several alternative values for the 'Y' unit are used.

The tinter is dispensed by setting the gauge at the nominated reading (according to the mixing instructions and the size of the container), opening the dispensing valve, and pressing down on the plunger. Some of the larger manual machines are able to have the dispensing level pre-set. This allows rapid multiple tinting of the same colour (and size), and such a machine is extremely useful in major warehouse tinting operations and in the larger stores that cater for the master painter.

Semi-automatic machines are similar to the machines described above but are powered typically by compressed air, and dispense the pre-selected quantity of tinter at the press of a button.

Automatic machines contain a microprocessor, which enables the machines to be interfaced with a computer, in which would be stored the mixing instructions for a total colour range. Simply entering the code for the colour is sufficient. The machine then determines the size of the paint can underneath the dispensing orifice, and delivers the correct tinters and quantities. Alternatively, the mixing instructions can be stored in the form of printed code bars, which are then read into the machine by means of a light pen. In this form, the machine need not be interfaced with a computer keyboard.

The semi-automatic and automatic machines are again best suited for multiple tints of the same colour in warehouses and depots.

Some types of machines are more accurate than others. Obviously, if a machine is to dispense the volume of tinter required, the pump calibration is important; so also is the rigidity of the pump assembly. For these reasons, machines with stainless steel pump assemblies are generally more accurate than their plastic counterparts. Accuracy also decreases as the size of the addition decreases. Whether these factors represent a serious commercial problem depends on variables such as the customers' colour match requirements, the strengths of the tinters used and the size of the paint can being tinted.

Some tinting machines have a tendency to allow the tinters to dry out on the nozzle of the dispensing pump assembly. This is, of course, also influenced by the formulation of the tinter. Machines that are used frequently, and are properly maintained, are unlikely to run into serious problems. Where machines are neglected and left unused for long periods of time, problems of blockage of the nozzle and seizure of the pump assembly become more common.

45.7.1 Other Equipment

Paint shakers are standard equipment in all but the smallest hardware store, and provide a degree of confidence that the paint and tinter will be properly mixed. Other equipment which is gaining acceptance, particularly with 20 L pails, is a hole cutter. This cuts a hole in the pail lid, through which the tinter can be added. The pail is sealed with a plastic stopper, the paint shaken, and handed to the customer. This saves the labour cost of opening a pail, possibly providing a new lid, and resealing.

45.8 REPRODUCIBILITY OF COLOURS

Some mention has been made in the previous section, of the effect of the tinting machine on the accuracy of the volume of tinter dispensed into the paint. Reproducibility of colour depends on all three elements of the tinting system: tinters, bases and tinting machine. Variation from batch to batch in both tinters and bases, coupled with variations in the volume of tinter added by the machine, must result in variations in the colours produced. It is therefore important to reduce variation in each of the components to a minimum, keeping in mind the cost of such stringent quality control. It is the marketing department's ultimate responsibility to weigh the factors and decide whether a given level of quality control is sufficient.

Keeping in mind the possible causes of variation, it becomes obvious why it is advisable to box tinted colours together before commencing painting, or at least start a new can of paint at a natural join, such as the corner of a room. Rarely will two tints be sufficiently close to be joined invisibly in the middle of a wall.

In addition there is the problem of line-to-line matching. Differences in gloss level, plus differences in the background tone produced by the different vehicles used in paint, introduce to a tinting system a further area of colour drift—or rather, colour difference. One way to minimise this is to take a particular line of paint and make the bases in that line act as the master standards for all other lines. For examples, a company might choose their premium semi-gloss latex paint as this master standard; all other lines are then referred back to this standard, when standards are to be issued.

45.9 CONCLUSION

Tinting systems are the life-blood of the decorative paint market in Australia. They are complex systems, being comprised of three distinct components—tinters, bases and tinting machine—and require a high degree of care in their design and implementation. Each one of the components introduces variations into the system; these variations can be controlled by careful quality assurance but cannot be eliminated entirely.

[*Editor's Note*: This chapter does not include formulations for colorants. Composition details are specific and confidential to each individual manufacturer.]

46 COLOUR MATCHING (USING COMPUTERISED TECHNIQUES)

[*Editors Note:* This chapter is an overview of a subject of considerable complexity. It is presented to give some insight into the subject; references have been included.—*JMW*]

46.1 INTRODUCTION

Colour measurement was introduced by the CIE more than fifty years ago, and slowly over the years it has been adopted by paint manufacturers for use at two quite separate stages during manufacture. The first stage is when a formulation is being devised for the first time, and this technique only became practicable with the development of the computer—first the analog computer (1958–65), which began to be superseded by the digital computer in 1963. The second stage is when a batch of paint has been produced, either by traditional methods or with the aid of computer match prediction, and it is necessary to decide whether the batch is a 'good commercial match' to standard. Instruments for this purpose were first marketed in the late 1940s and became very much more efficient in the 1970s when they were interfaced first to digital computers and later to microprocessors. Chapter 56 provides a review of the use of computers in surface coatings' applications.

46.2 COMPUTER MATCH PREDICTION

The colour of any paint panel is completely specified by a set of 16 *reflectance* factors, each of which is the amount of light at a specific wavelength reflected by the panel expressed as a fraction of that reflected by a perfect white, called the 'perfect reflecting diffuser'. Spectral reflectance factors are measured on an instrument called a *spectrophotometer*. If increasing amounts of a tinter are added to a white pigment, a series of reflectance curves can be obtained, the height of the curve changing at different rates at different wavelengths. Computer match prediction requires a mathematical function which will vary systematically, preferably linearly, with the concentration of the tinter; and when two or more tinters are present, the functions must be additive. Once these functions have been determined for all tinters including any black pigment, it then merely requires the solution of a number of simultaneous equations at each of sixteen wavelengths to obtain the same function as that of the colour to be matched.

The function most widely used in match prediction is that devised by Kubelka and Munk. It relates the reflectance at each wavelength to coefficients of absorption, K, and scatter, S. If pastel shades only are being manufactured, then the ratio K/S is adequate. This is related to reflectance by

$$K/S = \frac{(1 - r_\infty)^2}{2r_\infty}$$

where r_∞ is the reflectance of a completely opaque film and termed the reflectivity. If complete opacity cannot be achieved, r_∞ can be determined by applying a paint film of normal thickness to both a white and a black card and measuring the two reflectances. The reflectivity is then given by

$$r_\infty = a - (a^2 - 1)^{0 \cdot 5}$$

where

$$a = 0.5\left[r_W + \frac{r_B - r_W + r_G}{r_B r_G}\right]$$

r_W and r_B being the measured reflectance factors over the white and black backgrounds, respectively, and r_G the reflectance of the white background at each wavelength.

This simple function works well for paints when the content of white pigment is at least five times that of the other pigments present, as this ensures that the bulk of the scattering is due to the white pigment and therefore independent of the concentration of the other pigments which governs the co-efficient of absorption, K. The K/S function is not always linearly related to concentration, but several mathematical techniques such as *interpolation* and *polynomial curve fitting* can be used to overcome this.

The usual range of shades produced by a paint manufacturer, however, contains many in which the concentration of tinters is high enough for their scattering coefficients to be significant, and this requires a much more sophisticated set of equations known as the 'two-constant theory' which was not developed until the mid 1960s.

The first and probably the most important stage, if computer match prediction is to be reliable, is the preparation of calibration samples consisting of, for example,

(a) masstones of each white and coloured pigment;
(b) reductions of each coloured pigment with white in two ratios—for example, 10 : 90 and 20 : 80;
(c) reductions of each bright yellow, orange and red pigment with black in two ratios—for example, 98 : 2 and 99 : 1.

The reflectance factors at 400, 420, . . . , 680, 700 nm are then determined and stored in the computer. From such data, relative K and S constants can be calculated by different methods which have been published.[1–6]

Once the calibration data have been entered into the computer, match prediction is carried out by measuring the reflectance values of the target shade on the same spectrophotometer and under the same conditions as were used to measure the calibration samples. The pigments should then be specified to avoid the computer working systematically through every possible combination (thus wasting time trying to match a red shade with a white, two yellows, a blue and a black). What the match prediction program does is to vary systematically (iteration) the concentrations of each pigment and calculate the K and S values of each mixture until the values are as close as possible to those of the target shade. If this has been formulated with the chosen pigments, very similar values will be obtained. If the pigments are different, this will not be possible, although a near-perfect match can still be obtained. This will be *metameric* and it is achieved by an iterative algorithm aiming at identical *tristimulus values* under standard daylight (Illuminant D65). The degree of metamerism given by different formulations can also be calculated to ensure that it is not serious.

When the formulation is printed out, a small batch of the paint is prepared in the laboratory and applied in the normal manner. After drying it is compared visually or instrumentally (see Section 46.3) with the standard. If it is not a good commercial match, it is measured on the spectrophotometer, and from the reflectance factors a correction program will print out pigment addition details necessary to bring it on shade.

Some detailed match prediction programs using the two-constant theory have been published, and it would therefore be possible for a paint manufacturer to buy a spectrophotometer and for the staff to write a match prediction program for use in the company's general-purpose computer (or else by time-sharing a computer). Such an approach today would rarely be in the best interests of the manufacturer; a far better solution is to purchase or lease a complete system from one of the handful of companies that have specialised in this field and who can refer the potential purchaser to successful users of their systems. The match prediction programs used in some of these systems are written in a high-level language which makes it feasible for the user to modify the program if at any time this would seem to be desirable; in such cases, instructions on programming form part of the training course. Such computers invariably have spare capacity over that necessary for match prediction and other colorimetric calculations. This spare capacity can be used for providing varied management information, and additional terminals can be installed which give instant access even if the computer is controlling the spectrophotometer at the same time. The computer can simultaneously operate an automatic weighing machine.

The major feature of these systems is that the spectrophotometer is interfaced to the computer, making it unnecessary to enter data manually or via punched cards or tape. This results in a considerable saving in time in addition to that resulting from improvements in spectrophotometers: whereas in 1975 it would have taken at least half a minute to make a measurement, today it takes only a few seconds with no loss in either accuracy or precision but at a much lower cost. The calibration stage has also been considerably simplified: instead of having to adjust manually at each of sixteen wavelengths to obtain the absolute reflectances of white and black calibration tiles, interfacing requires only a measurement of each, every subsequent measurement being automatically corrected by the computer.

The savings arising once a system is installed are by no means confined to saving time in laboratory matchings of new shades. It is obvious that there are many different formulations giving a technically satisfactory match to most shades: only one, however, will be the cheapest, and the time required to discover this by traditional colour matching is so great that it is rarely practised. By incorporating pigment costs into a system, the cheapest formulation can be rapidly predicted and overall savings in pigment costs of at least 10 per cent can be virtually guaranteed.

Additional savings can also be achieved if the program includes an optimum pigment loading facility, because any excess over the minimum needed to give the necessary hiding power unnecessarily increases manufacturing costs. The problem is to predict the effect of adding a clear resin to a given paint, and several different solutions have been used in match prediction programs. The one that gives the best results and is the most flexible is based on the classical Kubelka and Munk analysis applied to films of known thickness over both a black and a white backing, namely:

$$R_B = 1/(a + b\cdot \mathrm{ctgh}\, b\cdot S)\lambda$$

$$R_W = [1 - R_G(a - b\cdot \mathrm{ctgh}\, b\cdot S)]/(a - R_G + b\cdot \mathrm{ctgh}\, b\cdot S)\lambda$$

where
R_B = reflectance over black
R_W = reflectance over white
S = scatter at specified thickness
a = (S + K)/S, where K = absorption at specified thickness
$b = (a^2 - 1)^{0.5}$
ctgh b · S = hyperbolic cotangent of b · S
R_G = reflectance of white backing

The values of K and S in the above equations are determined from the concentrations of the pigments combined with resin.

$$K = c_1 \cdot K_1 + c_2 \cdot K_2 + c_3 \cdot K_3 + c_W \cdot K_W + c_R \cdot K_R \lambda$$

$$S = c_1 \cdot S_1 + c_2 \cdot S_2 + c_3 \cdot S_3 + c_W \cdot S_W + c_R \cdot S_R \lambda$$

where $c_1 + c_2 + c_3 + c_W + c_R = 1$;
c_1, c_2, c_3, c_W & c_R denote the concentrations of colours, white and resin respectively;
and λ = discrete wavelength (10–15 nm bandwidth)

From these calculated reflectance values, the CIE tristimulus value Y (illuminant D65, 10° field) for both the black and the white backing can be calculated.

A rough approximation for hiding is given by the contrast ratio calculation:

Hiding when $\quad$ Y black/Y white $\geqslant 0.98$

An even more reliable method is to calculate X and Z as well as Y and then to apply a colour difference equation to the values over white and over black, preferably the optimised JPC79 equation mentioned in the next section. The numerical value of the difference which is just below the threshold of perceptibility is then used in the calculation of optimum pigment loading. Complex iterative routines are used to optimise the pigment concentrations relative to the resin addition to give any desired contrast ratio in seconds.

The values of K and S for each pigment and for the resin must be calculated in absolute terms for the film thickness under consideration. The usual expedient of calculating relative to a nominal scatter of 1 for the white paint at each wavelength in opaque films cannot be used. Experimental difficulties are encountered in the determination of these Kubelka and Munk constants, and both inverse hyperbolic cotangent equations and indirect methods must be used.

46.3 COLOUR DIFFERENCE MEASUREMENT

To quantify the colour difference between a specimen representing a batch of paint and the standard, it is first necessary to determine the XYZ tristimulus values of each. The ideal instrument for this purpose is a spectrophotometer, and the measurements of the reflectance factors at a minimum of 16 wavelengths between 400 and 700 nm should be made with the specular component excluded because in visual colour matching the specimens are always tilted to prevent the specular component entering the eye. The tristimulus values should then be calculated using the colour matching functions of the CIE 1964 supplementary standard colorimetric observer (often called the '10° observer') and the spectral power distribution of Illuminant D65.

Tristimulus values can also be determined with sufficient accuracy using much simpler and cheaper instruments called tristimulus colorimeters. Instead of measuring reflectance values at sixteen or more wavelengths, they measure reflectances at three, sometimes four, broad bands, the combination of the spectral power distribution of the light source, the transmission characteristics of the filters and the response of the photon counter being carefully chosen to give XYZ values for Illuminant D65 directly.

These two sets of tristimulus values are the starting point for calculating the colour difference between batch and standard and, until recently, the potential user was faced with a bewildering choice of colour difference formulae: more than twenty had been developed since the 1940s, and as many as 13 were in routine use. In 1976, however, the CIE published details of a colour difference formula called CIELAB, which is steadily becoming the preferred choice for every industry concerned with the colour of opaque objects. This is becoming adopted not merely because it has been recommended by the international authority, the CIE; their previous recommendation, CIE 1964 (U*V*W*), was almost completely ignored by industrial users. Its success is due to two factors which have been fully documented.

The first factor was that at the time of its publication, 1976, there was no colour difference equation that gave a higher correlation with the visual judgements of one hundred professional shade passers who made 55 000 assessments against standards of 240 different colours. Although most of these were of dyed textiles, there is no reason to suppose that the criterion of acceptability would have been any different from that of the average paint colourist.

The second factor was that the CIELAB formula permits any colour difference to be split into the three components of colour perception. This factor has turned out to be much more important than the correlation with visual pass/fail decisions, because even the most reliable formula does not permit a single value of ΔE to be chosen according to the closeness of match required. It is therefore necessary to establish colour-specific values of ΔE for a given end use, and these will vary by a factor of at least 3 : 1. Moreover, for a given colour, the acceptable tolerance if the difference is mainly one of hue is invariably less than if it is of the other two variables of perceived colour differences—lightness and chroma (saturation or purity).

The CIELAB formula uses the XYZ values of a specimen and those of the illuminant, $X_nY_nZ_n$, and converts them into L*a*b*, the Cartesian coordinates of a more uniform colour space, namely:

$$L^* = 116(Y/Y_n)^{1/3} - 16$$

when Y/Y_n is greater than 0.008856. Otherwise,

$$L^* = 903.3(Y/Y_n)$$

$$a^* = 500 \times [f(X/X_n) - f(Y/Y_n)]$$

$$b^* = 200 \times [f(Y/Y_n) - f(Z/Z_n)]$$

where $f(X/X_n) = (X/X_n)^{1/3}$ if $X/X_n > 0.008856$,
and $f(X/X_n) = 7.787(X/X_n) + 16/116$ if $X/X_n \leqslant 0.008856$,
with corresponding terms for Y/Y_n and Z/Z_n.

The colour difference between two specimens is then determined from the differences in L*, a* and b* by

$$\Delta E = [(\Delta L^*)^2 + (\Delta a^*)^2 + (\Delta b^*)^2]^{0.5}$$

In order to resolve any colour difference, ΔE, into the components of the visual perception of colour differences, the Cartesian coordinates of L*a*b* colour space are converted into the cylindrical coordinates L*C*h by

$$C^* = [(a^*)^2 + (b^*)^2]^{0.5}$$
$$h = \arctan b^*/a^*$$

the configuration of CIELAB space being as in figure 46.1.

This has the same configuration as that of Munsell space, which was devised purely on the basis of visual perception.

L* is a measure of lightness expressed on a 0–100 scale. If a batch has a higher L* value than standard, it is described as *lighter*; if its L* value is lower than standard, it is described as *darker*.

C* is a measure of chroma. If a batch has a higher C* value than standard, it is described as *stronger*; if its C* value is lower than standard, it is described as *weaker*.

L* and C* therefore quantify in CIELAB units two of the three variables of perceived colour difference. The third variable, the difference in hue, is quantified in the same units by

$$H^* = [(\Delta E)^2 - (\Delta L^*)^2 - (\Delta C^*)^2]^{0 \cdot 5}$$

The nature of the hue difference is deduced by rotating the chroma axis passing through the position of standard until it passes through the position of batch—the direction of rotation being such as to minimise the rotation necessary. This direction (clockwise or anti-clockwise) is noted. The axis is again rotated in the same direction from its position passing through standard until it cuts the first and second a* and b* axes. The nature of the hue difference is then given unambiguously by the order in which this occurs, as shown in the list below.

First	*Second*	*Hue difference*
Clockwise		
$(b^*)^-$	$(a^*)^-$	bluer (greener)
$(a^*)^-$	$(b^*)^+$	greener (yellower)
$(b^*)^+$	$(a^*)^+$	yellower (redder)
$(a^*)^+$	$(b^*)^-$	redder (bluer)
Anti-clockwise		
$(b^*)^+$	$(a^*)^-$	yellower (greener)
$(a^*)^-$	$(b^*)^-$	greener (bluer)
$(b^*)^-$	$(a^*)^+$	bluer (redder)
$(a^*)^+$	$(b^*)^+$	redder (yellower)

The first term will usually indicate the nature of the hue difference: if it is meaningless—for example, a batch is indicated as being redder than a red standard—then the second term in parenthesis will always provide the correct hue difference description.

Professional paint colourists often use a term to describe the visual effect of adding a trace of black to a coloured paint, although there is no general agreement as to what this term should be—*dirtier*, *duller* and *greyer* being the most common. The opposite effect is usually described as *cleaner*. The corresponding changes in CIELAB co-ordinates are as follows:

Dirtier, etc.—a decrease in both lightness and chroma to a similar degree.

Cleaner—an increase in both lightness and chroma to a similar degree.

Once the appropriate value of ΔE and, if necessary, separate values for ΔH^*, ΔL^* and/or ΔC^*, have been fixed, instrumental pass/fail decisions can be made with a much greater degree of consistency than occurs using the traditional visual method. Several analyses of the consistency of professional shade passers have shown that between 10 and 20 per cent of their decisions are wrong, and these are split more or less equally between accepting batches which should have been rejected and vice versa. Providing adequate care has been taken in setting the pass/fail values, wrong decisions will be virtually eliminated using instrumental methods.

The need for colour-specific pass/fail values may have been eliminated by the recent development of an optimised equation known as JPC79. Full details of this have been published[7] and prospective purchasers of instruments for measuring colour differences would be well advised to have them programmed to display JPC79 values as well as CIELAB values.

Whilst colour difference measurement is normally carried out by first measuring standard and then batch, this first step can be omitted if the instrument used is a spectrophotometer interfaced to a computer. Freshly prepared standards are measured several times and the average values stored in the computer. Whenever a batch is measured, the appropriate values for standard are re-called from store and used in the calculations. To ensure that the spectrophotometer has adequate stability for this purpose, standard black, white, red, yellow, blue and green ceramic tiles should be measured once a day and ΔE values for each calculated against stored data. The reliability of this method has been found to be as good as using a physical standard stored in a freezer at 0° C and better than the use of physical standards that have been subjected to the treatment they would normally receive in a paint factory (e.g. soiling and scratching).

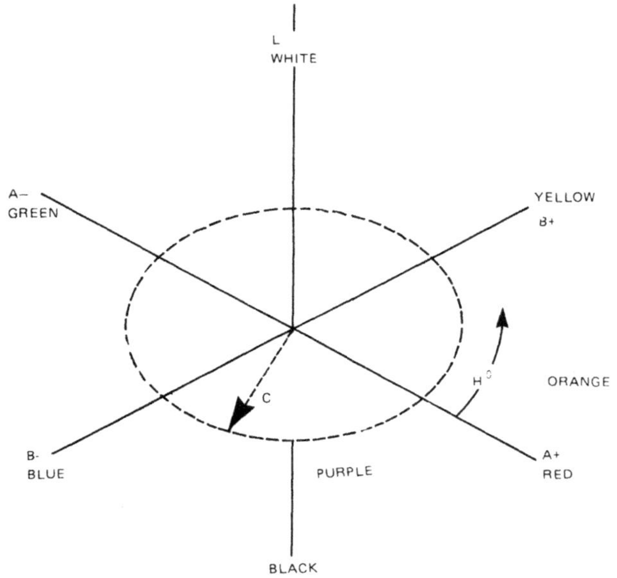

FIGURE 46.1
CIELAB space configuration

REFERENCES

1 D B Judd and G Wyszecki *Color in Business, Science and Industry* 3rd edn John Wiley & Sons New York 1975
2 SDC Colour Measurement Committee 'Selecting Surface Colour Measuring Instruments and Monitoring their Performance' *J Soc Dyers and Colourists* 1980 **96** 587
3 R M Johnston 'Color Measuring Instruments: A Guide to Their Selection' *J Color and Appearance* 1971 **1** 27
4 R T Marcus 'Long-term Repeatability of Color-Measuring Instrumention Storing Numerical Standards' *J Color and Appearance* 1978 **3** 29
5 CIE *Colorimetry* Publication no 15 1971; Supplement no 1 *Special Metamerism Index: Change in Illuminant* 1971; Supplement no 2 *Recommendations on Uniform Color Spaces, Color-Difference Equations, Psychometric Color Terms* 1978 Bureau Central de la CIE Paris (52 Blvd Malesherbes 75008 Paris)
6 R G Kuehni *Computer Colorant Formulation* D C Heath & Co Lexington Mass 1975
7 *J Soc Dyers and Colourists* 1980 vol 96 pp 372 418 486

47 TESTING AND QUALITY CONTROL

47.1 OBJECTIVES OF TESTING

The aim of testing may be one of the following:

(a) to determine the progress of some process (production control);
(b) to determine the change in properties as the ingredients are varied (research and development);
(c) to control batches within close limits to some standard (quality control);
(d) to determine whether a product is likely to be suitable for a particular use (marketing/quality assurance).

In the case of a product with a large range of possible uses, the number of tests necessary to ensure satisfaction in service becomes excessive. This often results in the testing operation being divided into two parts:

(i) quality assurance, product approval or type testing carried out on one sample, with an extensive range of tests over a prolonged period;
(ii) quality control or batch testing to ensure close similarity between a particular batch and the sample proved satisfactory by the above tests for marketing; quick tests are selected to indicate normality in major properties and to indicate that critical materials have been incorporated.

47.2 SELECTION OF PROPERTIES FOR TESTING

The properties usually chosen for testing are those considered of major importance to the user. This means that the specifier must have a complete knowledge of how the product is to be used. These properties are of six main types:

Paint properties: (a) Storage; (b) Application; and (c) Curing
Film properties: (d) Film appearance; (e) Film performance short-term; and (f) Film performance long-term

The paint properties are of first importance, because unless a satisfactory continuous film is produced, the performance properties will be substandard. Some film defects may be visually obvious in the film appearance, whilst others may not.

A considerable amount of film testing is carried out on films less than 7 days old; the assumption is made that the films will not change essentially in properties until the end of their useful lives (that is, until repainted). Unfortunately, this is rarely the case: films that dry by oxidation, in particular, continue to harden and shrink over several years (especially when subjected to UV radiation) with the result that they become more brittle and less extensible. Others degrade by loss of volatile material or chemical attack, so that a number of film properties should be tested after ageing for the expected period to repainting.

Often the most important function of a paint film is long-term protection of the substrate to which it is applied. This attribute is inconvenient to test because of the time scale over which the evaluation must be carried out, and it usually requires an adequate recording system to enable the results to be interpreted by persons other than those planning or starting the exposure series.

Ideally, the testing should be continued to the end of the useful life of the coating system. Interpretation of durability results at an early stage may be quite misleading: minor differences in decorative quality may have no relationship to ultimate protective value. Testing must be continued for more than half the expected life of the coating in order to ensure that any premature deterioration towards ultimate protective film failure is detected, even if only in its early stages.

A number of characteristics such as brushability are actually groups of related properties involving the flow performance at very different rates of shear together with early curing rates. Whilst it is often said that such testing is subjective, the aspect that is more so is the weighting given to each of the individual properties to arrive at a simple collective assessment.

47.3 SELECTION OF TESTS

A test may be selected because of its similarity with the actual use, because the tests are very sensitive or reproducible, because they provide a permanent record, or because they are cheap or rapid. Considerable experience is required to determine the most suitable tests for a particular purpose.[1]

For quality-control purposes, rapid simple tests are normally selected to provide a permanent record. Such tests cause the minimum interference with production whilst allowing easy control within close limits and ready auditing.

For research work or quality assurance, tests of high precision are required to produce a numerical record which allows statistical analysis and possibly an external audit.[2,3]

47.4 SELECTION OF EQUIPMENT

Important in the selection of suitable equipment for testing are:

(a) serviceability—robustness, ease of cleaning and servicing;
(b) range of operation;
(c) portability;
(d) necessity for controlled conditions (temperature, humidity, freedom from vibration or voltage fluctuations);
(e) speed of operation and operator time required (calibration, preparation of samples, reading, cleaning);
(f) training required for operator;
(g) low repeatability error;
(h) capital cost;
(i) standardisation with other laboratories.

47.5 CALIBRATION OF EQUIPMENT

If testing is being carried out for quality-control purposes in one production unit only, it might seem irrelevant to calibrate a piece of equipment, because all the results will be relative. However, unless the equipment has been calibrated, a serious problem results if it has to be replaced or the company expands its operations or merges with another. In such cases all the previous data may become valueless.

Calibration of equipment is essential if test limits or the results of testing are to be supplied to another body or manufacturing unit; it is also necessary for regular checking of the satisfactory operation of the equipment and to monitor the effect of wear over a prolonged period.

As part of the calibration exercise an equipment register should record the supplier, date of purchase, service contracts, each service and calibration, with the correction factors.

Advice on calibration of equipment is given in the appropriate Australian Standard test methods or may be obtained from the National Association of Testing Authorities, Australia.

47.6 PRECISION OF EQUIPMENT/OPERATOR

It is common to assume greater accuracy of a result than is statistically valid. It is as important to know what differences are significant as to know the actual test results. For some Australian Standard and ASTM tests, precision data, the repeatability (r) and reproducibility (R) values, have been determined by means of inter-laboratory tests on the same samples.

Checks should be carried out on equipment and operators to ensure that they fall within the precision expected of the test method. This is particularly relevant in the case of visual colour comparison, where the ability of individuals to differentiate between colours varies widely due to both inherited characteristics and acquired skill. For this purpose the Ishihara Test for Colour Blindness and the Federation of Societies for Coatings Technology of the U.S.A. Colour Matching Aptitude Set are available.

47.7 SAMPLING

Correct sampling is vital to any testing programme. Results are valueless unless the samples tested are representative of the material supplied. Most quality-control testing in the paint industry is based on the assumption that one sample is all that is required to define the characteristics of the whole batch. For this reason, adequate mixing before sampling and filling, and checks to ensure that containers filled at the start and towards the end of the filling operation are closely similar, are essential.

If it is shown that a sample is not representative of the batch, or that a batch is non-uniform, the quality-control system is invalid and all the testing carried out is suspect.

Retention of test samples for the maximum expected storage period enables differentiation between non-uniformity and changes in properties after storage.

47.8 CONDITIONS OF TESTING

Whilst some tests may be carried out under the prevailing conditions—for example, measurement of the viscosity of paint thinned for immediate spraying—most tests are carried out under artificial conditions because the conditions at the time of use are often unpredictable.

For testing carried out for quality-control purposes, the conditions will usually be controlled as much as possible to increase the precision of the results.

When testing is carried out for quality assurance or development programs, the conditions may be controlled to the most adverse under which the product is recommended for use.

The effect of variations in the environment on the results should be analysed statistically, particularly if the conditions cannot be controlled for testing. The conditions of testing should be recorded so that the significance of the results can be audited or a correction factor applied.

47.9 PANEL PREPARATION

Panel preparation is important for reproducible testing, because a number of tests can be affected by surface contamination or differences in adhesion. The moisture content of timber panels at the time of coating and the presence of surface cracks in timber and plywood (due to rapid changes in moisture content) play an important part in adhesion and long-term film integrity. The adsorbed layers of contaminants on the surface of steel affect the corrosion resistance of coatings to a great extent. Absorbent surfaces that have been wetted and allowed to dry may contain higher than normal levels of extractable materials. These surface layers must be removed to produce a true and reproducible substrate.

47.10 STANDARDISATION OF TESTING AND THE REPORTING OF RESULTS

A statement such as 'the drying time of a paint is 10 minutes' is meaningless to anyone other than the tester, and will become meaningless even to that person after a period of time, as the precise method used and the conditions under which the testing was carried out are forgotten. In order that a comparison can be made of results, either the detailed conditions and test method must be recorded in each case, or a standardised method used.

A standardised method is one in which the procedure is detailed and the conditions of testing defined. Such methods should be adequately documented, and given a reference number or distinctive name, so that only this reference need be quoted with the test result. The use of standardised methods enables direct comparison of results and quick recognition of differences in properties, either within the company (company standards), within the country (national AS standards) or internationally (ISO standards), to be made.

There are many national and international standards from which to choose; the list of test methods in Australian Standard AS 1580 is reproduced at the end of this chapter. Even non-standard methods are useful provided that the variations from the standard method are noted (see Appendix 47.2).

47.11 COST OF TESTING

The cost of testing is considerable and each laboratory should estimate the cost of the testing carried out. An assessment should be made of the cost and value of any testing program before commencement. This can be used to determine whether the resources would be better allocated to another project.[4]

47.12 QUALITY CONTROL

Quality is 'fitness for purpose'. To ensure this, companies employ scientifically trained personnel, almost all of whom are dedicated to this objective.

The market research personnel and technical sales representatives attempt to identify the customers' requirements. Product development personnel design products to satisfy the customers' needs and optimise the cost and balance of properties. Quality also involves the personnel who: establish a company standard; issue the master formulation; set raw-material standards; optimise the method of production; set quality control test limits; write label directions and data sheets. It again involves technical sales representatives to ensure that the product works in the customers' environment, who suggest modifications to application procedures and equipment to ensure complete satisfaction. The personnel who carry out any testing of batches of the product and competitive materials in the field (technical audit), and the investigation of complaints, are also a part of this team effort.[5]

Unfortunately, to many technical people, quality control is the poor relative of research and development work, and to others the mention of quality control brings notions of repetition and boredom. By contrast, quality control should be considered as one of the most technically important functions carried out by any company.

Without consistent quality of products being maintained by a manufacturing unit, all the good work that has gone before will have been to no avail. The company may have the most brilliant formulators, or the most capable sales staff; but without the ability to continually manufacture goods to a consistent standard of quality, their efforts will be ultimately wasted.

The maintenance of quality should be the prime objective of every employee within a company, from the guard who checks the load at the front gate, through to the managing director. It should be one of the chief aims of senior management to instill the employees of a company with a desire to manufacture only goods that completely meet the standards of quality laid down for their products. Every means that a company has at its disposal should be employed to further these aims. Books, films, in-house training and specialised college courses should be used wherever possible to assist the employees in carrying out this vital function.

Apart from the obvious legal reasons for manufacturing goods that are completely satisfactory, there are very sound economic arguments for ensuring continued quality. Goods that are obviously faulty have to be re-worked (or perhaps scrapped, with attendant disposal costs), and this involves considerable expenditure in terms of labour, machinery time and lost output.

In the case of a faulty batch of paint, the chemist's time is involved in determining the cause of failure and then devising some method to satisfactorily rectify the batch if this is possible, or converting it to some other less critical product. Machinery time and lost output must be costed, as well as factory labour plus the additional costs of recanning and re-labelling the batch.

Products that are not obviously faulty but are inconsistent in their properties as a result of loose quality-control procedures, create constant troubles in use and are the subject of many complaints which require investigation. The more serious of these may result in claims for compensation to repair and rectify. A few litres of paint spread over a large area can require considerable labour in removing the coating to the point where repainting can commence.

The greatest risk of poor quality control is the product and the company acquiring a poor reputation in the market place. This can have very serious consequences for the future growth and profitability of that company; it may take many years and countless batches of satisfactory products to counter this—provided that company is able to survive at all. Although very small companies may be able to exist by selling to new customers only, major companies must depend to a large extent on repeat business from the same customers. The cost to a company through the loss of a single customer as a result of a problem may be difficult to estimate, but this is why the quality-control system and staff are employed.

A company should view quality control as a profitable exercise, and it should be regularly reviewed to ensure that the budget is spent where it is most necessary. Neglect of quality control prejudices the future of the company, risking compensation claims that would be difficult to oppose, and wasting money originally expended on development of the product.

47.13 QUALITY CONTROL PROCEDURES

In order to ensure that a consistent product is produced by the factory, it is necessary to control all stages of manufacture, from the acceptance of raw material, through to filling and labelling.

47.13.1 Raw Material Inspection

Raw materials as delivered should be inspected by the receiving person to ascertain that the correct goods have been delivered. These goods should be held in a special receiving area to prevent inadvertent usage by the factory before testing. The goods are labelled with a 'hold' tag, pending laboratory examination for the appropriate properties to ensure uniformity from one delivery to the next.

The goods should then be labelled with an 'OK for use' tag, coded and safety-coded (if required) and made available for use. If the goods fail laboratory checks, then a 'reject' label should be affixed to the goods, which should be returned to the supplier immediately.

Raw material suppliers with the necessary technical facilities may be able to supply 'quality assurance' certificates ('SQA's') with each delivery of raw materials. This reduces the need for sampling and testing every delivery; however, audit checks may be necessary to avoid a drift in properties.

Samples of critical raw materials should be retained, with the date of supply, batch number, source and code number of the product.

47.13.2 In-factory Procedure

A batch process generally starts with the issue of a batch ticket numbered with an appropriate batch number and date. This may be a photocopy of an officially approved master card, or a computer-generated batch sheet. It is undesirable to issue original handwritten or typewritten batch sheets, as they can be subject to errors of transcription. In either case, the card should be checked to ensure correct transcription and to ensure that suitable raw materials are available to complete the batch.

Correct supervision of factory personnel during the manufacturing procedure is essential. It is the responsibility of management to ensure adequate training of staff in all operational areas, and technical staff have a major role here.

The ingredients for the batch should be weighed or measured carefully. The assembled goods should then be checked by a supervisor, before the charging of a mill or addition to a tank. It is most important that factory scales and fluid measuring devices be kept accurate and within tolerances by systematic checking. Service contracts with reliable contractors are a good means of maintaining standards. It is equally important for production staff to use measuring equipment appropriate to the quantity of material involved. An accuracy of $\pm\frac{1}{2}$ per cent is usually adequate.

After assembly of the batch ingredients, any deviations from the prescribed formula should be noted; where approved alternate raw materials are listed, then the actual material used should be entered. In some cases it may be necessary to record batch or lot numbers of raw materials on the batch sheet. Where critical items are added to a batch, this should be only under direct supervision, or by a person authorised to dispense those ingredients.

Whenever practical, the same manufacturing machinery and techniques should be employed for further batches. This helps reduce variations in properties which may occur with changes in processes. The formula card should nominate the manufacturing method and the precise machinery to be used.

Once the batch has been manufactured according to the batch formula, a representative sample should be taken to the control laboratory for technical evaluation.

47.13.3 Evaluation by the Control Laboratory

It is the responsibility of the control laboratory to ensure uniformity of product by the application of certain tests which are designed to give accurate and repeatable results. The types of tests which can be applied are many and varied, and it is not the intention of this chapter to examine the individual specific tests.

An experienced operator, or tests that are repeated often, allow heightened perception, giving greater control. A comprehensive list of tests and methods can be found in the Australian Standards (see Appendix 47.1), U.S. Standards, British Standards[6] and German Standards.

The tests selected are usually rapid sensitive tests, which are easy to reproduce. It is also essential that the tests applied are meaningful, and pertinent, to the end use of the product. For example, it would be no use determining the brushing properties of a batch of paint if it were to be applied by electrostatic spray.

The tests should also be chosen for suitability in detecting significant differences between batches. The word 'significant' in this instance may mean a relatively minor or quite major difference, depending on the property being measured.

An example of this 'significant difference' occurred with the manufacture of a larger than normal batch of flat latex paint designed for roller application by trade painters. The larger batch required a different manufacturing technique, although the equipment used was the same as normal and the formulation remained unchanged. The batch was duly manufactured, tested, passed as being suitable, and canned off. It was not too long before complaints started to come in that this batch lacked opacity—it required three coats to obliterate the substrate, whereas two normally sufficed. Initial checking showed the product to be normal for pigment content and the opacity appeared to be equivalent to previous production; viscosity was well within limits and the paint appeared normal in all respects. It took considerable detective work to isolate the problem, which was in part associated with roller skid. The roller was slipping slightly, rather than fully transferring the paint to the substrate, resulting in a thinner coat than normal. This problem occurred as a result of a changed manufacturing process and imprecise testing. Even though the viscosity was within limits, the rheology was different and this was not detected. The paint was rolled out on plasterboard as a routine test, but the method used did not differentiate between this batch and previous production. The differences were not detected because the tests employed were not sensitive enough.

Before completion and submission for final testing, certain checks are made while the batch is being processed. Early detection of problems can often initiate corrective action while the batch is still at the mill base stage. Removal of a contaminant by straining is obviously faster on the smaller volume involved; further grinding, with or without additional raw material, will be faster if treated on the lesser volume of mill base at the correct milling consistency and pigment loading.

When a production batch is brought into the control laboratory for checking, the batch sheet should be examined closely for changes and alterations and to ensure all the ingredients have been added and marked off. The bulk volume of the batch is checked against the theoretical yield and then the batch is measured for viscosity, density or other properties which will allow final adjustment. These adjustments should be fully documented on the batch sheet. Batches that require major additions should be viewed with suspicion and not readily accepted without thorough testing.

Whilst any number of tests may be applicable, fineness of grind, yield, viscosity, density and colour are regarded as quick, reliable tests which are indicative of whether the batch has been manufactured to formula. More detailed testing may be necessary for individual products.

Naturally subjective evaluation of the application properties and close examination of the wet and dry film by a skilled technician play a vital part in the overall assessment.

Once the batch is approved, the signed release should accompany the batch card, and filling should commence without delay. Should filling be postponed, then a recheck of key constants should be made. The laboratory should ensure that correct straining of the batch is undertaken. Spot checks may be made during filling if this procedure is prolonged, to ensure that the product remains homogenous throughout. Filled cans may be check-weighed during filling, to ensure correct filling level and as a double check on uniformity of product.

A sample should be taken early in the filling period, and a further sample towards the end.

These samples should be carefully labelled with the product code, date, batch number, colour and other required details, and the time when the sample was taken during filling. The samples should be returned to the control laboratory for careful recording and filing.

Checks should be made to ensure that the correct labelling, batch number, code, warning stickers, or other required information is on the label of the cans, and that they accurately describe the contents.

Upon completion of filling, the total number of litres canned off should be compared with the theoretical yield of the batch. Any variations greater than 5 per cent should be examined closely to determine the reason for the discrepancy. For example, a latex paint may become badly aerated during filling, resulting in underfilling and a higher yield.

It is essential that detailed records are kept by the control laboratory on the manufacture of each batch. Additions that are made to (and their effect on) the batch should all be documented. The actual batch sheet should be retained on file for reference should complaints or problems occur later. Retained samples should be filed and kept for the prescribed period. Retail products may have to be held for up to three years or longer, whilst 18 months may suffice for rapid turnover industrial products. Dried panels showing colour and finish should also be filed for future reference. Where practical, it is desirable for the results of all the tests to be quoted as a numerical value. The more subjective tests may be given a rating on a scale of 1 to 10 against a standard. The acceptable pass rating for the batch may then be nominated.

As with factory personnel, it is essential in the laboratory to have qualified supervisers working alongside testing staff. Supervision should include, among other things, the checking of calculations and methods. To avoid compromise on quality because of production pressure, the senior quality-control officer should have access to senior management for their opinion and directives should the aims of quality control and production output be in strong opposition.

A systematic spot-checking of products from the warehouse is beneficial, not only for maintaining a control on Q.C. testing, but also for detecting the start of some unwanted change of properties which may be developing on storage.

Equipment in the control laboratory should be kept in first-class condition and calibrated at regular intervals, to reduce the tendency for gradual drift in values. Cleanliness is also essential in ensuring reproducible results.

47.14 COMPLAINTS

A necessary part of all quality-control systems is a complaints register. Analysis of the results of the investigations should allocate the causes to one of six categories:

(a) defects not detected by quality control (requiring a reappraisal of the quality-control system for that product, and possibly others, and recall of the batch);
(b) defects that develop in the paint after some period of storage (requiring investigation of the formulation or raw-material specifications);
(c) the use of an unsuitable product for the purpose (there may or may not be a better alternative product);
(d) the use of the product under unsuitable conditions (there may be no alternative but to artificially improve the environmental conditions);
(e) application of the product over an unsuitable or improperly prepared substrate—insufficient attention to surface preparation is a very common cause of paint failure, in no way reflecting on the quality of paint used;
(f) cause unknown (this may be due to insufficient information or the unavailability of material for test); in this instance, no corrective action can be taken, and complaints from the same cause may recur.

47.15 IMPROVING STANDARDS OF QUALITY CONTROL

The preceding pages are not intended as a hard and fast method for quality control. The aspects covered are merely a broad-brush approach to the subject, and individual laboratories may vary this considerably. It is important, however, that quality control be seen in the correct perspective and be given the status that such a critical operation deserves.

47.15.1 Printed Matter

It is recommended that use be made of Australian Standards AS 1821–1823 Suppliers Quality Control Systems; and AS 2000, Guide to AS 1821–1823.[7]

A modest competence in statistical analysis is necessary to meaningfully design sampling procedures and interpret the results. A number of textbooks is available, and other Australian Standards may also be useful. Selected AS items are included in Appendix 47.1, located at the end of this chapter; AS 1580, which relates to paint testing, is listed in full.

The National Association of Testing Authorities (NATA), Australia, has issued a brief circular on quality-control testing of paints. This document is appended as Appendix 47.2. Copies of a batch sheet and test record card blanks are included in Appendix 47.3.

47.15.2 Training Courses and Conferences

Training courses on various aspects of quality control are available at many colleges of Technical and Further Education. Excellent courses and conferences are also conducted by the Australian Organisation for Quality Control (AOQC) who may be contacted in capital cities.

The NSW section of OCCA Australia held a symposium on 'Quality Control: A Profit Centre' in April 1982, and copies of these very useful papers are available from the Association.

It cannot be stressed too strongly that quality control is not simply the testing of materials against specifications; the quality control personnel must be involved in manufacturing and marketing decision-making and be fully aware of the need for the customer to receive the right product at the right time and right price.

REFERENCES

1 D F Crossing 'Architectural Coatings: Practical Performance vs Laboratory Testing' *OCCAA Proc and News* Jan–Feb 1975 14

2 A R Lansdown 'Method Evaluation' *Petroleum Review* June 1969 177–8

3 W E Craker and G M Deighton 'Application of Statistical Methods to the Subjective Comparison of Coatings' *J Coatings Technology* 1981 **53** (679) 55–64

4 Standards Association of Australia AS 2561–1982 'Guide to the Determination and Use of Quality Costs'

5 H M Werner 'An Overview of Quality Assurance' *J Coatings Technology* 1980 **52** (669) 53–6

6 British Standard BS 3900 'Methods of Test for Paints: General Introduction'

7 Standards Association of Australia AS 1821 to 1823–1975 'Suppliers Quality Control Systems'; AS 2000–1978 'Guide to AS 1821–1823 Suppliers Quality Control Systems'

8 J M Juran *Quality Control Handbook*; J M Juran and F M Gryna *Quality Planning and Analysis*; A F Cowan *Quality Control for the Manager*; E S Buffa *Modern Production Management*; M J Moroney *Facts from Figures*

9 Standards Association of Australia AS 1199–1972 'Sampling Procedures and Tables for Inspection by Attributes'; AS 1399–1973 'Guide to AS 1199'

APPENDIX 47.1

The following lists of Australian Standards and Test Methods for Paints and related materials were supplied by the Standards Association of Australia.

Standard No.	*Title*
1263–1972	Oil-based Putty
1433–1972	Paint Colours for Building Purposes
2009–1977	Glass Beads for Traffic Marking
2105–1978	Inorganic Zinc Silicate Paint
2204–1978	Zinc-rich Organic Priming Paint
2301–1980	Wood Primer, Solvent Borne, Brushing
2302–1980	Undercoat, Solvent Borne, Exterior/Interior
K108–1963	Metal Priming Paint, Anti-corrosive
K111–1968	Oil Gloss Paint for General Use
K113–1964	Enamelised Paint for Exterior Use
K11–1946	China Wood Oil (Tung Oil)
K122–1964	Latex Paints for Interior and Exterior Use
K126–1964	Full Gloss Enamel, Oil and Petrol Resistant
K127–1964	Undercoat for Oil and Petrol Resistant Enamels
K128–1964	Flat Enamel for Interior Use
K129–1969	Marine Underwater Systems for the Bottoms of Steel Ships
K130–1962	Decorative Thermosetting Laminated Sheet
K13–1929	Soya Bean Oil
K143–1962	Semi-gloss Enamel, Oil and Petrol Resistant
K145–1972	Red Lead Based Paint for Structural Steel
K146–1967	Road Marking Paints
K147–1964	Paint, Varnish and Lacquer Remover
K160–1966	Semi-gloss Enamel for Interior Use
K172–1970	Tar Epoxy Paint, Solvent Based
K179–1969	Semi-gloss Enamel—Low Fire Hazard type for Non-combustible Surfaces
K185–1968	Colours for Specific Purposes
K19–1971	Red Lead Pigment and Paste in Oil
K203–1970	Case Marking Paint
K211–1971	Zinc Chromate Primers for Structural Steel
K21 to K26–1930	Inert Fillers: Barytes, Blanc fixe, Calcite, Whiting, Kaolin, Talc
K28 to K29–1939	Benzol and Toluol
K2 to K6–1928	Linseed Oil: Raw linseed oil Refined linseed oil Pale boiled linseed oil Boiled linseed oil Linseed oil for varnish
K41.502.1–1965	True Red Lead in Red Lead Pigments
K59 to K65–1939	Pigments for Paints: Natural red oxide of iron for paints Manufactured red oxide of iron for paints Black oxide of iron for paints Purple oxide of iron for paints Natural sienna, raw and burnt for paints Natural umber, raw and burnt for paints Ochre for paints
K66 to K69–1939	Black Pigments for Paints: Carbon black for paints Bone black for paints Vegetable black for paints Lamp black for paints
K71 to K73–1940	Green Pigments for Paints: Green oxide of chromium for paints Lead chrome greens for paints Reduced lead chrome (Brunswick) greens for paints
K74–1941	Lead Chromes for Paints
K75–1941	Prussian Blue for Paints
K76–1941	Natural Sour (prime lactic) Casein for Glue Manufacture
K7–1941	Turpentine

AS 1580
List of Methods

Eleventh List of Methods
February 1981

No. of Method	*Date*	*Title*
101.1	June 1980	Air drying conditions
101.3	June 1980	Standard procedure for stoving
101.4	June 1980	Conditions of test, temperature controlled
101.5	June 1980	Conditions of test, temperature and humidity controlled
102.1	June 1980	Sampling procedure
103.1	June 1980	Preliminary examination and preparation for testing
104.1	July 1979	Recommended materials for test panels
105.1	June 1980	Pretreatment of metal test panels—Solvent cleaning
105.2	June 1980	Pretreatment of metal test panels—Sanding
105.3	June 1980	Pretreatment of metal test panels—Chromic acid dipping
106.1	July 1979	Preparation of timber test panels for outdoor weathering test
107.1	June 1980	Determination of wet film thickness from dry film mass
107.2	June 1980	Determination of wet film thickness from wet film mass
107.3	June 1980	Determination of wet film thickness by wheel gauge
107.8	July 1979	Determination of wet film mass (brush application)
108.1	June 1980	Determination of dry film thickness on iron and steel substrates (permanent magnet instruments)
202.1	June 1980	Density
202.2	June 1980	Density of water dispersed paints subject to foaming
203.1	June 1980	Skin formation
204.1	Aug. 1977	Fineness of grind
205.1	Feb. 1981	Application properties—Brushing
205.2	Feb. 1981	Application properties—Conventional air spraying
205.3	Feb. 1981	Application properties—Roller coating
205.4	Feb. 1981	Application properties—Airless spraying
208.1	Feb. 1981	Thinner compatibility
209.1	May 1975	Re-mixing properties
211.1	June 1980	Degree of settling
211.1	June 1980	Ease of manual reincorporation

No. of Method	*Date*	*Title*
402.1	Feb. 1981	Bend test
403.1	Feb. 1981	Scratch resistance
404.1	Feb. 1981	Recoating properties
405.1	July 1978	Determination of pencil hardness of paint film
407.1	Feb. 1981	Heat resistance—Slow cooling
407.2	Feb. 1981	Heat resistance—Thermal shock
408.1	April 1974	Adhesion (paint inspection gauge)
408.2	Dec. 1973	Adhesion (knife test)
408.3	Mar. 1976	Adhesion (Arco microknife)
408.4	June 1980	Adhesion (cross-cut)
452.1	May 1975	Resistance to humidity under condensation conditions
452.2	May 1975	Resistance to corrosion—Salt droplet test
453.1	May 1975	Resistance to petroleum spirit
454.1	May 1975	Resistance to mineral oil
455.1	May 1975	Resistance to water at room temperature
456.1	May 1975	Resistance to boiling water
457.1	May 1975	Resistance to outdoor weathering
458.1	May 1975	Resistance to wetting
459.1	Feb. 1976	Resistance to washing
460.2	Sep. 1974	Alkaline conditions (immersion test)
461.1	May 1977	Determination of resistance to yellowing (dark chamber)
481.1	May 1975	Assessment of individual defects of exposed films
481.2	Dec. 1975	Assessment of blistering of paint films
481.3	May 1975	Assessment of visible rusting
481.4	Dec. 1975	Assessment of corrosion of an underlying iron or steel surface
481.5	May 1975	Determination of durability, resistance to corrosion, and resistance to fouling of marine underwater paint systems
481.6	May 1975	Determination of durability of marine underwater paint systems under cathodically protected conditions
482.1	May 1975	Fastness to light
483.1	July 1973	Resistance to artificial weathering (carbon-arc type instruments)

No. of Method	*Date*	*Title*
212.2	Feb. 1981	Wet hiding power—Black and white chart brushout method
213.1	Feb. 1981	Relative dry hiding power
213.2	May 1975	Dry hiding power—Contrast ratio
214.1	May 1975	Consistency of paints measured by the Krebs-Stormer viscometer
214.2	May 1975	Consistency—Flow cup
214.3	Mar. 1978	Viscosity at a high rate of shear
215.1	May 1975	Levelling—Draw-down method
301.1	May 1975	Non-volatile content
301.2	May 1978	Non-volatile content by volume (volume solids)
302.1	May 1975	Pigment content
401.1	May 1975	Surface dry condition
401.5	June 1980	Hard dry condition—Sanding test
401.6	Nov. 1977	Hard dry time (mechanical thumb test)
401.8	Feb. 1981	No-pick-up time of road marking paints
501.1	Feb. 1980	Soluble lead content (gravimetric method)
502.1*	Nov. 1965	True red lead in red lead pigments
503.1	July 1973	Determination of water by the Dean and Stark method
504.1	April 1979	Metallic zinc content
601.1	May 1975	Colour—Visual comparison
601.2	Dec. 1975	Colour—Instrumental measurement of colour difference using the 'Colormaster' differential colorimeter
601.3	May 1975	Colour—Instrumental measurement of colour difference using the 'Color Eye'
602.1	Dec. 1975	Visual assessment of gloss
602.2	April 1974	Specular gloss
602.3	Sep. 1974	Visual assessment of gloss (Boller)
603.1	May 1975	Finish
604.1	Dec. 1973	45°, 0° Reflectance of white and pale coloured paints

* See AS K41.

AS 1580
Alphabetical List of Methods

Index
February 1981

Title	*No.*
Adhesion (Arco microknife)	408.3
Adhesion (cross-cut)	408.4
Adhesion (knife test)	408.2
Adhesion (paint inspection gauge)	408.1
Alkaline conditions (immersion test), resistance to	460.2
Application properties—Airless spraying	205.4
Application properties—Brushing (brushing properties)	205.1
Application properties—Conventional air spraying	205.2
Application properties—Roller coating	205.3
Artificial weathering (carbon arc), resistance to	483.1
Blistering of paint films, assessment of	481.2
Boiling water, resistance to	456.1
Colour—Instrumental measurement of	
Levelling, draw-down method	215.1
Light fastness—see Fastness to light	
Marine underwater paint systems, durability of under cathodically protected conditions	481.6
Marine underwater paint systems, durability, resistance to corrosion, and resistance to fouling of	481.5
Mineral oil, resistance to	454.1
No-pick-up time of road marking paints	401.8
Non-volatile content	301.1
Non-volatile content by volume	301.2
Outdoor weathering, resistance to	457.1
Pencil hardness of paint film	405.1
Petroleum spirit, resistance to	453.1
Pigment content	302.1
Preliminary examination and	

Title	*No.*
colour difference using the 'Colormaster' differential colorimeter	601.2
Colour—Instrumental measurement of colour difference using the 'Color Eye'	601.3
Colour—Visual comparison	601.1
Condition in container—Ease of manual re-incorporation	211.2
Conditions of test—Temperature and humidity controlled	101.5
Conditions of test—Temperature controlled	101.4
Consistency—Flow cup	214.2
Consistency—Krebs-Stormer viscometer	214.1
Contrast Ratio—see Hiding Power	
Corrosion of underlying steel surface, assessment of	481.4
Corrosion, Resistance to—Salt droplet test	452.2
Defects of exposed films, assessment of individual	481.1
Density	202.1
Density (for water dispersed paints subject to foaming)	202.2
Dry film thickness on iron and steel substrates (permanent magnet instruments)	108.1
Drying—see Hard dry, Surface dry, No-pick-up time	
Drying conditions—Air	101.1
Drying conditions—Stoving	101.3
Fading—see Fastness to light	
Fastness to light	482.1
Fineness of grind	204.1
Finish	603.1
Fouling—see Marine underwater paint systems	
Gloss—Visual assessment	602.1
Gloss—Visual assessment (Boller)	602.3
Hard dry condition (sanding test)	401.5
Hard dry condition (mechanical thumb test)	401.6
Hardness of paint film, pencil	405.1
Heat resistance—Slow cooling	407.1
Heat resistance—Thermal shock	407.2
Hiding power, dry—Contrast ratio	213.2
Hiding power, dry, relative	213.1
Hiding power—Wet (black and white chart)	212.2
Humidity under condensation conditions, resistance to	452.1
preparation for testing	103.1
Petroleum spirit, resistance to	453.1
Recoating properties	404.1
Red lead (true) in red lead pigments	502.1*
Reflectance of white and pale coloured paints, 45°, 0°	604.1
Re-mixing properties	209.1
Rusting, assessment of visible—see also Corrosion	
Sampling procedure	102.1
Scratch resistance	403.1
Skin formation	203.1
Soluble lead content	501.1
Settling, degree of	211.1
Surface dry condition	401.1
Test panels for outdoor weathering test, timber	106.1
Test panels, pretreatment of metal—Chromic acid dipping	105.3
Test panels, pretreatment of metal—Solvent cleaning	105.1
Test panels, pretreatment of metal—Sanding	105.2
Test panels—Recommended materials	104.1
Thinner compatibility	208.1
Viscosity at a high rate of shear	214.3
Viscosity—see Consistency	
Washing, resistance to	459.1
Water at room temperature, resistance to	455.1
Water determination—Dean and Stark method	503.1
Weathering—see Wooden panels	
Wet film mass (brush application)	107.8
Wet film thickness by wheel gauge	107.3
Wet film thickness from dry film mass	107.1
Wet film thickness from wet film mass	107.2
Wet hiding power—Black and white chart brushout method	212.2
Wetting, resistance to	458.1
Wooden panels for outdoor weathering test, preparation of	106.1
Yellowing (dark chamber), resistance to	461.1
Zinc content, metallic	504.1

*AS K41

APPENDIX 47.2 QUALITY CONTROL TESTS ON PAINTS

The National Association of Testing Authorities, Australia.

Most of the major paint manufacturers in Australia now hold NATA registration for tests on paints. In many cases, this interest in registration has arisen more from a need to issue NATA-endorsed test reports on their products to specific clients rather than as a means of monitoring the competence and validity of normal quality control testing.

In many cases, therefore, paint manufacturers have sought registration for research and development or service laboratories not directly associated with quality control testing had have gained registration only in terms of standard test methods (e.g. AS 1580, ASTM, BS 3900, etc.) or test methods specified by their clients.

Recent developments within the paint industry have prompted increased interest in NATA registration of quality control testing procedures. This circular outlines NATA's general requirements for registration of quality control tests on paints and should be read in conjunction with NATA's general requirements for registration expressed in the Chemical Testing handbook (January 1981).

Staff

NATA's normal requirements for staff apply. NATA recognises that paint testing is normally undertaken by unqualified, but trained, technicians working under the control of a knowledgeable and well-experienced paint technologist. In situations involving in-factory quality control laboratory units, effective supervision must be exercised from the main quality control laboratory.

Calibration

As a prerequisite for registration, NATA requires that all laboratories have their testing and measuring equipment calibrated in terms traceable to the standards of measurement held by CSIRO Division of Applied Physics.

It is normal for NATA to permit many items of working equipment to be checked in-house against a laboratory's reference equipment. This principle will be applied to checking of working equipment for quality control tests on paints.

Essentially, all laboratories holding NATA registration for tests on paints will be expected to have a basic range of calibrated equipment appropriate to the range of tests for which registration is held. Details of NATA's requirements for calibration of paint testing equipment are given in appendices 2 and 5 of the Chemical Testing handbook.

Equipment used in quality control testing may be cross checked against this calibrated equipment; this would apply to balances, wheel gauges, fineness of grind gauges, etc.

The verification of temperature distribution of ovens presents a particular difficulty. NATA is investigating the development of an in-house performance check for ovens used for determination of non-volatile matter; if this is successful, a circular will be issued in due course.

Because of the critical nature of the time/temperature relationship in stoving operations, ovens used for stoving of paint films must be calibrated by a NATA registered Heat & Temperature Measurement calibration authority.

Test Methods

For the purpose of NATA registration, AS 1580 test methods are to be preferred but registration will be granted for in-house quality control test methods provided that:

(a) either no AS 1580 method exists or there are valid reasons making the AS 1580 test method unsuitable for the particular product;

(b) in-house methods have been properly documented in a laboratory test method manual;
(c) if an in-house method produces a numerical result, the precision of the test method has been established;
(d) the laboratory does not classify the method as confidential and is willing to release details of it to NATA, GPC and customers if requested.

In the application of test methods, especially in relation to replication of results, NATA will recognise the normal industry practice for quality control. Single determinations will be accepted for such commonly determined properties as consistency, density, fineness of grind, colour and gloss. Duplicate determinations will be required for non-volatile content determinations.

Range of Tests

A laboratory seeking registration for quality control tests on paints must have facilities appropriate to the normal range of tests applicable to a quality control situation. The range of tests will naturally depend upon the nature of the products being manufactured by the organisation but the list that follows represents the basic framework:

Test Conditions

Routine conditions as specified in AS 1580 methods 101.1, 101.4 and 101.5 are acceptable; normally a small room fitted with a domestic or commercial airconditioner will satisfy this requirement.

Sampling

Method 102.1 is more appropriate to acceptance sampling rather than to production quality control.

NATA would require that sampling procedures appropriate to the manufacturing operations be established and documented in a laboratory test manual.

Non-volatile Matter

Methods 301.1 and 301.2 apply. If method 301.1 needs to be modified because of the nature of the products being tested, these modifications must be documented in the laboratory test manual.

Density

Methods 202.1 and 202.2 apply. In certain circumstances, a pressure cup might be appropriate for products susceptible to aeration as an alternative to 202.2.

Fineness of Grind

Method 204.1 applies. In accordance with Note 2 to clause 7.4, only one reading is necessary for routine testing.

Consistency

Any of the three consistency test methods currently in AS 1580 could be used. In addition, pending publication of corresponding AS 1580 test methods, in-house methods using the Brookfield viscometer or the Rotothinner will be acceptable, provided that the methods are documented in the laboratory test manual.

Normally, 25° C is the nominal test temperature for determination of consistency. However, some laboratories, for historical reasons, use other temperatures, such as 24° C or 27° C. NATA will accept this situation provided that:

(a) The test temperature is specifically noted in the test method contained in the laboratory test method manual.

(b) Any test report issued clearly indicates the temperature at which the consistency has been determined.

Drying Time

Methods 401.1, 401.5 and 401.6 are acceptable.

NATA would be willing to consider approval of in-house methods of equivalent reliability including tests performed on mechanically operated drying recorders.

As a prerequisite for registration for drying tests, some form of wet or dry film thickness measurement would be needed; methods 107.1, 107.2, 107.3 and 108.1 would be applicable.

Application Properties

Methods 205.1, 205.2, 205.3 and 205.4 apply. In-house methods of equivalent reliability will be acceptable.

Colour

Visual colour assessments may be made to method 601.1, Appendix A or Appendix B.

Instrumental colour difference measurements may be performed on instruments complying with AS 1580 methods 601.2 or 601.3 or any of the colour scales listed in ASTM D2244.

Gloss

Visual gloss determinations may be made to 602.1; instrumental gloss measurements may be made to 602.2.

Other Tests

In addition to the above, the product range may require inclusion of other specialist tests, such as facilities for stoving (103.1), pencil hardness (405.1), bend tests (402.1), scratch resistance (403.1) or infra-red reflectance.

Samples

Samples must be retained for a period of at least two years. While 500 mL samples are desirable (especially for Krebs-Stormer consistency and settling tests), 250 mL retained samples will be acceptable.

Laboratory Records

NATA's basic requirement is that records of test be made and retained in a clear, unambiguous, easily retrievable manner. In paint testing laboratories, test data are normally recorded either on job sheets or on product cards; either practice would be acceptable.

APPENDIX 47.3 SAMPLE BATCH SHEETS

Product Reference No.	Product name:			
SPECIAL INSTRUCTIONS:				
MANUFACTURING INSTRUCTION	Code No.	MATERIAL	KILOGRAMS	LITRES
		ADDITIONS TO BATCH: BATCH SIZE		
FORMULATOR........................				
FORMULA DATE / /	DIP VOLUME	THEORETICAL YIELD		

Batch Date / /	Batch No.		
Approved By	**Production Schedule**		
CHECK	Raw Mat. Batch No.		
		MILL No:	
		ASSEMBLED BY:	
		LOADED BY:	
		DATE IN MILL:	
		HOURS MILLED:	
		OK TO DROP:	
		FINISHED BY:	
		TINTED BY:	
		DATE TO LAB:	

CONTROL TESTING		
TEST	SPECIFICATION	RESULT
GRIND		
VISCOSITY		
DENSITY		
STRAIN THROUGH		
FILL INTO		

Mill Base Volume LITRES	Passed By:
	Date:

Filled By: Date / /

SIZE	200 L	20 L		4 L		1 L	500 mL	250 mL	
No. Required									
Actual									

APPENDIX 47.3 (cont.)

Product Reference No.	Product name:

FORMULA CHANGES DATE	DETAILS	INITIALS	FORMULA CHANGES DATE	DETAILS	INITIALS

REPLACEMENT MASTERCARD ISSUED ON: **ISSUE No.:**

Date												
Batch No.												
Yield												
Date												
Batch No.												
Yield												
Date												
Batch No.												
Yield												
Date												
Batch No.												
Yield												
Date												
Batch No.												
Yield												
Date												
Batch No.												
Yield												

APPENDIX 47.3 (cont.)

Date												
Batch No.												
Yield												

EXAMPLE LABELS	PACKAGING DETAILS
	Container Sizes
	Container Type
	Closure
	Label/Identification/Warnings, etc.
	Related/Replacement Products

APPENDIX 47.3 (cont.) SAMPLE TEST RECORD CARDS

PRODUCT REF. NO.	PRODUCT NAME	COLOUR	CARD NO.
SPECIAL INSTRUCTIONS			

DATE	BATCH NUMBER	INITIAL VISCOSITY	INITIAL DENSITY	FINAL VISCOSITY	FINAL DENSITY	GRIND	HIDING	DRYING TIMES		COLOUR			THEOR. YIELD	ACTUAL YIELD	ADDITIONS AND REMARKS	APPROVED BY

QUALITY CONTROL RECORD

Product: .... Ref. No.: ....

Colour Ref.: .... Master/s No.: ....

SPECIFICATION Date Issued: ....

Dispersion: .... Method of Application: ....

Viscosity: .... Flow Cup No. 4 at 25° C. Drying Time: ....

.... Krebs Stormer Units at 25° C. Thinners: ....

Density: .... kg/L. Covering Capacity: .... Gloss: ....

Special Tests: ....

....

Remarks: ....

....

....

48 THE SUBSTRATE AND ITS PREPARATION

48.1 INTRODUCTION

In the coatings industry, considerable research effort has been devoted to the development of a very broad range of coatings based on a great variety of vehicles or media of different chemical types. The chemical and physical composition of these coatings, their pigmentation, selection of solvents and rheology occupy much of the available skills and attention.

The technical personnel in the industry, being mainly chemically orientated by inclinations and training, tend to emphasise and consider the chemical structure of polymers and pigments; their importance and relevance to the profession and industries cannot be underrated. In this chapter, however, it is proposed to devote attention to the substrate itself—the surface to be coated—and to consider physical and chemical variations in these surfaces in relation to the effects these have on paint application and life.

Review of thousands of paint exposure evaluations over many years suggests that the nature and condition of the substrate immediately before coating is possibly the greatest single factor influencing the durability of the paint system.

While much has been done to improve coatings by raw material suppliers and formulators, practical performance can be optimised only by closer understanding of the requirements or peculiarities of the wide range of substrates to be coated. These variations impose special requirements, both in the correct surface preparation before coating and in the selection of the finishing system itself. Because of this interdependence, closer co-operation is needed between paint technologists, designers and manufacturers of such materials as wallboards and joinery. All too frequently, for example, new wallboards enter the market with little or no prior review by the coatings industry. These are, at times, subsequently the source of serious field problems due to unforeseen paint-holding defects. Many of the ultimate problems encountered tend to reflect on the paint industry and the painting trade, which must then jointly come up with some system which will work. How much better this could be done by prior consultation with the industry before release of new building materials!

All constructional materials, from brick to galvanised iron, present special painting problems; but where inter-industry co-operation exists, much progress has been made.

Expanding technology is continually creating new metal alloys and plastic surfaces which require both decoration and protection. New surfaces present a challenge to the paint chemist, who searches for new surface preparations and primers in order to satisfy these requirements.

48.2 COMMON REQUIREMENTS IN SUBSTRATE PREPARATION

In the following pages, some of the peculiarities of each substrate will be given—and perhaps in briefly indicating current views in this field, something of a practical nature can be achieved.

Among other factors, the shape, texture, colour, absorbency, cleanliness, chemical reactivity, dryness, flexibility and temperature of the surface will all play a significant role in influencing the performance of the coating system applied to it.

Adhesion must be achieved, and theoretical considerations do not appear yet to have fully resolved this important phenomenon. It does appear, however, that van der Waals forces existing between polar molecules are important, no doubt supplemented by mechanical, electrostatic and chemical bondings at times. Intimate contact with the surface and, hence, cleanliness of the surface is obviously critical.

48.3 CLASSIFICATION OF SUBSTRATES

Substrates may be classified in many ways. One division is by the substrate's porosity.

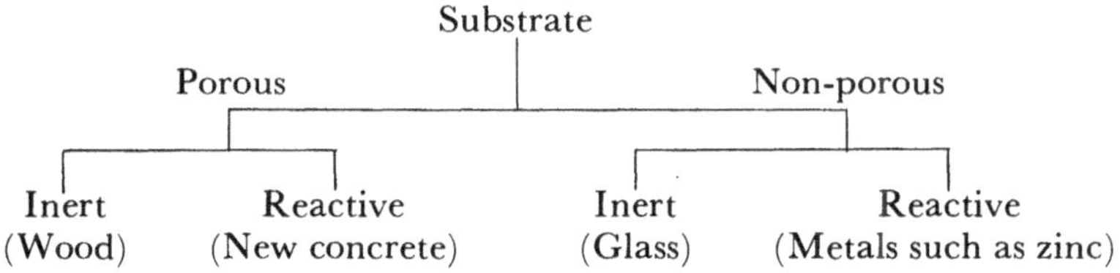

A broader practical classification could be as follows:

Type	*Common examples*
Metals	Steel, aluminium, zinc, copper, cadmium, chromium, lead, tin, magnesium
Timber	Many species
Masonry	Plaster, concrete, brick, stone
Wallboards	Masonite, Victor Board, Gyprock, Fibro, cement.
Plastics	Polythene, Polystyrene, Perspex.
Bituminous compositions	Malthoid, road surfaces
Previously painted	Various conditions
Glass and tiles	Glazed surfaces, ceramics
Paper	Packaging
Fabrics	Canvas
Leather	Furniture, equipment

In the sections below, the substrates and their preparation for coating will be grouped by type.

48.4 TYPES OF SUBSTRATES

48.4.1 Metals

In general, degreasing, removal of significant corrosion, micro-roughening and careful selection of primers is essential. Before paying more detailed attention to frequently used metals, the common methods of surface preparation utilised will be reviewed generally. Hand chipping or descaling may be required before any cleaning method is undertaken where heavy scale has built up on the article.

Chemical cleaning. Most frequently, corrosion build-up is removed with an inhibited phosphoric acid-based treatment. Such materials have the advantages of speed and their need for little equipment. Alkaline cleaners are generally employed to remove old surface coatings and frequently involve elevated temperatures and time-consuming dipping and soaking procedures. Citric acid solutions at about 10–15 per cent concentration are employed in baths at room temperature where rust removal can take place at a more leisurely pace and if importance of minimum metal removal is great.

Hand cleaning. Wire brushing is frequently used to achieve a minimum standard of cleaning before painting. It is important that the brushes are made from steel wire, as many are made from brass and may initiate electrolytic surface corrosion subsequent to treatment of ferrous metal. Power wire brushes tend to polish flat metal areas if used excessively, and this can result in adhesion difficulties.

Blasting cleaners. Sand, shot and bead blasting are particularly efficient ways of surface-cleaning metallic surfaces, with the added advantage of providing a surface texture which can be adjusted to suit the adhesion requirements of the metal/primer situation. Generally steel abrasives are most effective, and again the finer size range is most efficient, consistent with the size of anchor pattern required; for example, a heavy pattern is desirable for silicate-based zinc-rich primers, and a fine pattern for conventional primers. It should be noted that 20 000 to 120 000 pellets or 'hammers' per kilogram of shot will be applied to the task, depending on the abrasive size chosen. Blast cleaning equipment delivers shot on a kilograms per minute per horse power basis, so a great deal more useful action is normally associated with a size reduction in abrasive, provided that the anchor pattern is sufficient.

Glass bead blasting is a remarkably effective procedure for cleaning fine-machined metal parts with minimum surface disruption and metal loss, combined with attractive surface presentation. This can be important in many engineering applications or with the softer or more expensive metals.

48.4.1.1 Magnesium/Magnesium Alloys

These need to be cleaned with solvent or alkali at a pH about 11—the metal is rapidly attacked at lower pH's. These are chemically reactive substrates, and coatings should be non-acid forming. Wash primers are unsuitable and can cause gassing under the film. Half-second butyral clears are satisfactory.

Note: Mg (Rust) $\rightarrow$ Mg $(OH)_2$, which saponifies alkyds.

48.4.1.2 Copper–Brass–Bronze

Hydrogen sulfide as low as one part per million can cause severe tarnishing. Thermosetting silicone primers followed by acrylic topcoats are ideal. Good results are obtained also with air-dry acrylics plus benzotriazole. Generally, pigmented systems when applied over wash primers maintain good adhesion.

48.4.1.3 Cadmium (Plate) and Zinc

These substrates are usually in the form of very thin, dense, electroplated films. Paint films cured by oxidation usually lose adhesion rapidly, but good results can be obtained with etch primers and chlorinated rubbers. Both zinc and cadmium are associated with poor adhesion, probably because of a weak surface layer of friable oxide. Mechanical or chemical treatment removes this, and selected coatings will then adhere as well as to steel.

Galvanised iron with a zinc layer about 80 μm thick provides galvanic protection to steel and can cause adhesion problems unless carefully primed. Wash primers have provided good adhesion, and specially formulated non-saponifiable latex primers also perform very well.

Possibly because of the chemical reactivity of the surface, saponifiable coatings rapidly fail to adhere. Pretreatment will generally give a good key after efficient degreasing. Reliance on natural weathering seems to have little to commend it as a pre-painting requisite.

Electrolytically zinc-plated steel appears to offer some additional problems to surface coatings formulators, possibly because of a high degree of porosity, but latex primers often work well.

48.4.1.4 Chromium (Plate)

Little reference can be found to this substrate. Such surfaces offer little key to coatings. Surface abrasion is recommended before coating (e.g. shot blast or bead blast or at least power brush).

48.4.1.5 Lead

Lead is generally not a problem metal, because of good chemical stability. In many applications such as container tubes and flashings, flexibility of this substrate creates special problems calling for maximum adhesion and flexibility in the applied coating. Thorough degreasing of this metal is of critical importance but care must be taken to avoid destroying the surface by overenthusiastic surface preparation.

48.4.1.6 Tin (Plate)

Tin is generally a good surface to coat because of inherent chemical stability. Degreasing is critical—hot-dip tin plate generally has a thin film of palm oil, while electrolytic plate frequently is filmed with dioctyl sebacate. Adhesion can be further improved if the surface is lightly abraded with fine wet-dry paper which removes any tin oxide layer on hot-dip plate or the mixed chromium/tin oxide film sometimes found on electrolytic plate.

48.4.1.7 Aluminium

Aluminium is a most important structural metal. Special alloys have been evolved for many critical applications that were once thought unlikely for such a reactive metal—for example, in sea-going craft permanently exposed to salt conditions, and supersonic aircraft subject to elevated temperatures, high stress and vibration. Aircraft surfaces certainly suffer extremes of rapid temperature variation: at Mach 2–2.3 the aluminium skin of the 'Concorde' is reported to maintain approximately 130° C. Etch primers are recommended after careful surface cleansing.

48.4.1.8 Steel

Steel is possibly the most important substrate protected by surface coatings. Common mild steels are rapidly corroded in the atmosphere. Reactivity with oxygen in the presence of moisture is accelerated by electrolytes and the lack of chemical uniformity found in the metal itself.

Mill scale formed during high-temperature fabrication should be removed by blasting to white metal to ensure genuine coating contact with the metal itself. Alternative pickling or phosphating processes are also applied for the same purpose, although blasting is frequently favoured for critical applications. The subsequent protection of steel can depend on many factors (e.g. anticipated environmental factors, operating temperatures).

Where this important substrate can be thoroughly prepared and is to be coated, various types of zinc-rich coatings are widely favoured. If the preparation available is minimal, red lead or chromate primers are more suitable, as they wet the surface well and tend to inhibit further corrosion chemically. They do not have the advantage of cathodic protection, being a practical feature of the genuine zinc-rich primer; this is of value in aggressive locations, particularly if minor mechanical damage is likely in service.

Metal cleaning and pre-treatment is further covered in chapter 49.

48.4.2 Timber

Wood is unquestionably one of the most variable of substrates. Even within the one piece, major variations in density can be apparent, caused by various factors including knots and summer and winter growth.

Fluorescence microscopy indicates that it is mainly solvent that penetrates a wood substrate, with little actual penetration of the paint vehicle.

Adhesion and durability appear to be related to good wetting of the surface layer of wood cells—with different species of timber having a dramatic influence on the life of the most carefully selected coatings.

There is considerable room for improvement in the selection of the most suitable species for various building applications. In particular, the range of timber really well suited to exterior use would seem most limited and in need of close definition and recognition by the timber and building industries in Australia at least. Commercial milling and drying procedures for critical applications also appear to need some further review if this traditional building material and the final coated timber surface is to regain wide acceptance.

With co-operation between the timber and paint industries, vastly improved performance is technically achievable. It is pleasing to note progress is being currently made in this direction.

For full details of the correct preparation of timber refer to Australian Standard AS 2311–1979, 'Guide for the Painting of Buildings', Section 3, page 11.

48.4.3 Masonry

48.4.3.1 Plaster (fibrous)

This substrate is apparently not widely used outside Australasia.

Variable smoothness and density, water-soluble salts, dampness, alkaline reaction and high porosity all present special problems in coating. At times, the surface layer is notably weak and underbound and will create adhesion problems especially in latex systems unless treated; dilute phosphoric acid washes have been found effective. Lime plaster joints create special problems: in texture, alkalinity, relatively lower absorbency and the presence of calcium ions, which can flocculate colours in latex paints, causing pale areas over such joints.

48.4.3.2 Brick

Generally, brick is a fairly reliable substrate. However, some local types have a loosely bound surface, while others containing vanadium salts are prone to cause yellow staining under damp conditions—both of the brick surface itself and of subsequent coatings.

48.4.3.3 Stone and Terracotta Tiles

Generally, these offer no special problems as a substrate, being rigid dimensionally, chemically stable and at least microscopically rough.

48.4.3.4 Concrete

A firm concrete surface which is both cement-rich and well cured and at least microscopically rough, provides an almost perfect substrate. Most concrete work, however, tends to shrink on drying, leaving fine capillaries which readily reabsorb water on wetting and must also be regarded as a highly variable substrate. It is almost impossible to obtain a satisfactory bond to the friable, crumbly surface at times presented. Certainly, coatings need to have good alkali resistance, and sealers specifically formulated to resist alkalinity also can overcome excessive porosity.

With the rigidity and stability of such a surface, paints should achieve particularly long service lives when applied to well-made and well-cured concrete.

48.4.4 Wallboards

Generally, wallboards provide a sound substrate in themselves and most problems are associated with joints between sheets.

48.4.4.1 Masonite

Dark colour combined with high and variable liquid absorption rates renders this a difficult substrate for most coating systems. Standard oleo sealers have proved effective and also reduce the risk of water staining encountered at times with all latex systems.

48.4.4.2 Victor Board and Gyprock

This type of substrate is actually paper-faced gypsum boards. For many years cement sealers were specified to form a complete moisture barrier and to make the variable suction between board and joints more uniform. Generally, these boards accept paints well, especially when applied over specially formulated latex-based sealers.

48.4.4.3 Fibro Cement

Until about fifteen years ago, the high alkalinity of these sheets was a severe problem which limited choice of at least priming systems. Today, sheets are generally autoclaved and have little residual alkalinity. They thus represent a particularly stable substrate for both latex and conventional coatings, with a fair degree of porosity and good dimensional stability.

A simple experiment was conducted in an endeavour to permit realistic comparisons of rates of absorption of some simple paint liquids for timber, masonry and wallboards. A uniform bore glass tube was cut into equal lengths and mounted on the substrates with a silicone-based two-pack sealant so as to form a firm liquid-tight edge joint. The substrates themselves were as fresh as possible and fairly typical of commercial quality. At least two areas were checked for each liquid, and where results significantly varied further areas were chosen until a consistent performance was obtained.

All liquids were at approximately 10° C and were of standard current commercial quality. No attempt was made to allow for evaporation from the tubes, as this was found to be negligible under the conditions and time of the tests.

From figure 48.1 it may be noted the very wide range of rates of absorption among building materials most frequently simply classified as porous or absorbent. Mineral turps, possibly because of its low intrinsic viscosity, was most rapidly absorbed in all the cases examined, although in the case of brick, water was taken up as quickly. The extremely high rates of penetration noted in brick, fibrous plaster and Gyprock obviously call for special attention from the paint technologist. The hundred-fold gain in rate of penetration in Gyprock when 0.10 per cent of a non-ionic wetting agent was added to the water is interesting, but variation of pH from 7.0 to 9.5 produced no significant variation in absorption on this substrate.

48.4.5 Plastics

Some years ago many technologists believed that self-coloured plastics would rapidly reduce or largely eliminate the need for paint in many common applications. While this prediction has to some extent come to pass, plastic articles themselves are increasingly becoming substrates for new coatings.

A whole new field has developed, including moulded plastic furniture coated to give a natural wood look, plastic car parts painted to match the car, household appliances and toys. In fact, all major plastic types can be successfully coated, with the possible exception of Teflon.

48.4.5.1 Polyethylene

Adhesion to polyolefins is difficult to achieve. This is because of the lack of polarity or porosity of the surface and excellent solvent resistance. Surfaces can be prepared for painting by mechanical abrasion, or flame and/or chemical treatment. A brief impingement of oxidising flame results in slight oxidation of the surface layer, producing surface polarity which is most beneficial.

48.4.5.2 Polyesters

Polyesters are the most frequently painted plastics and present a rigid surface with few painting problems—provided intimate contact with the lightly abraded surface is obtained. Generally, wiping with water/alcohol/acetone mixtures is helpful in improving the surface cleanliness.

48.4.5.3 Polyvinyl Chloride (PVC)

PVC sheet may be coated with solution vinyl coatings low in plasticiser despite high flexibility requirements. Subsequent plasticiser migration from the sheet generally promotes adequate flexibility.

48.4.6 Bituminous Surfaces

Bituminous surfaces are most frequently involved in paving or roofing applications. Generally, slight solvent action is desirable but not essential to good adhesion, provided any powdery weathering residues are first removed. A fair degree of flexibility in the dried coating and the avoidance of strong solvents of low volatility are advisable because of the movement and softening of the substrate in hot weather and its inherent solubility.

48.4.7 Previously Painted Surfaces

Previously painted substrates are probably the most variable of all, the age, condition and composition of previously applied coatings having the most profound effect on the life of subsequently applied systems.

Less obvious variations in performance such as chalking rates and degrees of chalk fade can be significantly altered in repaints over old systems which may vary, for example, only in the grade of titanium dioxide used in them or in type of vehicle. This has been borne out in numerous exposure trials in Sydney.

Before recoating, thorough practical examination of the substrate is advisable. Adhesion can be assessed by tape testing after removing superficial chalk and grime. If the surface is not free of gloss, at least some abrasive treatment is desirable in most instances to increase the surface area for bonding and mechanical keying.

Despite reasonable precautions there are many instances of apparently sound, aged coatings deteriorating fairly rapidly after recoating, thus greatly reducing the anticipated life of the new system. Such failures are frequently attributable to the progressive loss of flexibility of the old system—particularly over timber after, say, three or four repaints with currently used coatings. It is widely recognised that latex paints have little tolerance for powdery chalked surfaces, and substrate preparation for such coatings must take this into account by either removing the loose layer or binding it with a more conventional-type coating before applying latex paints. Special care is needed with gloss latex paints as these are inherently very flexible and, while performing very well on new work, are likely to fail if applied over old oil/alkyd systems consisting of multicoats of brittle material due to excessive variation in flexibility.

In many instances, such as those cited above, it would be preferable to remove the coating before repainting. Various methods of paint removal are available and the choice would depend on the size and location of the substrate, as well as considerations of finance and time. The methods are briefly described below, with some being identical to those already described as being suitable for cleaning previously unpainted substrates.

48.4.7.1 Hand-Sanding, Mechanical Disc and Orbital Sanders

Many grades of papers are available, ranging from very coarse to very fine in carborundum, emery, alumina, glass and sand. Some are used dry and others wet or dry, depending on the grit, adhesive and backing used.

This method is commonly used in house painting and for furniture and is inexpensive. It is not often used for complete film removal, but will partially remove surface chalk and dirt and leave a good base for subsequent coats. Disadvantages include difficulty in reaching irregular surfaces, damage to soft substrates, high labour content and incomplete removal of paint.

48.4.7.2 Wire Brushing (Hand and Mechanical)

Wire brushes are available in coarse and fine grades and in many shapes to facilitate access to difficult areas. This method of paint removal is more suited to hard metal substrates and will damage soft substrates very easily. Labour content is high, but the equipment is inexpensive.

48.4.7.3 Burning Off—Blow Torch and Scraper

This method has been used for many years on house walls and is very effective, if carefully done. The earlier kerosene-fuel blow torch is gradually being replaced by liquified gas units with various nozzles to give the desired flame pattern. The method used is to soften the old paint and follow closely with a hand scraper. It is not intended to burn and char the old film. One disadvantage is a high fire risk, both for substrate and also for surroundings, unless burning paint droppings are carefully watched. Other disadvantages are relatively high labour cost, difficulty in scraping irregular surfaces, charring and distortion of substrates and unsuitability for removing some modern surface coatings which char, rather than soften, when heated. One incidental advantage with this method is that any mould or lichen on the surface is destroyed and the surface made sterile. It also ensures a dry surface substrate if coated without delay.

48.4.7.4 Solvent-based Strippers

At one time these were based on benzene, but because of toxicity and flammability this has now been replaced by methylene chloride or similar solvents. Likewise, paraffin wax was used as a thickener and evaporation retarder, but this leaves a wax deposit on the surface and, unless cleaned with further solvent, will retard drying of subsequent paints. Current strippers generally use thickeners derived from cellulose.

Small amounts of additive are included to give faster stripping action, longer solvent retention time even on vertical surfaces, and wetting agents to render residual stripper water rinsable. This method is suitable for cleaning small articles around the home but it is not practical or economical for stripping large areas such as walls. In addition, multiple applications are generally required for heavy film builds.

48.4.7.5 Grit Blasting—Wet and Dry

Many types of grit are available in various mesh sizes, and the type of grit selected depends on the film to be removed and the type of substrate. These units are used predominantly for industrial or maintenance applications and rarely for home cleaning. Strict regulations covering health hazards from flying grit make the unit more suited to fixed installation work than mobile units. Use of wet grit overcomes the dust hazard, but is more expensive because recovery is difficult. A major advantage with this method for ferrous substrates is that not only are old paint films removed, but also rust and scale, leaving a bright metal surface which is ideal for repainting.

48.4.7.6 Steam Cleaning

Steam cleaning is a relatively inexpensive method for factory maintenance, if steam lines are available, but if mobile units and skilled operators are required then it requires a fairly large job to be warranted. It is quite effective as a method on certain types of paints and substrates, but advice on its suitability should be obtained from steam-cleaning specialists before proceeding.

48.4.7.7 Water Blasting

A relatively new method, which relies on a water jet at very high pressure to remove the film of paint, Various nozzle types and pressures are recommended, depending once again on the film to be removed and the substrate type. Units may be purchased or hired and are quite small and mobile.

48.4.7.8 Chemical Strippers

These are a specialised field, and many types are available for industrial stripping baths where temperature and chemical concentration can be controlled. From the householder's point of view, chemical strippers available are generally of the caustic type and require careful handling and must be neutralised and thoroughly washed off the substrate. They are not very effective in a single application if applied as a wash to a surface, but work well if the article can be immersed in a bath of some sort.

They are relatively cheap materials, but because of removal difficulties and hazards associated with skin contact, are not often used.

48.4.8 Glass and Tiles

The smooth, non-porous surface of glass and tiles offers little mechanical or chemical foothold. Where such a surface can be slightly roughened by a light sand blast, good adhesion is ensured. If this cannot be done, careful surface cleaning and application of thin coats of a selected epoxy-based primer give generally acceptable adhesion.

48.4.9 Paper

Generally paper is a highly porous substrate with good paint-holding characteristics. Coatings are selected mainly for low penetration and good anti-blocking characteristics. Reports have been noted of the electrodeposition of colloidal dispersions on to suitably sized paper.

48.4.10 Rubber

Because of the flexibility and complex chemical nature of this substrate, rubber presents unusual difficulties to the coatings formulator. Adhesion is difficult to achieve, and discoloration of light colours over some rubbers is a problem. Coatings need high tensile strength, elongation and resilience, and minimum tendency toward embrittlement on ageing. Generally, special flexible two-pack polyurethane systems work satisfactorily under these conditions.

48.4.11 Fabrics

The coating of fabrics with conventional paint systems is not widespread. Generally, a coating penetrates the fabric readily, giving good adhesion. The fabric, if firmly fastened, will tend to reinforce the paint film and result in a hard wearing surface often used, for example, on decks of small craft. Pretreatment with anti-rot agents is desirable for natural fibres in damp locations.

48.4.12 Leather

A highly flexible, porous substrate of generally uneven texture. Coatings applied can lift the grain or cell wall structure unevenly, and it is frequently necessary to subsequently hot-press the coated surface to obtain uniform, smooth coatings. Aqueous acrylic and polyurethanes are widely used on this surface.

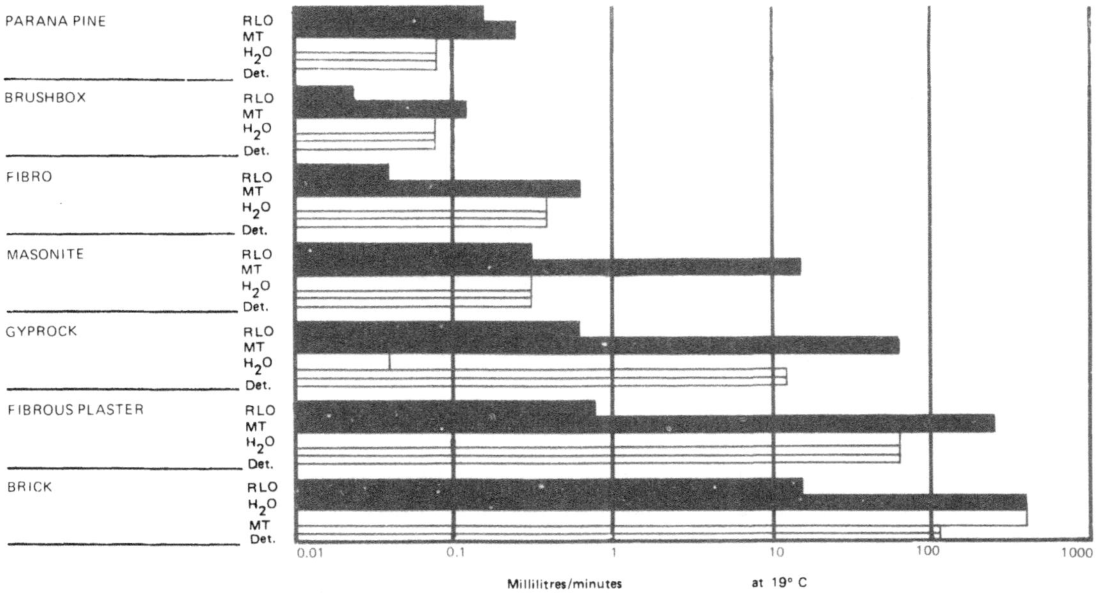

FIGURE 48.1
Some common absorbent substrates

48.5 CONCLUSION

It may be concluded that common substrates present a wide range of demands to the coatings formulator and applicator. In increasing knowledge of these surfaces and in the evaluation of new man-made alloys, boards and plastics as they are developed, the usefulness of coatings can be enhanced generally and the life and usefulness of many costly constructional materials increased greatly. Closer co-operation between the paint industry and those industries that produce materials ultimately requiring protection and decoration will benefit all concerned.

49 METAL CLEANING AND PRETREATMENT

49.1 INTRODUCTION

Prior to the application of organic coatings, all surfaces should be clean—that is, *free from liquid and solid contamination*, in a state where they provide a good 'key' and *maintain good adhesion under service conditions*, and be *resistant to degradation brought about by discontinuities in the film*.

In the case of metals, these requirements mean that oxides, swarf and other solid residues, grease, oil and other contaminants should all be removed. Further, the surface may need to be modified to enhance the adhesion of the organic coating and the corrosion resistance of the whole system.

The cleaning and pretreatment of metal should be considered not as a separate issue but as *part of the total finishing system*. The system selected should take account of the service conditions and required performance, together with an assessment of the most economic means of achieving the required result. Thus, although the optimum situation is for a surface to be perfectly clean, it is a fact that some finishes are able to cope better than others with certain amounts of contamination. Similarly, the inclusion of a *conversion coating* in the total system may be dictated by the performance standards, but an alternative paint system may be adequate without it.

Almost all conversion coatings are applied in aqueous solutions, so it is usually logical to carry out the prior cleaning step also using aqueous processes. Where no conversion coating is to follow, the use of aqueous cleaning may not be the best approach, especially if—as required for the subsequent application of non-aqueous paint—an energy-intensive drying step must follow.

The metals most commonly encountered in industrial finishing in Australia are steel, aluminium, zinc (mostly in the form of hot dip or electro-galvanised steel) and *Zincalume* (a 55 per cent aluminium 45 per cent zinc alloy coated steel of John Lysaght (Australia) Ltd). The cleaning and conversion coating requirements for these have both been well defined and they vary from metal to metal. Other metallic surfaces such as copper, brass, chromium, tin, and stainless steel are encountered as either a base metal or as electroplate. These require organic finishing in various contexts.

Cleaning and oxide removal of metals can be accomplished either by mechanical (abrasive) or chemical methods. The choice, as described before, is dictated by the total system requirement, plus economy and convenience. When chemical methods are used (generally by immersion or recirculating spray), the removal of oxide is usually a separate step best carried out after degreasing. A one-step process can cope if grease and oil contamination is light.

A further kind of surface phenomenon requiring attention is loosely described as *smut*. Smut may relate to alloying elements in the metal which are differentially affected by previous treatments such as acid or alkaline etching. The same term is often used to describe carbonaceous residues from heat treatment or similar processes. Sometimes chemical desmutting can be used; in some instances, cleaners analogous to paint strippers are effective (for example, for carbonised oils); but there are many occasions when only physical abrasion will suffice.

There is therefore a whole range of separate processes that may be involved in preparing a metal surface for the application of paint. Conversion coatings are relatively thin, inorganic coatings produced on metals by reaction with chemical solutions. The useful conversion coatings are adherent, inert and of high surface area per unit volume and thus improve the adhesion and corrosion resistance of paint systems. It is characteristic of solutions used to form conversion coatings to be reactive to the metal, and to be 'balanced' so that the reaction produces insoluble species, which bond on to the surface. For example, a typical zinc phosphating process is a solution of zinc phosphate that is just acidic enough to maintain the compound in solution. At the interface of the solution and the metal to be treated, the pH rises as acid is consumed by the dissolution of metal. In this boundary layer the (simplified) process is deposition of zinc phosphate which becomes supersaturated immediately adjacent to the metal surface. Commercially used solutions contain various additives such as activators and accelerators. Conversion coatings are described in greater detail in chapter 41.

In general, cleaning and conversion coating processes operate at elevated temperatures and generate sludge and wastes. A large proportion of research and development in the field of metal cleaning and conversion coating is devoted to decreasing the energy consumption and need for effluent treatment.

49.2 CLEANING

49.2.1 Metal Surface Preparation

Mechanical methods of preparation, which are commonly used to remove rust and scale from steel on large outdoor structures, are covered by relevant Australian Standards. Specialised equipment is available. In blasting, the *media*, where possible, are collected and re-used. *Wet blasting* uses water to minimise dust, and additives to the water are used to delay rerusting of the clean surface. *Wire brushes*, *needle guns* and *flame cleaning* are other non-chemical methods.

All the mechanical methods are used to some extent in industrial finishing. For example, *spot finishing* is widely used to remove local surface rust or other imperfections in mass production finishing. Blasting procedures are sometimes used: tubular steel furniture is one example, where the metal surface characteristically has smut, scale and rust and the system does not call for a conversion coating.

In the case of coated steels, mechanical abrasion of the surface would be used as a cleaning practice only in unusual circumstances. Because the life of a coated product is dependent on the metal coating thickness, mechanical abrasion can significantly reduce the coating thickness and hence the life of the coated steel.

49.2.2 Solvent Degreasing

Most oily residues are more easily removed by solvents than by any other method; even polymerised oils can be handled by preparations similar to solvent-based paint strippers. However, there are limitations with respect to toxicity, fire hazard, suitability for mass production processing, cost and compatibility with the total system. In practice, there are some situations, particularly in small- to medium-volume applications where no subsequent conversion coating is needed, where solvent degreasing has a definite place.

Vapour degreasing was developed utilising the properties of trichlorethylene, namely good solvency, convenient boiling point, non flammability, and high vapour density. It is therefore possible to design steam-heated tanks which maintain a zone of vapourised trichlorethylene above the boiling liquid. Cold workpieces lowered into the vapour collect condensed (and therefore pure) solvent which runs off taking the oily contaminants with it. Generally, a water condenser near the rim controls vapour loss, as does the provision of a close-fitting lid. The chlorinated solvent is stabilised to prevent the generation of hydrochloric acid—which would encourage rust—by degradation. With a large enough tank, a continuous conveyorised operation can be used. Unfortunately, vapour degreasing leaves almost all solid contaminants—even dust—on the degreased surface, so some auxiliary operation may be required.

Trichlorethylene not only produces toxic substances in contact with flames, such as cigarettes, but also has been increasingly recognised as being unacceptably toxic in its own right. However, alternative but similar solvents such as 1,1,1-trichlorethane have been adopted and vapour degreasing processes maintain a place in industry.

The point has already been made that the advantage of solvent degreasing compared to aqueous degreasing, namely the speed of subsequent drying, is of little consequence if an aqueous conversion coating is to follow; however, there are non-aqueous conversion coatings available. In fact, there is a rather elegant, if not widely used, process where degreaser, conversion coating and paint are all based on chlorinated solvents and in a three-stage conveyorised process, workpieces can be degreased (vapour), phosphated and painted (both immersion).

Manual, immersion or spray degreasing using solvents can be carried out using either chlorinated, hydrocarbon or various polar solvents. Manual solvent wiping is generally practised for small-volume production on large workpieces. If very clean rags are continuously applied, the results can be excellent, but inaccessible areas are a problem and the method has the disadvantage of being completely operator-dependent. In addition, the use of flammable solvents in this or other solvent degreasing processes is dangerous.

Immersion cleaning is used with various solvents. Technically, the best approach is where the solvent has been modified so that residues on the surface can be emulsified in a subsequent water rinse; the build up of contaminants in the tank is therefore not critical.

Formulations akin to cold paint strippers are used for very heavy duty cleaning—for example, the decarbonising of engine parts. Emulsifiable hydrocarbon solvents are used in a somewhat similar way: one particular example is used prior to manganese phosphating, in which the solvent cleaner has a more beneficial effect on crystal structure compared to the use of harsh alkaline cleaners. The solvent cleaners usually operate at ambient temperatures, in contrast to the hot alkaline cleaners.

49.2.3 Aqueous Cleaning

Aqueous cleaners include solvent emulsions at room temperature, detergent solutions, acid and alkaline solutions and combinations of some of these types. There are also available compositions that combine cleaning with etching, pickling or deposition of conversion coatings. Although, historically, soaps were important, and then sulfonated oils, modern cleaning formulations depend on the availability of a wide range of synthetic organic surface active agents whose properties depend on wetting, foaming and emulsifying characteristics and their hydrophilic/hydrophobic balance. Particularly in the case of alkaline cleaners, the optimum properties are usually obtained by combining the organic surfactants with inorganic 'builders', which add properties of, for example, saponification, sequestration and dispersion. Although there are good general-purpose cleaners, there is a need for many special-purpose compositions which may contain eight or ten different components.

49.2.3.1 Emulsion Cleaners

Emulsion cleaners are commonly based on oil-in-water emulsions of hydrocarbon solvents. When used at high concentrations they have some of the advantages of solvent cleaners without the same fire hazard. Such compositions are also used as 'in-process' cleaners with no rinsing, where the 'difficult' greases and solids are removed and the cleaner leaves soft oily residues which are easily removed later prior to painting. Emulsion cleaners are often effective at relatively low temperatures. They also have a favourable effect on the crystal form of conversion coatings such as zinc phosphate. Some types of contamination require warming to effect speedy removal by aqueous cleaners. In such cases, emulsion cleaners, or those with reasonably volatile solvents, are not very practicable.

49.2.3.2 Neutral Cleaners

Neutral cleaners, which may be based on surfactants with or without water-miscible solvents and inorganic 'builders', may be used for less 'difficult' cleaning jobs and also where etching of metals such as zinc or aluminium is not desired. A further advantage of neutral cleaners is that less pretreatment would usually be needed prior to the disposal of wastes and rinsings. The concentrated surfactant blends covered by this class of cleaner are more often used as boosters for other categories of cleaners such as the alkaline types.

49.2.3.3 Alkaline Cleaners

Alkaline cleaners are the most widely used for general applications. An extremely wide variety of formulations is available but the common ingredients are organic surfactants plus inorganic builders.

The organic component may include several non-ionic surfactants, possibly with some anionic types, chosen to give the metal wetting/soil removal and emulsifying properties dictated by the cleaning task, plus free rinsing. The inorganic component may include alkalies and/or alkaline salts such as carbonates, silicates, phosphates and borates. The alkaline salts provide saponifying and wetting properties; silicates combine these properties with good soil dispersing abilities; phosphates enhance many of the properties mentioned, and polyphosphates in particular are good sequestrants. Gluconates are also widely used as sequestrants both to minimise problems from hard water salts, and to assist etching cleaners to function in the presence of large amounts of dissolved metal.

There are no simple universal tests to indicate when a cleaner is 'spent' or, in many cases, when a cleaner is at the established concentration. In the case of alkaline cleaning baths, control is usually effected by determining the levels of free or total alkali. As the concentration of 'non-titratable' contaminants, such as oil increases, the effectiveness of the bath gradually diminishes. The time to dispose of the bath and start again is when the cleaning efficiency has reduced significantly and is not satisfactorily restored by moderately large amounts of replenishment. It is, in some cases, more economic and efficient to operate a cleaning bath at low concentrations, and dispose of these frequently, than to attempt a long bath life at increasing concentrations. Each cleaning task has different requirements depending on the type and quantity of soils, the metal substrate, the availability of equipment and the cleaning standard necessary for subsequent processing.

49.2.3.4 Acidic Cleaners

Acidic cleaners are used for specialised applications or where a combined cleaning action is sought, such as cleaning together with deoxidising, brightening, etching, or phosphating. Thus acidic cleaners may vary from a predominantly acid pickling composition (reinforced with solvent and surfactant) to mildly acidic phosphating solutions, typically at pH 5, together with surfactants. Soils of soap or fatty acid types are likely to form scums on the surface of strongly acidic solutions, in which case application by recirculating spray may be quite effective, but application by immersion may not.

49.2.4 Tests for Cleaning Efficiency

Testing ideally monitors (i) that the cleaning bath or medium is in the desired condition and (ii) that the metal surfaces after treatment are satisfactorily clean.

Unfortunately, immaterial of whether the system is solvent-based or one of the various forms of aqueous cleaning systems, the condition of the cleaner can, at best, be only partly characterised by simple plant tests. For example, checking the alkalinity of an alkaline cleaner gives only one parameter, while in all cases the degree to which the cleaner is 'spent' (by built-up soils) cannot be objectively rated. In practice, the best check on a cleaning system is how well it is cleaning.

The cleanliness of a surface can be assessed by wiping it with some clean material and examining both the material and the area wiped. A common test for aqueous cleaning is the degree to which the surface after rinsing remains 'water-break-free'. There are potential traps in these methods; for example, adsorbed surfactants can make an oily surface water wettable. In addition, some residues from 'dry-out' prior to rinsing may be extremely hydrophilic. Therefore, especially when the total system demands a high standard of cleaning, great care should be exercised in assessing results.

49.3 CONVERSION COATINGS*

Conversion coatings are so called to describe the process by which they are formed: conversion of the metal at the surface into a compound with desirable properties of *adhesion*, *high surface area* and *inertness*. Actually, according to strict definition, the term is for some coatings a misnomer because the compound is sometimes one of a metal other than that of the metal to be coated. Such examples include zinc and manganese phosphates on steel and chromium phosphate on aluminium.

Conversion coatings are formed, in common, from chemical solutions, reactive to the particular metal being treated, and balanced with respect to certain ingredients so that an insoluble compound builds up at the surface where the equilibrium is disturbed. After rinsing and drying the surface provides enhanced adhesion and corrosion resistance to most paint systems.

Some conversion coatings such as zinc and manganese phosphate are quite visibly crystalline. Many others are termed amorphous, often being hydrated gels. Among the 'amorphous' types are iron phosphate, chromate (on aluminium), and chromium phosphate.

The useful coatings, properly prepared, have good adhesion to the metal surface and, when dry, a large surface area. These factors account for their adhesion-promoting characteristics. The useful coatings are also relatively inert, and by covering a great proportion of the local electrodes on the surface, act as a barrier to corrosion. A suitable conversion coating thus provides enhanced adhesion to the subsequent paint, as well as corrosion resistance to the total system.

Although the coatings have a degree of porosity, the best corrosion resistance results generally from dense coatings with relatively few pores; the reactivity of the metal surface due to these pores can be further reduced using a final passivating rinse.

It is only since the early twentieth century that conversion coatings in the modern sense have existed. Phosphate processes developed at that time required long immersion times, often of several hours. Subsequently, studies of 'solution balance' and the discovery of accelerators have decreased coating times to minutes and, more recently, for industries such as coil coating, to seconds.

The developments in phosphates have been paralleled with other treatments, such as chromates for aluminium. In figure 49.1, the beneficial effect of certain pretreatments on the corrosion resistance of a one-coat acrylic enamel on steel is shown.

* Conversion coatings are discussed further in chapter 41.

49.3.1 Zinc Phosphate (see also chapter 41, section 41.5)

Zinc phosphate treatments are applicable to steel, zinc and aluminium, and have been widely used for decades in industries such as automotive and major domestic appliances; in fact, without pretreatments of this type, the mass-produced unitary-construction steel motor car body might not have been a practical possibility.

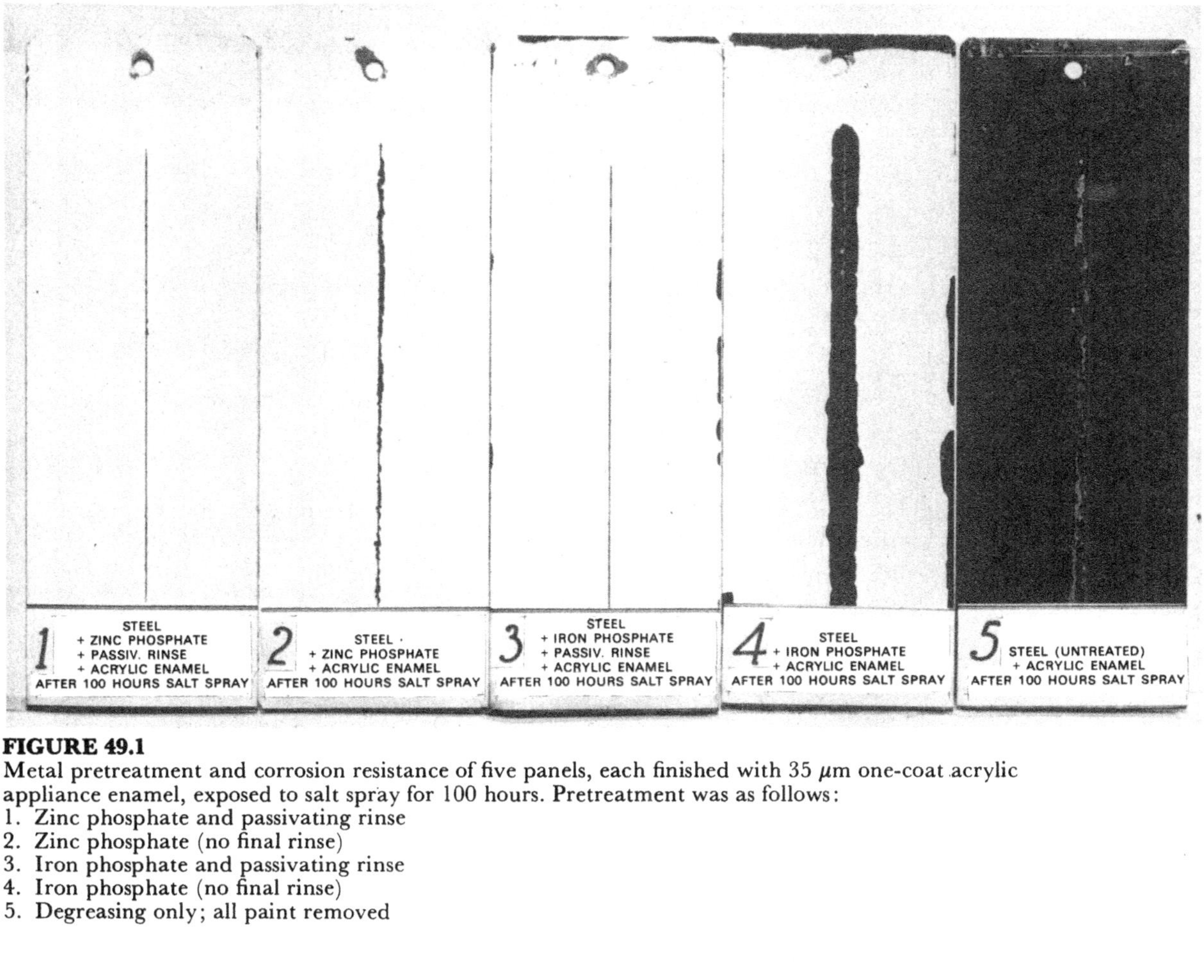

FIGURE 49.1
Metal pretreatment and corrosion resistance of five panels, each finished with 35 μm one-coat acrylic appliance enamel, exposed to salt spray for 100 hours. Pretreatment was as follows:
1. Zinc phosphate and passivating rinse
2. Zinc phosphate (no final rinse)
3. Iron phosphate and passivating rinse
4. Iron phosphate (no final rinse)
5. Degreasing only; all paint removed

The earliest zinc phosphate treatments were designed to treat steel only, in an immersion process. By today's standards, the process times were lengthy and the coatings rather coarse and heavy. Subsequently, process times and coating quality have been improved by the use of accelerators, improved 'solution balance' and application by recirculating sprays. The formation of finer, denser and more adherent coatings has also been found to be assisted by the prior application of conditioning chemicals such as colloidal titanium phosphate. Other formulation variations have extended the metals coated to include zinc and aluminium. Finally, continuing development in the field of final rinses has further upgraded the corrosion resistance.

Zinc phosphating solutions are generally made up as a water solution of 1 to 5 per cent of a concentrate, the level of zinc in the final bath being of the order of 1 to 5 g/L. The simplest concentrate consists of a close-to-saturated solution of zinc phosphate in phosphoric acid plus an oxidising agent, and a simplified mechanism of coating formation includes the following steps:

(a) the solution reacts with the metal at the boundary layer;
(b) the pH rises and zinc phosphate precipitates; and
(c) the accelerator 'oxidant', by depressing gassing and removing dissolved iron as insoluble ferric phosphate, maintains an active bath.

There are other oxidants used, such as chlorates; and other additives, such as nickel salts and fluorides are used particularly where zinc and aluminium are to be treated. The actual coating (on steel) usually contains some iron phosphate near the metal surface, in addition to the predominant zinc phosphate. A much simplifed reaction mechanism is:

(a) The metal surface reacts with the phosphoric acid, producing iron phosphate and reducing the phosphoric acid level:

$$3Fe + 2H_3PO_4 \rightarrow Fe_3(PO_4)_2 + 3H_2$$

(b) This effect encourages the formation of the insoluble tertiary zinc phosphate coating from the soluble primary zinc phosphate:

$$3Zn(H_2PO_4)_2 \rightarrow Zn_3(PO_4)_2 + 4H_3PO_4$$

(c) Sludge is formed by the production of ferrous phosphate due to the presence of nitrite accelerators:

$$H_3PO_4 + Fe_3(PO_4)_2 + 3HNO_2 \rightarrow 3FePO_4 \downarrow + 3H_2O + 3NO$$

A nitrite-accelerated bath requires the simultaneous replenishment of zinc phosphate concentrate and accelerator and testing of bath concentration ('pointage' to 'total acid') and nitrite concentration. Some baths require additional tests. Some zinc phosphate processes operate without an active accelerator such as this and, in operation, contain controlled amounts of ferrous iron in solution. Such baths, usually used for immersion processing, are sometimes described as running 'on the iron side'. The bath, typically containing zinc phosphate and nitrite, is capable of producing high coating builds of rather coarse crystalline form. Calcium–zinc variants of this system can, however, form quite fine coatings with excellent paint-bonding properties. Their temperature of operation, typically 80° C, may be 20–30° C higher than common nitrite accelerated counterparts.

49.3.2 Manganese Phosphate (see also chapter 41, section 41.5)

Manganese phosphate coatings are not so widely used today for paint bonding. The traditional manganese phosphate bath is similar to the high-temperature zinc phosphate process, operating 'on the iron side'. In the 1970s and 1980s, however, there has been limited usage of a manganese phosphate-containing process for spray pretreatment at quite low temperatures.

49.3.3 Iron Phosphate

Iron phosphate coatings are very widely used as a general industrial treatment. This system is also called 'alkali metal phosphating', as the bath is often based on chemicals such as sodium acid phosphates. The coatings are a hydrated iron phosphate/iron oxide, usually of rather low coating build (e.g. 0.2–0.8 g/m²).

While for good corrosion resistance it is desirable to use a chromic final rinse, the fact that cleaning and phosphating can be combined makes iron phosphate achievable in a three-stage process of clean-coat/rinse/final rinse. On the other hand the highest-quality iron phosphating processes are often accelerated with oxidising agent, as for zinc phosphate processes, and produced in at least five stages such as alkali clean/rinse/iron phosphate/rinse/final rinse.

The basic alkali metal phosphate composition is commonly modified with other chemicals to achieve various properties. In particular, fluoride modifications confer the ability to form useful coatings on other metals such as zinc and aluminium. These coatings do not always give the highest corrosion resistance but, if an effective final rinse is used, the resultant three or four stages provide a valuable multi-metal pretreatment for industrial finishing. The simplified (unaccelerated) process mechanism is:

(a) $2Fe + 3NaH_2PO_4 \rightarrow 2FeHPO_4 + Na_3PO_4 + 2H_2$
(coating)

(b) $4FeHPO_4 + O_2 \rightarrow 4FePO_4 \downarrow + 2H_2O$
sludge

The actual composition of some iron phosphate coatings has been found by analysis to be mixtures of Fe_3O_4 and $Fe_3(PO_4)_2.8H_2O$.

49.3.4 Chromium Phosphate

One of the earliest conversion coatings for aluminium was the chromium phosphate type, which in refined forms is still used today. The coating bath is an acidic mixture containing phosphate, chromate and fluoride, in which the fluoride acts effectively as an accelerator. The fluoride component promotes reaction with the aluminium substrate in which hexavalent chromium is reduced to the trivalent form and a hydrated amphous chromium phosphate coating is deposited on the surface. At high fluoride levels heavy green coatings are produced which have been used for decorative and protective coatings without subsequent painting. At the other end of the coating build scale, conversion coatings of the same mechanism provided the virtually invisible base for lacquers and decoration systems to launch the two-piece aluminium can industry. A very simplified form of the coating mechanism is as follows:

(a) $Al + 3HF \rightarrow AlF_3 + 3H^+$

(b) $CrO_3 + 3H^+ + H_3PO_4 \rightarrow CrPO_4 + 3H_2O$

The actual coating has been characterised as $Al_2O_3.2CrPO_4.8H_2O$.

49.3.5 Chromates*

Chromate coatings of many types are used on aluminium, as well as on zinc and certain other metals. A distinction should be made between coatings for paint bonding, and those for passivation of the surface, particularly on zinc: both effects need not necessarily be required. Chromate conversion coatings for aluminium often serve for both paint bonding and protection in their own right. There are many types of formulations, but one widely used high-quality process uses an acid bath containing chromate, complex and 'free' fluorides and a ferricyanide salt, the latter component acting as an accelerator. The amorphous coating is actually more oxide than chromate and has been characterised as $6Cr(OH)_3.H_2CrO_4.4Al_2O_3.8H_2O$.

Many so-called chromate coatings contain relatively little chromate, being substantially hydrated oxides of varying compositions. The presence of chromate, and the pH, affect the behaviour of a reactive metal to the conversion coating solution; and the characteristics of the coating, including its weight per unit area, depend on the various solution parameters. Particularly on zinc, where corrosion performance of the total paint system may be dependent on the chromate level, and the mechanical properties may be degraded as the coating build increases; the optimum coating is a compromise and requires careful definition and control.

---

* These are described in more detail in chapter 41, section 41.3.

49.3.6 Other Conversion Coatings

A number of special-purpose processes are used. Acidic compositions containing titanium or zirconium complexes are used to produce light 'complex oxide' coatings on aluminium for beverage containers. Alkaline chromate processes have been used on aluminium and tinplate. Alkaline compositions are used to form a 'complex oxide' coating on zinc; in conjunction with a suitable final rinse, this has been adopted to a significant degree for coil coating of galvanised steel.

As discussed previously, the basic mechanism involves a solution, which is reactive to the metal surface and balanced to provide the insoluble inert coating where the reaction takes place.

There are exceptions to this principle: certain unconventional approaches to, and new trends in, conversion coating. There are non-aqueous conversions coatings, exemplified by iron phosphate in chlorinated solvent-based compositions. There are also other very effective coatings composed of iron phosphate plus an organic binder, deposited by immersion or spray application of certain formulations. In these cases, precleaning is required, but no rinsing subsequent to the conversion coating stage.

The multi-metal, no-rinse coatings, finding increasing use as coil-coating pretreatment, are based on mixtures of chromates and organic binders. Coating formation involves the application by roll coat of the correct film thickness followed by drying under specified conditions.

49.3.7 Final Rinses

The inert conversion coating, which covers a reactive metal substrate, is normally incomplete and porous. Reactive sites are therefore present, which result in less than optimum resistance to corrosion under a paint film with any discontinuities. If these reactive sites can be passivated, then upgrading of the performance of the system results.

The final stage in most cleaning/conversion coating sequences is a passivating rinse to achieve the maximum level of corrosion resistance. The best of these materials are still chromate based, though there is much effort to find other formulations to decrease problems with disposal. Typically, the final rinse concentrates are made up as 0.1 per cent or less in water. The bath pH is generally within the range 2.5–4.5, and the contact times are short to avoid dissolution of the conversion coating. The selection of a suitable final rinse formula is very significant when the best corrosion resistance is being sought.

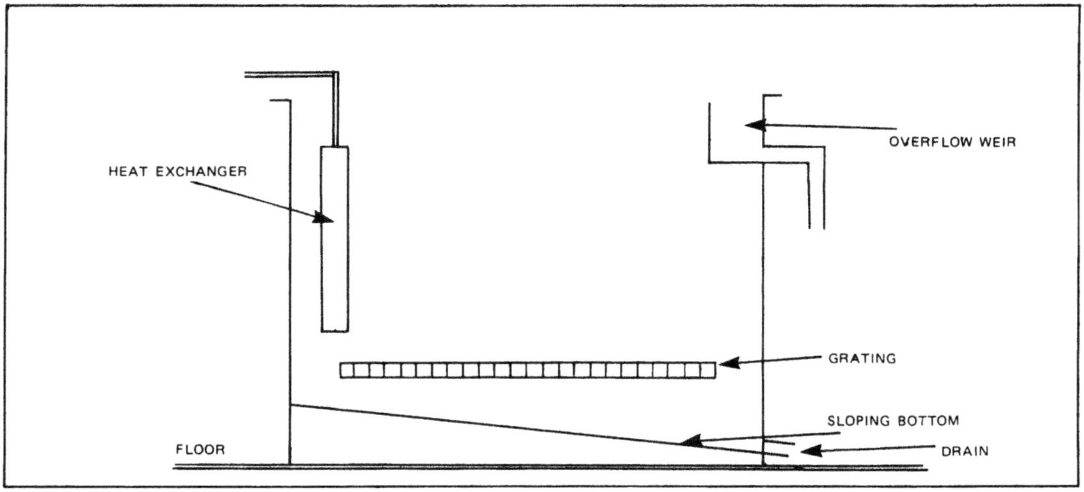

FIGURE 49.2
Typical dip tank design

49.3.8 Testing of Conversion Coatings

The determination of conversion coating efficiency includes the assay of the solutions from which they are applied, and the evaluation of the coatings themselves.

Solution checks involve determination of the concentration of the active ingredients; and of ingredient balance; and establishment that the build up of reaction products is not excessive.

With zinc phosphate the concentration may be indicated by total acid and oxidant (accelerator) levels. The balance is measured by determination of the ratio of total acid to free acid. Sludge reaction products vary: for a bath operating 'on the iron side', the reaction product is ferrous iron. Some simple iron phosphating baths are given no more than a regular concentration (total acid) check.

Chromium phosphate baths are checked for concentration of chromate and fluoride accelerator levels. The reaction products are dissolved aluminium and trivalent chromium of which the latter is checked. Other chromate systems use similar test methods.

On-line testing of the coatings themselves is often confined to a visual assessment of uniformity and coating appearance and integrity. Adhesion of the full system is checked regularly in critical situations.

More lengthy testing includes coating build, microscopic examination, and accelerated exposure testing of the conversion coating and/or the total finishing system.

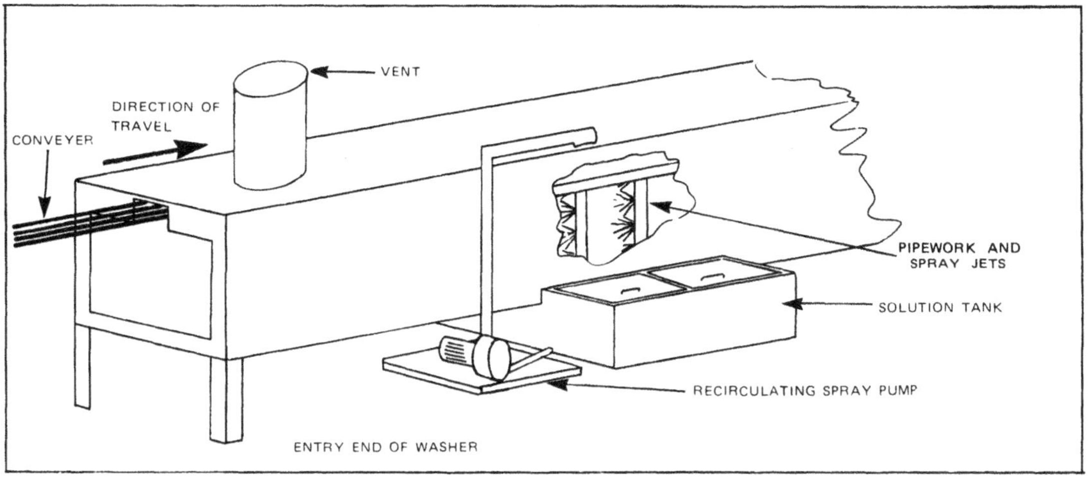

FIGURE 49.3
Features of spray washer

49.4 APPLICATION EQUIPMENT

In general, cleaning and the application of conversion coatings are carried out by immersion and/or recirculating spray. There are alternatives, such as the spray-to-waste process for low-volume, large assemblies, and roll-coat applications of conversion coatings prior to coil coating.

Although process temperature are reducing in response to the high cost of energy, most cleaning and conversion coating processes require heated solutions. Immersion processes, compared to spray applications, are characterised by higher chemical consumption, lower energy use, lower capital cost, longer contact times; the choice is therefore essentially dictated by production volume and economics. Most of the high-volume, mass-production industries are based on conveyorised finishing operations of which the pretreatment section involves multi-stage spray tunnels.

Although the design of tanks used for immersion pretreatment has been refined in various ways over the years, the equipment is simple. On the other hand, spray washer design is a matter of considerable expertise. These are shown in figures 49.2 to 49.4.

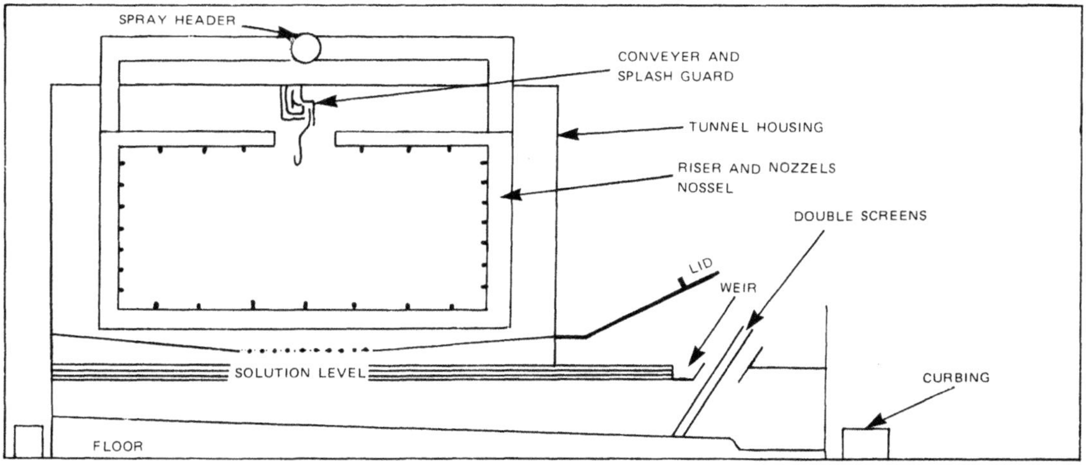

FIGURE 49.4
Cross-section of spray stage

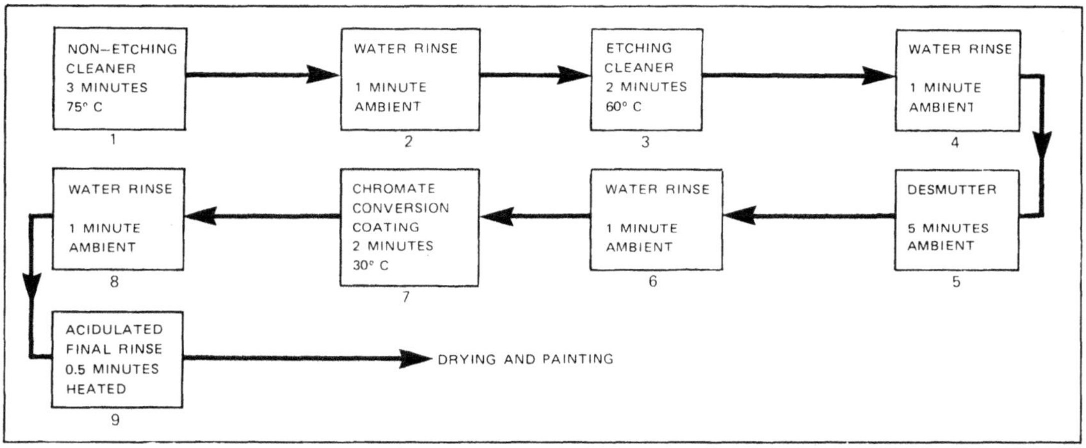

FIGURE 49.5
Immersion chromate treatment for aluminium

49.5 PROCESS STAGES

The number of stages in a process depends on the type of system required and the state of the metal prior to treatment. Thus, rust-free lightly oiled steel, calling for medium-quality iron phosphate, can be pretreated in three stages. In contrast, metal that needs deoxidising or desmutting or certain preconditioning treatments may need as many as nine or ten treatment stages, and even more if the advantages of multi-stage cleaning or cascade rinsing are to be enjoyed.

Figures 49.5 to 49.8 provide examples of selected process stages as used in industry.

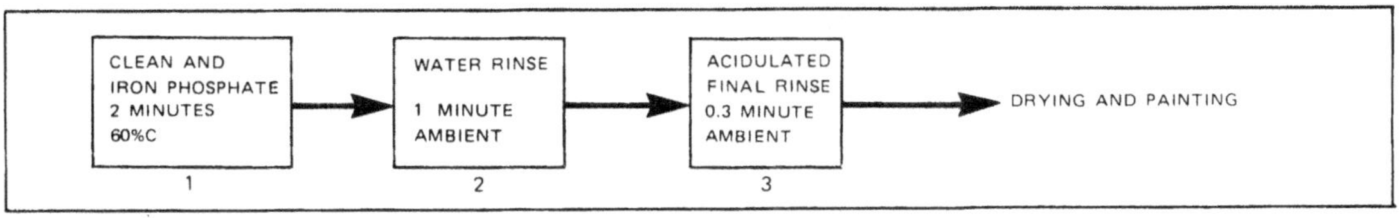

FIGURE 49.6
Spray iron phosphate treatment for steel

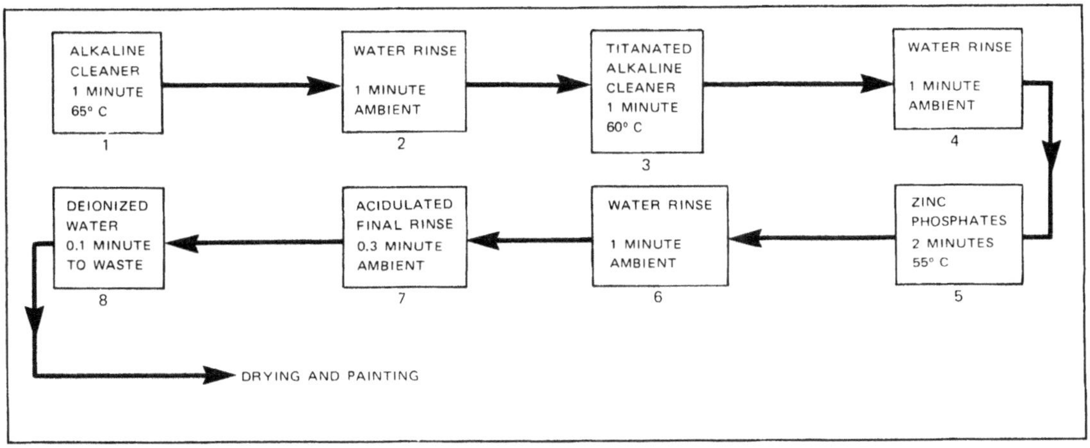

FIGURE 49.7
Spray zinc phosphate for steel and zinc

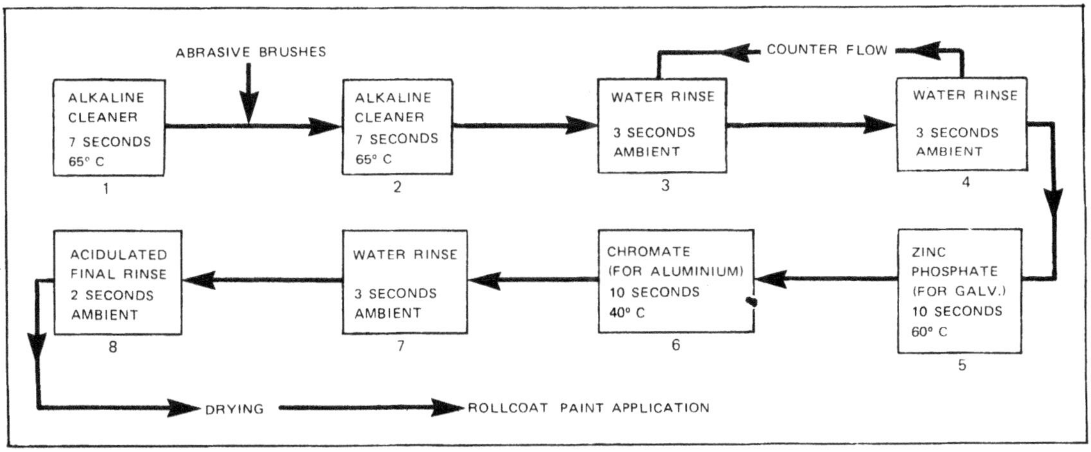

FIGURE 49.8
Multi-metal pretreatment for coil coating

49.5.1 Energy

Cleaning and conversion coating processes generally require heated solutions, and the energy required may be substantial when application is by circulated spray processes. This aspect has been subject to continuing scrutiny and there are few cases today where a spray applied conversion coating need operate much above 50° C. Cleaning processes have traditionally run at high temperatures and the greatest savings have been made here, with many modern compositions operating at 45° C and lower. Design of plant is a vital aspect of low temperature processing, and, in older plants, large energy savings might be achievable only at some reduction in quality.

49.5.2 Pollution Control

Cleaning and conversion coating may involve periodic disposal of the solutions used, and certainly do have a continuing requirement for rinsing. Increased control on effluents therefore affects the user.

Some success in formulation has been achieved. Biodegradable surfactants and phosphate-free cleaners are available where desired. Non-chrome conversion coatings are applicable in some areas, while for coil coatings the advent of 'no-rinse' conversion coatings offers a large step forward in the reduction of effluent.

Effective pollution control also requires efficient plant management and design. Good house-keeping, collection and treatment of wastes, and proper disposal are important. The volume of liquid wastes can be vastly reduced together with improved quality by the use of counterflow rinsing.

50 THE SELECTION OF DECORATIVE PAINTS

50.1 INTRODUCTION

Paint technology is not a study of individual products but rather the study of the interaction of substrates, primers, undercoats, finishing coats and the environment to which they will all be exposed. Any decision involving paint selection that does not recognise the importance of product selection and performance, as part of the total system, is likely to result in disappointing performance. Each coat of paint in a painting system has a job to do.

A *primer* or *sealer* may be required to:
(a) provide adhesion to the substrate;
(b) provide good adhesion for subsequent coats;
(c) regulate moisture movement (particularly important on wood);
(d) provide corrosion resistance on metals; and
(e) provide, in the case of sealers, alkali resistance on masonry.

An *undercoat* may be required to provide.
(a) adhesion to both primer/sealer and topcoats;
(b) film build;
(c) sandability;
(d) filling properties;
(e) opacity; and
(f) gloss holdout.

Topcoats give aesthetic appearance and the final service quality.

Because these coatings each have a specific job, a painting specification cannot simply leave some out. Some coatings, however, may perform more than one function. High-build, single-component coatings perform well in particular applications.

Multi-coat systems have an advantage over many high-build one-coat systems. In the one-coat system the product has to do everything—provide adhesion, build, aesthetic appearance and good wearing qualities. In the multi-coat system, individual products are designed to perform specific tasks on a cost-efficient basis.

It does not necessarily follow, however, that the more coats applied the better. There is an optimum film build for each system; exceeding this optimum results in unnecessary cost and may also result in system failures. Excessive film build can lead to stresses and strains being set up in the paint system, which can be relieved only by cracking.

50.2 FACTORS IN SELECTION

A wide variety of paints is produced by manufacturers to satisfy the differing needs of consumers. Factors that influence the selection of the right product for the job include:

(a) performance versus cost;
(b) application cost versus material cost;
(c) substrate protection versus aesthetic appearance; and
(d) whether the work is new, or a repainting job.

Examples illustrating paint selection of the basis on these factors are shown below.

50.2.1 Performance/Cost Considerations

A consumer may decide to paint a standard timber paling fence in order to improve the appearance of his or her property. For this application, the cost of using a premium timber finish may not be justified compared with a lower-cost product, which may have less coverage. In this example, the user wants to improve the appearance of the fence and is not concerned with its long-term protection (most paling fences fail under the ground at the supporting posts). The consumer would not be inclined to apply more than two coats, and probably will apply only one coat.

The same consumer may also be painting timber that forms part of the structure of the home. In this example, both improved appearance and substrate protection will be wanted, and here, therefore, premium products should be used, applying a minimum of two coats to a timber finish or three coats to existing paintwork.

50.2.2 Application Cost/Material Cost Considerations

The consumer may wish to paint a large area of previously unpainted interior wall space. Using a conventional brush or roller and a conventional paint, three coats would probably be required—or a high-build coating could be used, applying one heavy coat by airless spray. The cost of paint for the high-build coating will be higher than the cost of any of the individual conventional coatings, but substantial savings can be made in labour costs.

50.2.3 Substrate Protection/Aesthetic Appearance Considerations

Substrate protection, particularly if the substrate is metal, is enhanced by using thicker or multiple coatings of chemically resistant coating systems based on epoxy, chlorinated rubber or vinyl resins. These three coating types, while giving good protection, have rather poor aesthetic appearance, as they chalk freely. They make excellent coatings for industrial tank farms, for example, but their appearance is not acceptable for the exterior of commercial buildings or of domestic dwellings.

50.2.4 New Work/Repainting Considerations

This difference plays one of the major roles in a paint selection process. New work has all the advantages of a known substrate condition and the knowledge of the total paint system to be applied. Repaint work, on the other hand, may be over a sound paint system or over one that is failing; worse still, it may be over a previous system that looks sound but which harbours a hidden adhesion problem.

It is unwise to apply water-thinnable products* over solvent-thinnable systems and vice versa, particularly the overcoating of old solvent-thinnable systems with water-thinnable paints. Dra-

---

* Australian Standard AS 2311–1983 prefers the use of 'water-thinnable' to 'water-based', and the former term has therefore been used in this chapter. Because the terms are not interchangeable in the industrial coating field, we have not necessarily adopted the term throughout this book.

matic film failures can occur, for example, when dark-coloured gloss water-thinnable paints are applied over thick films or heavily exposed solvent-thinnable paints. Even though the old weathered solvent-thinnable coatings may look satisfactory, they can become very brittle and are unable to follow the movement of the dark-coloured gloss latex film as it flexes in hot and cold weather. The old paint will thus crack or fail in some other way under the new topcoat, causing the topcoat itself to fail prematurely.

50.3 THE CHOICE OF GLOSS LEVEL

(Refer Australian Standard AS 2311–1983, 'The Painting of Buildings')
Paints are available in a range of different gloss finishes generally known (in increasing level of gloss) as:

flat/matt; low gloss; semi-gloss; gloss; and full gloss.

In the selection of gloss level, the following points should be noted:

(a) The non-reflecting surface of flat/matt finishes minimises imperfections in the surface to be painted, while semi-gloss finishes and higher levels of gloss highlight such imperfections.
(b) Flat paints are not as resistant to cleaning as the higher gloss finishes, because they may show polishing or burnishing marks when rubbed.
(c) Matt and low-gloss finishes balance the factors of surface highlighting and suitability for cleaning, as shown by flat and semi-gloss coatings.
(d) Semi-gloss and full gloss finishes offer better wearing, cleaning and humidity resistance than their lower-gloss counterparts. Thus a semi-gloss or gloss paint would be preferred for kitchens and bathrooms or commercial food processing areas where ability to resist steam or cleaning activities to remove grease would be a desirable feature.
(e) Gloss or full gloss finishes also look better on kitchen cupboards, doors, guttering and exterior trims and are ideally suited for frequently handled areas such as hand railings, doors and garage doors. They are usually resistant to blocking.

50.4 THE CHOICE BETWEEN SOLVENT- AND WATER-THINNABLE FINISHES

In general use, largely for reasons of convenience, *latex paints** are gradually replacing solvent-thinnable finishes, but for some applications these latter '*enamel*' paints retain some advantages that cannot be matched; however, recent developments in acrylic polymer technology are already producing water-thinnable products with all the advantages of the traditional solvent-thinnable paints. Hence these innovations may in the near future change the perspective of the recommendations contained below.

50.4.1 Solvent-thinnable Finishes

The chief characteristics that influence selection of this class of coating are as follows:

(a) They have a higher, deeper gloss than water-thinnable paints, although improvements are constantly being made.
(b) They give better application—they do not foam, and they give better flow and levelling.
(c) They have greater resistance to abrasion, steam, water, grease, oil and chemicals and are therefore better suited to areas subject to high wear and tear.

* 'Latex paint' is the preferred term for a finish based on emulsion resins. Alternative terms such as 'plastic paint' should be avoided.

(d) The resin solutions (in contrast to the polymer dispersions used in latex paints) on which they are based penetrate reasonably well into mildly loose or powdery surfaces. This penetration factor means that solvent-thinnable undercoats tend to adhere better to such surfaces than their water-thinnable counterparts.

It is possible to achieve better flow and levelling properties in 'enamel' paints because there is a wider range of solvents that can be used to develop these properties. Water is a very intractable substance: the properties of water cannot be significantly modified, so water-thinnable paints themselves can be varied less readily to produce particular properties. Solvent-thinnable paints, on the other hand, can be based on any of a wide choice of hydrocarbon solvents which can be tailormade to give selective application properties.

Decorative paints based on alkyd resins (described in chapters 5, 6 and 7 in volume 1) dry first by solvent release and then, by an oxidative cross-linking reaction, develop high levels of resistance to moisture, abrasion and chemicals. This oxidative process, depending on the environment to which the paint film is exposed, will continue beyond the optimum stage, however, leading to a lack of flexibility and ultimate failure. Latex paint films develop useful properties due to the coalescence of polymer particles, and no cross-linking mechanism is involved.

50.4.2 Latex Paints

Water-thinnable paints are popular with both the professional painter and the 'Do-It-Yourself' painter, because they offer:

(a) low odour;
(b) quick and easy clean-up with water; and
(c) fast drying and re-coating.

For the purposes of this review, latex paints can be classified into two groups: those based on 'pure' acrylic emulsions, and those based on the variety of commercially available latices derived from vinyl acetate copolymerised with a range of acrylic monomers. (This technology is reviewed in depth in volume 1, chapters 16–20.) From the consumer's standpoint, this distinction is not always clear, as many copolymer-based paints are described on their containers as 'acrylic'. Styrene-acrylic resin-based paints can also be formulated to give 'acrylic' performance at one end of the PVC scale, and as low-cost very high PVC interior finishes at the other.

The pure acrylic resin-based group of paints are characterised by their excellent exterior durability, adhesion and resistance to chalking and dirt pick-up. They show excellent colour and gloss retention and do not yellow with age.

The copolymer-based range of finishes (some of which contain pure acrylic resins as a mixed binder) are suitable for many exterior situations as well as most interior applications. They are generally low in cost compared to the pure acrylic finishes but are not available in full gloss finishes.

Key factors that influence the choice between solvent- and water-thinnable can be summarised as follows:

(a) Solvent-thinnable gloss paints will outperform other paints in heavy wear areas.
(b) Paints based on pure acrylic emulsions give good performance in all areas, and have excellent wet adhesion properties.
(c) Other water-thinnable paints, which are less expensive than pure acrylic systems, are suitable for most interior decorative painting and many exterior jobs as well.
(d) There is little advantage in using solvent-thinnable paints giving flat/matt finishes when flat latex paints offer similar performance characteristics and are easier to use.
(e) Solvent-thinnable paints have a noticeable after-odour, take longer to dry and need turps

for cleaning up. Latex paints are almost odour-free, fast drying, and can be cleaned up with water.

50.5 RECOMMENDED FINISHES

In addition to the factors itemised above, the *Code of Practice for the Painting of Buildings*, produced by the Standards Association of Australia as document AS 2311–1983, provides complete specification details for new work and maintenance painting.

The following summarises the recommended finishes and procedures for common decorative paint situations. The qualifications regarding innovations in water-thinnable paint technology (see section 50.4) also apply to this section.

50.5.1 Interior Work

50.5.1.1 Heavy Wear and Tear Areas

A full gloss solvent-thinnable paint is usually recommended for heavy wear areas, especially those subject to moisture, steam and condensation and/or heavy cleaning. These paints are more resistant to dirt and grease, and have excellent washability characteristics.

If, however, a satin finish is required, possibly because of surface imperfections, semi-gloss solvent-thinnable paints will give good performance.

Where the previous paint surface is unidentified, pure acrylic latex paints may also be considered because of their excellent wet adhesion properties. The adhesion of the old paint system must first be established.

50.5.1.2 Moderate Wear and Tear Areas

It is usual to use a low-sheen or semi-gloss paint in these areas because they are more easily cleaned than flat paints and are more resistant to scuffing, abrasion and general wear and tear. The trend has been to latex paints for ease of use.

50.5.1.3 Light Wear and Tear Areas

Latex paints are usually recommended for these surfaces because of convenience and suitability, with a flat/matt or low-sheen finish depending on the surface.

A very flat paint will help to camouflage surface defects, such as joints in plasterboard, which can be highlighted by a high gloss coating. If a gloss finish is chosen for its fashion effect, the user should be prepared for imperfections to become more pronounced. With ceiling lighting, this is particularly true of ceiling areas and it is therefore customary (and cheaper) to use an ultraflat paint for ceilings.

50.5.1.4 Timber and Trims

Doors, bannisters, railings, window-sills, frames and tracks, skirtings and architraves are usually subject to constant handling, abrasion and cleaning. Gloss solvent-thinnable paints are therefore usually preferred in these areas: most latex paints are not as suitable and in some applications tend to retain dirt because of their themoplasticity. The same thermoplasticity can also cause adjoining surfaces to stick together in hot weather.

50.5.2 Exterior Work

Again, a gloss finish will give better performance than a flat finish but, in the case of exterior surfaces, the durability of solvent-thinnable paints is somewhat outweighed by their tendency to break down, at the end of their life, by cracking, flaking and, ultimately, peeling. Thus the preparation for repainting becomes more demanding than is the case with water-thinnable paints, which break down by chalking.

Gloss solvent-thinnable paints offer a deeper gloss initially than latex paints, but it drops away more quickly. Consequently, after a few years, latex paints retain a higher gloss and also tend to show less discoloration.

50.5.2.1 High Performance Areas

These are the areas exposed to the sun's ultraviolet rays, and to lashing rain and wind. These areas need a high performance paint to protect them, such as a pure acrylic-based latex paint.

50.5.2.2 Low Performance Areas

Sheltered areas—under eaves, verandahs or other overhangs—usually do not need such a high standard of protection. It is therefore possible to use a flat latex paint which is less expensive and offers the other benefits of water-thinnable paints.

50.5.2.3 Timber Trim

External timber trim—such as doors, windows, fascia boards and barge boards—are usually finished in gloss solvent-thinnable systems because of their high resistance to dirt and grime and the abrasion of regular cleaning. In some cases, a pure acrylic latex paint can be used, which does have the ability to 'move' with the timber when it flexes with changing humidity.

50.5.2.4 Metal

Guttering, downpipes and wrought iron normally need solvent-thinnable finishes for moisture resistance, but this is optional on adequately primed non-ferrous metals such as galvanised iron or *Zincalume*.

50.5.2.5 Roofs

Special paints are manufactured for use on roofs. The general rule is to use water-thinnable paints on asbestos cement and solvent-thinnable paints on metal surfaces. If water is to be collected for drinking purposes, particular care must be taken in selecting a painting system that will not contaminate and taint the water.

50.5.2.6 Fences

Generally fences should be treated in the same way as the body of the building they complement. Low-priced fence paints tend to give low-quality performance.

51 CORROSION AND PREVENTION

51.1 INTRODUCTION

Painting is the most common method of preventing the corrosion of mild steel which is the most widely used material of construction. Much of industry's plant, equipment and machinery is made of mild steel; examples include building frames, rolled steel joists (RSJ's), purlins, storage tanks, pipelines, structural steel generally, bridges and ships. Paint coatings are the only economic method of preventing rust and consequent deterioration, and at the same time providing a decorative appearance.

Protective coatings may differ from each other in many ways, but the most important variable is the resin binder. Pigments for opacity and colour are generally similar. Widely used resins include alkyd, styrene, epoxy, urethane, silicone, acrylic, phenoxy, chlorinated rubber, vinyl, phenolic and others. The selection of the type of paint is dependent upon the environment to which a structure to be protected is exposed, and on the materials of construction.

51.2 CORROSION MECHANISMS

Corrosion is an electrochemical phenomenon which occurs at the interface of the metal and the material or the environment with which it is in contact.

Mild steel is a heterogeneous substance containing, on an average, 0.2% carbon, 0.4% manganese, and traces of phosphorus, sulfur, silicon, nickel, chromium, molybdenum, copper, aluminium and tin; the remainder is iron so the surface consists of different chemical activities. The most reactive areas are *anodes*; *cathodes* are the less reactive. In the presence of oxygen from the air and water in any form, including normal humidity, iron immediately starts the process of oxidation. It propagates by an electric current passing between the anodes and cathodes by the transmission of electrons generated at the anodes. Their loss leaves the anode areas deficient in electrons and the iron there goes into solution as ferrous ions:

$$Fe \rightleftharpoons Fe^{2+} + 2e^-$$

The electrons, e, migrate through the steel to the cathode and react there in various ways which are dependent on the pH value and the availability of oxygen:

$$2H^+ + 2e^- \longrightarrow 2H \longrightarrow H_2$$

$$2H^+ + \tfrac{1}{2}O_2 + 2e^- \longrightarrow H_2O$$

$$H_2O + \tfrac{1}{2}O_2 + 2e^- \longrightarrow 2OH^-$$

The free hydroxyl ions (OH^-) from the cathode together with the ferrous ions from the anode form ferrous hydroxide next to the steel surface.

$$Fe^{2+} + 2OH^- \longrightarrow \underset{\text{ferrous hydroxide}}{Fe(OH)_2}$$

At the air-to-oxide surface, the reaction continues with additional oxygen:

$$4Fe(OH)_2 + O_2 + 2H_2O \longrightarrow 4Fe(OH)_3$$

The product, ferric hydroxide, may also be rewritten as $Fe_2O_3.6H_2O$ (hydrated ferric oxide).

The oxide layer that forms very rapidly is called *rust*, and it can be shown that the amounts of water and oxygen necessary to sustain rusting are extremely small. For example, an average rate of corrosion of steel of 70 milligrams per square centimetre per year would require only 11 mg of water and 30 mg of oxygen. The problem of rust as a corrosion product on iron and steel is that it is hygroscopic and is loosely adherent; it exhibits some slight solubility and tends to move away from the iron so that the formation of new rust can proceed. By contrast, the oxide film that forms on lead, nickel, cadmium, chromium, aluminium, zinc and copper is tightly adherent and hinders further corrosion.

Under conditions of differential aeration, where there exist varying availabilities of oxygen, ferric ions may also be formed.

$$Fe^{2+} \rightleftharpoons Fe^{3+} + e^-$$

Such conditions exist inside crevices and pits, and the primarily-formed dissolved ions Fe^{2+} and Fe^{3+} are hydrolysed to magnetite, Fe_3O_4, and ferric hydroxide, $Fe(OH)_3$, as follows:

$$3Fe^{2+} + 4H_2O \longrightarrow Fe_3O_4 + 8H^+ + 2e^-$$

and

$$Fe^{3+} + 3H_2O \longrightarrow Fe(OH)_3 + 3H^+$$

The essential point of the corrosion process is that it involves the movement of electrons (that is, the passage of electric current) through the metal, the circuit being completed through the water. If the water contains dissolved salts, such as salt from sea water or salt air, or ferrous sulfate from the reaction of sulfur pollution products with steel, then the transfer of electrons is made easier and the extent of corrosion is proportionally increased. In addition to increasing the conductivity of solutions, the chloride ion is especially aggressive and has a very damaging effect upon the corrosion of mild steel because:

(a) It keeps the corroding surfaces wet (being hygroscopic) for longer periods.
(b) The chloride anions (Cl^-) permeate and destroy any iron oxide that may have formed as a protective oxide film.
(c) It enhances the growth of pits that may have formed (e.g. in crevices) by 'self-stimulation'; the chloride ion migrates to the pit interior to maintain the electrical charge balance as electrons are transferred to cathode areas.

These are the reasons why corrosion is more severe under coastal and marine conditions than elsewhere.

51.3 CORROSION PREVENTION BY PAINT COATINGS

Carefully formulated and selected paint systems can and do prevent corrosion. Normally paint coatings are applied in two or three coats and, apart from a single coat of an inorganic zinc silicate, a one-coat paint system is rarely used and would require excessive care to ensure freedom from pinholes and minor discontinuities and to obtain an even thickness. The sequence of coats is called *primer*, *intermediate* and *topcoat*.

The *primer* is of the utmost importance to the paint system, as it is the tie-coat between the metal and subsequent coats. Primers should be:

(a) well adhering to the steel;
(b) anticorrosive; and
(c) a suitable base for the intermediate coat.

The adhesion to the steel is primarily a function of the resin binder in the primer, the anti-corrosive nature is largely contributed by the pigments, whereas the formation of a good base for the top-coats is partly a function of the binder, but also the ratio between the quantities of pigment and binder plays an important part. The choice of the type of binder for the primer is also governed by the choice of topcoats, which in turn is determined by the requirement for resistance to the environment. An all-purpose primer that can be overcoated with all types of topcoats would simplify the painting task, but this ideal is not yet available.

For a paint coating to stop corrosion from starting, it is evident from the previous reactions that this can be achieved by one of three methods, either alone or in combination:

(a) barriers to restrict the access of moisture, oxygen and salts;
(b) electrical methods;
(c) chemical inhibition.

51.3.1 Barrier Characteristics

As all coatings are permeable to both water and oxygen to some extent, it is therefore impossible to completely exclude them from reaching the steel surface. However, some coatings are more impermeable than others, and catalysed epoxies, coal-tar epoxies, solution vinyl and chlorinated rubber paint systems are used in wet or immersed environments. These are typical 'self-prime' materials which means that the intermediate coat or topcoat is used as first coat on the steel and an 'anti-corrosive' primer is not required; that is, they prevent corrosion by a barrier effect only.

These highly impermeable coatings rely on excellent adhesion to the surface and the maintenance of film integrity as water diffuses through the coating. For this reason it is essential that such coatings do not contain any water-soluble components; otherwise blistering with eventual rupturing and exposure of bare steel will occur due to osmosis, which is the passage of water through a semi-permeable membrance (the paint film) when different concentrations of salt exist each side of the paint coating. Equally damaging is the presence of salt from sea air or sea water and iron sulfate from sulfur pollution on the iron surface, which also are soluble and would cause osmosis. It is essential that steel be freshwater-washed to remove them or other soluble salts before any surface painting or preparation (including blast cleaning), as they can be blasted into the steel surface and become trapped beneath the paint coating.

51.3.2 Electrical Methods

Electrical methods of preventing corrosion involve minimising the flow of corrosion current so that, if negligible current flows, negligible corrosion results. Thus, in a corrosion cell:

$$I = \frac{E}{R}$$

where I is the corrosion current; E is the polarised potential difference between local anodes and cathodes; and R is the total electrolyte resistance, which must be much greater than the metal conductor resistance.

Thus the higher the value of R, the less will be the value of I, so that by making the electrolytic path of the current of high resistance, the movement of ions is impeded.

Paint resins with the highest electrical resistance are catalysed epoxies, phenolics, vinyls and chlorinated rubbers each with values in the order of 10^{10} ohms/square centimetre. The addition of coal tar further adds to the resistance, and the addition of extender pigments in the first coat such as talc, china clay, mica and iron oxide also assist in increasing the resistance. As mentioned above, the removal of soluble deposits on the surface is necessary because their presence will short-circuit the resistance of the paint film so that the value of I increases, possibly to the stage where rusting will occur.

Film thickness is variable and, as the diffusion of water through a paint film varies inversely with the film thickness, it is clear that it will take longer for moisture to diffuse through a thick than a thin film. Similarly the thicker the coating the higher will be the electrical resistance and, for immersion or buried surfaces, the dry thickness of paints would need to be between 250 and 500 μm.

An alternative method of electrically preventing corrosion of iron is to employ a metal more anodic (that is, less noble) than iron, such as zinc. Thus zinc-rich paint coatings will protect the steel; as the iron is no longer the anode of the electrical circuit, the zinc metal becomes the anode and the iron will not corrode. Zinc dust is discussed later.

51.3.3 Chemical Inhibition

Chemical inhibition refers to the use of special pigments which are employed in the primer coat. These anti-corrosive pigments function in varying ways, but broadly by:

(a) blanketing either the anode or cathode areas;
(b) interfering with either the anode or cathode reaction preventing the formation of ferrous hydroxide;
(c) in the case of zinc chromate, by oxidising the ferrous corrosion product to ferric oxide which plugs the anodic areas and increases the electrical resistance.

51.3.3.1 Types of Anti-corrosive Pigments

Red lead (Pb_3O_4). This is the oldest and probably the most widely known and most effective of the anti-corrosive pigments. Unfortunately, like all lead compounds, it is poisonous and State Health Departments are curtailing its use; there is a ban on its unlimited use in many countries, so its future is doubtful.

It functions by passivating areas to which water has pentrated by hydrolysing and blanketing these areas with insoluble lead hydroxide ($Pb(OH)_2$), which then increases the imperviousness of the film. Additional protection is provided in linseed oil primers—considered the standard by which others are judged—by the breakdown of the lead soaps such as lead linoleate, formed from the linoleic acid in the oil, to give various adsorption inhibitors such as lead pelargonate, which forms an impervious barrier over the surfaces.

A further reason why red lead is excellent as an inhibiting pigment is that the corrosive and damaging chloride and sulfate ions that penetrate to it are neutralised and rendered inactive by the formation of insoluble lead chloride and lead sulfate; the latter can still form soaps with linseed oil for continued protection.

Zinc chromate. This pigment is actually zinc potassium chromate ($K_2CrO_4.3ZnCrO_4.Zn(OH)_2.2H_2O$) and the chromate ion CrO_4^{2-} is an oxidising anion which oxidises any ferrous ions formed while the chromate itself is reduced; the ferric oxide and chromium oxide formed plug anodic sites and then contribute to the electrical resistance of the film. As zinc chromate is slightly soluble in water—0.11% by mass—it cannot be used in primers for continuously wet or immersed conditions; otherwise osmotic blistering will result.

Some industrial environments can produce ferrous sulfate which, if any penetrates to the primer, reduces chromate to chromium salts which are not anti-corrosive. Under such conditions zinc chromate is not as effective as red lead, unless the primer contains a high chromate content, so it must be present in sufficient quantities to be effective.

Zinc tetroxy chromate ($ZnCrO_4.Zn(OH)_2$). This chromate pigment is primarily used in two-pack etch (also called 'wash') primers which are based on polyvinyl butyral resin. A reaction between the zinc hydroxide and the phosphoric acid present in the primer forms a layer of zinc phosphate similar to that obtained in phosphate pretreatments. This process, together with the inhibition provided by the chromate ions, makes etch primers outstanding corrosion pretreatments for metals, including mild steel, galvanised iron, aluminium, zinc and aluminium metal sprayed brass, copper and stainless steel.

Zinc tetroxy chromate has low aqueous solubility, 0.003 per cent by mass, which is insufficient to cause blistering in sea water, but blistering is likely under freshwater immersion.

Zinc phosphate ($Zn_3(PO_4)_2$). This is a more modern anti-corrosive pigment which came into prominance in the 1970s as a result of a search for a non-toxic replacement for red lead. It provides protection more by a barrier effect than by the formation of any inhibitive ions.

Zinc dust. This is another anti-corrosive pigment with outstanding ability to protect steel and possessing very low toxicity. Zinc metal will protect steel galvanically: when the two are in contact, iron is the more 'noble' and is cathodically protected. The average particle size of zinc dust lies between 2 and 9 μm and zinc-rich paints are formulated to ensure contact between these particles and the steel surface. The formation of a layer of basic zinc carbonate and zinc oxide, together with basic zinc chloride in marine environments, provides the good performance of zinc dust primers. The conditions for their use are rather specific. In particular, they are not acid resistant; hence for specific acidic conditions, zinc dust primers are not suited.

51.3.3.2 Use of Inhibitive Pigments in Alkyds

The use of red lead and the zinc salts in 'conventional' paint systems, based on oil or alkyd media, has led to formulations for zinc chromate and red lead priming paints in Australian Standards K211 and K145, respectively. The appropriate topcoats are also based on alkyd resins and should conform to Australian Standard K113 'Exterior Enamelised Paint' or K126 'Petrol and Oil Resistant Enamel'.

A typical paint system based on alkyd resins is:

Primer. Two priming coats of red lead or zinc chromate, each 40 μm.*
Topcoat(s). Two topcoats of alkyd enamel, each 30 μm.
Film thickness. Total dry film thickness: 140 μm.

*For interior or mild exterior, the second priming coat may be omitted.

Alkyd paints are used for general interior and exterior atmospheric exposure in rural, industrial and marine environments and in industry on building frames, bridges, storage tanks, towers, girders, plates and structural steel generally. The use of micaceous iron oxide or aluminium pigment, both of which are lamellar in structure, in the topcoats will give improved exterior durability because of their superior weatherability. Because of the poor impermeability of alkyds, water penetration through the film under atmospheric exposure is such that a primer with an inhibitive pigment is essential.

51.3.3.3 Use of Inhibitive Pigments in Other Media

Alkyds are not suited to conditions of contact with acid or alkali, water-immersed or underground conditions or to constant splash or condensation, because the resin fails due to hydrolysis of the ester groups.

$$R{-}C{\overset{\displaystyle /\!\!/ O}{\underset{\displaystyle \backslash O{-}R_1}{}}} + HOH \rightleftharpoons R{-}C{\overset{\displaystyle /\!\!/ O}{\underset{\displaystyle \backslash OH}{}}} + R_1OH$$

Where resistance is required against chemical attack, or immersed conditions, underground, splashed or condensed water, or contact with chemical solutions, paint coatings that have high water impermeability such as the epoxies, solution vinyl copolymers, chlorinated rubbers, polyurethanes, silicones and acrylics are required. These same coatings are also used for maximum atmospheric durability, especially in chemical plants and industry generally. Where resistance against a special chemical or a combination of environments is required, the paint manufacturer should be consulted for a definite recommendation for the most suitable paint system, which must include the type of paint, film thickness necessary and degree of surface preparation and profile, time lapse between coats and optimum method of application.

51.3.4 Cathodic Protection

This is an electrical method of preventing corrosion of metals and is commonly used in the protection of steel in underground pipes as used for gas or water, in tanks, and in marine environments including shipping, steel wharf piles and off-shore drilling rigs.

The requirement is to introduce a *new* electrode into the electrochemical corrosion cell and provide sufficient electromotive force (EMF) so that the new electrode functions as the anode and the structure to be protected becomes the cathode. Two methods are used: *galvanic* and *impressed current*.

51.3.4.1 Galvanic Protection

This involves the use of sacrificial anodes. A metal, M, in the presence of an electrolyte will dissociate,

$$M \longrightarrow M^{n+} + ne^-$$

and the formation of ions is the equivalent of a flow of electrical current from the metal into the electrolyte, as shown in figure 51.1.

In a corrosion cell, the electrode at which current flows into the electrolyte and which *generates* electrons is the anode, while the electrode which consumes the electrons is the cathode; the anode therefore corrodes.

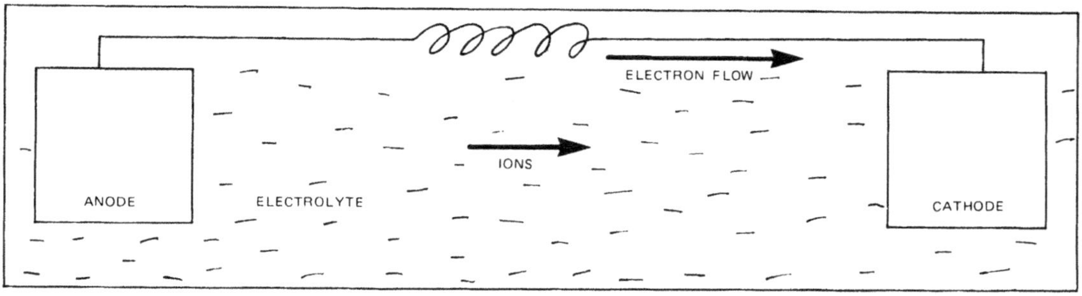

FIGURE 51.1
Metal dissociation

When the metal M is in equilibrium with its ions, the Nernst equation provides:

$$E_M = E^0_M - \frac{0.059}{n} \cdot \log(M^{n+}) \text{ volts @ } 25° \text{ C.}$$

Values of E can be calculated for each metal. With the activity of M^{n+} constant the values of E^0_M obtained from a concentration of a 1 Normal solution gives the *electrochemical series*; when the metals are immersed in sea water the single potentials are different and this series is known as the *galvanic series*. Metals with large positive potentials are called 'noble' (gold, platinium) because they do not readily dissolve, while the more reactive metals are called 'base'. Hydrogen, by convention, is zero; values at 25° C are, for iron, −0.44 volt, zinc −0.76 volt, aluminium −1.67 volts, and magnesium −2.37 volts.

Iron is therefore more noble than zinc, aluminium and magnesium, and these metals are used as sacrificial anodes to protect steel structures. By coupling iron with zinc, a simple primary cell is established in which iron is the cathode and is protected.

51.3.4.2 Impressed Current

In this method the supply of electrons is provided by a DC battery or generator—the negative terminal joined to the structure being protected and the positive terminal joined to an anode (scrap iron, graphite, etc.) buried or immersed in the conducting medium adjacent to the metal being protected. By forcing the structure being protected to consume electrons, it is thus made the cathode. Figure 51.2 compares this method diagrammatically with that of the galvanic method.

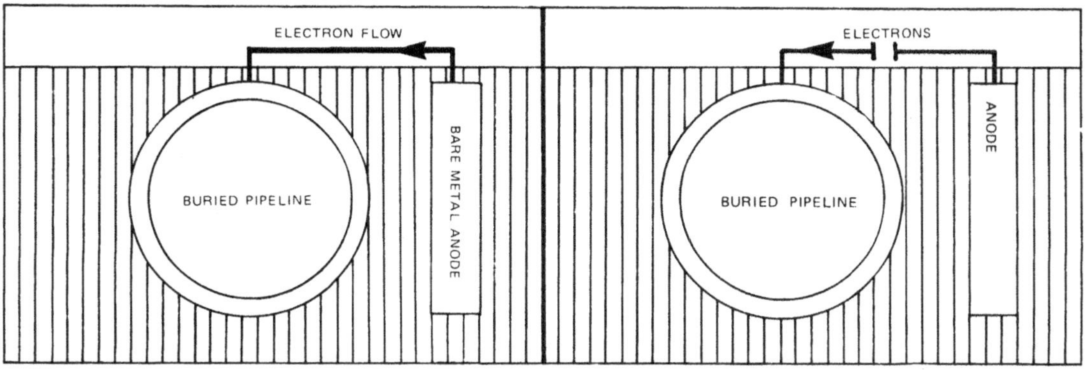

FIGURE 51.2
Galvanic (a) and impressed current (b) methods of protection

Cathodic protection (CP) is used in conjunction with a protective coating, since the current required on bare steel would be prohibitive and CP is used to protect bare steel at those areas where damage occurs or the paint film is incomplete. Since the cathodic reaction produces hydroxyl ions, the paints must be alkali resistant, such as the epoxies, vinyls or chlorinated rubbers.

51.3.4.3 Cathodic Protection by Zinc-rich Primers

Cathodic protection of steel can also be provided by the application of a zinc-rich coating. Such a product is characterised by a very high degree of pigmentation, typically 90 per cent zinc dust in the dried film. The essential property is that the paint film shall be electrically conductive and be in electrical contact with the underlying steel. These properties can be obtained only with a high proportion of zinc particles that are in contact with each other and with the steel. Such a paint will normally be applied at a dry film thickness of 65–90 μm.

Basically there are two types of zinc-rich coatings which derive their classification from the binder or resin in which the zinc dust is incorporated. They are:

(a) organic primers, based on resins similar to those in most paint coatings including polystyrene, epoxies, urethanes, acrylics, phenoxy, chlorinated rubber, vinyls and phenolics.
(b) inorganic primers, in which the binder may be based on alkali silicates, such as potassium silicate, or on an alkyl type, such as ethyl silicate.

Both types can be applied alone as a single coat, and often this is standard procedure; there are many instances of single-coat applications performing satisfactorily. Because zinc is amphoteric, zinc-rich coatings should not be used outside the pH range 5–9. Topcoats, if used to extend the service life, must be compatible. Only non-saponifiable types, such as vinyls, epoxies, chlorinated rubbers and acrylics, should be used. Oil-containing binders, such as alkyds and epoxy esters, should not be applied directly over zinc-rich coatings, as poor adhesion will eventually result.

51.4 SPECIFIC FORMS OF CORROSION AND THEIR CONTROL

There are several other forms of corrosion, not quite so obvious or well known, and their causes and prevention are described below.

51.4.1 Acid Attack

Acid attack occurs on steel and the less noble metals at atmospheric temperatures through direct displacement reaction; that is, the metal is dissolved by the acid producing metal ions and hydrogen gas. Use is made of this in industry to dissolve rust from steel by pickling, and *filming inhibitors* are added to the pickling bath to prevent direct attact on the steel. These inhibitors are aliphatic and aromatic amines.

The only method of preventing acid attack using protective coatings is to separate the metal from the environment by the application of an acid-resistant barrier. There are also some very acidic salts such as ammonium sulfate and urea (both used as fertilisers) and ammonium nitrate (used as an explosive) which can cause rapid and severe corrosion of, for example, iron, concrete, aluminium and galvanised iron.

51.4.2 Galvanic Corrosion

When two metals or alloys are in electrical contact and immersed in a conducting solution (electrolyte), the more active metal will dissolve first and during the process of sacrificing itself will prevent the less active metal from corroding. The more active metal is the anode, the less active the cathode. Cathodic protection, described in section 51.3.4.1, employs this principle.

Conversely, if the steel is coupled with a less active metal, such as copper, the steel becomes the anode and sacrifices itself to protect the copper; the steel corrodes more rapidly by galvanic corrosion than if it were by itself. Whether a metal in contact with a second metal will corrode depends on many factors, among which are the electrical conductivity of the solution, the presence of air or an oxidising agent, and the relative tendency of the metal to go into solution. For example, galvanic corrosion will be slight in neutral distilled water compared to the galvanic corrosion in sea water. Galvanic corrosion of the metals commonly used in water heating systems is not significant because the air is removed from the water on heating and is vented from the system. The presence of oxygen in the electrolyte plays an important part.

The amount of galvanic corrosion with a less noble metal will depend on the *areas* of the two metals in contact, and therefore the potential or voltage existing between the two metals in a given environment is governed by the distance apart of the two metals in the galvanic table (see table 51.1). If the anode is small and is coupled to a large area of a metal cathode, rapid

TABLE 51.1
Galvanic series of metals

Anodic end (Greater tendency to corrode)	
1. Magnesium	15. Copper
2. Zinc	16. Bronzes
3. Aluminium	17. Nickel–silver
4. Cadmium	18. Copper–nickel alloys
5. Steel	19. Monel
6. Cast iron	20. Silver solder
7. Stainless steels (A)	21. Nickel (P)
8. Lead–tin solders	22. Iconel (P)
9. Lead	23. Stainless steels (P)
10. Tin	24. Silver
11. Nickel (A)	25. Graphite
12. Iconel (A)	26. Gold
13. Nickel–chromium alloys	27. Platinum
14. Brasses	*Cathodic end* (Less tendency to corrode)

(A) Active metal surface
(P) Passive metal surface

corrosion of the small anode will take place, resulting in perhaps catastrophic failure of the structure. For example, an iron valve in a copper vessel will rapidly corrode and be eaten away. Conversely a brass or copper valve in an iron tank will probably result in little or no corrosion because this is an example of a large anode and a small cathode.

Having dissimilar metals in contact is generally avoided; this is a design weakness and the deliberate joining of dissimilar metals is usually carried out only in cathodic protection. If dissimilar metals must be joined, they should be suitably insulated from each other. However, galvanic corrosion can occur on the surface of a single metal because some areas are more active than others and will try to corrode more rapidly than the less active areas with which they are in contact. Some instances of this are as follows, and in these instances suitable corrective action should be taken, such as more careful surface preparation or double priming of these particular areas.

(a) Steel is strongly anodic to millscale.
(b) New steel is anodic to old steel, and this occurs where a new section is joined to an old portion by welding or bolting.
(c) Highly stressed surfaces are always anodic to those unstressed; examples are pipe bends, rivetted areas, welds, bolted areas, and areas that have been hammered or otherwise fabricated. Where the jaws of a pipe wrench bite into the pipe surface, this strained area will rust because it is anodic to the surrounding area. Also the head and point of a nail are anodic to the shank.
(d) Where a rough metal surface is adjacent to a smoother surface, the rougher surface is anodic to the smoother one.

51.4.3 Protective Film of Corrosion Products

The formation of a corrosion product, such as an oxide film, on the surface of some metals and alloys will hinder and prevent further corrosion by forming a tight skin on the metal surface. Such metals are said to be *passivated*. These include lead, nickel, cadmium, chromium and aluminium. The addition to steel of small quantities of chromium and nickel gives *stainless steel*, and the formation of a surface layer of chromium oxide is the reason for the corrosion resistance of this metal. However, if an ample supply of oxygen is not available to some of these metals, they can become *active* and corrosion can continue; this is why the galvanic table (Table 51.1) lists some metals as both active and passive, representing their position in an ample supply of oxygen—with minimum corrosion—and with little or insufficient oxygen to maintain the protective oxide film, respectively. Industry makes use of the fact that a protective film can be formed on the surface of a metal and examples follow:

(a) Pretreatment processes result in the formation of protective films of iron and zinc phosphates (and other phosphates) on the surface of steel, aluminium and galvanised iron. In the formation of the phosphate film some of the metal is, however, corroded by the phosphoric acid treatment solution.
(b) The protective film of aluminium oxide on aluminium is artifically increased by the process called anodising to such an extent that corrosion of anodised aluminium under atmospheric conditions is virtually nil. This oxide film can be made in many pleasing colours by the inclusion of dyes in the solution in which the film was deposited. The process is described in chapter 41.
(c) Lead is used as a lining for vats or vessels to contain varying concentrations of sulfuric acid. Very rapid corrosion takes place initially which leads to an insoluble layer of lead sulfate on the surface, which prevents further attack on the underlying metal.
(d) Mild steel vessels are used for storage and transport of concentrated nitric acid, a strong oxidising acid. Under such strongly oxidising conditions, the oxide film formed is protective; the iron is considered passive, dissolving only very slowly through the build-up of oxide.

    Likewise, magnesium in hydrogen fluoride produces a very protective insoluble magnesium fluoride film which prevents further corrosion.

51.4.4 Corrosion Inhibitors

The severity of corrosion of a metal immersed in an electrically conductive solution (electrolyte) in many cases depends on the solubility of the corrosion products at the cathode, at the anode, or at both. If the corrosion products are soluble, they go into solution as they are formed and leave a fresh metal surface available to further attack. If the corrosion products at either a cathode or an anode are not soluble, they form on the surface of the metal and present a barrier to further corrosion. Some chemicals have the ability to combine with, or to coat over, the reaction products on the metal surface or even on the metal itself and to form a film which slows the corrosion rate. Such chemicals are called inhibitors.

Inhibitors are added to the water used in industry, involving millions of litres per day, for purposes such as cooling towers, condensers, heat exchangers, boilers, washers and recirculating systems. These may be constructed of many metals including mild steel, aluminium, brass, copper and lead.

Inhibitors may function by preventing either the anodic or cathodic reaction. They include chromates such as potassium chromate or dichromate, sodium nitrite, calcium bicarbonate (which is already present in 'hard' waters), alkali phosphates such as sodium hexametaphosphate ('Calgon'), alkali silicates and organic inhibitors, examples of which were mentioned in section 51.4.1.

There is significant industrial effort involved in manufacturing and selling proprietary chemicals for inhibiting corrosion and the appropriate companies should be contacted for advice. It should be emphasised that the use of a protective coating is not always the best approach; the use of an inhibitor may be by far the easiest and most effective method of preventing corrosion.

51.4.5 Crevice Corrosion

Corrosion is more severe in crevices. In outdoor situations, water from rain or dew remains in cracks long after other portions of the structure have dried.

The use of skip welding and other forms of poor design, such as structural steel placed back to back, cause crannies where corrosion is more severe than on other parts of the structure and it is not possible to completely protect these areas.

Variations in the availability of oxygen at different parts of a metal result in a corrosion cell with separated anodes and cathodes on the same surface. The exclusion of air from any part of the metal system can lead to localised corrosion in those areas where oxygen is least available.

$$2H_2O + O_2 + 4e^- \longrightarrow 4OH^- \quad \text{cathode reaction}$$

$$E_{O/OH} = Eo_{O/OH} - \frac{0.059}{4} \cdot \log \frac{(OH)^4}{(O_2)} \quad @\ 25°\ C$$

If the oxygen concentration is reduced, the more negative is the value of E and the potential difference is therefore anodic to other areas. This occurs in crevices and areas of poor design which can interfere with air circulation. Defectively coated parts may also accumulate dirt and other contaminants. Severe pitting results, which is particularly dangerous as the damage inflicted is out of all proportion to the actual amount of metal wastage for example, on the floors of storage tanks, where pitting forms under localised porous deposits of corrosion products and foreign matter. This mechanism is called *differential aeration.* It can be shown that oxygen causes corrosion and, in this instance, lack of oxygen causes more severe corrosion.

Stainless steel and aluminium owe their corrosion resistance to the formation of an oxide film on the surface. This oxide film prevents further corrosion because it is so chemically resistant. However, such metals are subject to severe corrosion in such places as crevices where there is not sufficient oxygen to maintain the protective oxide film in repair; such areas will normally be out of sight, and a structure can be severely damaged before it is realised that the normally corrosion resistant materials are being destroyed.

51.4.6 Sulfur Corrosion of Steel

51.4.6.1 Hydrogen Sulfide

Hydrogen sulfide corrodes steel by direct reaction which results in the formation of iron sulfide and hydrogen.

$$Fe + H_2S \longrightarrow FeS + 2H \longrightarrow H_2$$

When the hydrogen is first formed, it is in the form of atomic hydrogen; the hydrogen sulfide retards the transformation to molecular hydrogen and the nascent atomic hydrogen can diffuse into the metal. This is a significant problem in the oil and gas industry, where hydrogen sulfide occurs in oil and coal. If the steel has a hardness above Rockwell C20, it is susceptible to hydrogen embrittlement which is the loss in ductility of steel caused by the entrance of hydrogen into the metal and, if sufficient stress is present, cracking will result. High-strength steels, because of limited ductility, are more susceptible to hydrogen cracking than are low-strength steels; and while hydrogen will enter the steel in both instances, failure may not occur in low-strength steels.

Hydrogen sulfide corrosion is particularly severe on the interior of steel sewerage digestors and sour crude oil storage tanks and pipelines. The best barrier found so far against hydrogen sulfide is a coating system based on tar epoxy resins.

Hydrogen sulfide will discolour paints that contain lead by the formation of black lead sulfide on the surface, but this generally has no deleterious effect on the corrosion resistance of the product.

51.4.6.2 Bacterial Corrosion

Sulfur and its compounds are an essential part of the metabolic processes of the most important bacteria involved in microbial corrosion. Together with animals and plants, these micro-organisms play an important part in the successive large-scale transformation of sulfur in nature. Part of the transformation is to inorganic sulfates.

In the absence of oxygen, where normally little or no corrosion can be expected to occur, sulfate-reducing bacteria, the most common of which is *Desulfovibrio desulfuricans*, convert the sulfate to sulfide which reacts with ferrous irons to form iron sulfide, which is a corrosion product. The effect of this corrosion is normally in the form of pitting, which can be quite severe—for example, in the bottom of crude oil carriers with high-sulfur oil.

51.4.6.3 Atmospheric Sulfur Pollution

The atmospheric corrosion of steel because of sulfur pollution is accelerating with increased industrial pollution from the burning of coal, oil and other products. The sulfur oxides formed by combustion combine with water in the atmosphere to form sulfurous and sulfuric acids, resulting in direct acid attack on steel and causing corrosion which can be increasingly severe as the sulfuric acid concentrates because of evaporation. The iron sulfate which is formed as a corrosion product is soluble in water and is leached away by rain, exposing further metallic iron to be further attacked by new acid.

Alkyd paints are fairly resistant to the mild attack of normal industrial pollution, but in areas of severe pollution improved surface preparation and more chemically resistant coatings are needed to satisfactorily protect steel work. In industrial plants, increasing use is being made of epoxy, chlorinated rubber and vinyl coatings for best protection.

51.4.7 Stress Corrosion

Stress corrosion cracking is a phenomenon which may result in the brittle failure of alloys, normally considered ductile, when they are exposed to the *simultaneous* action of tensile stress and a corrosive environment, neither of which when operating separately would cause major damage. Stress can be residual from cold deformation or applied stress from an external load. The mechanism is a complicated mixture of electrochemical, mechanical, stress sorption and microstructural processes.

It is important to note that for almost every alloy or metal there is a specific corrosive medium for the inducement of stress corrosion cracking. If stress is not present, the degree of corrosive attack will be small; likewise if the corrosive medium is absent, the attack is insignificant.

Examples of stress corrosion cracking include that which occurs in carbon steel due to hot nitrate solutions, austenitic stainless steels in hot chloride solutions, and copper alloys in moist ammonia. Generally, cracks occur on the interior of vessels or pipes where they cannot be seen and do not become apparent until perforation occurs with leakage and sometimes catastropic failure. The remedy is not to use these alloys in the susceptible environment specific to them.

51.4.8 Corrosion Fatigue

Corrosion fatigue is the simultaneous action of cyclic stress and a corrosive environment which may result in cracking. Failure often starts at corrosion pits that act as stress raisers, and the corrosion lowers the endurance limit of the steel or alloy and fatigue accelerates the corrosive attack.

Corrosion fatigue failures are particularly likely to occur, for example, near restraints, attachment welds and nozzle welds. At elevated temperatures corrosion fatigue is generally enhanced by oxidation; the failure results from oxide formation and fatigue. As with stress corrosion cracking, sudden failure occurs and the equipment fails rapidly.

The remedy for corrosion fatigue is engineering design, including a change of materials to those that are not susceptible, proper location of anchor points, supports to minimise tensile stresses, reduced vibrations or stress relieved weld joints or cold-worked materials.

52 INDUSTRIAL COATINGS: APPLICATION AND CURING METHODS

52.1 INTRODUCTION

With current coating systems it is essential for paint technologists to understand the various application, pretreatment and curing methods available in order to obtain the maximum benefit and economy from the coating. For example, whilst it might be preferred to coat appliances such as refrigerators and washing machines with an automatic electrostatic system, a complete system may cost in the order of $60 000 to $250 000, and this system would often have production capacity far in excess of demand. Therefore, some manufacturers utilising conventional hand spray guns or semi-automatic electrostatic spray units will always be found. Whilst these methods may be outdated and consume more paint, the end product is still quite satisfactory.

Not only is it necessary to understand application techniques from the formulation aspect, but also—as customers generally approach their paint suppliers for advice on new or alternative application methods—technologists should be able to advise coating users on the various application methods and what system is most suited to their needs.

It is usually agreed that formulation technology has at times outstripped available application methods, and it can often take some time for suitable production systems to be designed—as in the case of electrodeposition.

The earliest applications of paint were made by finger painting; then by very crude brushes. It has also been established that in some cave drawings of 20 000–25 000 years ago, the coatings were 'blown on' through pieces of hollow bone. This would be the first evidence of spray painting.

These early applications were applied for purely decorative purposes. The main advances in technology, which have occurred in the past century, have been made in response to the need for coatings that are not only decorative but also protective and durable.

52.2 BRUSH APPLICATION

In general all coatings for brush application are air-drying, with the most common application areas in the home and building construction industries.

Brush application is one of the oldest methods of applying paint and, although with today's need for higher production rates using minimum labour, a very large proportion of paints applied by brush will still be found.

Brushes are designed with more air spaces between the bristles at the tip of the brush than at the stock and will therefore hold more paint in the lower portion of the brush. They are generally made from horse hair and recently from special nylon fibres, although the best brushes are made from Chinese hog bristles. These brushes have splayed or split ends, allowing a far greater contact of the brush with the surface and producing a more even spread of the paint.

In painting structural steel one very important advantage of applying a protective primer by brush is that any slight surface contamination by oil and rust will be dispersed by the brushing action. This in turn will result in better adhesion and surface penetration of the primer than spray-applied coatings over these surfaces.

The main advantages of brush application are:

(a) versatility;
(b) suitability for use under variable conditions;
(c) the operator has full control of the quantity of paint being applied;
(d) low wastage of paint;
(e) little contamination of nearby areas.

The main disadvantages are that it is relatively slow and more costly in labour than many of the other methods of paint application.

52.3 SPRAY APPLICATION

52.3.1 Conventional Spray Application

The spray gun was first used about 1907 by furniture manufacturers and later used by the automobile industry. Spray application was expanded with the development of nitrocellulose lacquers in the early 1920s.

The basic principle of spray application is to atomise the paint into a fine spray and to direct the spray onto the object to be coated. Supplementary equipment such as a source of compressed air, flow control valves, filtration units for the removal of dirt, oil and water from the air supply, and containers for the paint supply are also needed. Paint supply containers vary in size from the 250 mL cups attached to the gun to remote 500 litre pressure pots at some distance from the spray guns.

Conventional hand spray guns are still the most common form of industrial paint application in Australia, with major uses in the automobile and furniture industries.

The main advantages of conventional spray guns are:

(a) the speed of application;
(b) the degree of control that operators have over the film thickness and nature of the finish deposited. This is typified in furniture finishing,where skilled spray operators can allow for both colour and grain variations of the timber substrate by varying the quantity of coating deposited on the substrate.

Pulling back the trigger first opens the air valve to allow 'dusting' air. As the trigger is further retracted, it unseats the needle valve in the fluid nozzle and allows paint to leave the gun and be mixed with compressed air in a tubular stream. This is called 'first-stage atomisation' and takes place immediately outside the air nozzle.

Second-stage atomisation follows within 10 mm of travel, and it is here that the paint is converted from a tubular stream to a flat spray pattern, by means of a blast of compressed air from each side of the air nozzle horns.

The main parts of the spray gun are:

(1) Gun body
(2) Trigger
(3) Air valve
(4) Needle valve
(5) Fluid nozzle
(6) Air nozzle
(7) Fan control knob
(8) Fluid control knob
(9) Needle packing

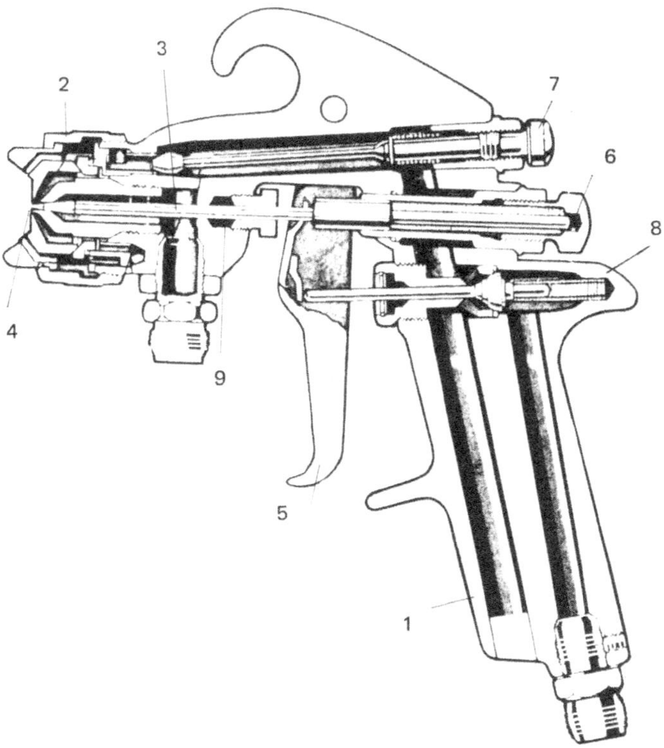

FIGURE 52.1
Typical spray gun

The width of this pattern is controlled by the fan control knob, which restricts the air flow as it is screwed in, reducing the fan width. Thus a wetter and more concentrated pattern results.

When a spray gun is used with an underslung cup, the paint is drawn from the cup by the vacuum formed by the compressed air passing over the tip of the fluid nozzle. This method, called 'suction feed', is suitable for thin-viscosity fluids, generally in the region of up to 26 seconds in a BS No. 4 Flow Cup.

Heavier fluids are 'pressure fed' by means of a pressurised cup, pressure feed container, or paint pump, and here the size of the fluid nozzle is important to fluid flow.

The air nozzle is used to direct compressed air to the paint stream, to atomise the paint and form a spray pattern. The air nozzle can be of either internal or external mix and can be either syphon or pressure feed. Factors to be considered in selecting air nozzles are the volume of air, the paint feed system, paint characteristics and the fluid flow.

The fluid nozzle controls the paint flow and directs it into the air stream. It also acts as a seat for the needle valve to close off the flow.

The nozzles are usually of stainless steel to avoid corrosion, or of tungsten carbide to avoid extreme wear from abrasion.

The needle valve is used to shut off the paint flow and to regulate the paint flow. Its size is determined by that of the fluid nozzle. Some needles are adjustable for wear and for the correct amount of dusting air.

The trigger operates the air valve and opens the needle valve. The air valve must open before the needle valve.

The fluid control knob restricts the movement of the needle, thus controlling the paint flow.

The fan control knob controls the air to the horns of the air nozzle, thus controlling the fan pattern—from a wide line to a small circle.

The air valve controls the air flow through the gun. It must open before the needle valve and close after the needle valve to ensure that compressed air is present to atomise the first and last drops of paint. Low-pressure spray guns do not have air valves. Air bleeds constantly through these guns, maintaining equal pressure throughout the system.

For satisfactory spray application, the following points should be observed:

(a) an adequate clean air supply for the spray gun;
(b) the correct air cap and fluid tip—the choice of cap and tip is governed by the type of paint, the spraying rate and the width of paint 'fan'.
(c) the viscosity of the paint;
(d) atomising pressure at the spray gun and pressure pots;
(e) the correct solvent balance of reducer and paint, to suit ambient conditions.

Although many new forms of spray applications are used today, it is interesting to note that 'hammer finishes' and certain metallic and 'toning' paints are best applied by the conventional spray gun.

52.3.2 Automatic Spraying

Automatic spraying units in widespread use include the horizontally transversing guns and various reciprocating and fixed systems incorporating special spray gun heads. They differ from manual guns in that the trigger is replaced by an air cylinder activated by an air valve.

With the transverse units, the guns are mounted on an overhead gantry which moves to and fro while an open mesh conveyor belt passes the articles beneath. This method is suited only to comparatively flat articles or sheets. The automatic coating of *large* flat areas or sheets is more economical by using either curtain coating or rollercoating.

Automatic coating machines are mainly used for continuous painting of small parts. These units consist essentially of a small conveyor passing articles through a suitable pretreatment system and then on to a rotary table where the articles are coated by several fixed guns, and finally cured in an oven.

There are many other complex automatic systems consisting of a series of moving guns designed to automatically follow the surface contours of large, irregularly shaped articles. These are in common use in automobile factories.

The main advantages of automatic spraying are:

(a) definite saving of skilled labour;
(b) strict control of application and paint thickness;
(c) variations of gun settings, which are easily made;
(d) the machines are very compact and occupy the minimum of floor space;
(e) they offer 'dust-free' painting conditions.

As with all automatic processes, automatic spraying will show cost advantages only over very long production runs.

Robot painters: Robots offer optimum application over long periods and can be used to apply toxic coatings. These machines generally consist of four components: an electronic control unit, a hydraulic power unit, a manipulator arm and an independent programming arm. The robot should ideally have the flexibility of human motion in painting; it should be able to match the articulation of shoulder, elbow and wrist and to some extent the flexing of the back and the bending of the knees. To program such a machine, a skilled painter moves the lightweight counter-balanced arm through the desired pattern.

52.3.3 Hot Spray Application

Hot spraying consists of manually or automatically spraying coatings that are heated by passing the paint through a heat-exchanger to about 60–70° C in the spray unit before application. This allows for higher application solids and faster production rates.

In conventional cold spraying, the application viscosity is obtained by adding 25–100 per cent reducer; in hot spraying, reduced viscosity is obtained by heating. With most hot spray coatings, heating reduces the viscosity by 25–30 per cent.

With cold spraying, one-third of the solvents present in the coating evaporate before the wet paint reaches the article, producing a drop in temperature of 10–15° C below ambient. With hot spraying, almost three-quarters of the solvent present will evaporate by the time the atomised droplets reach the article, producing a greater drop in temperature. However, because of the higher initial temperature of the paint, the final temperature at the workpiece will be slightly above ambient.

By using the hot spray process, less energy is required for atomisation, halving the consumption of compressed air.

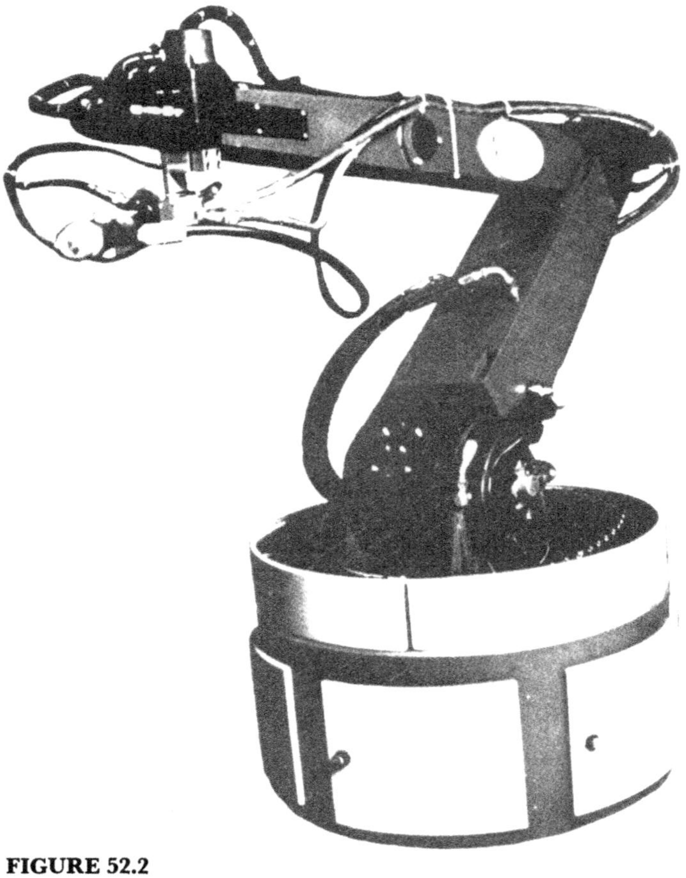

FIGURE 52.2
A robot spray painter

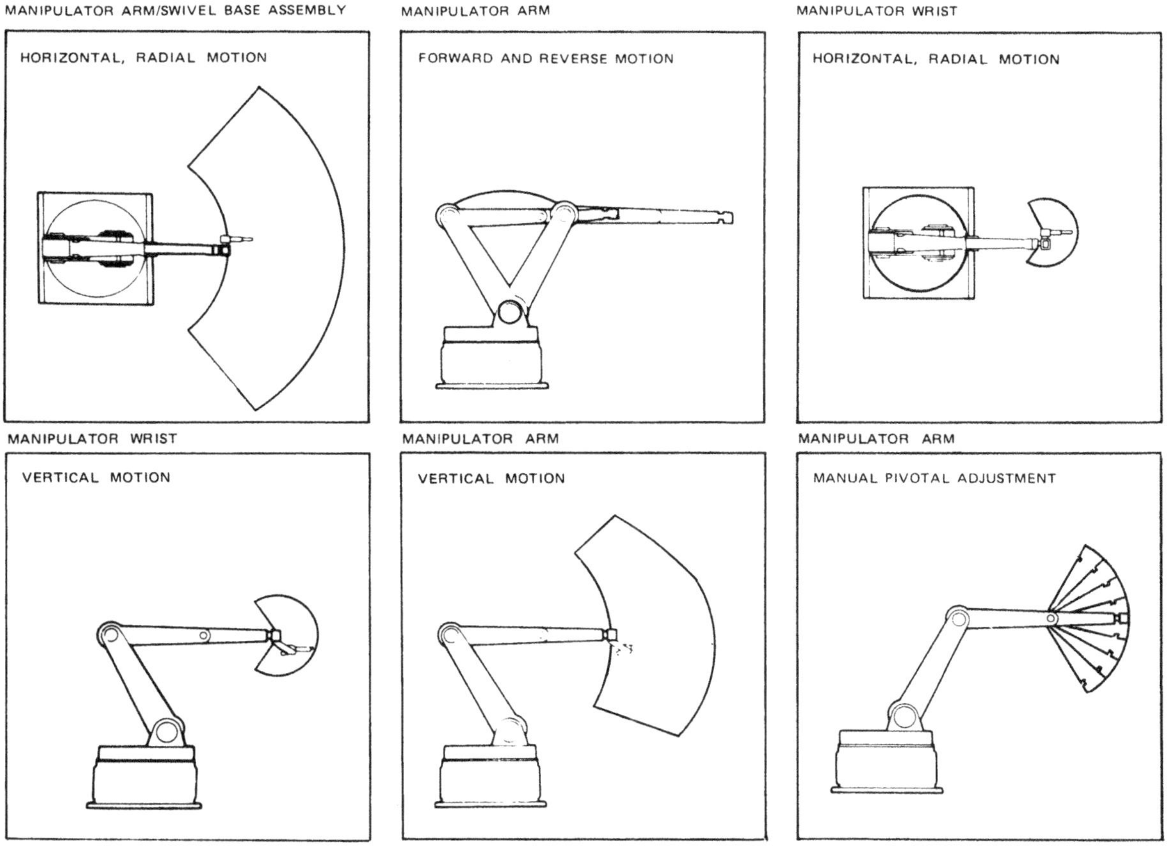

FIGURE 52.3
Limits of movement of robot painter illustrated in Figure 52.2

The advantages of hot spraying are:

(a) higher solids application;
(b) articles coated at higher than ambient temperature;
(c) reduced atomising pressure;
(d) better surface wetting, resulting in improved adhesion, flow and durability.

Manual hot spraying is mostly used for the application of nitrocellulose-based furniture lacquers; the higher temperature reduces the tendency of these lacquers to 'bloom'.

A major use of the hot spray technique is in the appliance industry—for the application of thermosetting acrylic enamels.

52.3.4 Airless Spraying

One of the drawbacks of conventional air spraying is the considerable volume of overspray and 'bounceback', resulting in noticeable paint losses to atmosphere.

Airless spraying minimises overspray, enabling high film builds on large surface areas in relatively short times.

The advantages of airless spraying are:

(a) less overspray;
(b) ease of painting surfaces inside a structure;
(c) greater paint economy—up to 20–30 per cent of conventional spraying;
(d) very fast application—at least twice as fast as conventional spray;
(e) the possibility of applying very high film builds;
(f) less pollution.

Airless pumps are driven either electrically or by compressed air. Pressures of up to 30 000 kPa are exerted as the paint discharges through a small tungsten carbide tip, used to minimise wear. The tungsten tips range in size from 0.18 to 1.8 mm and are arranged to give a fan pattern width of from 10 to 85°.

Airless units are capable of spraying paints at container consistency, and thinning is not required. Smaller units will handle the normal decorative coatings—gloss enamel, PVA acrylic, sealers and undercoats—and are generally a one-gun unit.

Larger units will handle multiple guns or heavier coatings up to and including high-build or solventless epoxy coatings and, with the correct nozzle tip, will apply up to 300 μm thickness in one pass of the gun.

Airless spraying is ideal for the decorator painter for broad areas; steel fabricators for items sprayed in situ; ship painting, tank painting, roof painting, etc. It is a popular method for painting outdoor structures such as bridges.

Portable units are widely used by master painters for commercial building work. Single-gun units are capable of covering 84 square metres of wall or ceiling area in about 15 minutes with a coating of latex paint sufficient to give complete coverage in one application.

Many airless equipment suppliers offer an accessory that extends the gun-body to spray-tip distance up to 1 metre to give greater reach and accessibility. Such units are called 'pole guns'.

Another innovation is the airless pump-fed automatic paint roller, enabling considerable time savings over conventional paint rollers.

52.4 ELECTROSTATIC PAINT APPLICATION

To meet the need for faster and more economical paint application, Ransburg developed, in the late 1950s, a system achieving more than 90 per cent paint deposition on the object being coated. With conventional spray application, up to 75 per cent of the paint used fouls the spray booth or the surrounding atmosphere.

The use of both manual and automatic electrostatic painting is one of the major advances in paint application. This process now finds widespread use in all industries where mass production dictates efficient paint application.

52.4.1 Principles of Electrostatic Spraying

In the conventional air spraying process, the materials are atomised by means of converging air jets, and the resultant stream of paint droplets and air is directed at the object by movement of the gun. Only one face of a stationary article may be coated by moving the gun in one plane.

In the electrostatic process, a controlled flow of paint is fed into a rotating insulated disc or bell. Centrifugal force takes it to the outer edge of the bell, where it would normally fly off at a tangent in coarse droplets or strings. However, an electrostatic charge of about 100 kV is applied to the insulated bell and causes the paint droplets to become 'charged'. This process of charging serves a dual purpose. Firstly it assists in the atomisation of the paint and produces small particles,

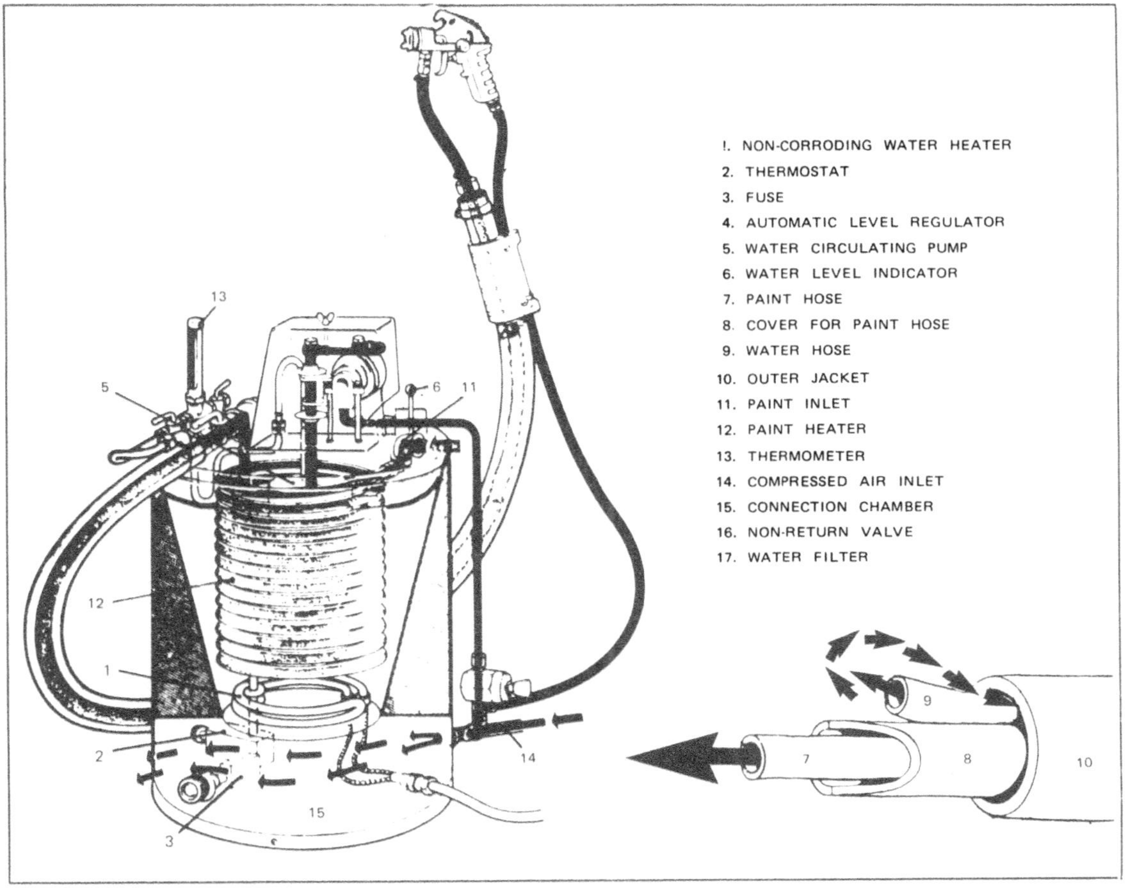

FIGURE 52.4
Typical hot spray units

and secondly these charged particles are strongly attracted towards the nearest earthed object in their path and impinge upon all sides of it.

A tube of up to 50 mm in diameter may be sprayed from one direction only; because there is no blast of air, overspray is eliminated, all particles being drawn to the object.

In order that the object may be earthed, it must be capable of conducting an electric current. All metals, most papers, concrete, plaster and some timbers have sufficiently good conductivity to be sprayed satisfactorily, providing they are securely connected to earth.

A finely variable DC voltage up to 90 kV generates an electrostatic field between the object and gun. The charged paint particles move along the field lines to the earthed object. Having unipolar charge, the flying particles repel each other and then 'wrap' themselves uniformly around the antipolar surface. As the field lines also envelop the object, paint particles cannot fly straight past, but reliably deposit on the rear surface as well. This is the well-known electrostatic 'wrap-around' effect.

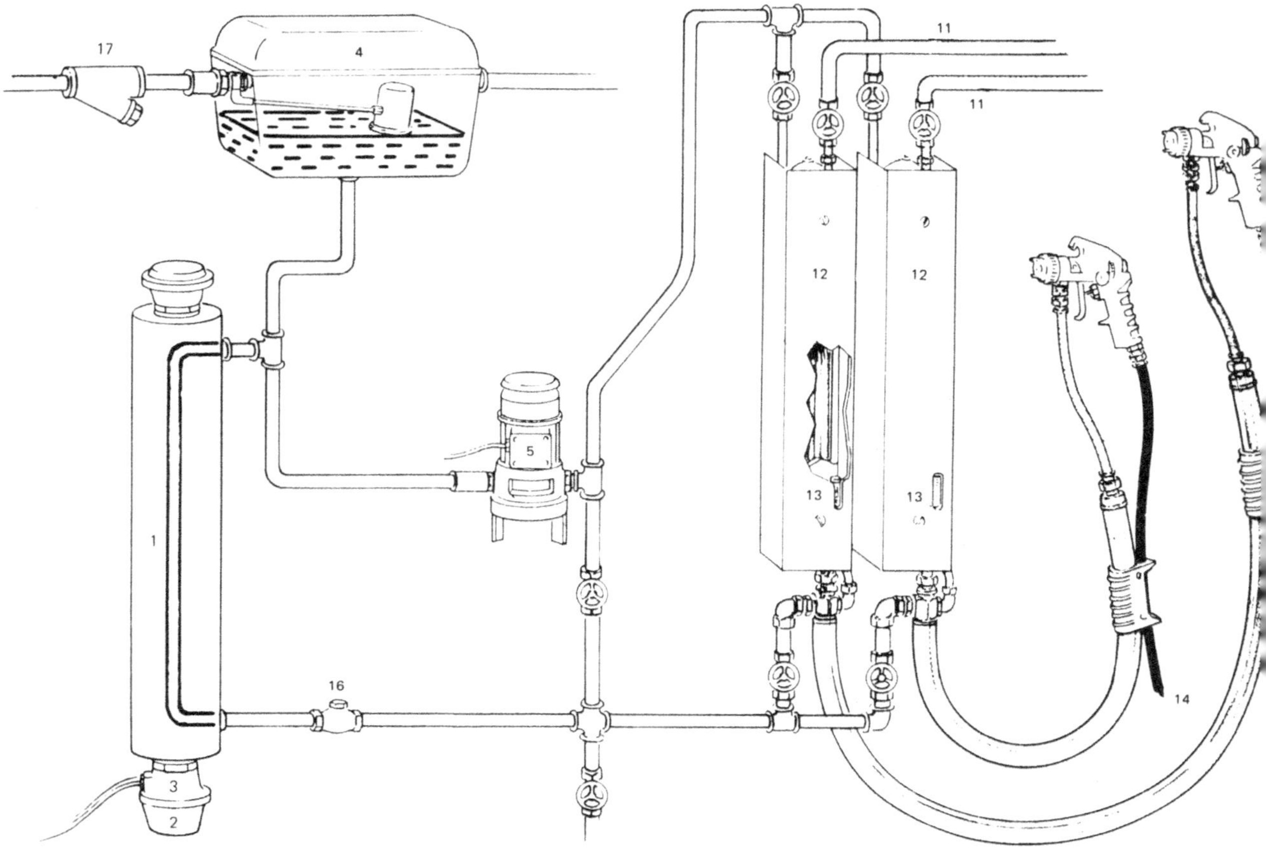

FIGURE 52.4
Typical hot spray units (see page 725 for legend)

52.4.2 Industrial Applications

Both disc and bell systems are widely used. Installations vary from the small stationary unit to the more complex multiple-gun systems fitted to reciprocating arms. The articles being painted are usually transported on a continuous monorail conveyor through the coating zone and then through a curing oven.

Complex automatic systems are best suited for long runs of similar articles with minimum colour changes and with careful loading of the conveyor line. Although low-speed reciprocating discs are still used, many plants are converting to high-speed discs to achieve even greater economies.

It is possible to vary the speed and stroke of the disc and to use varying sizes of discs. Once these variables have been established with a given work loading, no further adjustments are necessary.

One disadvantage of electrostatic spraying is that it will not completely coat the inside surfaces of hollow vessels and a manual reinforcing coat must be applied to these areas.

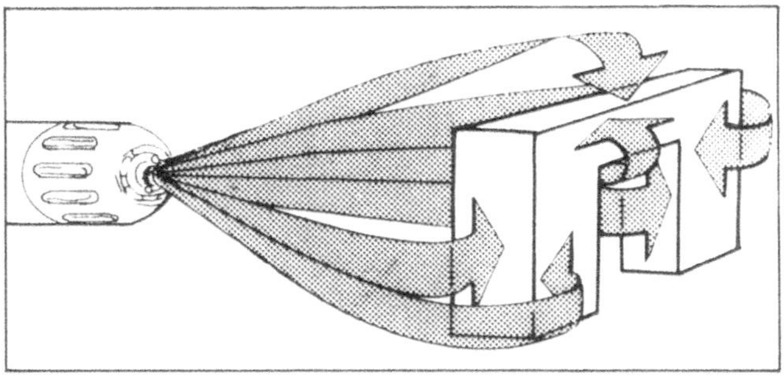

FIGURE 52.5
Movement of charged paint droplets

52.4.3 High-speed Electrostatic Disc Systems

The development of more efficient high-speed disc applicators was brought about by the increasing need to apply high-solids and water-based industrial coatings.

Most major appliance producers throughout the world are switching from the slower 1000–4000 rpm reciprocating discs to the new high-speed 20 000–30 000 rpm units.

The advantages of high-speed electrostatic disc systems:

(a) they are capable of applying water-based and high solids (up to 80 per cent) industrial coatings;
(b) they are capable of applying coatings at high viscosity at room temperature (i.e. 80 s BS No. 4 Flow Cup); because the speed of rotation is infinitely variable, paint can be applied as received without adjusting viscosity;
(c) there is no need to adjust the resistivity of the paint; voltage is infinitely variable up to 150 kV;
(d) they give uniform application with as little as 5 μm variation in film thickness;
(e) they have up to four times the paint output of conventional discs or bells;
(f) they give better atomisation;
(g) they supply better penetration into recessed areas and better wrap-around.

Despite these advantages, high-speed disc systems are not a complete answer to the application difficulties of waterborne and high solids coatings. In the case of waterborne coatings, the user is limited in the solvents that can be used. If there is a problem in 'wetting-out' the disc or avoiding dry spray, it cannot be solved by the introduction of 'slow' organic solvents as is often the case with solvent-based coatings.

Similar problems are seen when applying the high solids coatings. There are a number of installations that have been successfully spraying higher solids in the 60 per cent volume solids range. But generally when the solids exceeds 70 per cent, heating equipment is added in order to achieve best results. With some high solids coatings, there are problems in properly wetting the disc surface and there are definite limitations as to what additional organic solvents can be added to overcome these surface tension problems.

52.4.4 Automotive Painting

Several car producers have installed fully automated electrostatic multiple-gun systems capable of applying high solids and water-based coatings while having sufficient flexibility and high volume delivery to avoid the use of costly reciprocators.

Electrostatic attraction is combined with a gentle air flow that carries the paint to the work surface, providing a very even, consistent coating. Metallic coatings, which are very popular, are readily applied with more uniform results than with manual spray; this is particularly important where thermosetting acrylics are used, as excessive or uneven film build can result in solvent boil. The strong wrap-around effect almost eliminates paint overspray, and the spray pattern remains intact right to the target.

Control systems give accurate memory programming for virtually any conveyor requirement by means of a synchronous memory system. These memory systems record style, size or complete contours of irregular shapes passing on a conveyor at random. Strategically placed photoelectric cells differentiate between parts. The memory read-out initiates the spraying cycle when the part is at the proper location for each atomiser.

52.4.5 Electrostatic Hand Guns

Following the advent of the automatic electrostatic process, a later development was the electrostatic hand gun for manual use. The first manual gun to find widespread acceptance consisted of a bell-type rotating atomiser which was situated at the end of a plastic spray gun 300 mm long so that the the charged bell was remote from the operator. The gun housed an electric motor which rotated the bell at approximately 600 rpm.

In order to render the hand gun safe from accidental electric shock, the potential used was approximately 90 kV, and the amperage was considerably reduced compared to that of the automatic process. The potential for the hand gun was generated from the mains supply via a constant voltage transformer and voltage pack, and was transferred to the rotating bell by means of a special cable and a contactor brush touching the outer rear edge of the rotating bell. Both the high voltage supply and the bell motor were actuated by a toggle switch at the rear of the gun.

The paint supply from a conventional pressure pot was carried up to the gun in a special static electricity dissipating hose. The rate of delivery of the paint was regulated by the air pressure on the pot, as the paint trigger on the gun was of the on/off type. The paint was fed up to the inside edge of the rotating bell by a small nylon feed pipe.

The high negative potential on the bell was concentracted at the sharp outer edge. Atomisation of the paint, by the electrostatic field so produced, took place. The paint particles were then transferred to the object—the negatively charged particles were thus attracted to all sides of the object. The atomised paint particles adopted a curved path which depended upon the relative strengths of the centrifugal force.

There are now many other types of electrostatic hand guns, all relying on the same fundamental principles but differing in such factors as potential, means of generating potential, amperage, shape and means of rotation of the bell, and speed of bell rotation. This wide range of available guns has much extended the range of paints that can be applied by this means.

52.5 DIP APPLICATION

Dip application is one of the most economical means of coating large numbers of articles. It can be either manual or automatic.

The most suitable coatings for dipping are those that have the ability to cure well at varying film thicknesses. It is more common to see the use of thermosetting coatings on an automated conveyorised system, although several smaller installations still exist using manual dipping and air-drying coatings.

Obvious advantages of dipping are the low cost of equipment, low paint losses, high production capacities and the ability to coat internal and external surfaces.

The most important aspect of the dip process is the correct design of the hooks or jigs for holding the articles at the optimum drain angle. The jigging of articles should be arranged with sufficient space between the articles; otherwise 'solvent washing' can occur from the large volume of solvent vapours acting on the uncured paint film.

The most common form of dipping is the conveyor system used in conjunction with a hump-backed forced-draught stoving oven. The articles are suspended from a conveyor, which is diverted for the dipping operation down through 45°, allowing the articles to be lowered into the dip tank, after which a 45° rise withdraws them from the dip tank. The coated articles then pass over a draining area where the surplus paint is allowed to drain from the article (this paint is generally returned back to the dip tank after filtration). In order to reduce the 'drips' from the bottom of the articles, it is possible to use an electrostatic 'de–tearing unit' to produce a more uniform appearance.

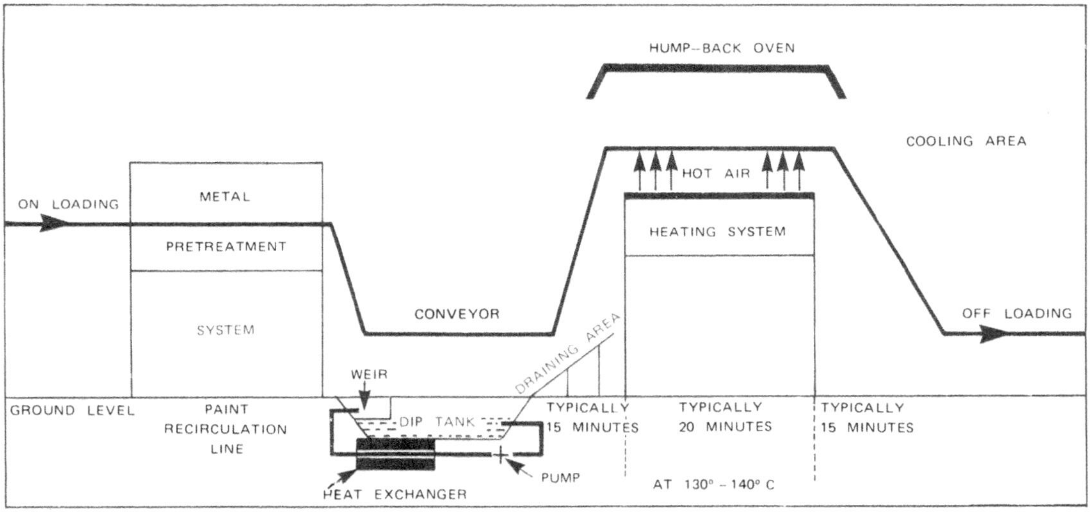

FIGURE 52.6
A typical dip system

Considering the large volumes of paint contained in dip tanks, the following points are important:

(a) To obtain uniform results, a regular check of viscosity and solids should be made.
(b) To reduce solvent losses, dip tanks should have a minimum surface area.
(c) The system should include adequate filtration and recirculation of the surplus paint from the draining area.
(d) The temperature should be kept constant.
(e) Articles should be thoroughly cleaned before entering the dip tank.

A dip system would probably have a time schedule similar to the following: after dipping, the articles would be drained for about 15 minutes, then stoved for 20 minutes at 130–140° C, followed by 15 minutes in a cooling section assisted by cooling fans. This curing cycle is typical for the application of an alkyd-melamine stoving enamel. Other coating types may require far higher curing temperatures for longer periods. (Other factors affecting the cycle are the conveyor speed and the thickness and quantity of the metal articles on the conveyor.)

Many conveyor systems use a high-intensity infra-red heat source for the curing cycle. These ovens cure similar coatings at much shorter schedules.

One of the many dipping systems in current use is Leyland's patented process, the 'Roto-Dip'. As the name suggests, the object (a car body) is rotated through both the dipping and draining operations, which give superior flow without tears or sags. It is generally used only for the application of primer coats and provides penetration of paint to the inaccessible areas that are most liable to corrosion.

Various conveyor systems are used in dipping whereby the conveyor has a spasmodic movement: as it stops over the dip tank, either the article is lowered or in some cases the dip tank is raised, at a controlled rate giving uniform painting.

Most dipping plants are designed for specific purposes and vary according to the size and shape of the articles to be coated. Small objects such as paint-brush handles and wooden utensils are usually coated by the 'slow dip process' where the withdrawal rate is controlled. The coatings are generally very high viscosity lacquers that can often provide adequate film build in one application over a suitable sealer.

Another variation—commonly used for coating wooden broom handles, mop handles, pencils and similar articles—removes excess lacquer from the surface of the objects as they are withdrawn from the dip tank, by passing them through either a neoprene or felt squeegee. This can be either a manual or automatic operation, producing uniform film thicknesses over long lengths and at fast production rates.

A similar method is used for coating wires. A coil of wire is passed through a tank of insulating varnish and then through a squeegee or a metal die, resulting in a smooth even coating. After coating, the wire is passed through a curing oven and then recoiled at the other end of the line. This process is continuous, to produce the correct film build.

The electrical industry is an important user of the dipping process, coating electrical units such as transformer ballasts and electrical motor windings. The electrical component is usually dipped in a low-viscosity phenolic insulating and bonding varnish to achieve the maximum penetration and impregnation of the coating.

Many recent installations for dipping use aqueous stoving paints for such applications as refrigerator condenser coils, aluminium extrusions and underbody automotive parts.

52.5.1 Paint Requirements for Dipping

Dipping is one of the simplest methods of paint application. However, with incorrect paint formulation, it can be one of the most difficult to control. The ideal formulation should meet the following criteria:

(a) high opacity to give complete hiding over low film build areas;
(b) suitable rheological properties to minimise sags, tears and 'running away' from sharp edges;
(c) the correct solvent balance to avoid solvent boil or sagging;
(d) good anti-settling properties;
(e) good stability to withstand aeration, oil and other contaminants.

These requirements can usually be met by using a slightly higher pigmented version of a conventional industrial coating with a fairly active blend of medium to slow boiling solvents. In non-airconditioned plants it is customary to use a different solvent blend for summer and winter.

52.5.2 Vacuum Impregnation

A more sophisticated method of dipping is sometimes used for coating armature and stator windings with high solids thermosetting insulation coatings. The process is known as vacuum impregnation and is carried out by placing the windings in an autoclave and reducing the internal pressure to a near vacuum before admitting the varnish. After a suitable impregnation period, pressure is increased to complete the coating operation. Some units have provision for the contents to be heated while under vacuum, giving even better penetration.

Widespread use of vacuum impregnation is made for treating timber with various preservatives and fungicides to improve its durability under severe exposure conditions.

52.5.3 Flow Coating

Flow coating is a variation of the dipping method. Instead of immersing the article in a large reservoir, the surface is flooded from squirting nozzles. This is not a spraying operation in the strict sense, because the paint is not atomised to fine particles, the liquid velocity is low and very little solvent vaporises in the spray pattern. An important advantage over conventional dipping is the elimination of the serious problem of dip tank maintenance; hence more modern paints can be used which have only limited dip tank stability. For example, primers based in part on desirable resins such as the epoxy type lend themselves well to flow coating. As in dipping, proper draining is critical.

In precision flow-coating systems, the initial few minutes of draining is done in a solvent-saturated atmosphere in a drain chamber of a size that is consistent with the speed of the conveyor and the size and shape of the objects coated. The remaining time before stoving is under conditions of ventilation. Solvent balance of the paint must be precisely controlled to suit the conditions of drain, flash-off and reflow in subsequent baking.

The paint drainings can be thinned back to viscosity specifications and returned to the supply reservoir; hence there is little waste.

A good flow coating operation requires careful control of the following variables:

(a) composition, viscosity and temperature of the paint;
(b) pressure control at the pumps to maintain optimum pattern and flow rate of the paint through the nozzles;
(c) temperature and ventilation of the drain areas; if the film formation process is sensitive to water vapour, humidity control is essential;
(d) cleanliness of atmosphere and paint—for example, dust-free air is desirable because dirt can promote 'pinholes' or 'fish-eyes' in thin films. Also the paint collected in the drain pans should be filtered before returning to the supply tanks. Periodically the flow coating machines should be cleaned either manually or by flushing the system with solvent.

Flow coating has most of the advantages of dipping—for example, areas inaccessible to spray painting can be coated more or less uniformly. The method is more suited to objects that can be coated by spraying only at the expense of considerable overspray. Under most conditions, unless the work is very large, flow coating is faster than dipping and requires less floor space and floor load. However, as in dipping, flow coating is a low-precision operation with respect to control of film thickness, appearance and protective properties of the finish.

52.6 CURTAIN COATING

Curtain coating is a method of application that finds widespread use in the furniture industry, although it has been used for some time in industries such as confectionery and ceramics.

This process was invented by a Swiss bicycle engineer who later sold the patent rights to Ulrich Steinemann who produced the first commercial machines.

Curtain coaters are used for objects with flat or slightly curved surfaces and those with raised decorations and mouldings. For some applications the conveyors and feed tables are tilted up to 30° from the horizontal to provide uniform coatings.

Essentially curtain coating consists of the rapid horizontal movement of the object through a vertical falling curtain or 'waterfall' of liquid coating material. The volume of paint flowing in the curtain is controlled by careful micrometer adjustment of the slot in the bottom of the head tank. Pump speed is set to maintain a continuous curtain, and the conveyor speed is set to provide the correct film build.

It is possible to apply very thin uniform films by curtain coating. On the other hand high-build two-pack polyester coatings can also be applied by the use of twin heads set about 75 cm apart. With these catalysed finishes, two films are applied consecutively with sufficient mobility left in the applied double film for the accelerator and catalyst to mix with the polyester to provide satisfactory cure.

The conveyor speed can be varied from 24 to 122 metres per minute, thus offering very fast production rates. Curtain coating installations can be part of automated production lines incorporating in-line sanding, dust and fume extraction units, and curing ovens.

Formulation of materials for satisfactory curtain coating performance generally presents no problems, providing the material has sufficient flow and has a suitable surface tension. The surface tension is important to allow the material to form a continuous uniform curtain, free from waves and areas of varying thickness. Coatings that tend to foam or bubble will require the addition of a suitable antifoam.

The main advantages of curtain coating are speed of application and economy in the use of labour and materials, although installation costs of automated systems are fairly high.

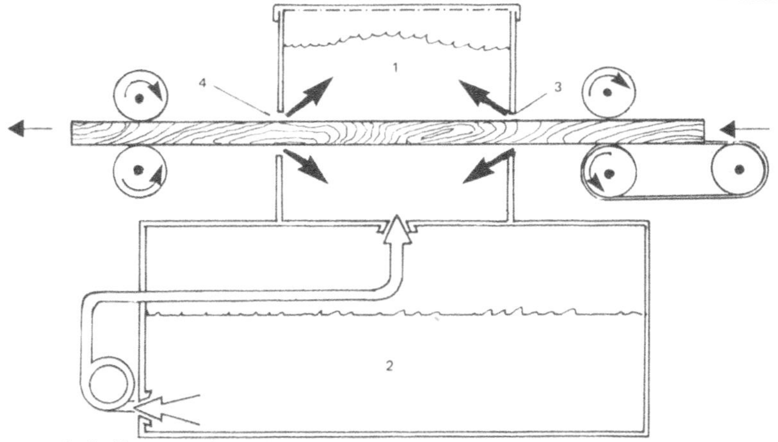

In the Vacuum coating applicator, the coating material is pumped into the application chamber (1) from the main reservoir (2). There it forms a coating bath that the boards pass through. Air is pulled into the application chamber through the board entry (3) and exit (4) ports.

FIGURE 52.7
Vaccum curtain coater

52.6.1 Vacuum and Pneumatic Coaters

Several woodworking and furniture companies are successfully using coaters which flood boards and then strip off excess coating material—rather than applying a pre-determined amount of coating to the board. The excess is recycled through the machine, thus avoiding any loss of material. Two types of these coaters exist, one operating on a vacuum principle and the other on a pneumatic principle.

The Paint-O-Matic coatings applicator, made by Paint-O-Matic Willits (California, U.S.A.) works on a vacuum principle. The object to be coated is roller-fed through an application chamber filled with coating material. A constant vacuum in the chamber not only keeps the material in the chamber but also strips off excess as the object leaves the chamber. Because the object is immersed in a bath, all exposed surfaces are coated. When using coatings of low viscosity (e.g. stains), boards can be handled and stacked immediately after coating. Excess coating is removed by air rushing over the board through the exit port at speeds of up to 650 km/h. The operator adjusts the thickness of the coating by regulating the vacuum. Without the need for drying after staining, less expenditure on energy is necessary.

The vacuum coater was originally designed for the application of stain. Since then, the machine has been developed to apply primer, sealer and topcoats—all water-based coatings.

The Pneumatic Coater uses a jet of air to strip off excess coating material. The object to be coated—a board or panel—is first flooded with paint by a curtain coater in a layer thick enough to obscure all texture. Then the board travels 3–4 m to allow it to absorb the paint. It passes under an 'air-knife' which blows off the excess into a collection tank. The distance it travels to the air-knife is sufficient to allow enough paint to soak in to coat the board evenly. The amount of paint delivered to the board is a function of air pressure, air-knife angle, distance of the knife from the board, and any change in the surface of the board feed.

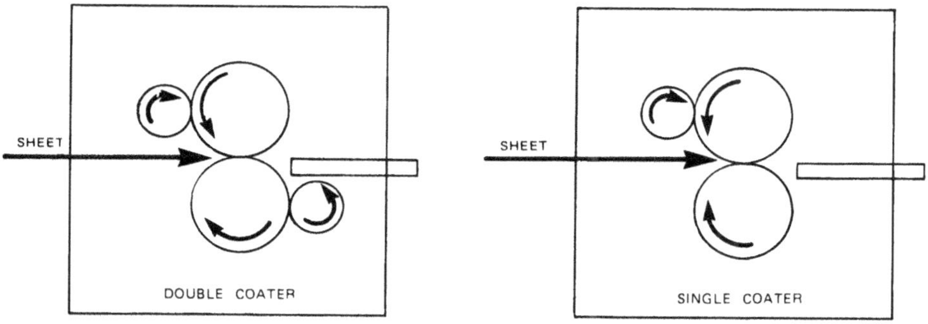

FIGURE 52.8
Roller coating

52.7 ROLLER AND COIL COATING

Roller coating is an efficient and economical means of coating sheets or strips such as plywood, hardboards and paper boards for packaging. Modified coaters are used for coating flexible tubes and containers.

The most significant use of roller coating, however, is for coating continuous steel and aluminium coiled strip at speeds of up to 180 metres per minute. In many installations, both sides of the strip are coated simultaneously. The coil coating industry is one of the most important consumers of paint and coating products, as most plants operate on a 24-hour, six-day basis.

Roller coating is the method used almost exclusively in the metal container and tinplate industries where final products are formed from the coated sheets. The machines are of the single-coater or double-coater type, using three or four rollers consisting of a doctor roller, applicator roller and feed roller. The coating material is held in the space between the doctor and applicator rollers, and the final coating thickness is adjusted by the clearance between these rollers. The coating is transferred to the flat sheet as it passes between the applicator and the feed roller. The machines are generally set up with automatic feed systems, including pretreatment and final curing ovens, on a continuous basis coating up to 100–150 sheets per minute.

Similar machines are used for coating plywood, particle boards and hardboards on conveyor production lines, applying fillers and other materials by the reverse rollercoating technique. The shearing action of the roller forces the coating into the pores of the timber, providing an excellent basis for subsequent coats. Using patterned or embossed rollers, various surface effects can be obtained by coating a base coat first, then overcoating with a contrasting colour.

52.7.1 Continuous Coil Coating

Coil coating is a high-speed process for prefinishing aluminium or steel strip with a thermosetting enamel, for subsequent fabrication into many varied end products.

The coil of aluminium, steel or tinplate is unwound and passed through a five-step pretreatment section. It is then dried and passed through a roll leveller and into the primer paint room where a two- or three-roll paint applicator applies the required wet film thickness. The strip is then heated to the required temperature of the semi-cured primer, cooled and passed into the finishing paint room where the topcoat is applied. The strip continues into the finishing oven where the system is fully cured, then cooled, recoiled and stored in readiness for fabricating into the finished article.

Coil coating has a number of advantages over other methods of paint application, but the main advantage is the fast production speed it offers.

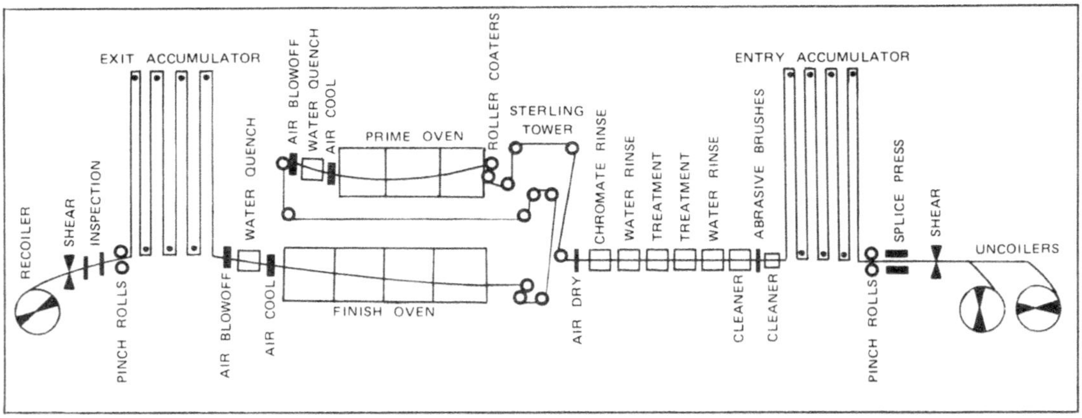

FIGURE 52.9
Typical coil coating line

52.7.2 Paint Requirements for Roller and Coil Coating

When formulating enamels for the coil coating industry, the polymer type governs the cured film properties. The formulator should know the degree of durability required, the flexibility and adhesion performance expected, whether the enamel is of sufficient hardness to withstand recoiling, slitting and rollforming, whether it can be supplied as a matt or semi-gloss finish if specified, and so on.

The flow requirements of a rollercoating enamel differ a great deal from those of other industrial coatings. During application, the enamel must be applied evenly and level rapidly under the influences of surface tension and, to a lesser extent, gravity. The enamel should exhibit a slight thixotropy. A dilatent paint would not wet the application roll properly, causing bare spots (ghosting) and an uneven 'raggy' surface upon application. A high degree of thixotropy would also cause problems in poor levelling. There must be no change in rheology due to solvent evaporation; thus medium to high boiling solvents must be used in the formulation. Because of their slow evaporation, these solvents, once the paint is applied, take on the role of maintaining the flow properties. Low-viscosity (low molecular weight) silicone fluids are sometimes used to improve levelling qualities.

It must be remembered that at the high production speeds encountered with coil coating, any paint fault can result in a great quantity of unusable coated stock.

Foaming can be a problem. It can cause air bubbles to be trapped in the surface of the cured paint film, causing blisters or bumps which can act as sites for corrosion. The use of baffles in the pick-up paint tray will assist in keeping this problem in check.

At present most enamels are cured by heating the metal substrate. The formulation must have a fairly broad margin of undercure and overcure, as coil gauge may vary slightly and temperatures may fluctuate slightly throughout a production run.

When the strip is recoiled, it is still quite warm. Special care must be taken to formulate to avoid pressure marking or 'blocking' in the coil. This problem may be caused by the cured film being soft or by waxy slip agents migrating from the surface of the film. The reverse side enamel—called the 'backing coat'—can also cause pressure marking and in some instances it will actually 'pick off' the topcoat in isolated areas. It is therefore good practice to formulate a backing coat of similar cure properties (mainly gloss level, hardness and adhesion).

As with industrial baking enamels applied by other means, many different polymer types are suitable. In rollercoating the emphasis is usually on decoration, whilst in coil coating it is usually on extreme exterior durability.

Typical coatings in common use in the coil industry are based on the following polymer systems:

Primers: Epoxies

Finishing coats: Alkyd amine
Vinyl—solution and plastisol types
Acrylics
Silicone acrylics
Polyesters
Silicone polyesters
Fluorocarbons

Environmental problems in the coil coating industry have caused a trend to water-borne coatings, mostly of an acrylic emulsion type. These exhibit excellent combinations of durability, formability, applied cost and non-polluting characteristics.

52.8 POWDER COATING

Powder coating involves the application of 100 per cent solid pigmented powder to the articles to be painted. The powder coating is 'fused' to a continuous film in an oven curing cycle or applied to a preheated article. Powder coatings may be either thermoplastic or thermosetting, and their formulation and manufacture is covered in detail in chapter 42.

52.8.1 Fluidised Bed

Fluidised bed is a technique analogous to dipping of wet coatings. Thermoplastic powder that has been reduced in size to about 100–200 μm is held in a dip tank with a porous base. Low-pressure air is applied to the base and 'fluidises' the powder. The article to be coated is preheated to a little over the melting point of the powder and then dipped briefly into the bed of 'fluidised' powder. The powder adjacent to the hot metal surface melts and coats the metal. The coated article is now withdrawn, and generally the coating is fused in an oven with a temperature 5–10° C higher than the melting point of the powder. It is sometimes possible, if the mass of metal is sufficient and with adequate preheat temperature, to fuse the coating without further heating.

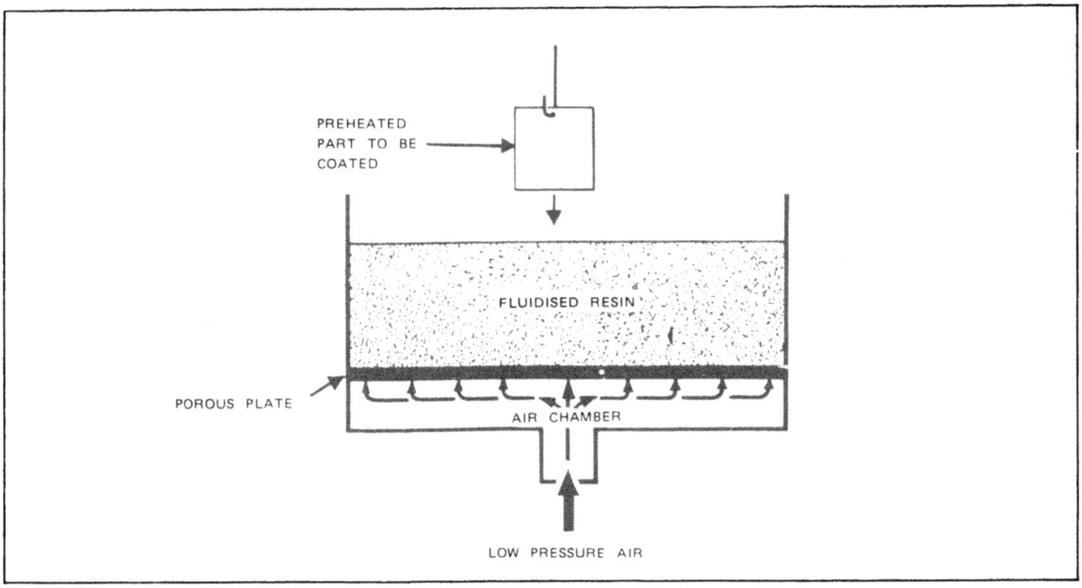

FIGURE 52.10
Fluidised-bed coating

The film build depends on the gauge of the metal, the pre-heat temperature, the time taken from pre-heat to the bed, and the dwell time in the bed. The final film is usually quite thick, about 250 μm. If an article is made up of both thin and thick metal sections, different film builds are produced on each.

In some automatic systems the bed of powder is raised and lowered as the articles enter the top of the dip tank. This is done by turning the air supply on or off.

Because no solvent is present in the powder coating to aid the wetting of the metal surface, all forms of powder coating require very clean substrates in order to prevent 'pinholes' or 'cissing'. Recommendations are for grit blasting or chemical pretreatment.

52.8.2 Electrofluidised Bed

This is a modification of the standard fluidised bed, in which an electrode charged to a high potential is situated above the perforated base of the dip tank, charging the powder as it is raised by the air supply.

With this method, the article is earthed and cold when lowered into the fluidised bed. It becomes coated by a layer of electrostatically held powder because of the induced opposite charge on the earthed article.

A lower film build generally results because of the good insulating characteristics of the powders. After a certain film thickness has built up, the approaching particles are repelled—they carry the same charges, and no leakage occurs back to earth. The coated article is withdrawn and fused as before, and care is taken not to disturb the loosely held coating until it is firmly fused.

Although this method can yield more uniform film builds, it does suffer (like all electrostatic processes) from poor coverage on recessed areas.

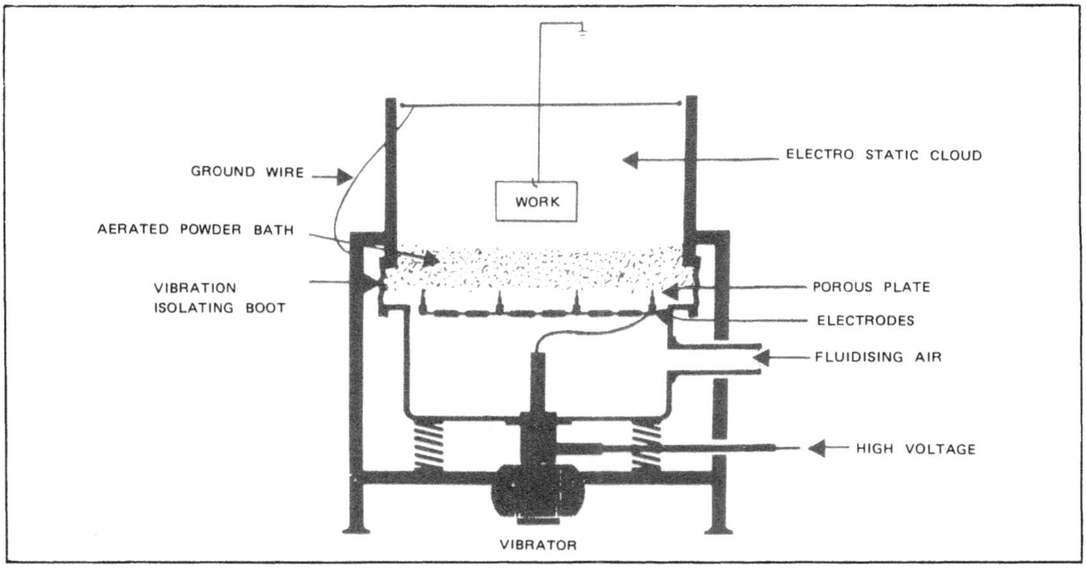

FIGURE 52.11
Electrostatic fluidised bed

52.8.3 Electrostatic Powder Hand Guns

In early designs of electrostatic powder hand guns, the powder was contained in a vessel with a porous base through which air was forced at low pressure; this gave a similar effect to the fluidised bed process. More recent developments include vibrational or Venturi systems, and pulsating flow methods in which a rotating wheel forces the pressurised powder outlet to open and close rapidly. The aerated powder is transferred to the gun itself either by simple air pressure or by using a Venturi-type pump. The rate of powder transfer to the gun can therefore be closely controlled.

Some guns use the principle of a rotating head powered by an air motor in order to distribute the powder towards the article. The rotating head can be a bell, in which case the powder coming from the shaft of the gun hits a deflector which spreads it out over the inside of the bell from which it leaves by the sharpened edge. Other models utilise a slotted mushroom-shaped rotating head which causes the powder to be uniformly distributed towards the article being sprayed. Whatever the design of the rotating head it is charged to a high, often negative, potential by means of a charged probe or brush. The powder, on leaving the rotating head, has this high potential and is therefore attracted to the earthed article being coated. Because these powders are generally good electrical insulators, on reaching the earthed article only a part of their charge leaks to earth and the residual charge holds the particles in position. After a certain thickness of charged powder is built up on the article, further particles are repelled because they carry the same charge and so the process tends to be self-limiting.

In the nozzle type of gun, the charge is generally attained by passage of the powder over a charged electrode. The potential in all these guns is obtained by similar means to those used for paint electrostatic spray. In general a negative potential is used with the thermosetting epoxy powders, whereas some thermoplastics (such as nylon) give superior results with a positive charge. Some guns are therefore designed to spray with either charge.

It is usual to use a cold earthed article, but if a thicker film is required the article can be preheated; more powder will adhere because the first layers will melt. Guns are available that do not require compressed air, the powder being propelled up to the gun head by a motor-driven vane in the powder container.

Powder is more expensive than conventional paint, so it is imperative to collect any overspray and excess material. With electrostatic spray application of paint, overspray is slight; but with powder, because of the reduction of attraction after a certain thickness has been obtained, overspray will occur. Consequently, methods are available for recovery and re-use of surplus powder.

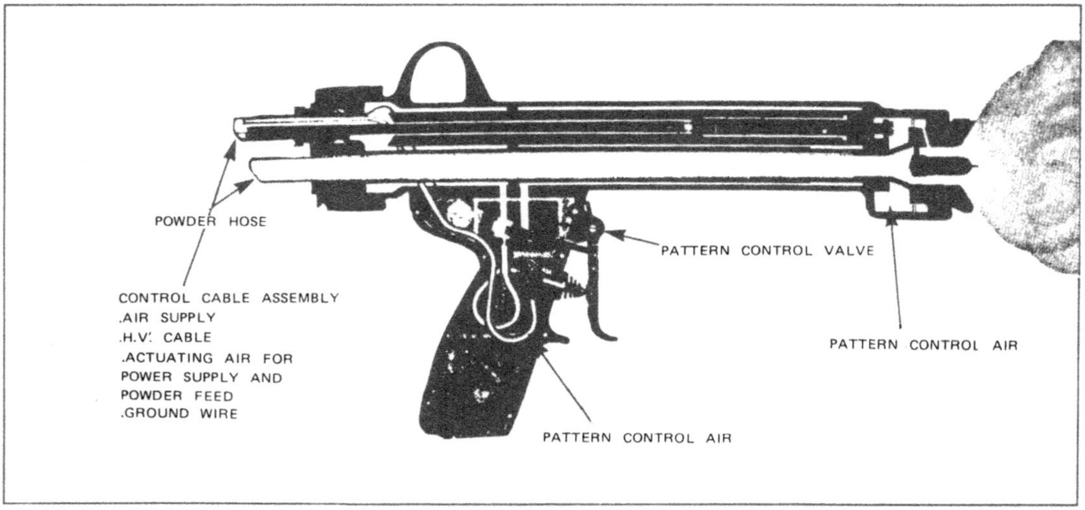

FIGURE 52.12
Electrostatic powder handgun

52.8.4 Automatic Electrostatic Powder Coating

The most fundamental automatic systems employ fixed guns positioned either side of a moving conveyor. More complex installations utilise reciprocating gun units.

In one typical installation, five automatic electrostatic guns are mounted on the cross-bar of a vertical stroke reciprocator and are used to apply powder coatings to corrugated building sheets. A cam device provides automatic movement of the guns to follow contours and maintain spacing from guns to target surface. A tracking method allowing movement of the entire reciprocator is necessary when curved rather than flat sheets are to be coated.

Other automatic systems, commonly referred to as multi-gun systems, are widely used for pipe coating. Pipe lengths are generally connected end-to-end and passed as a continuous train through a cylindrical coating chamber. Automatic electrostatic guns are mounted in a helical pattern through the sides of the chamber. The mounts allow for target distance adjustments when pipe diameters change. Line speed, rotation speed, powder delivery, gun placement and fan pattern must all be properly adjusted to eliminate a striping or barber-pole effect and to maintain tolerances.

Advantages of the fixed gun technique include less complex control equipment, minimum maintenance and down-time and, in some instances, more consistent film thickness control.

Another automatic method of powder coating is the 'Cloud Chamber' process. A cloud of charged powder rotates slowly in a tunnel through which the earthed article passes on an overhead conveyor. The cloud is produced by several distribution heads within the tunnel, and these are set in such a way as to maintain a rotating cloud with the assistance of low-speed directional air streams. Clouds of powder rotating in opposite directions produce a more uniform coating on complex shapes.

52.8.5 Flame Gun

The flame gun method is used mostly for applying thermoplastic powders such as nylon. The powder melts as it passes through the flame at the end of the gun. The article is heated by the flame and the hot molten plastic adheres to its surface. This method does not generally require separate pre-heat or post-fusion stages, although it does suffer from fairly high overspray losses and requires very skilled operators.

52.8.6 Powder Recovery and Safety Aspects

While a powder is inherently safer to use than volatile liquid coatings, the powder to air ratio must be kept at a safe level when in the presence of a potential source of ignition. This is best accomplished by continuous system operation during the working period, thereby avoiding any settling or excessive collection of powder within the booth or application area where ignition could occur.

During operation, the air flow balance through the system must be maintained to regulate booth entrance velocity, to prevent the powder from drifting out of the booth or cabinet. This is best accomplished with adjustable access and profile openings. It is also necessary to maintain a system air velocity that will keep the powder from settling anywhere within the system. Stagnant areas have an adverse effect on efficiency, safety and quality of finish.

52.9 ELECTRODEPOSITION OR ELECTROCOATING

This method of paint application was first patented by Cross and Blackwell in 1930. The first commercial application was in 1962, for coating auto wheels.

Today the most significant use of electrodeposition is for application of automotive primers because of the excellent corrosion protection and very even film thickness.

Electrodeposition is a very similar operation to electroplating as there is an anode and cathode immersed in an aqueous medium through which a current is passed. The early commercial systems were anodic electrocoat, where the article to be coated was made the anode in the 'plating' process. These systems are still in commercial operation. However, recent technology in use in the automotive industry is based on cathodic or cationic electrocoat, where the article to be painted is made the cathode in the process. This system produces primers with significantly improved corrosion resistance.

Although the base resin dictates certain final film properties, electrodeposition-applied primers result in adequate protection to all normally inaccessible areas—such as high spots on the base metal, weld area seams and inside internal box sections—giving greatly improved performance over conventionally applied primers.

Electrodeposition also has widespread use in the application of primer and finish coats to domestic appliances, electrical goods and components, lawnmowers and tractors, aluminium extrusions, metal shelving and toys and many other articles.

52.9.1 Principles of Anodic Electrocoating

The article to be coated is immersed in a tank containing a suitably controlled aqueous paint system. An electric current is applied to the tank and the article to be coated is made the anode. The cathode can be the tank itself or, in the case of an insulated tank, separate cathodes are inserted. When the current is applied the negatively charged paint solids are attracted to the article and 'plate out' on its surface. The coated article is then removed from the tank, washed free of excess paint, then stoved.

The separate electrolytic processes take place in the following order:

Electrolysis. This is essentially the separation of the paint into negatively charged paint solids and positively charged solubilising agents and other additives. These negative ions discharged at the anode will generate protons—acidity, or low pH value.

Electrophoresis. This is the movement of the paint solids to the anode and their deposition on the work under the influence of an electric current.

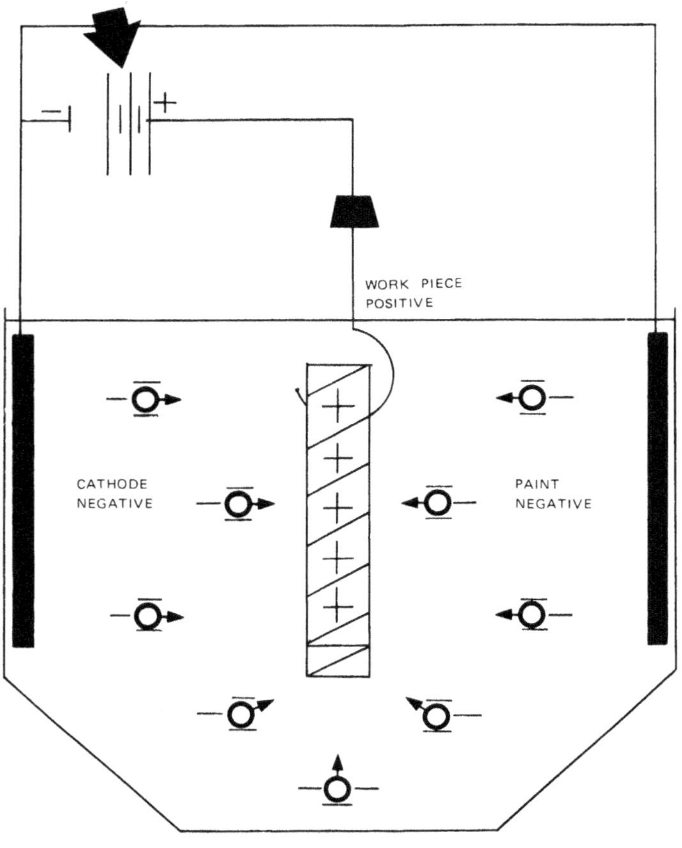

FIGURE 52.13
Principle of anodic electrocoating

Electro-osmosis. This is the movement of the water and soluble ingredients out of the film. Due to the passage of the current, the water within the paint film is squeezed out by osmotic dehydration. Thus on removal from the dip tank, the film has a water content usually below 8 per cent and is quite firm and dry to touch. Unlike other dipping methods, there are no runs or sags and coated components can enter the curing oven without the usual 'flash-off' period.

During this operation, paint solids cover every conductive surface, regardless of shape, with a uniform film whose thickness can be controlled very accurately. As soon as this thickness is achieved, the resistance built up by the coating slows down further deposition and the article is withdrawn from the tank.

During the process of film deposition, the film thickness increases rapidly at first then more slowly. With the dehydration of the film, the electrical resistance of the film increases sharply towards the end of the deposition. The applied current will therefore diminish until insufficient protons are being produced at the anode to reduce the pH to cause resin precipitation and the deposition virtually ceases.

In actual practice the areas of the anode (workpiece) closest to the cathode are coated first, and as the film build increases and builds up resistance, the more remote areas are then coated.

This whole coating operation usually takes 1–3 minutes.

Throwing power. The 'throwing power' of a paint is a term used to describe the ability of the paint to coat recessed and shielded areas. It depends upon the electrical resistance of the film during deposition and the electrical conductivity of the liquid paint. Large particle size emulsions tend to produce coatings of higher film build but with lower throwing power. Solution coatings generally produce lower film builds with better throwing power. To obtain the maximum throwing power it is quite common to have the voltage increased to a level just below the 'rupture voltage value' of the coating.

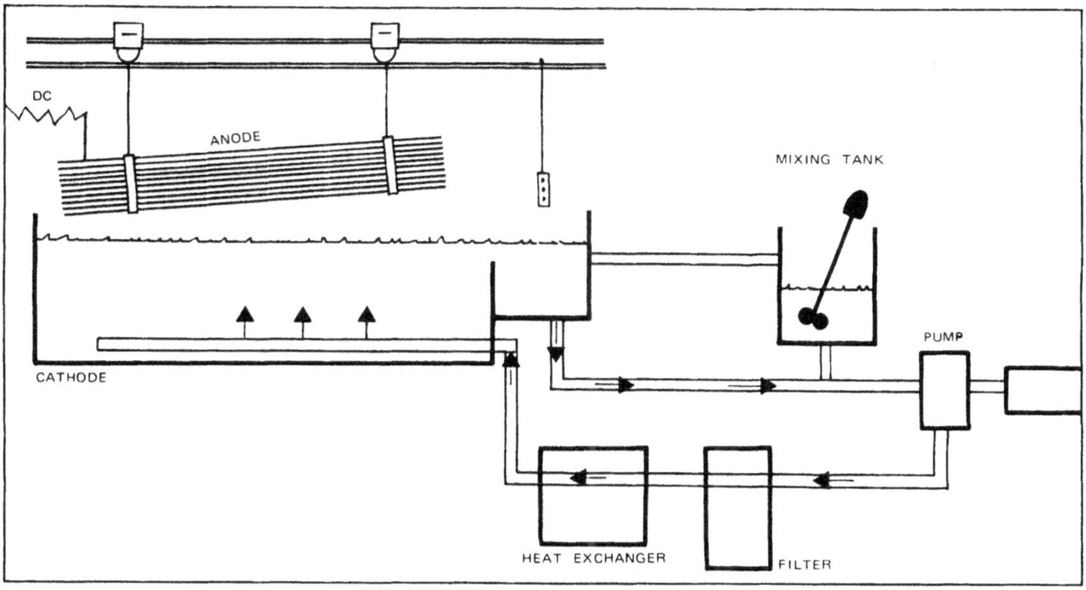

FIGURE 52.14
Simplified diagram of installation for electrocoating extrusions

Rinsing. At the end of the electrodeposition operation, the dipped article also becomes coated with an additional film of adhering paint called 'drag-out'. The drag-out contains a high proportion of water and is more porous than the plated coating. Therefore, it is usually rinsed off. In the past, the drag-out and rinse were discarded, resulting in paint wastage and disposal problems. Most lines include some means of re-using this paint by returning it to the tank. Rinsing does not affect the plated paint because of its excellent adhesion even before curing. As stated, the coating feels dry to touch and can even be rubbed gently. The rinsed article is cured in a conventional oven.

52.9.2 Control Systems

During the coating operation, the pH rises considerably and, unless controlled, can reduce the deposition. The two main systems used for paint control are the base deficient make-up system and the membrane system.

52.9.2.1 Base-deficient Make-up System

The 'Ford-Glidden' process is currently used in several very large plants for auto body priming. The process consists of adding acidic or base-deficient paint which in turn reacts with and dissolves in the excess of base present in the dip tank. As the stability of concentrated acidic paint is rather limited, it is common with this system to use separate additions in a pre-mix tank of highly concentrated aqueous pigment dispersion and acidic resin solution. These individual 'feedstocks' are added to small quantities of the main paint and after vigorous agitation are fed back into the main dip tank.

52.9.2.2 Membrane Systems

The resin and pigment particles deposited on the anode are removed from the electropaint bath, but the corresponding amount of base released at the cathode immediately redissolves in the paint. The consequences of continued paint deposition under these conditions would be a decrease in the ratio of resin/pigment particles to base particles. The pH would correspondingly rise, leading eventually to loss of control. The use of simple cathode boxes circumvents this difficulty.

A cathode box is a structure, one side of which is closed only by a simple membrane; inside the box is the metal cathode. In operation, the box is filled with water at about pH 10–11. The cathode box is immersed in the paint bath so that the liquid levels of the box and the bath are equal. Base ions can easily pass through the membrane into the cathode box but, after liberation at the cathode, cannot pass back to the paint bath because there is no great movement of liquid across the membrane and no electric field causing them to move out of the box. The net effect is that the resins plus pigments are removed from the paint bath at the same rate as the base, so that the pH remains constant.

By using this technique it is possible to use fully solubilised feed material to replenish the tank.

52.9.3 Cationic Electrodeposition

A further development in electrocoating has been the introduction of cationic electrocoating equipment. Cationic electrocoat offers significantly greater corrosion resistance than the best anionic electrocoat, especially in respect to throwing power and thin-film corrosion resistance. Such systems are reported to give an increase of 50 per cent in the throwing power and up to three times increase in corrosion protection.

As most cationic systems operate at up to double the solids of anionic systems, better tank circulation and mixing are essential.

In order to obtain a smooth surface it is important to have a good rinse system to remove the non-deposited film from the object—a car body—as soon as it leaves the dip tank. Between rinses, the car body is tilted to facilitate drainage. The final rinse is with deionised water.

The pH of the bath is controlled by using anolyte cells. The cells enclose the anodes and are constructed from high-density polyethylene with an ion-selective semi-permeable membrane on the side facing into the bath. This membrane allows acetate ions produced by the cathodic reaction to pass through as they are attracted to the anode, but the acetic acid resulting from the anode reaction cannot pass back through the membrane into the bath.

Cationic tanks give not only improved corrosion resistance but also better tank stability and general performance and at lower cost.

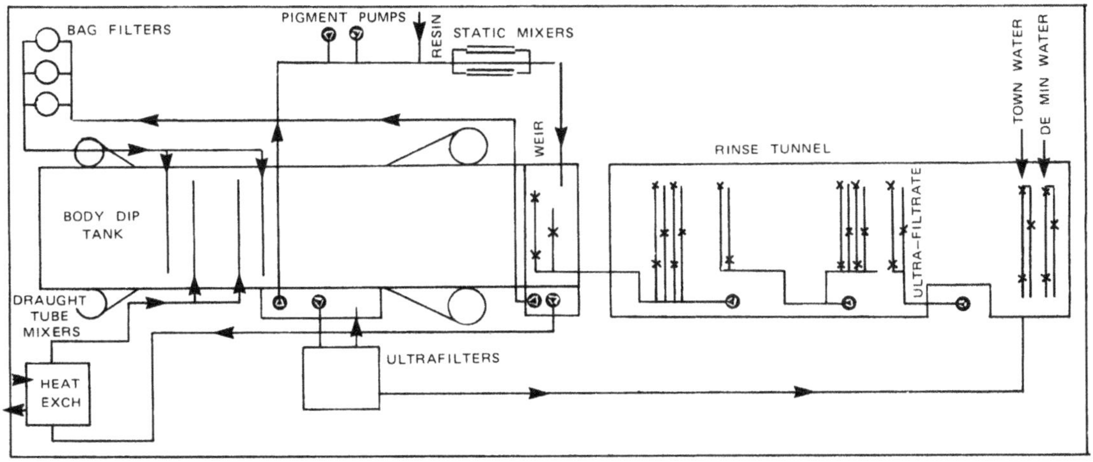

FIGURE 52.15
Cathodic electrocoat system

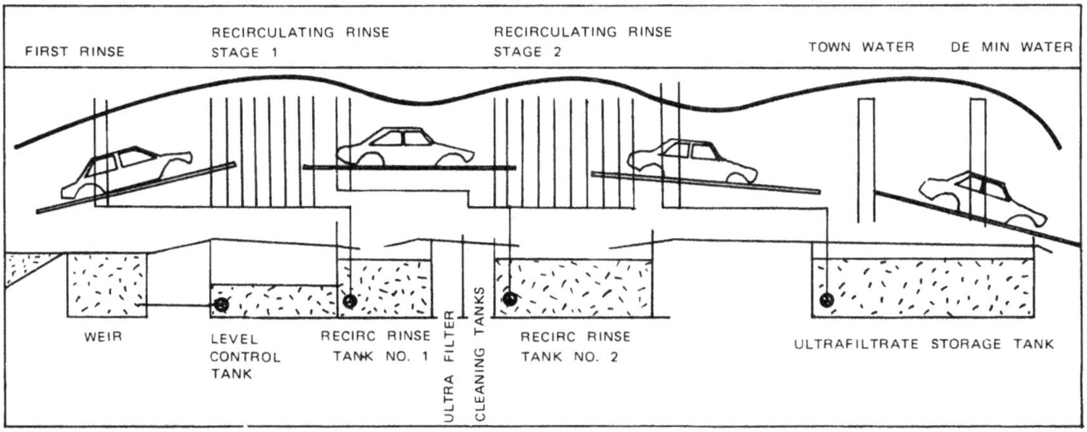

FIGURE 52.16
Spray rinse system

52.9.4 Advantages of Electrocoating

(a) The main advantage is the uniform film build produced on all surfaces, even in recessed areas, sharp edges and tapped holes.
(b) It results in superior adhesion and corrosion resistance of the uniformly dense film which is in fact 'forced' against the substrate during painting.
(c) The deposited film contains about 8 per cent of water, requiring no 'flash off' before curing.
(d) These are fewer fire hazards and less air pollution, because the main solvent used is water.
(e) It can be a fully automatic and mechanised process, utilising more than 95 per cent of the coating.

52.9.5 Disadvantages of Electrocoating

(a) Because of the high installation cost, the system is suitable for only very large numbers of similar articles.
(b) Formulation and control of the paint are extremely critical, and the paints are very sensitive to contaminants.
(c) High gloss finishes are difficult to produce.
(d) With anodic deposition, metals tend to dissolve into the coating, producing discolouration.

52.9.6 The Future of Electrocoating

There is no doubt that electrocoating systems are now firmly established in industry, and the method should continue to gain wider acceptance in many new areas in the future. Although its major use is in the automotive industry, its use will no doubt be extended for coating appliances and aluminium extrusion.

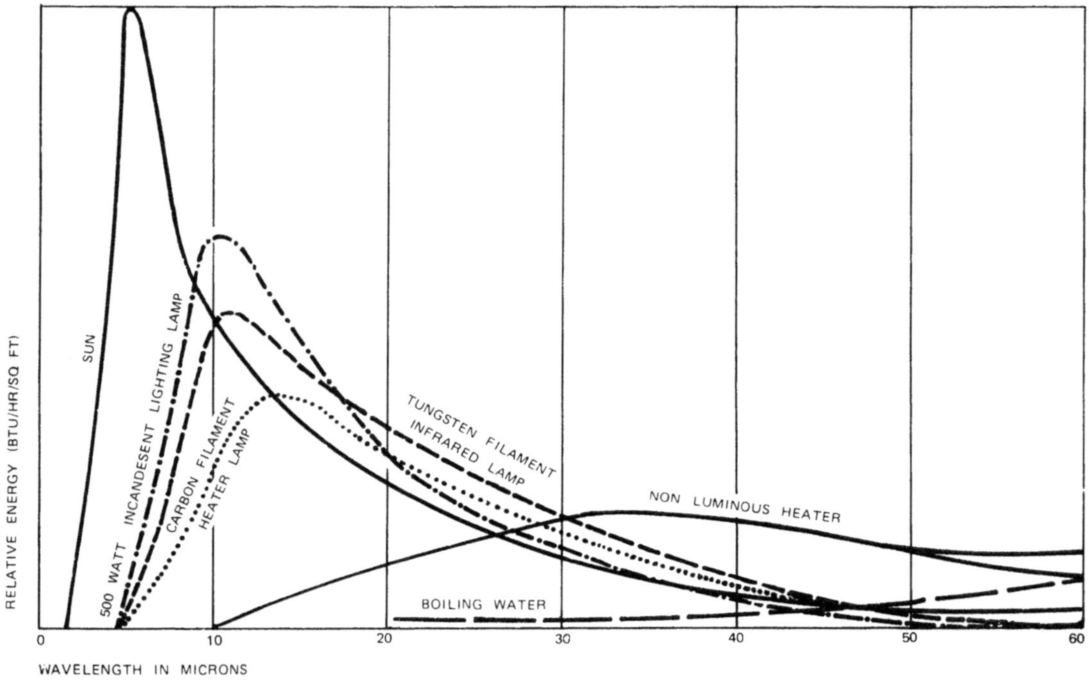

FIGURE 52.17
Energy distribution of infra-red sources

52.10 OVEN CURING METHODS

52.10.1 Radiant Heat Ovens

Radiant heat ovens are usually conveyored using an overhead monorail. The oven section comprises infra-red emittor panels arranged around the workpiece. The power source is usually electricity or gas—electric panels generally being preferred for their higher efficiency.

Heat losses can be reduced by having reflector plates to divert the heat back to the work piece. The radiant panels have an exceptionally quick heat-up time, reaching operating temperatures from cold in a very short time.

The best effects are obtained from infra-red panels when the rays strike the work perpendicular to the surface to be dried; articles with a flat smooth surface are ideally suited to this method of drying.

Infra-red ovens no longer use the fragile glass lamps, but usually incorporate a series of robust sheathed wire elements or silicon carbide rods. Widespread use is also made of tungsten filaments in inert gas-filled quartz tubes.

It is often assumed that a forced-draught convection oven uses only convection as the means of heating, and that infra-red ovens rely solely on radiation. However, the modern high-temperature forced-draught oven utilises a considerable amount of radiation from the hot walls of the oven. In an infra-red oven, the air and the walls become hot and thus hot air convection plays a significant part in the paint stoving. In order to gain full use of this convected heat, the ovens are usually enclosed on the top to prevent hot air loss, and on the sides to reduce cooling draughts.

The radiating sources can operate in the region of 260–800° C although some units reach up to 2000° C. It is important when working with infra-red ovens to measure surface temperature of actual coated panels, as different colours and coatings are capable of absorbing different quantities of radiation. This temperature measurement can be made by using either thermocouple devices or thermopapers attached to the article by heat-resisting tape.

With infra-red ovens, the following variables can be adjusted to achieve uniform results: conveyor speed; conveyor loading; radiation intensity; and oven cycle time.

Most electric infra-red ovens incorporate controls to allow the intensity of the radiation to be varied from 0 to 100 per cent of the capacity of the unit. This allows for greater versatility in respect to conveyor loading and metal gauges used.

52.10.2 Forced Air Circulation Ovens

Industrial convection ovens are usually in the form of simple box ovens for batch loading or continuous loading where the work flows continuously through at a predetermined speed, often suspended from an overhead conveyor as in the familiar 'camel-back' oven.

The air is heated by any conventional heating medium and is then blown into the work chamber. The main advantage of this system for drying is that it is not limited to straight line travel, the air being guided or 'splashed' into the interior surfaces of irregular shapes.

The oven can have a low roof, or it may be as much as three storeys high with the conveyor moving forward on the lower level and back again at a higher level. The greater the cross-section of the oven, the greater are the problems of uniform heat distribution and ventilation. Preferred designs are as long as possible and as small in cross-section as is feasible. Most improvements in ovens are in the direction of providing better ventilation.

When evaporating water, the vapour tends to accumulate above the wet surface. Unless this vapour is constantly blown away and replaced with dry air, the atmosphere immediately above the wet surfaces becomes saturated and this slows or even stops evaporation. When drying paint or any finishing materials containing flammable solvents, it is important that the vapour concentration is kept four times below the lower explosion limit.

52.10.2.1 Heat Sources

As previously mentioned, the heating source may be gas, electricity, light fuel oil, hot oil or steam.

There are two types of heating systems for fuel firing—direct and indirect. The direct method heats the air directly; thus products of combustion enter the oven. This is accomplished by locating the burner in the air stream. The indirect method heats the air using a heat exchanger; thus the products of combustion do not enter the oven. Of the two methods, direct heating is more widely used because of its efficiency and lower initial costs. Indirect heating would be used if the products of combustion are harmful to the finish of the work. Efficiency of indirect heating (approximately 50–75 per cent) is much lower and the initial cost is higher, largely because the heat exchanger is of stainless steel and refractory construction.

52.10.2.2 Air Circulation

There are two systems of air movement over the heating medium. One is the pull method, where the fan is located downstream of the heat source. The second is the push method, where the fan is located upstream of the heat source. Of the two systems, the pull method is preferred because the heated air can be mixed in the air-circulating fan to achieve a uniform temperature of air entering the work chamber.

The air circulation pattern, flow and distribution around the workpiece are important—the shape of the object being dried usually dictating the flow pattern. Generally a horizontal cross-draught is used, but vertical draught and end-to-end air circulation can be employed. In addition, multi-zoning of oven chambers can be considered.

52.10.3 Paint Problems with Oven Curing

52.10.3.1 Gas Checking

This problem may occur in the smaller, inefficient plants. It is usually observed in the form of dull patches on an otherwise glossy surface. In extreme cases it can show as a complete surface 'wrinkle pattern'.

The most common cause of gas checking or accelerated surface curing are from small pockets of acidic fumes that collect in the curing oven. The acidic fumes could originate from: acid pickling baths; or electric welding, giving oxides of nitrogen; or chlorinated solvents degreasing units. If the source of contamination cannot be removed or traced, it is often possible to alter the paint formulation to overcome the problem.

52.10.3.2 Solvent Washing

Solvent washing is caused by articles in the oven being at different temperatures. The solvent from the warmer piece evaporates and it condenses on the cooler article, washing the paint from parts of the article.

The following can help prevent this problem:

(a) the use of a faster solvent, so that the solvent evaporates before the articles enter the oven;
(b) increasing the 'flash-off' time by slowing down the conveyor;
(c) rearranging the packing of articles on the conveyor to allow greater air space between them;
(d) increasing the circulation of air within the oven to prevent solvent condensation.

52.10.3.3 Solvent Boil

Solvent boil, or 'popping', is caused by a rapid heat build-up, curing the surface of the paint before all of the solvent has left the coating. Thermosetting acrylic materials are very prone to this—their fast solvent release causes the surface to skin—and it is therefore necessary to adjust the solvent balance of the formulation to suit the particular stoving conditions.

Solvent boil can usually be eliminated by one or more of the following methods:

(a) changing the solvent composition of the coating;
(b) increasing the 'flash-off' time by slowing down the conveyor;
(c) the use of lower build by lowering the application viscosity;
(d) reducing the rate of heat-up in the oven;
(e) slowing down the cure of the coating.

Solvent boil is most commonly caused by a heavy film build in some sections of the articles.

52.10.3.4 External Contaminants

Another source of trouble with industrial paint application is 'cissing' and dirt contamination. Many plants use silicone-based grease and lubricants which can be readily distributed by careless operators or by air circulation throughout the oven and plant.

Dirt and dust are sometimes caused by a build-up of oxidation products in the oven and by bad air filtration on the incoming air. Ovens should be periodically cleaned, as should hooks and conveyors to minimise dirt and dried paint particles falling on the freshly painted articles.

52.11 ULTRAVIOLET CURING

Ultraviolet light greatly accelerates the formation of free radicals and curing is thus more rapid—for example, less than 1 minute at room temperature. It is mostly used in conjunction with fast coating systems (such as curtain coating), thus enabling curing speeds to equal coating speeds.

The two main sources of UV energy are low-power actinic fluorescent ultraviolet lamps and high-pressure quartz mercury vapour lamps.

Because of the health hazard associated with high-pressure mercury vapour lamps, careful installation and operating procedures are necessary.

The oil crisis of the mid 1970s intensified the search for alternative energy sources. UV curing systems underwent considerable improvements and are now finding widespread use with furniture, board coating, paper board (graphic arts) and printed circuit boards.

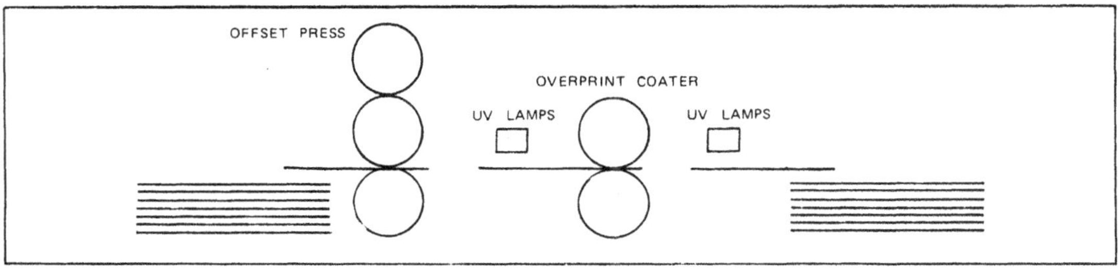

FIGURE 52.18
UV cured overcoated offset printing

53 SURFACE COATINGS DEFECTS

A defect in a surface coating can be the result of any one of a number of causes and may therefore have a corresponding number of remedies. Also, several defects may occur simultaneously, further complicating the technologist's task. This chapter classifies defects into broad groups, gives alphabetical listings of defects within each group, and provides relevant cross-references. The classifications are:

1. Defects in the liquid paint
2. Defects during application
3. Defects during drying/curing
4. Defects in the dry film

53.1 DEFECTS IN THE LIQUID PAINT

Aeration. 'Incorporation of bubbles of air in paint during stirring, shaking or application'. This can lead to *foaming* during application of the coating, or *cissing* or *cratering* during drying. Aeration can be controlled by the addition of proprietary defoamers in the case of latex paints, or bubble-release agents in the case of solvent paints. Aerated paints will exhibit subnormal density values, which provide an easy test for this defect at the manufacturing stage.

Bodying. See *viscosity increase.*

Caking. See *settling.*

Can corrosion. This may be caused by incorrect pH of latex paints, or incorrect choice of ingredients leading to acidic by-products on storage. The remedy is careful selection of can lining, or perhaps the addition of anti-corrosive agents to the paint, or improved formulation and adjustment.

Coagulation. This refers to the premature coalescing of emulsion resin particles in the paint. This is also termed 'breaking' of the emulsion. Excessive stirring, solvent addition or addition of coalescing agents may be the cause. Because universal colorants contain solvents (typically glycols), they may have the same effect if added too quickly or without post-stirring. There is no truly effective remedy for coagulated paint. Straining (followed by addition of further latex) may partially recover a batch, and permit blending off.

Curdling. See *gelling* and *coagulation.*

Gassing. This is *aeration* due to a chemical reaction within the liquid paint during storage. It can result in explosion of cans, with resultant hazards to health and property. The action of water on aluminium or zinc-based paints, or acid on calcium carbonate will give this defect. In the case of aluminium paints, an air vent in the lid (covered with paper sticker) is a wise precaution. Such paints may also be held in bulk until gassing testing is completed. The formulator should also consider including a small addition of a water or acid scavenger in the paint.

Gelling. 'Deterioration of a paint or varnish by the partial or complete changing of the medium into a jelly-like condition'. The cause of this condition may be a chemical reaction between certain pigments and vehicles (such as zinc oxide and an acidic vehicle) or between atmospheric oxygen and oxidisable or polymerisable oils in the vehicle. A paint that has gelled to a livery mass that will not disperse on stirring, even with added solvent, in un-recoverable.

Livering. See *gelling.*

Settling. 'Separation of paint in a container in which the pigments and other dense insoluble materials accumulate and aggregate at the bottom'. The law of gravity applies to paint, as does Stokes' law. An increase in consistency will help, as will a thixotropic rheology. Various additives are available to do this.

Skinning. 'The formation of a tough, skin-like covering on liquid paints and varnishes when exposed to air'. A skin sometimes forms across the surface of a paint during storage in sealed or unsealed, full or partly filled containers. If the skin is continuous and easily removed, it is not as troublesome as a slight, discontinuous skin which may easily become mixed with the remainder of the paint.

The formation of skin is due to oxidation and polymerisation of the medium at the air–liquid interface. Anti-skinning agents, usually volatile anti-oxidants, are generally added to paint to prevent skinning. A proportion may be lost, by evaporation, if the batch is left for an excessive time before filling. Because air (oxygen) is generally necessary, the best way of preventing skinning is to keep the air away as much as possible. When skin is encountered in a full container, the seal of the lid may be the cause.

Thickening. See *viscosity increase.*

Viscosity decrease. Many cases of reported low viscosity are the result of failure to stir thoroughly, or over-reduction. This may be checked by determining the density, or solids content, on a sample and comparing it with the original figure recorded for the batch. Other causes of viscosity drop are:

(a) enzyme attack on cellulose thickeners in latex systems, because of insufficient preservatives or contaminated ingredients; this is usually accompanied by putrefaction (odour) and possibly pressure build-up.
(b) failure of gellant or soap in solvent systems to maintain gel, due to chemical or physical causes.
(c) changes in pigment orientation on storage, such as dispersion of a partially flocculated system.

Viscosity increase. A slight increase in viscosity during storage of a paint is not uncommon, but rapid or excessive thickening is either because of instability of the medium or because of a reaction between the pigment and the medium known as 'feeding'. This can sometimes be corrected by adjusting the drier content, or by the use of anti-skinning agents, stronger solvents or certain chemical additives.

53.2 DEFFCTS DURING APPLICATION

Bubbling. See *Aeration*.

Cobwebbing. 'The formation of fine filaments of partly dried paint during the spray application of a fast-drying paint'. This can be caused by incorrect solvent blend in the coating, or by spraying too far from the article. The remedy in each case is obvious.

Coverage. 'The spreading rate, expressed in square metres per litre'. Poor coverage is a defect related to either *sticky application* because the *viscosity* of the paint is too high, resulting in too much paint being applied; or to an absorbent substrate. In the latter case, reduction with the appropriate thinner for the first coat only, will provide the remedy.

Drag. See *sticky application*.

Fly-off. 'The throwing-off of particles of paint from a paint roller'. This is a particular instance of poor rheological control.

Foaming. This is the formation of a stable gas-in-liquid dispersion, in which the bubbles do not coalesce with each other or with the continuous gas phase. It is a defect commonly encountered in application by roller, particularly with latex paints. The remedy is the addition of an anti-foam agent in the manufacture of the paint, and/or a reduction in the speed of the roller. See also *aeration*.

Frothing. See *aeration*.

Frying. See *lifting*.

Lifting. 'The softening, swelling and wrinkling of a dry coat by solvents in a subsequent coat being applied'. Usually, it is the action of strong solvents which cause this effect. Coatings that dry by oxidation are particularly prone to lifting. It is important to observe the recommended recoating times nominated by the supplier, as the rectification involves sanding and recoating.

Shrivelling. See *lifting*.

Spatter. See *fly-off*.

Sticky application. Sticky application of a paint may be caused by a number of factors:

(a) the choice of formulation ingredients.
(b) excessively rapid loss of solvent by evaporation during application.
(c) excessive viscosity, possibly because of bodying of the paint on storage or because of application at very low temperatures.
(d) high temperature either of the air, surface or paint itself.
(e) highly absorbent substrate or poor-quality brush.

Poor application characteristics can be controlled to a certain extent during production by viscosity adjustment and choice of a suitable solvent balance, providing the basic formulation is correct. Errors in raw-material weighing and contamination can obviously affect application properties.

Streaking. 'The formation of irregular lines or streaks of various colours in a paint film caused by contamination of, insufficient or improper incorporation of colorant'. This can be remedied by adequate incorporation of the colorant, or by the use of additives (as detailed in the drying defects *floating* and *flooding*).

Working up. See *lifting*.

53.3 DEFECTS DURING DRYING/CURING

Bleeding. 'Discolouration caused by migration of components from the underlying film'. Substrates that can cause problems are those coated with tar- or bitumen-based materials, paints made on certain red and yellow organic pigments (which are partially soluble in solvent), some wallpapers, and timber stains that contain soluble dyes. The remedy is to use a specially formulated sealer or an aluminium paint.

Blooming. 'The formation of a thin film on the surface of a paint film thereby reducing the lustre or veiling its depth of colour'. See also hazing. This defect occurs mostly in stoving enamels (particularly blacks) in gas ovens. For further treatment of this defect, refer Hess (1979). Lacquers also exhibit this defect, especially when used with low-quality thinners under certain ambient conditions. See also *blushing*.

Blushing. 'The formation of milky opalescence in clear finishes caused by the deposition of moisture from the atmosphere and/or precipitation of one or more of the solid constituents of the finish'. This defect is generally associated with quick-drying lacquers. The rapid evaporation of solvent causes the cooling of the substrate and the consequent condensation of moisture. The remedy is to adjust the evaporation rate of the solvents used, or pre-heat the article being coated.

Bridging. 'The separation of a paint film from the substrate at internal corners or other depressions due to shrinkage of the film or the formation of paint film over a depression or crack'. Undercoats or primers that do not have adequate filling properties will give rise to this defect. Poor surface preparation is another cause. The remedy is to provide adequate surface preparation, and apply an undercoat with good filling properties. A lower application viscosity may also be helpful.

Brush marks. 'Lines of unevenness that remain in the dried paint film after brush application'. Brush marks and ropiness are associated with poor flow and sticky application. These defects are more often encountered in highly pigmented products and in certain latex paint formulations. Too-rapid recovery of consistency in a thixotropic system will also cause these defects. The remedy may be the addition of a flow promoter, reduction in consistency or modification of the rheological properties.

Cheesy film. 'The rather soft and mechanically weak condition of a dry-to-touch film but not a fully cured film'. See *drying problems*.

Cissing. 'The recession of a wet paint film from a surface leaving small areas uncoated'. This is a consequence of improper wetting of the substrate by the paint. Frequently it is an aggravated form of pinholing. Where cissing is due to high surface tension inherent in the coating, specific proprietary additives can be used to remedy the situation. Examples are cellulose acete butyrate for polyurethane lacquers, and anionic or non-ionic surfactants for latex paints.

Clouding. See *hazing*.

Cratering. 'Residual effect of burst bubbles'. Cratering describes the formation of small bowl-shaped depressions in a paint or varnish film. For remedy, refer to *aeration*.

Crinkling. See *wrinkling*.

Curtaining. See *sagging*.

Drying problems. Where the defect relates to rapid drying, refer to sticky application. In the case of slow drying, there are two general defects.

(a) slow drying due to poor formulation or environmental conditions—for example, the drier blend may be incorrect, or the atmosphere too cold or damp;
(b) loss of dry on storage or after addition of universal colorants. The loss of dry here is attributable to absorption of the metal drier onto the pigment particle, or possibly by precipitation in an insoluble form.

The remedies are various. Adjustment of the drier blend can achieve improvements in both areas above. Control of conditions under which painting occurs will greatly assist the first area. Limitations on the amount of universal colorant used will safeguard the second area.

Fat edge. 'Accumulation of paint at the edge of a painted surface'. See *sagging*.

Filling. See *bridging*.

Floating. 'Separation of pigment which occurs during drying, curing or storage which results in streaks or patchiness in the surface of the film and produces a variegated effect'. Close examination will reveal Bénard (hexagonal) cells. This is due to differences in pigment concentration between the edges and centres of the cells, caused by convection currents in the drying film. Thixotropic paints will minimise this defect and the use of proprietary materials may also assist.

Flooding. 'An extreme form of *floating* in which pigment floats to produce a uniform colour over the whole surface which is markedly different from that of a newly applied wet film'. Again, thixotropic systems or specialist additives will provide a remedy.

Flow. 'The ability of a paint to spread to a uniform thickness after application'. See *brush marks* as a special case of poor flow. The remedies there apply here also.

Grain raising. 'The swelling and standing-up of wood fibre resulting from the absorption of water or solvent(s)'. This can be avoided by priming the timber or composite board with a good-quality primer/sealer (not water-based) before application of the finish coats. Where grain raising has occurred, the remedy is to sand back to a level surface, prime and finish coat.

Hazing. Loss of gloss after drying. It is usually caused by application of a gloss paint on a ground coat that has not hardened sufficiently; or excess driers in the final coat; or partial solution of organic pigments in the paint. See also *blooming*.

Hold-out. This is a defect in an undercoat, where the topcoat sinks into the undercoat due to high porosity. The undercoat or sealer should be tested after manufacture for this property. It will usually be necessary to apply a further coat of good-quality undercoat, followed by topcoat, to correct this defect in situ.

Mudcracking. 'Visible irregular cracking in thick films of paint caused by shrinkage tension during drying'. Generally but not always associated with latex paints, particularly highly pigmented materials. The defect can be overcome by the addition of further coalescing agent.

Ropiness. See *brush marks*.

Runs. See *sagging*.

Sagging. 'Excessive flow of paint on vertical surfaces causing imperfections with thick lower edges in the paint film'. Poor application technique and the condition of the substrate are major causes of this fault; however, the rheology of the paint can influence results obtained. Many paints currently sold have thixotropy deliberately built in, to facilitate the application of heavier coats with less tendency to sag. The rate of recovery of viscosity after application is the key to reducing sagging. In production, paints can be checked for poor flow or sagging tendency by various types of sag index blades or combs. These deposit tracks of paint of varying film thickness, and the resultant tracks, if left in a vertical position, give an indication of the flow behaviour to be expected in use.

Sink-back. See *hold-out*.

Sleepiness. See *hazing*.

Stippling. See *brush marks*.

Tackiness. 'The degree of stickiness of a paint film after a given drying time'. See *drying problems*.

Tears. See *sagging*.

Texture. See *brush marks*.

Wrinkling. 'The development of wrinkles in a paint film during drying'. This defect is closely associated with *drying problems*. Its cause is the surface of the film drying too rapidly before the underlying layer has firmed up. Correct balance of metal driers and solvents will cure this defect. Excessive film thicknesses may also be a factor. See also *hazing*.

53.4 DEFECTS IN THE DRY FILM

Ageing. 'Degeneration occurring in a coating during the passage of time and/or heating'. See also *checking*, *cracking*, *crocodiling*, and *embrittlement*:

Alligatoring. See *crocodiling*.

Bitty film. 'A film containing bits of skin, gel, flocculated material or foreign particles, which project above the surface of the film'. See also *seeds*. Where this is encountered, it is not always obvious whether the contamination is in the paint or has resulted from dust in the atmosphere. Some causes of bitty films due to contamination of the paint are:

(a) failure to strain the paint properly during the filling process;
(b) contamination with dust either during filling or during application;
(c) partial skinning of the paint;
(d) incomplete re-incorporation of settled pigment;
(e) post-seeding of paint because of an insoluble resin fraction.

The remedy for bitty films is to allow to hard-dry, sand back and repaint with clean material in a dust-free environment.

Blistering. 'Isolated convex deformation of a paint film in the form of blisters arising from the detachment of one or more of the coats'. This is often the consequence of faulty surface preparation, leading to poor primer–substrate adhesion. Dark coatings are more prone to this defect than light coatings. The only effective remedy is removal of the surface coating, careful preparation of the substrate, and repainting with the correct materials. Painting under very hot ambient conditions should be avoided.

Blowing. See *popping*.

Bronzing. 'The formation in a paint film of a characteristic red or yellow metallic lustre that is visible only at certain angles of illumination'. This phenomenon seems to be correlated with the presence of a substantial amount of material with a particle size below about 0.1 μm. See also *blooming*.

Chalking. 'Change involving the release of one or more of the constituents of the film, in the form of loosely adherent fine powder'. This is generally a result of the gradual breakdown of the binder because of the action of the weather. Careful selection of pigment types and levels and the use of more durable binder types retard the process. In flat white paints, chalking will enable the finish to be self-cleaning. A chalked surface requires washing down, or sealing with a penetrating sealer, before repainting.

Checking. 'Breaks in the surface of a paint film which do not render the underlying surface visible when the film is viewed at a magnification of 10 × '. Slight checking is not a serious defect, as it indicates a relieving of shrinkage stresses in a paint film. See also CRACKING.

Cleanliness. See *bitty film*.

Colour differences. Refer to chapter 45, on universal colorant systems.

Covering power. See *opacity*.

Cracking. 'Formation of breaks in the paint film that expose the underlying surface'. This is the most severe of a class of defects which include *checking*, *crocodiling* and *embrittlement*. These phenomena do not necessarily indicate that anything was or is wrong, if they are the natural consequences of normal *ageing* of the film. The process of breakdown leading to cracking in the paint film is essentially shrinkage of the film. Much of the cracking noted over timber is caused by splitting and grain opening and not defective paint. The following should be avoided, if cracking is to be reduced:

(a) applying paint before the previous coat is sufficiently hardened, particularly when the undercoat has been applied too thickly;
(b) applying a hard-drying finish over too elastic an undercoat.

In addition, careful selection of the raw materials used in the manufacture of the paint, and careful choice of the type of painting system used, will enable the surface coatings technologist to give maximum life for a particular application. The only effective remedy for cracked paint is total removal and repainting.

Crocodiling. 'The formation of wide criss-cross cracks in a paint film'. Here the cracks are pronounced, and expose the underlying paint films. See also *cracking*.

Dirt collection. 'The presence of matter adhering to the surface of or embedded in a film but not derived from the film'. This generally refers to atmospheric dust deposited on the film. The degree to which this can be removed by washing is related to the gloss and hardness of the film. These in turn are dependent on the formulation. See also *chalking* as a remedy for dirt collection.

Discolouration. 'Any change in the colour of a film as a result of exposure, including that due to chalking and dirt collection'. See *chalking, dirt collection, fading, ageing, mould, staining* and *water spotting*.

Efflorescence. 'A deposit of salts that remains on the surface of masonry, brick or plaster after the evaporation of water'. This should be washed off and the surface primed with a cement-proof paint to prevent *saponification* of subsequent paint films.

Embrittlement. This can occur where the curing process continues throughout the life of the coating —for instance, alkyd enamel drying by oxidative cross-linking. See also *cracking*. Reduction of drier levels may be helpful.

Erosion. 'Attrition of the film by natural weathering which may expose the substrate'. See also *ageing* and *chalking*. This is normal in any paint system. It becomes a defect only if it occurs within the expected lifetime of the coating. In this case, the cause may be incorrect choice of binder and pigment types, or poor quality control.

Fading. This can be caused by poor lightfastness of the pigments used, or by *chalking*. The use of the cheaper, low-fastness red and yellow organic pigments can represent a serious problem in exterior-quality surface coatings.

Flaking. See *peeling*.

Hiding power. See *opacity*.

Metameric match. 'Close colour conformity under a particular illumination which changes to an appreciable colour difference under other light sources'. Chapter 45 provides some background. Metamerism is a function of the pigments used to achieve the colour. This is not generally a serious problem in decorative systems, but assumes major importance in automotive and industrial systems. The defect can be avoided only by careful formulation of the colours, and strict control of pigment substitutions.

Mould. The growth of mould is associated with dampness, either of the substrate or of the surrounding atmosphere. It is recognised by black or vari-coloured spots or colonies which may be on, in, or beneath the paint film and can occur on almost any type of building material. The growth may penetrate the underlying plaster or brickwork and become difficult to eradicate. On new work there should be little risk of mould growth if normal precautions are observed in relation to adequate ventilation. Mould is prevalent in premises that have been left unoccupied and closed for a considerable period or where there has been persistent damp penetration. There have been many articles published on the effect of mould inhibitors, but as yet no perfect solution has been found. Some are very effective for an initial period but, being fugitive, eventually disappear from the film. Others impart unacceptable colour or toxicity to the film, or rapidly leach out because of high water solubility. Paints based on certain types of zinc oxide are good mould inhibitors, but other properties such as opacity, gloss, storage stability and durability usually suffer.

Nibs. 'Small pieces of foreign material, pieces of skin, coagulated medium, etc., which project above the surface of an applied film, usually a varnish'. See *bitty film*.

Opacity. 'The ability of a paint to obliterate the colour difference of a substrate'. Insufficient opacity, or failure to cover adequately, may be a consequence of insufficient covering pigment in the formulation; where poor opacity is claimed on products of known good quality, the following causes may apply:

(a) over-reduction or overspreading;
(b) pigment settlement not re-dispersed;
(c) poor application techique.

Over-reduction can be established by comparing the result obtained with that given by a freshly opened tin. Overspreading is frequently encountered, particularly in latex paints (which are generally characterised by ease of application).

Peeling. 'Loss of adhesion resulting in detachment and curling out of the paint film'. Peeling is essentially a manifestation of poor adhesion, either between the paint and the substrate, or between successive coats of paint. Peeling may occur as a result of a number of different factors:

(a) a substrate that is powdery, damp, dirty or unduly absorbent at the time of painting;
(b) ingress of moisture into substrate, because of structural defects;
(c) inadequate rubbing down of one coat before applying the next;
(d) painting at a time when the surface was wet with rain or dew.
(e) composition of different coats being incompatible—for example, painting over surfaces previously painted with Kalsomine (distemper) or similar loosely bound paints that have not been effectively removed or treated.

The effects of poor adhesion may not be apparent until something occurs to disturb the film, such as the action of heat or light or ageing, repeated wetting and drying, the exudation of resins, or the crystallisation of salts beneath the film. The only remedy is complete removal of all peeling and flaking paint, and repainting.

Pinholes. 'Minute holes in a dry film which form during application and drying of paint'. See *cissing*.

Popping. 'A small bubble-like defect in a paint film, resulting from the expansion on hydration of extraneous material in a plaster substrate'. Avoid the use of water-based paints over fresh plaster walls or over surfaces where popping has occurred previously. The craters formed in plaster can be filled with quick-set plaster compound, the area sanded and repainted.

Rust damage. See *staining* and refer to chapter 51 on corrosion control.

Saponification. 'A defect resulting from binder degradation by alkali'. The introduction of latex paints (which are more resistant to saponification) and of autoclaved asbestos cement products (which are much less alkaline) have nowadays reduced the problem considerably. 'Saponify' means 'to turn into soap', and the reaction that takes place between a highly alkaline surface and the alkyd in the paint film is similar to that between caustic and oil or fat used to produce soap. Under damp and humid conditions, free alkali present in walls can cause saponification of enamel films. When saponified, the paint becomes soft or sticky, showing loss of gloss and darkening or other discolouration. Most cases of saponification are found on cement-rendered walls and lime plaster joints. The use of unwashed beach sand in cement rendering must always be avoided, as the presence of salts in reaction with lime in the cement produces the alkalinity needed to cause saponification. The remedy is total removal of the coating, and repainting with alkali-resistant paints.

Seeds. 'Undesirable particles or granules other than dust, found in paint or varnish'. See *bitty film.*

Sheariness. This defect is observed as patches of higher sheen in an otherwise uniform (usually flat) finish. This can be caused by poor substrate *bridging* or by *sticky application*, and is frequently observed in conjunction with *brush marks.* Reformulation using alternative extenders, thickeners or resins is usually necessary.

Staining:

Sulfide staining. With the dominance of lead-free paints, sulfide staining has become a rare phenomenon. Its presence—more prevalent with alkyd flats than other finishes—is evidenced by brown to purplish-black stains, which are readily removed by dilute hydrogen peroxide. These stains are caused by a reaction between small amounts of lead compounds in the paint and traces of hydrogen sulfide in the atmosphere to form black lead sulfide. Such staining may also occur on lead-free alkyd or latex paint that had been applied over a previous, lead-containing paint system, by leaching of the lead through the new system.

Other stains. Other types of stains can occur on painted surfaces and most can be traced to either the substrate or atmospheric contamination. Stains can be caused by water, tannin, rust, mould, soot or dyestuffs on the substrate, or by cooking fumes, tobacco smoke or industrial pollution. The paint can contribute, since most paints contain trace amounts of lead or mercury compounds as driers or fungicides. Yellowing of paint films can also be included in this category and is influenced by the degree of lighting to which paint is exposed. For enamels the darker the area the more the yellowing. The degree of yellowing is also influenced by the type of oil used to make the alkyd resin. Safflower, soya and sunflower oils are relatively non-yellowing, when compared with linseed. The advent of latex paints and solution acrylics has improved the colour retention of commercially available coatings.

Water spotting. 'Spotty appearance on a dry paint film caused by the drying out of droplets of water'. In the case of latex-based paints, the water-solubles are leached from the film and leave a shiny mark when the water dries. Hosing down will help. This is generally a transient problem which will improve as the coating is repeatedly subjected to rain.

A completely different effect is the surface damage caused by water on excessively thermoplastic automotive finishes in hot weather. In extreme cases, cutting and polishing is needed to renovate the paint surface.

FURTHER READING

A standard text for further reading is:

Hamburg H R and Morgan W M eds *Hess' Paint Film Defects: Their Causes and Cure* 3rd ed Chapman & Hall London 1979

54 ANALYSIS OF POLYMERIC MATERIALS

[*Editors Note:* The analysis of coating materials is a subject of enormous scope, involving the use of a range of analytical procedures applied to a wide variety of organic and inorganic materials including cured and uncured polymer systems. It is, however, of considerable relevance to the modern coatings technologist, and the approach here is to provide an overview to give some insight into the theoretical and practical aspects of the subject. Since this chapter is of necessity in summary form, extensive references have been included.—*J M W*]

54.1 INTRODUCTION

Surface coatings are normally compounded mixtures of a variety of materials that are themselves not single compounds. The analysis of such products is complex, and the results are largely dependent on the skill and experience of the analyst. An outline of the procedures for analysis is presented, and reference is made to original works which detail the individual analyses of interest.

Coating analysis is time consuming and thus costly, and optimisation of resources is essential. The analytical chemist should be aware of the purpose of a request for analysis before commencement of work, as much effort may be expended to provide a mass of data which may not be of value in solving the problem at hand.

Few modern texts specifically concern the analysis of coating materials,[1,2] and it is frequently necessary to refer to works associated with other fields where common materials are used.[3–7] New raw materials and intermediates continue to be developed and used in coatings, and accordingly new methods of analysis are required. A familiarity with the chemical literature is necessary, which is simplified by the availability of various abstracts and review journals.

The need to perform paint analyses is varied, and may be associated with manufacturing problems, customer servicing, monitoring competitors' products or even health reasons. Some areas of possible concern to the paint analyst are:

(a) *Whole paint*—for quality control, to evaluate competitors' products, to identify unlabelled drums.
(b) *Paint flakes*—for example, to ascertain recoating compatibility.
(c) *Faulty production batches*—to identify the source of the problem in a faulty production batch, such as contaminants or a deviation from formulation. This may be in response to customer complaints which may have legal implications; here the analyst's results may be used as evidence in courts of law.

(d) *Environmental monitoring*—at the manufacturing site, to trace airborne monomers, solvents or particulates, in the interests of workers' safety.

54.2 INITIAL SEPARATION OF MAJOR COMPONENTS

Generally, a paint consists of three major component groups; solvents, resins and pigments/extenders. Each group has a particular role to play and may be the subject of analysis in order to establish its composition and proportion. They usually must be separated prior to identification and this can be achieved by the following means.

54.2.1 Drying

Typically, a 1 g sample (as a thin film) is dried at 105° C for 3 hours. The dry mass gives the non-volatile content (or 'per cent solids') and is an indication of the proportion of solvent and/or plasticiser present when weighted to constant mass. See Appendix 1.
Variations of temperature or time are sometimes made for specific products, for example in order to match the actual conditions used for oven drying on industrial production lines.

Further procedures are detailed in many specifications[8] for routine use. In special cases—for example, extremely volatile solvents or highly viscous materials—modified procedures have been developed. The volumetric solids are frequently of interest and again are extensively specified.[9]

54.2.2 Centrifugation

Whole paint samples can be centrifuged (at 3000–15 000 rpm for about 30 minutes), usually after thinning with appropriate solvents such as MEK or THF, and the supernatants (solvents and resins) removed. After solvent washing and re-centrifuging two or three times, the residue is then dried at, say, 105° C for 2 hours, leaving the pigments and extenders. An Australian standard specification[10] describes the procedure in considerable detail.

There is sometimes difficulty in removing very fine or low-density pigments such as carbon black. Problems can also be encountered with water-based coatings, where the dispersed resin may 'drop out' on centrifugation. This may be overcome either by drying a film on a Teflon plate, then extracting the dried paint with MEK or THF using a mechanical shaker or sonic bath before centrifugation; or by careful pre-mixing of the paint with water and other solvents such as ethylene glycol.

54.2.3 Filtration

If only a small amount of pigment-free resin is required, the sample can be solvent-thinned and passed through a 0.5 μm filter (Millipore) in an attachment on a small syringe. Sufficient material can be obtained for spectroscopic or chromatographic investigation. *Note:* The pigment and extender must be coarse enough to be retained within the filter pad.

54.2.4 Ignition

Heating, usually in a muffle furnace, to temperatures above 600° C, destroys all organics present (i.e. dyes, resins), leaving the inorganic pigments and fillers. However, decomposition of carbonates such as whiting and loss of water of crystallisation (e.g. from gypsum) can occur.

54.2.5 Distillation

Removal of the solvent or solvent mixture is a prerequisite to analysis, and several procedures have been employed. Solvents have been separated from the polymer traditionally by:

(a) dry distillation alone;
(b) dry distillation in the presence of a considerable amount of inert material (i.e. dry sand) such that an underbound residue results and the solvents are not entrained;
(c) dry distillation in the presence of a high boiling plasticiser; or
(d) steam distillation.

The first three methods are unsuitable for use with gas chromatography, as the procedure is much more sensitive than the older chemical and/or physical methods. Decomposition products of the resin and organic pigments, if present, are produced, and these are observed on the chromatograms. Volatile decomposition products of low molecular weight produce peaks early on the chromatogram, and with high relative response factors are more serious than the tails produced by the less volatile materials.

Steam distillation has been used for the recovery of hydrocarbons (as in solvent-based finishes), but it is unsuitable for other materials, as many alcohols, esters, ketones and ether alcohols are wholly or partially soluble in water.

The quantitative injection of a viscous solution with a conventional microlitre syringe is virtually impossible, while alteration of column performance may be experienced because of the deposition of polymeric materials on the support. Decomposition of polymer may also occur in the injection system with the production of extraneous peaks. Accuracy of injection and the production of charred residues in the system may be substantially overcome by the use of a suitable syringe and a chromatograph fitted with a precolumn[11] but the problem of spurious peaks remains.

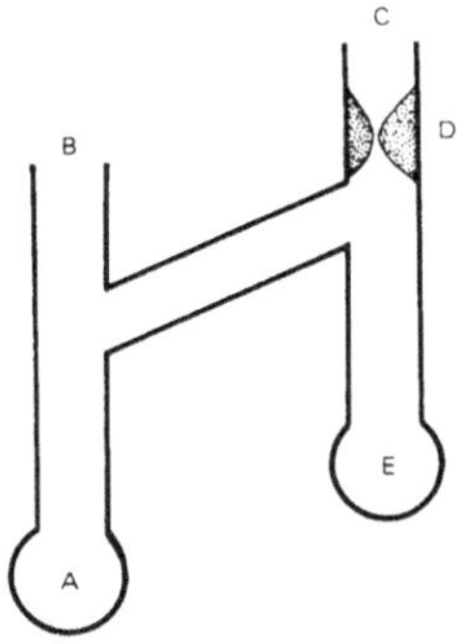

FIGURE 54.1
Apparatus for isolation of solvent

Simple micro vacuum distillation using the apparatus developed by Haslam *et al.*[12] as shown in figure 54.1 has been used with considerable success for the removal of solvents from coatings. The apparatus may be of any size, but samples of 1–5 g are suitable for most purposes. The sample is transferred to section A, using a coarse syringe or Pasteur pipette and ensuring that the sealing point B is not contaminated. The sample is frozen and point B sealed. The system is evacuated and point C sealed. The sample-containing section is removed from the coolant and the receiver E is cooled. Part A is immersed in boiling water and the solvents are distilled off and collected in the receiver.

54.3 SOLVENT ANALYSIS BY GAS CHROMATOGRAPHY

Whitham[13] in 1956 supplemented the usual chemical and physical tests for solvent mixtures with gas chromatography by first examining the sample, then fractions obtained by the use of the fluorescent indicator absorption technique developed by Ellis and Le Tourneau.[14] A fluorescent indicator absorption technique for the determination of hydrocarbon types in liquid petroleum products is available as an ASTM specification.[15]

Haslam and his co-workers[12] used gas chromatography on two coloumns of different polarity and infra-red spectroscopy to resolve complex solvent mixtures. This method may be readily used by an experienced analyst, because probably less than six of more than a hundred potential constituents would be present in a given solvent blend. In the presence of hydrocarbon fractions, a multi-peak hydrocarbon chromatogram is obtained as a background, and identification may be more difficult. The use of infra-red is of particular value, as it may readily show the presence or absence of a particular carbonyl or absorption due to aromatics or ethers.

Haken and McKay[16] have reported a method of analysis using the sequential application of solubility tests, supplemented if necessary with several functional group reactions. The components of hydrocarbon fractions may be identified if necessary after removal of the simple chemical types by further chemical and extractive tests, by the use of selective stationary phases. Or the fraction as a whole may be identified by comparison with a library of 'fingerprint' chromatograms;[17] such a library should be regularly updated, as the solvents are not sold on a composition basis (which thus tends to vary).

54.3.1 Isolation of the Solvent

The analysis is most conveniently conducted with about 8 mL of solvent obtained from a coating using several of the expendable glass tubes shown in figure 54.1. The examination of smaller samples is possible and simply necessitates re-examination of a single sample. The aqueous extract is treated with 85 per cent sulfuric acid and then with concentrated sulfuric acid.

Example. Three 2.5 mL solvent samples were each mixed and vigorously shaken with a similar quantity of water, 85 per cent sulfuric acid and concentrated sulfuric acid, respectively, in 5 mL stoppered graduated cylinders. The samples, after shaking, were allowed to stand, and on separation the residue was examined. The treatment with 85 per cent sulfuric acid was conducted under running water. Under these conditions, aromatic and unsaturated hydrocarbons are essentially unaffected, and trials on larger samples have shown that reproducibility of about 1 per cent is normally encountered. The effect of the solubility tests on the various solvents for which retention data are available has been demonstrated.[16]

An essentially similar scheme using a reaction gas chromatograph[18] allows a reliable quantitative analysis to be made with a few microlitres of extracted sample or a gas sample with three reactive column packings used in an abstraction unit. The columns used were 10 per cent sodium metabisulfite, to remove ketones, 3 per cent boric acid to remove alcohols and, 3 per cent lithium hydroxide to saponify esters, with removal of the corresponding alcohol with a boric acid column connected in series.

Separation of the oxygenated solvents from a thinner mixture may also be carried out using a liquid chromatography column[19] based on the fluorescent indicator procedure of Ellis and Le Tourneau.[14] The thinner is absorbed on a silica gel column and then eluted by passing methanol down the column. The fractions are monitored by the position of the dye front—the hydrocarbons eluting before the dye, with the oxygenated solvent following the dye. The procedure required 0.4–0.5 mL of solvent, with the two fractions being resolved by subsequent gas chromatography.

54.3.2 Basic Analytical Procedure

Gas chromatography can be used routinely to obtain quantitative data in a reasonable time by the following method:

(a) inject appropriate reference standards and measure retention times and response factors;

(b) inject the unknown sample;
(c) compare the retention times to identify sample components, and integrate the peaks and apply response factors to calculate percent composition;
(d) confirm results so obtained by running a sample blend made up with solvent standards.

The problem of inadequate resolution (peak overlapping) may necessitate changing the applied oven temperature gradient, using longer columns, or different stationary phases. Frequently, the analyst needs to use two or three column types to confirm the composition of a solvent mixture, because different solvents may have very similar retention times on some columns but not others (due to, for example, relative polarities and boiling points).

It is sometimes desirable to separate the hydrocarbon portion of the solvent mixture in order to simplify the chromatogram. This is accomplished by mixing a known volume of the solvent mixture with 80–85 per cent sulfuric acid. Any oxygenated solvent (ethers, esters, etc.) will dissolve in the acid, leaving the hydrocarbons as a readily removed separate top layer which can be further identified by gas chromatography, or by its infra-red or nuclear magnetic resonance spectrum to determine whether it is aliphatic or aromatic.

54.4 POLYMER ANALYSIS

Before identification, the binder in a whole paint should be removed using the methods previously described. Paint flake analysis may require an edge-view microscopic examination in order to determine whether more than one coat is involved. Top and base coats may be carefully extracted by spotting THF on to the surface and then removing it by Pasteur pipette, followed by centrifugation. They may also be gently abraded with fine sandpaper and the material removed, then solvent extracted and centrifuged. The separated resins can be identified or characterised by the following methods.

54.4.1 Polymer Analysis by Infra-red Spectroscopy

In polymer analysis, infra-red spectrophotometry is generally used first. There are three basic types of instrument commercially available:

Optical null
Ratio recording
Fourier transform

The optico-electronic differences between them will not be discussed here. Although there may be important operational differences in spectral expansion facilities, scan times and low transmittance sensitivities, spectral interpretation procedures are the same for all three.

Analysis is usually carried out on a dried film cast on a sodium chloride, potassium bromide, or caesium iodide plate that is transparent over the spectral range of the instrument (normally between 4000–6000 and 4000–200 wave numbers).

Water-based systems have the disadvantage that the casting plates are usually soluble in water. Three methods of overcoming this problem include:

(a) Using water-resistant plates such as silver chloride or KRS-5. These are expensive and have other disadvantages—see table 54.1.
(b) Spotting a drop of the emulsion on to a disposable potassium bromide disc, which is then carefully dried at about 60° C.
(c) Evaporating the water and reconstituting the binder in an organic solvent such as THF.

Reflectance techniques, attenuated total reflectance (ATR) and multiple internal reflectance (MIR), can be used to study at least partial composition of coatings surfaces (and also surface contaminants). Complete paint flakes may also be studied this way, but after the interference of pigment absorption it makes spectral interpretation difficult. High gloss films, being richer in binder at the surface, give the best results. The MIR method is the more sensitive; reflection is quite small—see figure 54.2.

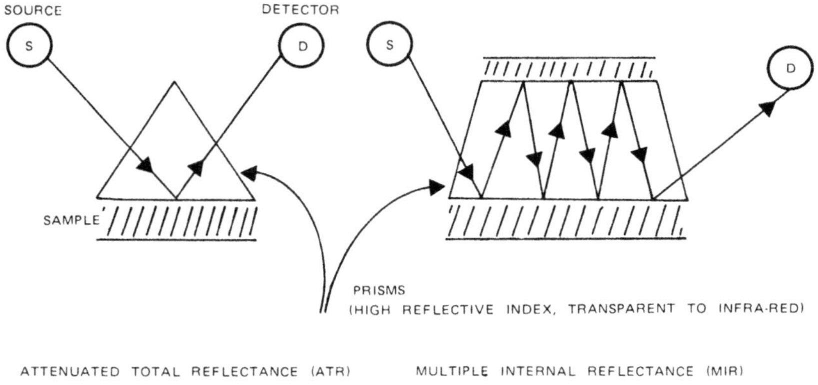

FIGURE 54.2
Two reflectance techniques

There is no systematic procedure for identification of an unknown. In the simplest case, the analysis consists of comparison with a library of reference spectra which ideally would have been produced on the same instrument. The size of the reference library is of major importance, and in an industrial laboratory spectra of new materials should be constantly added as samples become available. As the size of the library increases so does the time required to attempt to achieve a match and here subdivision of the reference spectra is necessary. It is relatively simple to achieve a partial identification, even by a relatively unskilled operator by the use of absorption bands due to major functional classes, and this may serve as a basis for subdivision. For a simple comparative examination, the absence of a particular functional feature may be all that is necessary to achieve the required result. Several libraries of spectra of materials relevant to coatings have been prepared[20–24] and all are of considerable value to the analyst. These should be regarded as an extension to the analyst's own library, which in a coatings laboratory should soon surpass the commercial compilations.

A knowledge of the origin of the absorption bands in a spectrum is necessary for more detailed interpretation, and here also the relative intensity and the shape of various bands must be considered because the contributions from all functional groups of equivalent concentrations may vary greatly.

With small molecules, infra-red assignment is conveniently developed from a study of paraffins; and with polymers, polyethylene similarly provides an appropriate basis for study. The spectrum of the polyolefin is superficially simple and expectedly similar to that of a simple paraffin. In both cases the most intense absorption is that in the carbon–hydrogen stretching region, which is evident with spectra from low resolution spectrometers as a broad absorption band centred at about 2940 cm^{-1}. This is largely due to the methyl and methylene groups but also to the weaker absorbing C—H groups. In the spectrum, the next most intense band occurs near 1460 cm^{-1} due to CH_2 deformation; a weaker band near 1380 cm^{-1} is due to C—CH_3 vibrations. The final major bands occur as a doublet at 720 and 730 cm^{-1} due to rocking motions of pairs of methylene groups in the chain.

The spectra of vinyl polymers may be considered as a first approximation as a composite of absorption bands due to the polymer chain, which are basically weak, and of bands attributable

TABLE 54.1
Properties of window materials

Material	*Useful range* (cm^{-1})	*Reflection Losses* at 1000 cm*$^{-1}$	*Solubility g/100 mL @ 20° C*	*Relative cost*	*Physical properties*
Sodium chloride NaCl	>4000 to 625	7.5%	40	1.0	Soluble in water. Robust but easy to grind and polish. Cleaves well. Fogs slowly.
Potassium bromide KBr	>4000 to 400	8.5%	70	1.4	Soluble in water. Slightly softer that NaCl but still easy to polish and cleave. Fogs slowly. Slightly more expensive than NaCl but has the advantage of an extended wave-number range.
Calcium fluoride CaF_2	>4000 to 1000	5.5%	Almost insoluble	4.3	Almost insoluble in water and resists most acids and alkalis. Must not be used with solutions of ammonium salts. Does not fog.
Barium fluoride BaF_2	>4000 to 750	7.5%	Insoluble	6.3	Chemical characteristics similar to CaF_2. Must not be used with solutions of ammonium salts. Very sensitive to thermal and mechanical shocks. Does not fog.
Caesium iodide Cs	>4000 to 180	11.5%	80	14	Soluble in water. Soft and is easily scratched. Does not cleave. Fogs slowly.
Irtran-2 polycrystalline ZnS	>4000 to 700	12%	Insoluble	15	Insoluble in water but is slightly attacked by HNO_3, H_2SO_4 and KOH solutions. Hard and mechanically stable, it can withstand a wide range of temperature changes.
Silver chloride AgCl	>4000 to 450	19.5%	Insoluble	1.8	Insoluble in water but soluble in acid and NH_4Cl solutions. Corrosive to metals and alloys. It is ductile and resistant to thermal shocks. Darkens on prolonged exposure to UV light.
Silver bromide AgBr	>4000 to 280	25%	Almost insoluble	1.6	Almost insoluble in water. Soft and easily scratched, it will cold flow. Darkens on exposure to UV light. Suffers from high reflection losses.
KRS-5 TlBr/TlI	>4000 to 250	28%	0.1	11.4	Slightly soluble in water, soluble in bases but insoluble in acids. Soft and easily scratched, it will cold flow. High refractive index makes it widely used as MIR crystal. Highly toxic.
Infrasil SiO_2	>4000 to 2850	NA	Insoluble	NA	Insoluble in water, soluble in HF solutions, slightly soluble in bases. Mechanically very hard, does not cleave.
Polyethylene (high density)	625 to 10	NA	Insoluble	0.6	Insoluble in water and attacked by few solvents. It is soft and tends to swell, making it difficult to clean. The powder can be pressed into dises. Low-cost window material for far infra-red.

* Reflection loss at two surfaces NA Not applicable.

to the pendant groups. The spectrum of polyvinyl acetate demonstrates the point, as the major bands are those due to the stretching vibration of

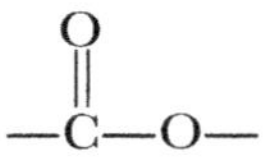 (i.e. 1240 cm^{-1})

and  (i.e. 1740 cm^{-1})

while the bands in the C—H stretching region are very weak. The spectrum of polyethylene–vinyl acetate copolymer, where the doublet due to the long paraffin chain is minimal, shows the effect of the copolymerisation. The position of the —C—O— band is characteristic for many esters but for aliphatic esters it is almost stable, although the C—O single band is more variable and of use such that propionates may be differentiated from acetates by the position of this band.

The carbonyl-stretching frequency is of considerable interest and in association with several other bands should allow the type of carbonyl functional group to be established.

Carboxylic acids in addition to carbonyl absorption show absorption due to hydrogen bonding between the hydroxyl and the carbonyl groups; carboxylate ions show characteristic absorption frequencies that are lost on liberation of the free acid; anhydrides all show two carbonyl absorption frequencies; aldehydes exhibit carbonyl absorption together with characteristic C—H stretching vibrations near 2700 cm^{-1}; while ketones have a strong carbonyl band but none of the others indicated here.

As absorption bands occur according to well-defined selection rules, the spectra are more complex as the number of atoms present are increased—although with polymers to a first approximation, this complexity is dictated by the repeating unit.

The spectrum of polystrene is widely used for instrument calibration, and the increasing complexity or number of bands in the spectrum is evident as compared with a polyolefin. Considerable information is frequently obtained from aromatic spectra, as characteristic absorption occurs with substitution by polar substituents. The carbon–hydrogen stretching frequencies of hydrogen atoms on the aromatic ring are higher (i.e. greater than 3000 cm^{-1}) than experienced with aliphatic compounds; C═C skeletal in-plane vibrations show strong bands near 1600 cm^{-1} and medium-intensity bands near 1580 and 1450 cm^{-1}.

The out-of-plane carbon–hydrogen deformations provide strong bands in the region 1000–650 cm^{-1}. With mono-substituted aromatics the band position is in the range 770–730 cm^{-1}, and a further strong band is observed 710–690 cm^{-1}. Compounds having four adjacent ring hydrogen atoms (*o*-disubstitution) absorb at similar frequencies to mono-substituted compounds. Aromatic rings with three adjacent hydrogen atoms (i.e. 1,3- and 1,2,3-substitution) show a frequency shift to the 810–750 cm^{-1} region and a second medium-intensity band in the region 725–680 cm^{-1}.

A further shift to higher frequencies (i.e. 860–800 cm^{-1}) occurs with two adjacent hydrogen atoms, as with 1,4-, 1,2,4- and 1,2,3,4- substitution, while a further shift occurs with a single isolated hydrogen atom. With 1,3-, 1,2,4-, 1,3,5-, 1,2,3,5-, 1,2,4,5 and 1,2,3,4,5- substituted compounds, absorption occurs in the 900–860 cm^{-1} region. With 1,3,5- trisubstitution, a second band is observed in the range 730–675 cm^{-1}. These and other bands in aromatic compounds are of considerable value for identification.

A more systematic approach can be made to polymer classification by tracing the pathways outlined in table 54.2.

Systematic studies of frequency assignment with many classes of compounds have been made and are discussed in various works, one of the most important being that of Bellamy.[25] With polymer analyses, infra-red spectrophotometry should be the first analytical tool used, as a skilled

TABLE 54.2
Polymer classification scheme

Carbonyl band near 1725 cm^{-1}: Present				Carbonyl band near 1725 cm^{-1}: Absent			
Bands near 1950 and 1490 cm^{-1}				Band near 1590 and 1490 cm^{-1}			
cm^{-1}	Present (aromatic)	cm^{-1}	Absent (aliphatic)	cm^{-1}	Present (aromatic)	cm^{-1}	Absent (aliphatic)
1540 1220	Polyurethanes	1430 1235	Poly (vinyl acetate)	3330 1220 910–670	Phenolics	3330 1430 1100(b)	Poly (vinyl alcohol)
1300 1230 725	Isophthalate alkyds and polyesters	1430 690(b)	Poly (vinyl chloride-acetate) and poly (vinylidene chloride-acetate)	1430 1110–1000(s)	Phenylsiloxane	2940(vs) 1470 1380(w) 730–720(d, s)	Polyethylene
1230 1120 1075 740 705	o-Phthalate alkyds and polyesters	1265 1240 1190–1150(s)	Polymethacrylates	1235 1180 825	Bisphenol expoxides	2940(vs) 1470(s) 1380(s) 1160 970	Polypropylene
1265 1110 867 725	Terephthalate alkyds and polyesters	1250(b) 1190 1150 (s)	Polyacrylates	814 780 700	Polyvinyltoluene	2260	Polyacrylonitrile
1235 1175 826	Bisphenol epoxy esters	1110–1150(b)	Cellulose esters	760 700	Polystyrene	1640 1540	Urea-formaldehyde and polyamides
813 781 700	Vinyltoluene esters					1640 1280 834	Cellulose nitrate
778 700	Styrenated esters					1540 826	Benzoguanamine-formaldehyde
						1540 813	Melamine-formaldehyde
						1430 690(b)	Poly (vinyl and vinylidene chlorides)
						1265 1110–1000(s)	Methylsiloxane
						1229–1150(d., vs)	Polytetrafluoroethylene
						1110	Poly (vinylethers and acetals), cellulose ethers
						1250–1110 1000–910	Polychlorotrifluoroethylene

b = broad, s = strong, vs = very strong, w = weak, d = doublet.

worker will obtain what information is possible quickly and accordingly at little cost. With mixtures, differences in the spectrum may often be made by liquid--liquid extractions. The film examined may be extracted and if significant bands are reduced or eliminated, similar treatment of a larger quantity of the material will provide adequate extracted material for a spectrum to be provided.

An example of some importance concerns the orthophthalates which may be present both as plasticisers and resins. Characteristic absorption evident as weak twin bands near 1600 cm^{-1} occurs in both types of compounds. Clarification is frequently achieved by extraction of the coated specimen by careful refluxing with methanol, ether or n-hexane where monomeric phthalates are dissolved away to allow further study of the polymer. The plasticiser may be examined using infra-red spectrophotometry. However, the spectra of the various dialkyl phthalates are very similar, and definite identification of the particular alkyl group is doubtful. Further, many phthalates are currently produced from alcohol fractions such that the mixture of phthalate esters will not be resolved by infra-red spectrophotometry. In this case the use of gas chromatography is indicated, which has the further advantage of utilising the extract from a small halide plate without the necessity of extraction of a larger sample.

The use of liquid–liquid extraction frequently for prolonged periods with refluxing is common for the determination of stabilisers and anti-oxidants in polymer products.[26] The additive materials are normally used at very low levels and are not apparent in the spectrum of the total polymer products; but after extraction, residues with complex spectra are evident. Mixed additives are frequently encountered, and further separation after extraction using column, thin layer or paper chromatography is necessary to allow identification of the individual components.

While the detection of isocyanate groups is readily carried out using the intense absorption that occurs near 2270 cm^{-1}, the identification after conversion to a polyurethane is much less certain because the spectrum might often readily be identified as a polyester. The simple technique of Kaczaj[27] is often of value and simply involves a partial pyrolysis of the surface of the cast film or disc. An almost red hot spatula is rapidly wiped across the polymer surface, causing some degradation with liberation of isocyanate which is readily apparent from the intense absorption at 2270 cm^{-1}. The halide or disc usually cracks with this treatment but is a small price if the identification is changed from a polyester to a polyurethane. Polyurethanes, however, frequently require chemical techniques for reliable analysis, and representative procedures are described in later sections of this chapter.

The infra-red spectra of close homologues are very similar, and the identification of mixtures is difficult and where possible the sensitivity is poor. If volatile, such homologues are readily separated and quantitatively estimated using gas chromatography; a wide variety of chemical reactions have been applied to polymers to render them amenable to gas chromatography. A reduction in molecular weight is necessary, and vigorous chemical reaction or thermal degradation —i.e. pyrolysis—is necessary to fragment the molecule. Many examples of thermal degradation have been reported and are detailed in various reviews.[4-7] The techniques are readily applied, but the results are very variable and frequently are suitable only for qualitative analysis. Chemical degradation is much more time consuming, but the results achieved are often essentially quantitative. Both types of procedure are discussed using various polymeric products used in the coatings industry.

54.4.2 Pyrolysis

The fragmentation or reduction in molecular weight of polymeric materials to facilitate analysis has been described using various chemical procedures, but degradation may also be achieved by thermal means—that is, pyrolysis—where the volatile products formed are determined by gas chromatography (GC) or gas chromatography–mass spectroscopy. Pyrolysis gas chromatography is simpler and quicker than chemical degradation, but the analytical precision is much poorer (despite various claims in the literature).

In this technique, the resin is thermally degraded and the volatile decomposition products are flushed through a GC column with carrier gas to give a characteristic 'fingerprint' chromatogram. The 'unknown' chromatogram is again compared to a reference library. Although only a small sample size, perhaps 10 μg, is needed, instrumental parameters must be strictly controlled to ensure reproducibility.

The pyrolyser accessory is typically a temperature-programmable probe which can be fitted at the chromatography injection port. A very rapid temperature rise to 500–900° C is used to degrade the polymer.

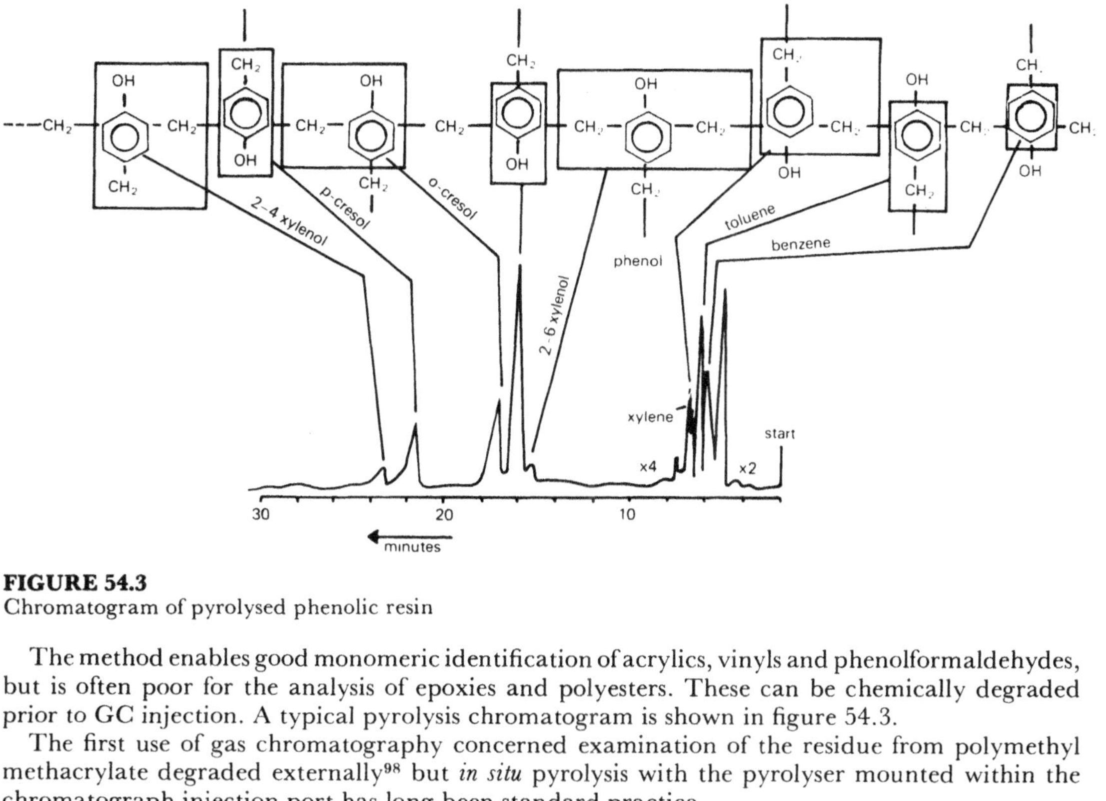

FIGURE 54.3
Chromatogram of pyrolysed phenolic resin

The method enables good monomeric identification of acrylics, vinyls and phenolformaldehydes, but is often poor for the analysis of epoxies and polyesters. These can be chemically degraded prior to GC injection. A typical pyrolysis chromatogram is shown in figure 54.3.

The first use of gas chromatography concerned examination of the residue from polymethyl methacrylate degraded externally[98] but *in situ* pyrolysis with the pyrolyser mounted within the chromatograph injection port has long been standard practice.

Several thousand reports of polymer pyrolyses have been published, and many of these are included in a number of reviews.[3-7] This chapter is restricted to a number of polymers of importance to the coatings industry and will indicate those materials where the technique is considered to be advantageous.

The available pyrolysis equipment may be classified as operating either by a pulse or by continuous mode. The first type, which is the one of importance, may be illustrated by heating the specimen on a filament; the second, by admission of the sample into a pre-heated zone. The advantages of pyrolysis are evident, as only a few micrograms of sample are required. The chromatographic peaks may often be characterised by their retention behaviour, or the effluent may be further examined as by a direct coupled mass spectrometer. The pyrolysis chromatogram may also be used for 'fingerprint' identification. Much more care is necessary for fingerprint comparison than was suggested by the early workers, but the technique has been extensively developed for forensic work.[99]

Pyrolysis has been used for both qualitative and quantitative analysis, for differentiating copolymers and homopolymer blends of the same composition and for the elucidation of polymer microstructure.

Acrylic polymers were first examined by pyrolysis gas chromatography in 1959[100] when depropagation of the poly n-alkyl methacrylates with near-quantitative yields of monomer was achieved together with chain scission of the polyacrylates with low yields of monomer. At about the same time[101] the pyrolyses of acrylic polymers were shown to be dependent on their structure. The pyrolysis of a copolymer and a homopolymer blend, each containing 80 per cent methyl methacrylate and 20 per cent methyl acrylate, produced essentially similar methyl methacrylate yields but the acrylate yield from the copolymer was more than twice that obtained from the homopolymer blend.[101]

Detailed quantitative assays of polyacrylic systems have been reported by McCormick[102] using stepwise degradation on a helical filament. From plots of monomer yields with temperature, identification of the sample as a copolymer or mixture was possible. A less-cumbersome procedure using a radio frequency pyrolyser with three ferromagnetic elements of increasing Curie Point was developed by Haken and McKay.[158]

The temperature at which the pyrolysis is carried out is of prime importance, as the higher the temperature used the higher the yield of gaseous products. The gaseous products are largely decomposition products of the primary products of the degradation, and the greater the amount of these products the less characteristic is the chromatogram produced. The pyrolysis temperature used depends on the stability of the particular polymer system and is generally within the temperature range of 600 to 900° C. Minimisation of secondary reaction products is achieved when the pyrolysis time is short, and thus pyrolysers are designed so that the pyrolysis temperature is achieved very rapidly after application of the electrical energy to the pyrolyser.

Recent studies with polyacrylates have been concerned with examination of the large number of products produced. Early workers suggested that the major products were gaseous with low yields of monomer and other volatiles. However, it is now established that a series of saturated and unsaturated oligomers are the major degradation products. The complex chromatogram produced complicates the analysis, but provides much data for fingerprint analysis. A number of studies have considered automotive finishes, and pyrolysis gas chromatography has been developed for forensic purposes for identification of unknown coating fragments.[199,197-111]

Polystyrene which degrades to form a substantial monomer yield was one of the early systems studied by pyrolysis.[4] Polystyrene is seldom used as a sole film former because of its inherent brittle nature, and if in pure form it is more suitably examined by infra-red spectrophotometry. Copolymers with butadiene, acrylonitrile and acrylics are commonly used, and all have been examined using pyrolysis gas chromatography producing fragments largely appropriate to the individual components. Examples of the qualitative and quantitative examination of styrene copolymers are included in tables 54.3 and 54.4.

The pyrolysis of copolymers of vinyl acetate with homologous acrylate, methacrylate and maleate esters have been reported.[112] Acetate–acrylate–maleate–fumarate copolymers have also been examined. Vinyl copolymers containing vinyl chloride and vinyl acetate provide products appropriate to the two individual monomers such that acetic and hydrochloric acids are the major products.[113]

Condensation polymers used in coatings have been examined by pyrolysis with varying results; phenolformaldehyde resins, which are difficult to analyse, have provide some outstanding results with very characteristic pyrolysis chromatograms; polyesters have produced disappointing results; and products containing vegetable oils are very poor. A chromatogram provided in an early

TABLE 54.3
Qualitative or fingerprint analyses of polymers and copolymers

Polymer or copolymer	*Reference*
Acrylonitrile-vinyl acetate	119
Acrylonitrile-α	119
Alkyl methacrylates	105–106, 120–121
Cellulose	122–127
Cellulose acetate	90–91, 123, 127, 128
Cellulose nitrate	129
Chlorinated polystyrene	10, 130
Chlorinated polyvinyl chloride	131–132
Coumarone indene	13–14, 133–134
Cyclised polyisoprene	15, 135
Epoxy	16–21, 117–118, 136–139
Furfuryl alcohol resins	140
Hydroxyalkyl methacrylates	105
Neoprene	123
Nylon 6	141
Perfluoroalkylene-aromatic imide	25, 142
Phenol formaldehyde	26–38, 114, 137, 142–154
Polyacrylic acid	155
Polyalkyl acrylates	121
Polyacrylonitrile-styrene	119, 156, 157, 159
Polyamides	150
Polybutadiene	159–160
Polybutene	98, 161–162
Polyisobutylene	48–50, 158, 163–164
Polybutyl methacrylate	51, 165
Polycarbonates	52–54, 166–168
Polycyclopentadiene	135
Polyesters	34, 55–62, 116–150, 169–170, 172–176
Polyethylene	159, 162, 178–180
Polyethyl acrylate	66–67, 100, 182
Polyindene	68–69, 183–184
Polyisoprene	45, 59, 70, 98, 123, 159, 185
Polymethylene	169, 186
Polymethyl acrylate	100, 187
Polymethyl-2-cyanoacrylate	188–189
Polymethyl methacrylate	52, 91, 129, 150, 154, 184, 190–191
Poly-α-methylstyrene	68, 77, 183, 192
Polyphenyl ether	78, 193
Polypropylene	79–81, 159, 161–162, 194–196
Polystyrene	34, 40, 43, 59, 63, 68, 69, 79, 82, 83, 150, 156, 159, 173, 178, 183–184, 194, 197–198
Polystyrene-divinyl benzene	34, 43, 79, 88, 150, 159, 173, 183–184, 194, 197–199
Polytetrafluoroethylene	84, 87, 200–201
Polyurethanes	34, 85–86, 150, 202–203
Polyvinyl acetate	45, 79, 98, 194
Polyvinyl acetate-acrylic acids	92, 112
Polyvinyl alcohol	9, 79, 93, 129, 194, 205
Polyvinyl butyral	94, 200
Polyvinyl chloride	9, 34, 42, 80, 83, 95–98, 129, 150, 158, 195, 198, 207–210

TABLE 54.3 (cont.)

Polymer or copolymer	*Reference*
Polyvinyl chloride-butyl acrylate	211
Polyvinyl chloride-methyl methacrylate	119, 212
Polyvinyl chloride-polyvinylidene chloride	79, 93, 194, 205
Polyvinyl propionate	79–83, 194–198
Polyvinylidene chloride	79, 100, 194, 213
Propylene-vinylcyclohexane	101, 214
Silicones	50, 102–104, 159, 215–217
Automotive finishes	99, 107–111, 222
Bitumen	112–113, 225–226
Flame-retardant finishes	5–11, 122–125, 129–131
Naturally occurring resins and pitch	114–115, 226–227
Thermosetting acrylics	116, 228
Urea formaldehyde	98,210
Hydroxy alkyl starch	117, 229
Starch	118,230

German trade brochure is shown in figure 54.3 and shows the powerful nature of pyrolysis with a suitable system.[114]

Results with polyesters have been poor, and chemical degradation is far superior to pyrolysis. Examination of polyester resins from cured laminates was carried out at 760° C and gross differences in composition were readily apparent. The presence of maleic and fumaric acids produced identical results, while ortho- and isophthalic acids were readily distinguished—as were ethylene glycol, diethylene glycol and propylene glycol.[115]

Polyesters based on diethylene glycol' 1,2-propylene glycol and 1,4-butanediol with maleic anhydride and succinic acid were pyrolysed at temperatures between 550 and 750° C. The gaseous products were collected in a cold trap and subsequently examined by gas chromatography on an activated charcoal column; the volatile products were examined on a dodecyl phthalate column. Examination of volatiles for the operation is of little value, while examination of the liquid products showed qualitative differences in the ratios of the two dibasic acids and in the glycols present; to be of any real value, considerable refinement of the technique would be necessary.[116]

Epoxy resins have been subjected to pyrolysis, as have epoxy resins cross-linked with polyamine, with a considerable number of products being produced—the major product being toluene—while little evidence of the amine was apparent. The very wide variety of minor compounds produced provide little basis for identification.[117–118]

A very large number of polymer and copolymer systems have been examined using pyrolysis gas chromatography, and tables 54.3 and 54.4 show examples of qualitative or fingerprint analysis and of quantitative studies on materials relevant to the coatings industry.

54.5 GENERAL RESIN ANALYSIS

54.5.1 Nuclear Magnetic Resonance (NMR) Spectroscopy

Briefly, NMR involves measuring the energy absorption due to nuclear spin behaviour of a substance in a varying magnetic field. Proton or ^{13}C NMR spectra of polymers can give structural configuration and composition data. The former are often poorly resolved and difficult to interpret because of high solution viscosities, but may be useful for analysing fairly simple copolymer systems. The more costly ^{13}C NMR is frequently more useful. Again, comparison of the 'unknown' spectrum to those in a spectral reference library is most beneficial.

54.5.2 High-performance Liquid Chromatography (HPLC)

HPLC and size-exclusion chromatography (SEC) instrumental techniques can be used for determining the molecular weight distribution of polymers and for separating polymeric mixtures for further analysis. The methods are analogous to gas chromatography in that the sample is eluted through a separatory column—the carrier, however, is a solvent or solvent blend. SEC separates on the basis of molecular size, while other HPLC methods largely utilise differences in polarity to effect separation.

54.5.3 Other Gas Chromatography Methods

As earlier mentioned, resins can sometimes be identified by chemical degradation and subsequent GC injection. For example, alkyds may be analysed for oil type or polyol composition. The rapid identification of the oil in an alkyd may be made by the following procedure:

(a) Saponify 100 mg of the resin with tetramethyl-ammonium hydroxide solution (25 per cent in methanol) by heating for about 10 minutes on a steam bath. This liberates the fatty acids.
(b) Add some dimethylformamide and then methyl iodide methylating reagent to give the volatile fatty acid methyl esters.
(c) Inject the supernatants into the GC, which should be fitted with diethyleneglycol succinate (DEGS) columns, and compare the resulting chromatogram to those of standard oils similarly treated. In many cases, direct injection after (a) at a fairly high injector temperature can facilitate pre-column methylation of the fatty acids without the addition of other reagents.

54.5.4 Thermal Analysis

Various characteristics of a polymer can be observed by heating it over a broad temperature range (typically from sub-zero to decomposition) and recording enthalpic changes against a thermodynamically stable reference standard.

Differential thermal analysis (DTA) provides a means of doing this and can give information on glass transitions, melting and associated crystallisation effects and thermal degradation.

Differential scanning calorimetry, a closely related technique, also measures the rate of energy absorption by the test sample relative to that of a reference material, but employs a slightly different instrumental set-up.

54.5.5 Qualitative 'Spot' Tests

There are many chemical tests designed to generally characterise a resin. The need for such simple and quick wet methods largely evolved before the widespread availability of the instruments used today.

54.5.5.1 Rosin Presence (Liebermann-Storch Test)

A 100–200 mg sample of resin is heated in 15 mL of acetic anhydride, filtered and acidified with a drop of 80 per cent sulfuric acid. A violet colour develops immediately if rosin is present, due to the reaction of abietic-type diene acids.

54.5.5.2 Bisphenol Epoxies (Swann Test)

100 mg of resin is dissolved in concentrated sulfuric acid to give a colour approximating that of 0.1 M potassium dichromate solution. A stirring rod is then dipped into the solution and streaked across a filter paper. If bisphenol-type epoxy resins are present, a bright purple colour develops within a minute, and eventually turns blue. Further confirmation can be gained by adding a couple of drops of 40 per cent formaldehyde solution to the acid solution, when a brick red colouration appears.

54.5.5.3 Sodium Fusion Test

This well-documented method offers a fairly simple means of determining the presence of nitrogen (polyamides), halogens (PVC), sulfur, or phosphorus, which may be useful in broad polymer classification.

54.5.6 Kappelmeier Saponification

This procedure is applicable to oil-based polyesters and enables their chemical breakdown to the major constituents. The separated vehicle is refluxed for 1½ hours or more with ethanolic potassium hydroxide, in benzene. On cooling, potassium ethyl phthalate is recovered, and the filtrate extracted with diethyl ether to yield solvent thinners and unsaponifiable matter. The remaining aqueous fraction can be taken to pH 2 with concentrated hydrochloric acid and ether extracted again to recover the fatty acid components for further analysis.

54.5.7 Solubility Tests

Paints and their resins can sometimes be characterised by solubility testing with a sequence of solvents. For example, acrylic automotive coatings may be classified by the scheme shown in figure 54.4.

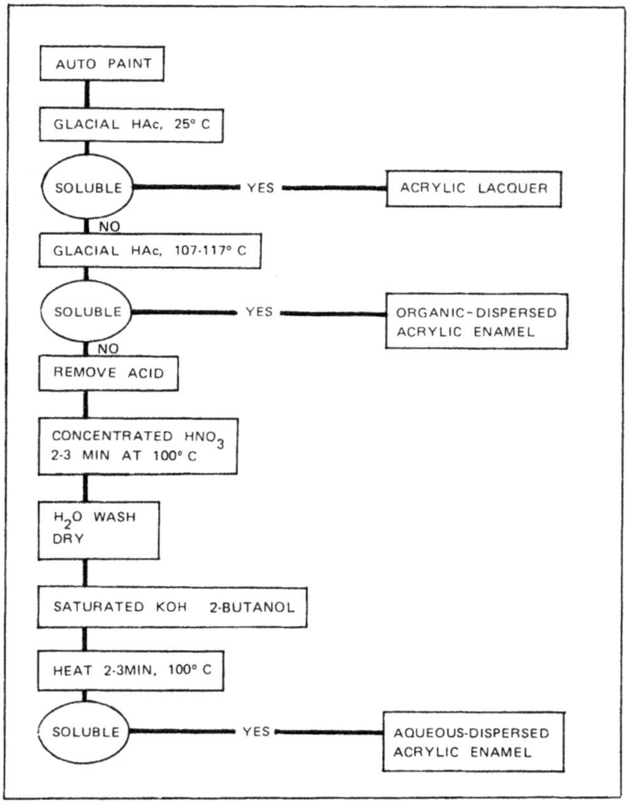

FIGURE 54.4
Solubility tests to identify an automotive coating

54.6 SPECIFIC RESIN ANALYSIS

54.6.1 Carboxylic Esters

Carboxylic esters are used in coatings as solvents, plasticisers, additives and resins. Where two or more acidic and/or hydroxyl components are present, techniques supplementary to infra-red spectrophotometry are needed for detailed analyses. Hydrolysis with cleavage of the acyl-to-oxygen bond occurs with nucleophilic substitution, to yield the alcohol and acid salts as described below.

54.6.2 Alkyd Resins and Linear Polyesters

Although an extensive range of resinous products has become available since the introduction of alkyd resins, they remain as the major film former of the coatings industry. The linear polyesters are of more recent development and are finding increasing use. Oil-modified resins are essentially dimers or trimers of molecular weight 2000–3000, but the linear polymers are of substantially higher molecular weight.

The linear polyesters also differ from the alkyd materials in that the reactants are substantially difunctional, of shorter chain length than the fatty esters, while dibasic acids other than orthophthalic are commonly used. This type of material finds use in convertible industrial enamels in addition to the long-established use in fibreglass reinforcement.

All of these products require chemical cleavage, the traditional hydrolysis reaction being commonly employed. A modification of the Kappelmeier procedure is an established procedure.[28] Alkaline saponification is followed by gas chromatographic determination of both acid and hydroxyl components. The polyols are recovered and examined as acetyl or more commonly trimethyl silyl (TMS) derivatives. The acids are liberated from the alkali salts and examined as methyl or TMS esters. The technique of on-column chemical reaction for the formation of the TMS derivatives has been developed by Esposito.[29] Aqueous or alcoholic solutions of the liberated acids and/or alcohols are injected into the chromatograph and immediately followed by an injection of the silylating reagent. Depending on the silylation reagent, procedures are recommended for both qualitative and quantitative analysis.

The analytical scheme shown below is for an idealised alkyd resin consisting of orthophthalic acid, glycerol and fatty acid.

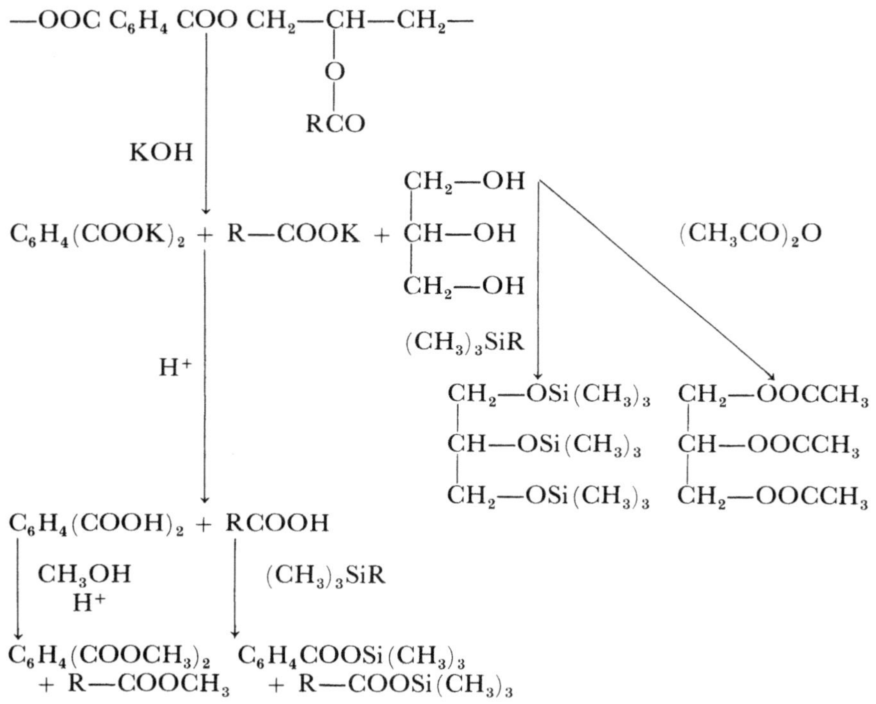

A second procedure for the analysis of alkyd type resins involves liberation of the polyols by reaction with alkyl or aromatic amines. Butylamine[30–31], benzylamine[32], phenylethylamine[33–34] and hydrazine[35] have been used with polyester and polyurethane foams. Extended reaction periods —up to 42 hours—have been reported; the amide formed with the constituent acids is not suitable for analysis without further reaction.

Alkyds have also been subjected to transesterification using catalysis with lithium methoxide[36], a method subsequently adopted as an ASTM specification.[37] A semi-quantitative modification by Percival[38] required prolonged methanolysis for 18–42 hours with sodium methoxide. The transesterification rate is increased by sealing the resin in glass tubes with a large excess of methyl acetate, sodium methoxide as catalyst, and heating for 1 hour at 175° C.[39] The same procedure was used in the first report of chromatographic separation of the three isomeric phthalic acids[40]. It is not possible to differentiate between maleic and fumaric acids, as both form dimethylmethoxysuccinate by addition at the double bond—a compound, however, that is readily eluted during subsequent gas chromatography.

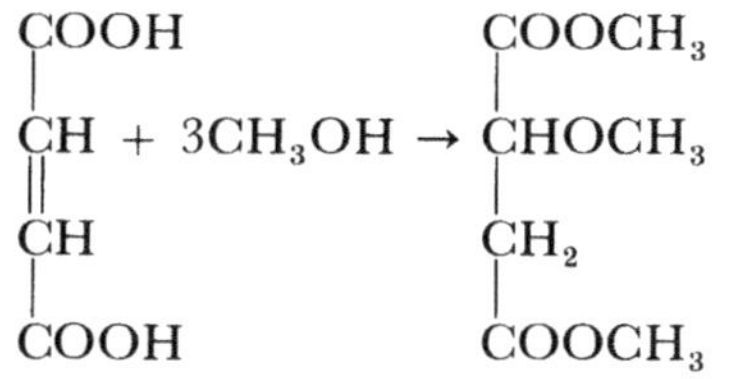

The resistance to hydrolysis of alkyds or polymeric esters is increased with the presence of isophthalic acid and greatly increased with terephthalic acid. Simple solution hydrolysis becomes

TABLE 54.4
Quantitative analysis of copolymer series

Copolymer systems	*Reference*
Adipate polyesters	172
Cellulose esters	231–233
Cellulose ethers	2, 5, 231, 234
Hydroxyethyl cellulose	6–7, 235–236
Chlorobutyl rubber-natural rubber	5, 234
Chlorobutyl rubber-styrene-butadiene	5, 234
Elastomers	3–5, 8, 232–235, 237
Ethyl acrylate-acrylate esters	102, 103, 238
Ethyl acrylate-methacrylate esters	102, 103, 238
Ethyl acrylate-styrene	12, 102, 125
Ethylene-methyl methacrylate	13, 239
Ethylene-methyl acrylate	13, 239
Ethylene-vinyl acetate	14–15, 240–241
Ethylene-ethyl acrylate	14
Ethylene-propylene	16–22, 114, 232, 242, 244, 246
Ethylene-propylene-dicyclopentadiene	20, 211
Ethylene-isobutylene	23–24, 247–248
Ethylene-butene (hexene-octene)	4, 20, 211, 233
Ethylene-styrene	25, 249
Ethylene oxide-propylene oxide	4, 13, 233, 239, 250
Ethylene-trioxane	27, 181
Isoprene-vinyl toluene	28, 252
Isoprene-vinyl xylene	28, 252
Methyl acrylate-methyl methacrylate	103, 101, 254
Methyl acrylate-styrene	256
Methyl acrylate-α-methyl styrene	256
Methyl methacrylate-methacrylic acid	29, 101
Methyl methacrylate-ethylene glycol methacrylate	29, 101
Methyl methacrylate-ethyl acrylate	9–10, 30–31, 102–103, 254–255
Methyl methacrylate-n-butyl acrylate	9, 10, 30, 102–103, 254
Methyl methacrylate-n-propyl methacrylate	10, 25, 30, 103
Methyl methacrylate-ethyl methacrylate	9–10, 30–31, 102–103, 254–255
Methyl methacrylate-n-butyl methacrylate	9–10, 30, 102–103, 254
Methyl methacrylate-n-hexyl methacrylate	9, 33, 102, 253
Methyl methacrylate-n-pentyl methacrylate	9, 102
Methyl methacrylate-vinyl chloride	34, 156
Methyl methacrylate-vinylidene chloride	35, 258
Methyl methacrylate-ethyl acrylate-2-ethylhexyl methacrylate	11, 238
Methyl methacrylate-styrene	36–41, 156, 107–111, 189
Nylon 6	171
Natural rubber-styrene-butadiene	43, 170
Phenol formaldehyde	41, 47–52, 124, 131, 158, 204, 243, 251, 220, 189
Styrene-acrylonitrile	271
Styrene-butadiene natural rubber ethylene-propylene-terpolymer	177, 234
Vinyl acetate-vinyl chloride	156, 197
Vinyl acetate-n-butyl acrylate	112
Vinyl acetate-butyl maleate	112
Vinyl acetate-2-ethylhexyl acrylate	112
Vinyl acetate-2-ethylhexyl maleate	112

unsatisfactory, and more vigorous conditions employing pressure or molten reactants are required to achieve a result in a reasonable time. All three phthalate isomers exhibit characteristic absorption in the infra-red, but attempts at quantitative analyses are generally poor.

Alkali fusion of carboxylic esters[41] with gas chromatography of the alcohols liberated has been reported with several classes of esters including alkyl phthalates and terephthalates. The samples were fused at 240–320° C for 30 minutes with 99–100 per cent conversion, and the procedure should be equally applicable to polymeric esters.

54.6.3 Polyacrylic Esters

Both acrylate and methacrylate esters exhibit characteristic infra-red absorption bands, but the reliable analysis of copolymers is limited and further identification techniques are required. The esters, particularly the methacrylate esters, are very resistant to hydrolysis such that solution hydrolysis is not applicable. The use of the Zeisel Reaction with chromatography of the liberated alkyl halides has been known for three decades and the reaction is shown below.

$$\underset{\displaystyle -\!\!\!\left(\text{CH}-\text{CH}_2\right)\!\!\!-_x}{\overset{\text{COOR}}{|}} \xrightarrow{\text{HI}} \underset{\displaystyle -\!\!\!\left(\text{CH}-\text{CH}_2\right)\!\!\!-_x}{\overset{\text{COOH}}{|}} + \text{RI}$$

Haslam and his co-workers in 1958[42] described the analysis of a copolymer of methyl methacrylate containing 5 per cent ethyl acrylate—a comonomer concentration not evident by infra-red spectrophotometry. The polymer (20 mg) was reacted with phenolic (hydriodic) acid and the alkyl halides trapped for chromatographic separation using a short (1.8 m) column packed with dinonyl phthalate. A modified procedure[43] determined the total alkoxy content by titration of iodine liberated from the alkyl halides with sodium thiosulfate. A second reaction employing gas chromatography allowed separation of the propyl and butyl esters. It is not clear why the total alkoxy is needed, as it is simply a summation of unknowns and of little value. In conjunction with a series of empirical relationships, analysis of a terpolymer of methyl methacrylate–ethyl acrylate–butyl acrylate was carried out. While determination of the pendant groups is clear, attachment to the backbone is not obvious. The alkyl groups of maleate, acrylate or methacrylate esters are estimated together. Recovery and infra-red spectrophotometry of the carboxylic acid containing polymer will indicate the presence of individual carbonyl stretching frequencies of the various esters. Higher resolution and slow scanning speeds (capabilities usually not available with the lower-cost bench spectrophotometers) are required. The technique of band enhancement by reduction of the internal energy of the molecules, developed by Werner and Haken[44] a decade ago, with scanning at liquid nitrogen temperature may also be applied to advantage.

The Zeisel cleavage has also been applied to polymers by Anderson and his co-workers[45] and the effect of stereoregularity and of hydroxyl-containing monomer units on the reaction has been observed.

Care must be exercised in interpreting the results of the reaction, as ethyl iodide is formed by reaction of hydriodic acid with hydroxyethyl acrylate or methacrylate; this is analogous to the reaction of hydriodic acid with ethylene oxide derivatives.[46] With hydroxypropyl esters, isopropyl iodide and propionaldehyde are formed.

The Zeisel reaction is not applicable to all acrylic systems: many thermosetting acrylics include etherified melamine resins, etherified methylolated acrylamides, and phenol formaldehyde condensates, all of which interfere and must be removed before analysis.

While the polyacrylic esters are resistant to simple hydrolysis, the longer chain acrylate esters —octyl (2-ethylhexyl), decyl (n-C_{10}), and lauryl (n-C_{12}), have been saponified with alcoholic potassium hydroxide in a pressure vessel at 160° C for 4 hours, the liberated alcohols being readily separated by gas chromatography. Alkali fusion of polymethyl, n-butyl and isobutyl methacrylate and a copolymer of polymethyl acrylate with the monobutyl esters of polymethyl vinyl ether and maleic acid have been reported with near quantitative conversion.[47]

A quantitative procedure for the determination of the alkoxylation level of acrylamide acrylic interpolymers has been developed by Anderson and his co-workers[48] using alcohol exchange followed by gas chromatographic examination of the reaction products.

Etherified alkylated thermosetting acrylamide interpolymers are shown below.

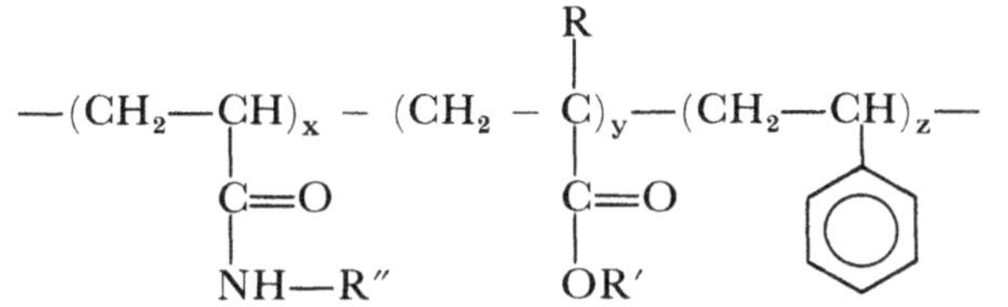

where R = H or CH_3
R′ = H, CH_3, C_2H_5, C_4H_9 or C_8H_{17}
R″ = H, $CH_2OC_4H_9$ or CH_2OH

Reaction with 2-ethyl hexanol (1 per cent m/v) and p-toluene sulfonic acid (1 per cent) as catalyst was effected by refluxing for 4 hours before gas chromatography was carried out.

The etherifying alcohols were exchanged according to the reaction shown below.

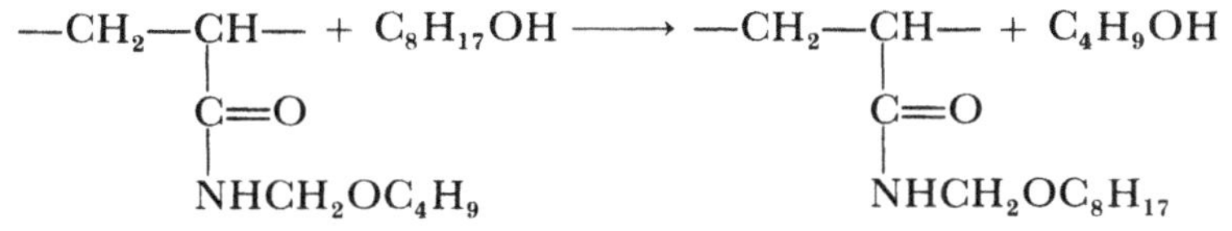

In polymers with the same alkyl ester and alkyl ether groups, the relative amounts of each can be assessed by carrying out both Zeisel cleavage and alcohol exchange determinations.

54.6.4 Polyvinyl Esters

The examinations possible using infra-red spectrophotometry may be refined using chemical means. The determination of acetyl groups in vinyl acetate–vinyl chloride copolymers, of propionyl and butyryl groups in vinyl esters and of acetyl groups in cellulose esters has been carried out by saponification of the polymer, acidification, and gas chromatography of the free organic acids.

Aydin and his co-workers[49] have determined the individual acids in vinyl acetate—vinyl propionate and ethylene–vinyl acetate copolymers by fusion with 4–10 times its mass of p-toluene sulfonic acid at 160° C for 2–2.5 hours. The free acid was determined using gas chromatography. The same workers[50–51] have shown that in the presence of acrylic esters, acetic acid is split from the acrylate ester, but gas chromatography reveals that some esterification occurs and alkaline hydrolysis was recommended for the system. The alcohols and acids formed are determined by gas chromatography and titration, respectively. This method is simple, reproducible and precise. The polymer sample and a few millilitres of 2 N NaOH are sealed in a small ampoule and heated at 100° C for 3–15 hours. After cooling, *t*-butanol (as internal standard) is added to the alkaline solution and the alcohols are determined by gas chromatography. Then the alkaline solution is heated in order to evaporate the alcohols; the remaining acids can be determined by titration or gas chromatography. This method can also be used for polymers in the form of powders, dispersions, emulsions or films. Gas chromatography has the great advantage that mixtures of different alcohols and acids can be determined simultaneously. It is therefore possible to analyse copolymers containing four different base units, such as a copolymer of methyl acrylate, *n*-butyl acrylate, vinyl acetate and vinyl propionate.

Williams and Siggia[52] have reported the quantitative analysis of vinyl esters—that is, polyvinyl acetate, polyvinyl propionate, ethylene–vinyl acetate copolymer and polyvinyl acetate–n-vinyl pyrrolidone copolymers—and of cellulose esters by liberation of the carboxylic acid by hydrolysis with molten orthophosphoric acid.

54.6.5 Cellulose Esters

Cellulose esters are susceptible to hydrolysis, and many procedures reported for other esters are applicable: the simple saponification of alkyd resins developed by Esposito and Swann[53]; the transesterification with sodium methoxide–methanol–methyl acetate of Jankowski and Garner[54]; the alkaline fusion of Whitlock and Siggia[55]; and a method developed for high molecular weight cellulose acetate butyrate and propionate esters, requiring methanolysis with absolute methanol containing 10 per cent boron trifluoride. The samples sealed in glass tubes were heated at 140° C for 3 hours, with subsequent gas chromatography.[56]

54.6.6 Polyurethanes

Chemical cleavage has been applied to linear polyurethane materials using acidic[57] or alkaline hydrolysis[58–62] or aminolysis[63–64].

A linear polymer prepared from hexane-1,6-diisocyanate and butane-1,4-diol was quantitatively hydrolysed by boiling with 50 per cent sulfuric acid for 2 hours.[57] The products of the hydrolysis are butane-1,4-diol, which is largely dehydrated to tetrahydrofuran, and hexane-1,6-diamine, from which the reactant isocyanate is prepared by reaction with phosgene. The products are identified by gas chromatography, the diol as the TMS ether derivative and the diamine as the trifluoroacetyl (TFA) derivative.

Alkaline conditions have been reported by Mulder[58] by heating the sample with 2 per cent aqueous sodium hydroxide in a pressure vessel at 180–200° C for 6 hours. The hydrolysis products were examined by infra-red spectrophotometry, ion-exchange, thin layer, and gas chromatography.

Polyester urethanes and unmodified polyesters have been hydrolysed by heating at 350° C under nitrogen with potassium hydroxide in water. The diamines and polyhydric alcohols were separated using diethyl ether, and the free diamines and the acetylated alcohols were determined by gas chromatography.[59] Dawson and his co-workers[60] similarly effected hydrolysis using polyurethane foam and a solution of potassium hydroxide in water heated overnight at 150° C in a stainless steel pressure vessel. Fijolka and his co-workers[61] conducted the hydrolysis of cross-linked polyurethanes by treatment with 20 per cent hydrochloric acid or 1 N ethanolic potassium hydroxide at 160° C for 16 hours under pressure; the individual homologous products were separated by gas chromatography after preliminary separation with a cation exchange resin. The composition of ester-type polyurethanes has been determined by Klacel and Svoboda[62] by preliminary alkaline hydrolysis.

The determination of small amounts of trimethylolpropane in polyurethane foams and in the ester component before reaction has been carried out[63] using a modification of the aminolysis procedure of Esposito and Swann.[30,32] A sample of the polymer was refluxed with phenylethylamine at about 200° C for 3 hours. After refluxing, the reactants were cooled, acetic anhydride added, and further refluxing for 1.5 hours was carried out. The reaction products were extracted with chloroform, which after extraction with water was injected into the chromatograph.

The aminolysis of elastic polyurethanes has been conducted by reaction with 100 per cent morpholine at 125° C for 25 hours. With polyurethanes based on polyethers, the reactant polyether is liberated, while the reactant diisocyanate reacts with morpholine to form a urea. The glycols were determined by gas chromatography. Polyurethanes based on polyesters were reacted in the same manner, the dicarboxylic acids liberated being examined by gas chromatography of the dimethyl esters.[64]

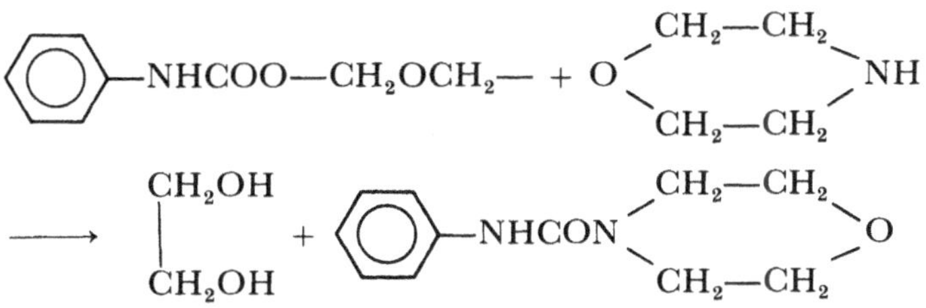

54.6.7 Polyethers

A variety of procedures have appeared[65–72] for the analysis of oxyethylene and oxypropylene groups in alkylene oxide polymers used in the preparation of polyurethanes and their intermediates. Cleavage is again the key to the analyses; and phosphoric acid[65], hydrogen bromide[66], 2M potassium hydroxide[67–69], and a mixture of p-toluene sulfonic acid and acetic anhydride[70–72] have been used.

Molecular weight distribution of low molecular weight polyols—usually propylene oxide adducts of polyhydric alcohols used in polyurethanes—have been determined by gas chromatography of the trimethylsilyl ethers.[73] The polyol was dissolved in acetone and shaken in hexamethyldisilazine, after which trimethylchlorosilane was added. The mixture was allowed to stand for 5 minutes and after further shaking was centrifuged and samples taken for examination. The temperature was limited to about 300° C by the stability of the stationary phase, and polyols of molecular weight of about 1000 could be eluted.

Withers[74] has shown the examination of polyethers without prior chemical reaction. The separation of polyethylene glycols with 13 oxyethylene units and of molecular weight 590 has been achieved by chromatographing the trisilyl ethers.

54.6.8 Methylolated Melamines

The alkoxy content of etherified methylolated melamines has been determined by alcohol exchange with 2-ethyl hexanol by refluxing for 1 hour. The temperature was then increased, and 15 mL of distillate was collected for gas chromatographic examination.[75]

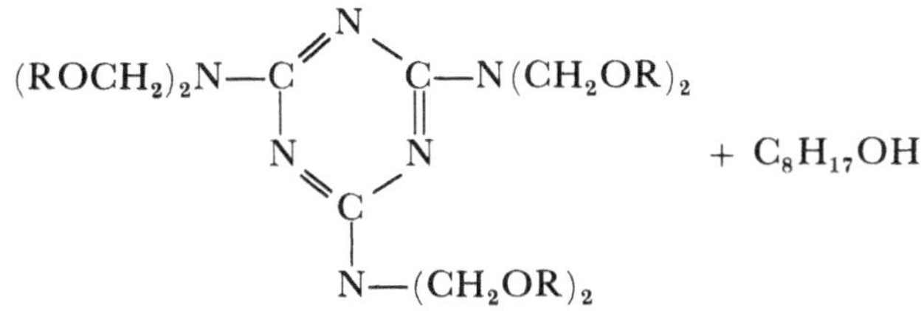

where $R = CH_3, C_2H_5, C_3H_7, C_4H_9$.

54.6.9 Polyamides

Fatty polyamides find considerable use as the major cross-linking agents for epoxy systems. Examination by infra-red spectrophotometry provides characteristic spectra, but trade products of significantly different composition produce similar spectra and chemical cleavage into the acidic and amino constituents as employed with Nylon materials is necessary.

An analytical scheme proposed by Frankoski and Siggia[76] utilises alkali fusion for 30 minutes at 240° C. The amines are estimated by gas chromatography either as TFA derivatives on a packed column or as free amines on a porous polymer column. The fatty acids present as salts are converted to methyl esters for estimation. The dimer acids are similarly converted to esters and separated by gel permeation chromatography to provide peaks for the $C_{18(monomer)}$, $C_{36(dimer)}$ and $C_{54(trimer)}$ components.

Pendant amide groups on the backbone of polyamides and polymeric nitriles have been analysed by alkali fusion, as outlined earlier. Polyacrylamide and polyacrylonitrile liberate ammonia quantitatively.

$$-(CH_2-\underset{}{CH}(CN))- + H_2O \rightarrow -(CH_2-CH(CONH_2))- \xrightarrow{KOH} -(CH_2-CH(COOK))- + NH_3$$

The alkali fusion reagent provides the water to convert the nitrile to the amide and, upon further reaction, to ammonia and a carboxylic acid salt.[76] Alkali fusion of the monobutylamide of poly(methyl vinyl ether-co-maleic acid) gave the expected n-butylamine content. The nitriles also give characteristic infra-red absorption, and acrylonitrile-containing materials are readily identified by the band near 2200 cm^{-1} which, however, is weak and quantification is restricted. Polyacrylonitrile or other nitrogenous polymers are also readily determined quantitatively by nitrogen determination either on a micro or macro scale, providing the polymer is known.

54.6.10 Polyimides and Poly(amide-imides)

Polyimides are a class of thermally stable polymers or copolymers, usually prepared from dianhydrides and diamines or diisocyanates, which find limited application as films and wire enamels and in the fabrication of components for use at high temperatures.

Poly(amide-imide) polymers have both the amide and imide linkages. Like the polyimides, they have favourable thermal, mechanical, chemical and electrical properties. They are used in laminating varnishes, high-temperature enamels and adhesives. More than qualitative analysis of these products requires chemical degradation, and alkali fusion has been successfully used. Reaction for 30 minutes at 250° C produces the volatile amines and the tri and/or tetracarboxylic acids as salts, both functional species being converted to derivatives to allow separation by gas chromatography.[77]

54.6.11 Epoxide Materials

The epoxy resin oligomers from DP 0 to DP 24 have been examined using liquid chromatography with gradient elution on a 1.2 mm × 2 m column packed with Bondapak C_{18}/Corasil. A water to tetrahydrofuran gradient was used. Size exclusion studies were also reported using standard Waters Poragel columns.[78]

Gel-permeation chromatography—that is, size exclusion[79]—has also been applied to simple epoxy resins which are reaction products of bisphenol A or its tetrabromo-derivative with epichlorohydrin or the diglycidyl ether of bisphenol A. The procedures employed tetrahydrofuran as solvent with ultraviolet detection using Merckogel OR 6000 columns. Pure oligomers were used to establish calibration graphs for the quantitative characterisation of resins. Cross-linked styrene polystyrene columns have also been used by Schulz and Raubach[80] with dimethyl formamide as the solvent.

54.6.12 Epoxide Oils

The cleavage reaction of paraperiodic acid has been applied to the analysis of epoxide oils, with direct cleavage of the epoxide functions and production of aldehydes and methyl azelaldehyde. The reaction was carried out in aqueous dioxane at room temperature with stirring for 15 minutes.

The aqueous products of the reaction were extracted with low boiling petroleum ether and after concentration were examined by gas chromatography.[81]

The epoxy glycerides of C_{18} acids have been examined by gas chromatography after reaction of the epoxy groups with ketones in the presence of boron trifluoride to produce 1,3-dioxolone derivatives. The glycerides were dissolved in isooctane and the boron trifluoride and ketone added, and after shaking the mixture was allowed to stand for 2 hours. The reaction was terminated by the addition of 10 per cent aqueous sodium chloride and the reaction products injected into the chromatograph.[82]

54.6.13 Polysiloxanes

The polysiloxanes exhibit characteristic and intense absorption in the infra-red, and the double band at 1020–1100 cm^{-1} due to the Si-0-Si structure is difficult to mistake. Bands due to CH_3-Si- and C_6H_5-Si- are well known, and assignments for polysiloxanes have recently been extensively reviewed.[83] A wide range of substituents find application and for detailed analyses fragmentation of the molecules is necessary. The chemical cleavage agents available are shown in table 54.5.

54.6.14 Cellulose Derivatives

Cellulose is a linear polymer consisting of glucose units bonded through β-linkages and containing three hydroxyl groups; it is capable of forming ester and ether derivatives, the esters having been discussed earlier. The infra-red spectra of many cellulose derivatives are not particularly characteristic, with many broad bands, and auxiliary analysis may be necessary. While the cellulose molecule may be degraded and the fragments identified, the major relevant interest concerns the type and amounts of substituent groups, all of which may be estimated after chemical reaction.

Methoxyl groups in partially methylated cellulose have been estimated[93] after hydrolysis and conversion to the methyl glucosides with methanol and hydrochloric acid. Products with from 5.3 per cent to 45.2 per cent methoxyl groups (i.e. fully methylated) have been examined with separation of tri- and di-O-methyl glucosides.

The series of glucose ethers formed are: 2,3,4,6-tetra-O-methyl-D-glucose (from chain ends);

TABLE 54.5
Organosilicon cleavage reagents (84–92)

Organic group	*Cleavage reagent*	*Reaction products*	*Conditions (°C)*
Dimethyl	Boron trifluoride	Dimethyl fluorosilane	70
Ethyl	Boron trifluoride	Ethyl trifluorosilane	70
Vinyl	Phosphoric acid	Ethylene	80–600
	Sulfuric acid	Ethylene	250
	Alkali	Ethylene	
Phenyl	Potassium hydroxide	Benzene	120
Alkoxy	Boron tribromide	Corresponding alcohol	130
Phenyl	Boron tribromide	Benzene	130
Ethylene oxide	Potassium bisulfate	Acetaldehyde	260
Propylene oxide	Potassium bisulfate	Propionaldehyde	260
Phenyl	Potassium bisulfate	Benzene	260
Isopropylene oxide	Potassium bisulfate	Isopropionaldehyde	260
Glycerin	Potassium bisulfate	Acrolein	260
Ethylene oxide	Potassium persulfate		260
Propylene oxide	Potassium persulfate		260
Terminal $(CH_3)_3Si$-		$(CH_3)_3SiOH + (CH_3)_3SiOSi(CH_3)_3$	200

2,3,6-tri-O-methyl-D-glucose; 2,6-di-O-methyl-D-glucose; 2,3-di-O-methyl-D-glucose; 3,6-di-O-methyl-D-glucose; 6-mono-O-methyl-D-glucose; 3-mono-O-methyl-D-glucose; 2-mono-O-methyl-D-glucose; D-glucose; all of which may be separated by gas chromatography.

The determination of ethoxyl groups in ethyl cellulose has been carried out by reaction of the cellulose with 30 per cent m/m aqueous chromic acid for 3.5 hours at room temperature. Complete reaction was indicated by solution of the sample.[94] Ethoxy groups have also been determined in hydroxy ethyl cellulose by hydrolysis of the sample by refluxing with 1 N sulfuric acid for 4 hours. The hydrolysate was deionised and the trimethylsilyl derivative formed.[95]

A Zeisel-type procedure has been reported for the analysis of cellulose ether groups. The polymer was reacted with hydriodic or hydrobromic acid, and monoiodoalkanes and alkenes or monobromoalkanes and dibromoalkanes were formed.[96]

54.6.15 Starch Derivatives

Starch is similar to cellulose, except that the glucose units are bonded by α-linkages at the C_1 and C_4 position and some chain branching occurs at the C_6 position, and thus a similarity of reaction should be expected. The modification of starch with ethylene oxide produces a hydroxyethyl product of commercial importance. Hydrolysis with sulfuric acid (as described in section 54.6.14) has been used with identification of the TMS derivatives.[97] Reaction between ethylene oxide and the hydroxy-ethyl groups produces side products with poloxyethylene side chains, which complicate the chromatograms produced but which may be identified by comparison with known derivatives.

54.7 PIGMENT/EXTENDER ANALYSIS

Pigments and fillers are recovered from the paint by ashing, or centrifugation and solvent washing (with paint flakes, care should be taken to first remove residual substrate material) and analysed by either instrumental or wet chemical methods.

54.7.1 Infra-red Spectroscopy

A general idea of the extender line-up can be obtained quickly by IR spectroscopy using a potassium bromide pellet or a Nujol mull. A KBr (or other alkali metal halide) pellet is a highly compressed disc of dry KBr (about 100 mg) containing a dispersion of finely ground sample (about 2 mg). A Nujol mull is a fine dispersion of sample in paraffin oil.

It is often difficult to obtain good resolution of mixed inorganics. Ashing the sample to remove organic pigments, or acid washing to remove whiting and other acid solubles, may assist in reducing the complexity of the spectrum.

54.7.2 X-ray Techniques

X-ray fluorescence and diffraction methods can also give valuable information on the inorganic species present, including sulfides, halides and phosphates, but again data acquired from mixtures can be difficult to interpret and quantify.

54.7.3 Emission/Absorption Spectrometry

Information on the cations present in the sample can be gained from both emission spectroscopy and absorption spectroscopy. They are most useful in conjunction with IR or X-ray techniques.

54.7.4 Colorimetric Methods

There are many chemical analyses that can be applied to extenders and pigments. They are generally complicated and tedious, and hence have been largely replaced by instrumental methods. There are some specific analyses which involve a combination of wet chemistry and instrumentation.

An analysis for titanium dioxide in a paint is often important for costing purposes. There are several means of doing this, including the Jones reductor, cupferron-complex gravimetric and polarimetric methods. Perhaps the most straightforward technique is via wet ashing and colorimetry, as follows:

(a) About 300 mg of paint is wet ashed over a bunsen with concentrated sulfuric acid and ammonium persulfate initially, and then with a little concentrated nitric acid.
(b) The reaction mixture is diluted with distilled water and a small amount of hydrogen peroxide added, to form a titanium peroxide complex.
(c) The absorbance of this solution is read at 408 nm on a colorimeter and compared graphically against a standard curve.

54.8 IDENTIFICATION OF MINOR ADDITIVES

The minor constituents of a surface coating are sometimes the subject of analysis. The small quantities present and the large variety of additives in a particular class, can often make analysis and identification difficult. Usually, isolation of the components can be made by solvent extraction or chromatography, and subsequent identification by spectroscopic means.

Surfactants and plasticisers can be separated by extraction with appropriate solvents. For example, phthalate plasticisers can be removed with hexane, diethyl ether, or methanol reflux, and identified by gas chromatography on neopentylglycol succinate (NPGS) or silicone SE-30 columns. Alkyl and aryl sulfonate surfactants can be extracted with petroleum spirits and classified by IR spectroscopy. Several collections of the IR spectra of common plasticisers exist. Minor additives can also be separated with HPLC/GPC and identified by NMR or colorimetrically.

Trace metals, such as from driers, can be analysed by atomic absorption spectroscopy. This technique is often employed for the determination of lead levels in epoxy food can linings. Inductively coupled plasma and complexometric–spectrophotometric techniques may also be employed.

54.9 ENVIRONMENTAL MONITORING

In the interests of worker safety, paint manufacturing companies increasingly will need to monitor their production sites for airborne monomers, solvents and particulates. Their efforts in this area are being spurred by union pressures and government legislation. The paint chemist is normally required to perform these analyses.

Volatile organics are generally collected by either passive absorption on to activated discs, or pumping air through a tube containing activated charcoal over a set time period. Desorption is generally achieved by washing with carbon disulfide. This solution is injected into the gas chromatograph and the peaks identified in the usual manner.

REFERENCES

1 J S Long and R R Myers *Treatise on Coatings Vol 11 Parts I and II Characterisation of Coatings* Marcel Dekker New York. Various Chapters
2 J K Haken *Gas Chromatography of Coating Materials* Marcel Dekker New York 1974
3 M S Stevens *Characterization and Analysis of Polymers by Gas Chromatography* Marcel Dekker New York 1969
4 G M Brauer in *Thermal Characterization Techniques* P E Slade Jr and L T Jenkins eds Marcel Dekker New York 1970
5 V G Berezkin V R Alishoyev and I B Nemirovskaya eds Gas Chromatography of Polymers, *J of Chromatography Library* Vol 10 Elsevier Amsterdam 1977
6 C E R Jones and C A Cramers eds *Analytical Pyrolysis* Elsevier Amsterdam 1977
7 R W May E F Pearson and D Scothern *Pyrolysis Gas Chromatography* Chem Soc, London 1977
8 *Australian Standard 1580 Method 301.1* May 1975
9 *Australian Standard 1580 Method 301.2* May 1978
10 *Australian Standard 1580 Method 302.1* May 1975
11 J K Haken and T R McKay *J Gas Chromatography 4* 132 1965
12 J Haslam A R Jeffs and H A Willis *J Oil Col Chem Assoc 45* 325 1962
13 B T Whitam in *Vapour Phase Chromatography* D H Desty ed Butterworths London 1956 p 395
14 E H Ellis and R L Le Tourneau *Anal Chem 25* 1269 1953
15 *ASTM Specification D1319* ASTM Philadelphia 1970
16 J K Haken and T R McKay *J Oil Col Chem Assoc 47* 513 1964
17 J K Haken and T R McKay *Aust Oil Col Chem Proc & News 1*(5) 6 1964
18 J K Haken and V Khemangkorn *J Oil Col Chem Assoc 54* 764 1971
19 M F Dante *J Paint Technology 47* 606 1975
20 R A Nyquist *Infrared Spectra of Plastics and Resins* Dow Chemical Co Midlands 1961
21 R B Du Vall *Infrared Spectra of Plasticisers and Other Additives* Dow Chemical Co Midlands 1966
22 *Adhesives and Sealants* Sadtler Research Laboratories Philadelphia 4 volumes
23 D O Hummel *Infrared Analysis of Polymers, Resins and Additives* Wiley-Interscience New York Vol 1 1969 Vol 2 1971
24 Chicago Society for Paint Technology *Infrared Spectroscopy—Its Use in the Coatings Industry* Federation Soc Paint Tech Phil 1969
25 L J Bellamy *Infrared Spectra of Complex Molecules* 3rd ed Chapman and Hall London 1975
26 T R Crompton *Chemical Analysis of Additives in Plastics* 2nd ed Pergamon Press Oxford 1977
27 J Kaczaj *Applied Spectroscopy 21* 180 1967
28 *ASTM Specification D 563* ASTM Philadelphia 1952
29 G G Esposito *Anal Chem 40* 1903 1968
30 G G Esposito and M H Swann *Anal Chem 33* 1854 1961
31 *ASTM Specification D24256* ASTM Philadelphia 1969
32 G G Esposito and M H Swann *Anal Chem 34* 1173 1962
33 R E Wittendorfer *Anal Chem 36* 930 1964
34 F H de la Courte, N J P Van Cassel and J A M Van der Valk *Farbe und Lack 75* 218 1969
35 H D Dinse and E Tucek *Faserforch Text Tech 21* 205 1970
36 G G Esposito and M H Swann *Anal Chem 34* 1048 1962
37 *ASTM Specification D2455* ASTM Philadelphia 1969
38 D F Percival *Anal Chem 35* 236 1963
39 R J Rawlinson and E L Deeley *J Oil Col Chem Assoc 50* 573 1967
40 G G Esposito *Anal Chem 40* 1903 1968
41 S Frankoski and S Siggia *Anal Chem 44* 507 1972
42 J Haslam J B Hamilton and A R Jeffs *Analyst 83* 66 1958
43 D L Miller E P Samsel and J G Cobler *Anal Chem 33* 677 1961
44 J K Haken and R L Werner *Appl Spectrosc 22* 345 1968
45 D G Anderson *et al Anal Chem 43* 894 1971
46 S Siggia *J Amer Oil Chem Soc 35* 643 1958
47 J S Cobler and E P Samsel *SPE Trans 2* 145 1962
48 D G Anderson *et al Anal Chem 47* 1008 1975
49 O Aydin B J Kaczmar and R C Schulz *Angew Makromol Chem 24* 171 1972
50 O Aydin and R C Schulz *Makromol Chem 176* 3537 1976
51 R C Schulz and O Aydin *J Polymer Sci* Sympos No 50 497 1976
52 R J Williams and S Siggia *Anal Chem 49* 2337 1977
53 G G Esposito and M H Swann *J Paint Technol 43* 60 1971
54 S J Jankowski and P Garner *Anal Chem 37* 1709 1965

55 L R Whitlock and S Siggia *Separation and Purification Methods 3* 299 1974
56 M Wandel and H Tengler *Gummi Asbest Kunstoffe 19* 141 1966
57 E Schroder *Plaste Kautsch 9* 121 186 1962
58 J L Mulder *Anal Chem Acta 38* 563 1967
59 I Ligotti G Bonomi and R Piacentini *Rass Chim 17* 137 1965
60 B Dawson S Hopkins and P R Sewell *J Appl Polym Sci 14* 35 1970
61 P Fijolka R Gnauck and G Schulz *Plaste Kautsch 19* 751 1972
62 Z Klacel and P Svoboda *International Polymer Science and Technology 2* 3 57 1975
63 R E Wittendorfer *Anal Chem 36* 931 1964
64 R N Mokeeva Ya A Tsarfin and V D Kharchenkova *Soviet Plastics* no 9 79 1972
65 H D Graham and J L Williams *Anal Chem 36* 1345 1964
66 A Mathias and N Mellor *Anal Chem 38* 472 1966
67 R J Kern and J Schaefer *J Am Chem Soc 89* 6 1967
68 J Schaefer R J Katnik and R J Kern *Macromol 1* 101 1968
69 J Schaefer R J Kern and R J Katnik *Macromol 1* 107 1968
70 M H Karger and Y Mazur *J Amer Chem Soc 90* 3878 1968
71 K Tsuji and K Konishi *Analyst 96* 457 1971
72 K Tsuji and K Konishi *Analyst 99* 54 1974
73 G E Corbett W Hughes and R G Morris-Jones *J Appl Polym Sci 13* 1297 1969
74 M K Withers *J Gas Chromatogr 6* 242 1968
75 D G Anderson D A Netzel and D J Tessari *J Appl Polymer Chem 14* 3021 1970
76 S Frankoski and S Siggia *Anal Chem 44* 2078 1972
77 D D Schleuter and S Siggia *Anal Chem 49* 2343 1977
78 W A Dark E C Conrad and L W Crossman *J Chromatography 91* 247 1974
79 D B Braun and D W Lee *Angnew Makromol Chem 57* 111 1977
80 G Schulz and H Raubach *Plaste Kautsch 24* 325 1977
81 G Maerker and E T Haeber *J Am Oil Chem Soc 43* 97 1966
82 J A Fiorita M J Kanuka and R J Sims *J Chromatogr Sci 7* 448 1969
83 J Helflys J Schraml and M Horak *Handbook of Organosilicon Compounds* Marcel Dekker New York 1973
84 J Franc *Czech Patent 154* 436 15 8 74
85 C L Hanson and R C Smith *Anal Chem 44* 1571 1972
86 G W Heylmun and J E Pikula *J Gas Chromatogr 3* 266 1965
87 G W Heylmun R L Bujalski and H B Bradley *J Gas Chromatogr 2* 300 1964
88 V M Krasikova and A N Kaganova *J Anal Chem* USSR *25* 1212 1970
89 V M Krasikova A N Kaganova and V D Lobkov *J Anal Chem* USSR *26* 1458 1971
90 V M Krasikova V P Milishkevich and A N Kaganova *J Anal Chem* USSR *29* 1028 1974
91 E R Bissell and D B Fields *J Chromatogr Sci 10* 164 1972
92 J Franc and K Placek *Coll Czech Chem Commun 38* 513 1973
93 C E Lott and K M Brobst *Anal Chem 38* 1767 1966
94 H Jacin and J M Slanski *Anal Chem 42* 801 1970
95 A A Karnishin *Zh. Anal Khim 23* 1072 1968
96 G Bartelmus and R Ketterer Fresenius Z *Anal Chem 286* No 3/4 161 1977
97 J Franc and K Placek *J Chromatogr 48* 295 1970 *67* 37 1972
98 W H T Davision S Slaney and A L Wragg *Chem Ind* London 1356 1954
99 W D Stewart, *JAOAC 59* 35 1976
100 E A Radell and H C Strutz *Anal Chem 31* 1890 1959
101 J Strassburger *et al Anal Chem 32* 454 1960
102 H McCormick *J Chromatography 40* 1 1969
103 J K Haken and T R McKay *Anal Chem 42* 1251 1973
104 G G Esposito *Anal Chem 36* 2183 1964
105 G G Esposito and M H Swann *J Gas Chromatography 3* 282 1965
106 S Paul *J Coating Technology 52* 47 1980
107 N C Jain C R Fontain and P L Kirk *J Forensic Sci 5* 102 1965
108 W D Stewart *J Forensic Sci 19* 121 1974
109 B B Wheals and W Noble *J Forensic Sci 14* 23 1974
110 R W May *et al Analyst 98* 364 1973
111 P Perros *JAOAC 58* 1150 1975
112 J C Daniels and J M Michel *J Gas Chromatography 5* 437 1967
113 R S Lehrle and J C Robb *Nature 183* 1671 1959
114 Bodensee Perkin Elmer Co *GmbH Germany Bulletin GC202*

115 C C Luce *et al Anal Chem 36* 482 1964
116 C Beleinski J C Rosso and F Lalau-Keraly *Rec Aerosp 114* 51 1965
117 L H Lee *J Poly Sci A3* 859 1965
118 L H Lee *J Appl Poly Sci 9* 1981 1965
119 C N Cascaval and I A Schneider *Rev Roumaine Chim 20* No 4 1975
120 J E Guillet W C Wooten and R L Combs *J Appl Polym Sci 3* 61 1960
121 K H Duebler and E Hagen *Plaste Kautsch 16* 169 1969
122 I Rusznak *et al Kororiszt Ertesito 16* No 7–8 185 1974
123 S K Yasuda *J Chromatogr 27* 72 1967
124 J Voigt *Kunstoffe 51* 18 1961
125 C W Stanley and W R Peterson *SPE Trans 2* 298 1962
126 A Lipska and F A Wodley *J Appl Poly Sci 13* 851 1969
127 J Kammermaier *Kolloid Z 209* 20 1966
128 K Kato and H Komorita *Agric Biol Chem 32* 21 1968
129 K Ettre and P F Varadi *Anal Chem 35* 69 1963
130 S Tsuge H Ito and T Takeuchi *Macromolecules 2* 277 1969
131 S Tsuge T Okumoto and T Takeuchi *Bull Chem Soc Jap 43* 3341 1970
132 J Mitera and J Michal *Chem Prum 26* No 8 417 1976
133 F Sontagg *Brennst Chem 47* 264 1966
134 B G Luke *J Chromatogr Sci* 11 435 1973
135 J Zachoval *et al Sci Papers Inst Chem Technolog* Prague *C18* 37–44 1972
136 D P Bishop and D A Smith *Ing Eng Chem 58* 8 32 1967
137 I Tanikawa *et al J Jap Soc Col Mat 42* 349 1969
138 J Q Walker and R J Morgan Paper presented at the 166th ACS Natl Mtg Chicago 11 Aug 27–31 1973
139 C Waysman D Matelin and C L Duc *J Chromatog 118* 115 1976
140 J H O'Neill *et al J Gas Chromatogr 1* 28 1963
141 J Reardon and R H Barker *J Appl Polym Sci* 18 1903–17 1974
142 J L Cotter *Org Mass Spectr 5* 851–55 1971
143 J Martinez and G Guichon *J Gas Chromatogr 5* 146 1967
144 G E Fisher and J C Neerman *IEC Prod Res Dev 5* 288 1966
145 C Landault and G Guichon *Anal Chem 39* 713 1967
146 V T Brooks *Chem Ind* London 1090 1960
147 W Sassenberg and K Wrabetz *Fresenius Z Anal Chem 184* 423 1961
148 J Zulaica and G Guiochon *J Polym Sci Pt B 4* 567 1966
149 W M Jackson and R T Conley *J Appl Poly Sci 8* 2163 1964
150 E Hagen *Plaste Kautsch 15* 711 1968
151 M Tsuge T Tanuka and S Tanaka *Jap Analyst 18* 47 1969
152 H Mosimann and W Weber *Schweiser Arch Angew Wiss Tech 36* 402 1970
153 D Braun and J Arndt *Kunstoffe 62* 41 1972
154 M Yamao and Y Iida *Bunseki Kagaku* 21 1602–08 1972
155 A Frank and K Wunscher *Chem Ztg Chem App 91* 7 1967
156 R S Lehrle J C Robb *J Gas Chromatogr 5* 89 1967
157 J Nematollaha W Guess and J Autian *Microchem J 15* 53 1970
158 J Chih-an Hu *Anal Chem 49* 537 1977
159 J Zulaica and G Guichon *Bull Soc Chem Fr* 1351 1966
160 T Shono and K Shinra *Anal Chem Acta 56* 303 1971
161 E M Barrall R S Porter and J F Johnson *J Chromatogr 11* 177 1963
162 J Voigt *Kunstoffe 54* 2 1964
163 Y Tsuchiya and K Sumi *J Poly Sci Pt A 7* 813 1969
164 H M Cole *et al Rubber Chem Technol 39* 259 1966
165 F A Lehmann and G M Brauer *Anal Chem 33* 673 1961
166 A Davis and J H Golden *J Gas Chromatogr 5* 81 1967
167 B M Kovavaskaya *Chem Zvesti 18* 13 1964
168 S Tsuge T Okumoto and T Takeuchi *J Chromatogr Sci 7* 253 1969
169 E W Cieplinski *et al Fresenius' Z Anal Chem 205* 357 1964
170 K Tsuge J Ando and N Okubo *J Soc Rubber Ind* Japan *42* 851 1969
171 H Senco S Tsuge and T Takeuchi *J Chromatogr Sci 9* 315 1971
172 F Farre-Rius and G Guichon *J Gas Chromatogr 5* 457 1967
173 C E R Jones and A F Moyles *Nature 191* 663 1961
174 J S Parsons *Anal Chem 36* 1849 1964

175 A G Nerheim *Anal Chem 35* 1640 1963
176 F Sadowski and H Kanfeld *Farbe Lack 69* 597 1963
177 D J O'Neil *J Compos Mater 2* 502 1969
178 F W Willmott *J Chromatogr Sci 7* 101 1969
179 V G Berezkin I B Nemirovskaya and B M Kovarskaya *Zavod Lab 35* 148 1969
180 T Abo and T Watababe *Kogyo Kagaka Zasshi 74* 885 1971
181 K H Burg E Fischer K Wiessermel *Makromol Chem 103* 268 1967
182 D Noffz and W Pfab *Fresenius' Z Anal Chem 228* 188 1967
183 P Fuchs and C Szepesy *Chromatographia 1* 310 1968
184 F Sontag *Fette Seifen Anstrichm 70* 417 1968
185 P G M Van Stratum and J Dvorak *J Chromatogr 71* 11 1972
186 Y Tsuchiya and K Sumi *J Polym Sci. Pt B 6* 357 1968
187 J K Haken D K M Ho E Houghton *J Polymer Sci* Chem *12* 1163 1974
188 S K Yasuda *J Chromatogr 51* 261 1970
189 Y Sibasaki *J Polym Scipt* A 1 *b* 21 1967
190 A Barlow *et al Polymer 8* 523 (1967)
191 G Bagby R S Lehrle and J C Robb *Polymer 9* 284 1968
192 D H Grant E Vance and S Bywater *Trans Faraday Soc 56* 1697–1703 (1960)
193 M T Jackson Jr and J Q Walker *Anal Chem 43* 74 1971
194 D Noffz W Benz and W Pfab *Fresenius' Z Anal Chem 235* 121 1968
195 Y Tsuchiya and K Sumi *J Appl Chem 17* 364 1967
196 M Dimbat in *Gas Chromatography* 1970, R Stock ed Inst Petroleum London
197 F Spagnolo J *Gas Chromatogr 6* 609 1968
198 S Tsuge T Okumoto and T Takeuchi *J Chromatogr Sci 7* 250 1969
199 R H Wiley and F E Martin *J Macromol Sci Chem Al* 635 1967
200 R D Collins P Fiveash and L Holland *Vacuum 19* 113 1969
201 W Coleman L Scheel and C Gorski *Am. Ind. Hyg Assoc J 29* 54 1968
202 T Takeuchi S Tsuge and T Okumoto *J Gas Chromatogr 6* 542 1968
203 H Szymanski C Salinas and P Kwitowski *Nature 188* 403 1960
204 R Vukovic and V Gnjatovic *J Polym Sci* Pt A 1 *8* 139 1970
205 A N Genkin and N A Petrova *J Chromatog 105* No 1 25–32 1975
206 D E Hillman and H Wells *J Oil Colour Chem* Ass *52* 727 1969
207 W G Geddes *Europ Polym J 3* 267 1967
208 E A Boettner G Ball and B Weiss *J Appl Poly Sci 13* 377 1969
209 M M O'Mara *J Polymer Sci* Al 9 No 5 1387–1400 1971
210 S A Leibman D H Alstrom and P R Griffiths *Appl Spectros 30* No 3 355–57 May/June 1976
211 J Van Schooten and K Evenhuis *Polymer 6* 561 1965
212 R S Lehrle *Lab Practice 17* 697 1968
213 G M Galpern *et al Zavod Lab 32* 931 1966
214 K P Kornieyeva *et al Plasticheskie Massy 1574* No 3 75–77 1974
215 G Garzo and F Till *Talantha* 10 583 1963
216 J Franc K Placek and F Mikes *Coll Czech Chem Commun 32* 2242 1967
217 M Blazso G Garzo and T Szekely *Chromatographia 5* No 9 485–92 1972
218 L D Turkova and B G Belenky V Y Sokomol *Soedin Ser A 12* 46 1970
219 Y Shibasaki *J Polym Sci* Pt A 1 *5* 21 1967
220 H Kambe and Y Shibasaki *Chem High Polym 21* 65 8 1964
221 A S Barlow R S Lehrle J C Robb *Polymer 2* 27 1961
222 R Saferstein and J M Chao *JAOAC 56* 1234–1238 1973
223 C E R Jones and G E T Reynolds *J Gas Chromatogr 5* 25 1967
224 D Poxom R Wright *J Chromatog 61* 142 1971
225 Z Ramljak D Deur-Siftar and A Solc *J Chromatog 119* 1976
226 K Takiura A Yamaji and H Yuki *Yakugaku Zasshi 93* 776 1973
227 K Takiura A Yamaji and H Yuki *Yakugaku Zasshi 93* 769 1973
228 N Macleod *Chromatographia 5* 516 1972
229 H Tai R M Powers and T F Protzman *Anal Chem 36* 108 1964
230 D J Bryce and C T Greenwood *Appl Polym Symp 2* 149 1966
231 W D King and D J Stannors *TAPPI 52* 465 1969
232 B Groten *Anal Chem 36* 1206 1964
233 E W Neumann and H G Nadeau *Anal Chem 35* 1454 1963
234 A Krishen and R G Tucker *Anal Chem 46* 29 1974

235 H Tai, R M Powers and T F Protzman *Anal Chem 36* 108 1964
236 D J O'Neil *Anal Letters 1* 499 1968
237 R O B Wijesekera and M R N Fernando *J Chromatogr 65* 560 1972
238 J K Haken and T R McKay *J Chromatogr 80* 75 1973
239 K J Bombaugh C E Cook and B H Clampitt *Anal Chem 35* 1834 1963
240 E M Barrall II R S Porter and J F Johnson *Anal Chem 35* 73 1963
241 T Oku Moto T T Keochi and S Tsuge *Kogyo Kagaku Zasshi 73* 702 1970
242 D Duer-Siftar *J Gas Chromatogr 5* 72 1967
243 A Krishen *Anal Chem 44* 494 1972
244 M Dimbat and F T Eggertsen *Microchem J 9* 500 1965
245 V R Alishoyev *et al Vysokomol Soedin Sera 11* 247 1969
246 V M Androsova N M Seidov and T M Shabayev *Zavod Lab 34* 668 1968
247 E Hagen and G Hazkoto *Plaste Kautsch 16* 21 1969
248 G Hazkoto and E Hagen *Muanyag Gumi 7* 210 1970
249 K Jobst and L Wuckel *Plaste Kautsch 12* 150 1965
250 R N Mokeyeva and Y A Tsarfin *Plaste Massy 3* 52 1970
251 D Braun and R Disselhoff *Angew Makromol Chem 23* 103 1972
252 K V Alekseyeva L P Khramova and I A Strelnikova *Lavod Lab 36* 1304 1970
253 I Takeuchi T Ukumoto and S Tsuge *Jap Analyst 18* 614 1969
254 R L Gatrell and T J Mao *Anal Chem 37* 1294 1965
255 E C Ferlauto *et al J Appl Polym Sci 15* 445 1971
256 J K Haken D K M Ho *J Chromatog 126* 239 1976
257 J E Guillet W C Wooten and R L Combs *J Appl Polym Sci 3* 61 1960

55 TECHNICAL SERVICE IN THE SURFACE COATINGS INDUSTRY

55.1 INTRODUCTION

Technical service is an important and integral part of the marketing of both raw materials and finishes within the surface coatings industry. Through technical service, the benefits and most appropriate use of a company's products are defined to assist in making the sale. By examining the basic functions in a typical technical service operation, the important position of this function in marketing may be understood. The main activities are:

(a) advising on product use and handling
(b) relating customer needs to product features
(c) assessing future requirements
(d) servicing complaints
(c) assisting in the preparation of literature.

There are three broad areas of technical service activity within the surface coatings industry:

(a) in the marketing of raw materials and equipment, in which emphasis on product benefit is paramount;
(b) in the marketing of retail paints, in which consumer assistance is important;
(c) in the marketing of industrial finishes, in which involvement in the manufacturing process is of considerable significance.

55.2 RAW MATERIALS TECHNICAL SERVICE

In this area it is interesting to look firstly at the recent development of the chemical industry in Australia and to consider the impact on user industries.

Since the Second World War there has been huge growth in the worldwide chemical industry. In Australia, major advances have taken place in the past twenty years, and further large-scale developments are projected for the immediate future.

Funds employed in the chemical industry in Australia have grown from A$163 million in 1960, to $463 million in 1969, and to $1022 million at the end of 1978.

The impact of this growth is a huge increase in the number and variety of chemicals available. The coatings industry has proved no exception and has sought to improve performance, reduce costs and formulate products that were previously technically or economically impossible to make in this country.

A review of the changes in the raw materials used by the surface coatings industry—and other chemical industries—illustrates the effect of the greater availability of chemical raw materials.

Formerly the surface coatings industry used predominantly natural products such as oils, gums and earth colours, with only a few synthetic chemical products such as white or red lead and Prussian blue. Today a very broad range of chemicals are used. Typical examples are the synthetic resins—alkyds, urethanes, epoxies, PVA and acrylic emulsions, for example—synthetic pigments, surface active agents and a wide variety of other so-called 'auxiliary' chemicals. From this growing list, many examples can be found to illustrate the valuable role of technical service in the surface coatings industry.

For example, since its introduction to the industry in the 1930–40 period, titanium dioxide has become the totally dominant white pigment used in surface coatings. This has been because of its features of high opacity, bright colour, durability, stability, and very low toxicity. In some instances the value of these features is not immediately apparent to the end user. There are many cases where the role played by technical service has been to demonstrate these features in a practical way and thus enable the paint manufacturer to achieve improvements in the performance of the company's products. This does not imply that the paint manfacturer lacks the technical knowledge or resources to carry out such work himself; but rather that technical service staff, who have specialised product knowledge, are able to co-operate with the user's technical team in the development process. The specialised technologist may also advise on the specification and introduction of new products by keeping the customer informed of new market trends where appropriate.

A similar example to the developments in surface coatings can be found in the evolution of the detergent industry. It is not so long ago that soap powders, built up with alkalis such as sodium carbonate or silicate, were the major raw materials in this field. Today detergent and laundry products designed for many different specific uses are available in the marketplace; examples include the spray-dried powder and liquid detergents, liquid surface-cleaners, fabric softeners or conditioners, bleaches and hair-care products. These are derived from a wide variety of new chemicals such as alkyl aryl sulfonates, alkyl sulfates, non-ionics, polyphosphates, perborates, optical brighteners, enzymes and many others. All are products from what is now a highly diversified chemical industry.

In the USA and elsewhere there has been public concern to reformulate detergents because of the eutrification of waterways caused by the high phosphate content in laundry effluent. It is the role of technical service to advise on formulations which will overcome such problems, hopefully ahead of any legislation that might eventuate, and this is a good example of the importance of the technical service function in predicting future needs. Government regulation in the protection of the environment is increasing rapidly, and technical service must become more and more involved in assisting the users of raw materials to adjust to these changes.

Many more examples could be found to show the changing market demands for an ever widening range of chemicals to meet the needs of future applications. The availability of this wide variety of chemical raw materials enables the manufacturer to meet this changing market and, at the same time, poses the more complex problem of selecting the optimum products to suit specific formulation needs.

The technical service facilities offered by chemical manufacturers have been developed to help the end users in the selection and use of the appropriate raw materials or products, and have thus become a vital factor in the total marketing effort.

The functions of technical service may now be discussed in greater detail.

55.2.1 Advice on the Use and Handling of Products

This familiar aspect of technical service includes provision of information on how to use the product, often by recommending starting-point formulae, and outlining precautions to be observed in storage and handling (particularly if the product is hazardous). For this, and for the second function, it is essential that the personnel concerned have a thorough knowledge of the product, its properties and applications.

55.2.2 The Relationship between Customer Needs and Product Features

By assessing the needs of the user with regard to his production facilities, product range, the segment of the market to which he is selling, and relating these to features of their own products, technical service personnel can assist in the selection of the most suitable material and the optimum method of use for the intended purpose.

A critical analysis of needs related to features is extremely valuable to the end user and to both the seller and the buyer. From the end user's point of view, this technical service can lead to increased efficiency in manufacture, improvement in cost/performance, reduction in the time to formulate a new product or to modify an existing item, and incorporation of new performance features into an existing product.

For the sales personnel of the chemical manufacturer, it highlights the benefits of purchasing a particular product and thus improves their sales presentation. In addition, where circumstances permit, technical service staff can assist the sales staff of the end product manufacturer to improve their own product knowledge and thus function better.

55.2.3 Assessment of Future Needs

Because technical service personnel are in close contact with the market and constantly assessing the requirements of the market relative to their products, they are in an ideal position to predict what will be required to meet the future needs. The feedback of intelligence from the marketplace is a valuable source of information to help product development and production personnel to establish their forward plans.

Technical service is not the sole source of this information: obviously sales staff are also in a position to contribute; both complement the information obtained by formal market research in making the decisions for future products. The technical service division, by virtue of its close technical involvement with customers, offers a valuable means of collecting such data.

55.2.4 Servicing Complaints

It is important that complaints are attended to promptly and efficiently. Where these involve a technical problem, technical service can materially assist in resolving the complaints in the shortest possible time. Although nobody welcomes complaints—neither the supplier nor the user—it should be realised that they can provide useful information and, properly handled, can improve the goodwill of the user. Thus for a number of reasons, the function of technical service, in assisting with complaint reactions, is of significant importance.

55.2.5 Assisting Literature Preparation

Technical literature is an essential and valuable tool in the marketing of chemical products. Because of their close contact with the end user, the people involved in technical service can make a valuable contribution to the subject matter in technical literature. Similarly, in the preparation of advertising copy, the technical serviceman can provide useful practical background information and suggestions.

In any negotiation leading to the decision to purchase a new or alternative product—whether it be a relatively short discussion over a desk or an extended procedure involving detailed technical consideration and product evaluation—the following five steps occur: attention; interest; desire; conviction and action.

The first three steps can be initiated by a number of means including direct contact, advertising, information on overseas developments or competitors' activity.

Although technical service may participate in these initial stages, its most important function is to provide the data that lead to the essential conviction step in the negotiation. This is the evidence that gives the assurance in the buyer's mind that the choice under consideration is correct. It is not a novel idea, but buying and selling are reciprocal functions: the supplier needs to sell the product profitably and the purchaser seeks continuity of supply of the item at a price relative to performance which will enable him to manufacture his own product competitively. In arriving at a mutually satisfactory solution to these reciprocal requirements, technical service fulfills a basic requirement in increasing sales.

55.3 TECHNICAL AND CUSTOMER SERVICE IN THE RETAIL SECTOR

Although an important part of the formulation of retail paint products involves testing to eliminate field and in-service problems, and quality paints are subjected to rigorous quality control tests, there will still be a number of complaints received from, and technical service required by, the customers. It is, in fact, uneconomical to provide the high levels of quality assurance testing necessary to eliminate all adverse consumer feedback.

It must be realised that quality paint manufacturers produce and sell, to non-technical consumers, a complex technical product. It is applied to a variable and often inadequately prepared substrate, sometimes over a non-uniform and undetermined coating or series of coatings. The coating is also applied by various application techniques and allowed to film-form under a variety of conditions of temperature and humidity. Thus adequate control of the *use* of surface coatings in the retail sector is virtually impossible.

Consequently, paint companies must regard complaints as inevitable. When these cannot be effectively resolved in writing or by telephone, it is customary that a skilled technical service representative calls and services that enquiry. When this occurs, it is common practice to log the complaint into a register with the following information:

(a) Reference number
(b) Date of enquiry
(c) Name of client, address, telephone number
(d) Geographic area of enquiry
(e) Product (if applicable)
(f) Classification of defect (as detailed below)
(g) Type of defect (also detailed below)

Defects may be broadly classified into three broad groups:

(a) In-can defects
(b) Application and film formation defects
(c) Performance defects

The defect classes may be further sub-divided as follows:

(a) Skin, seed, lumps
Consistency problems
Container faults

(b) Colour
Drying
Coverage/Opacity

Finish (e.g. sheariness, poor coalescence, mud cracking)
Gloss level
Colour acceptance; flotation
Application, roller matting
Foaming
Odour

(c) Cracking, crazing, flaking, peeling
Discolouration, chalking, staining, bleeding, efflorescence, colour change
Blistering
Water sensitivity, wet adhesion
Washability
Mould
Saponification

A more detailed discussion of defects is given in chapter 53.

It is imperative that all enquiries are handled objectively. All clients must be treated with respect, since feedback on retail products (which may detail any one of a number of commercial problems with that product) can come only from consumers. A well-maintained servicing system can highlight potential problem areas with minimum delay, inconvenience and cost if an open-minded attitude prevails.

Feedback from clients is also useful in aiding technical people to write or modify label instructions.

TABLE 55.1
Suggested representatives' kit

Moisture meter
pH papers
Paint inspection gauge
Magnifying glass
Polaroid camera with flash
Various solvents
Reagents for spot tests (such as oxalic acid, bleach, peroxides, etc.)
Paint scraper or scalpel
Sample envelopes
Paint stirrer

It is important when servicing complaints for representatives to carry a kit. (see table 55.1).

It is also essential that a representative possesses several fundamental prerequisites to aid him in his dealings with the problem and the client:

(a) He should be convinced that his products are controlled under a tight and effective system of quality assurance. This provides him with *confidence* in the range of products he is representing.
(b) He should possess full knowledge of the *recommended use areas* for each product, and be assured his products are adequately labelled with full directions for surface preparation, specification for use, application procedures and special warnings if applicable. Examples of these are: 'Do not apply at temperatures less than 10° C'; 'Do not wash or scrub film within seven days of application'; and 'Provide adequate ventilation during application and drying'.
(c) He should have adequate *field experience* so that the problem may be easily identified; his experience should enable him to assess the reasons for the complaint almost without exception.

(d) He should be very open-minded about the problem and be sensitive in his communication with the client.
(e) He should be prompt in his response to the complaint, because the client has expectations in this regard. Delays tend to increase the anxiety of the client, which can affect the communication between the parties. Ideally the matter should be finalised on the spot rather than merely reported for action by a series of other people.
(f) He should be aware of any shelf-life limits that apply to the product, and the relationship, if any, between the batch number and date of manufacture (DOM).

55.4 TECHNICAL SERVICE IN THE INDUSTRIAL SECTOR

The service of surface coatings for industrial use differs from retail technical service, in that most industrial coatings are uniquely formulated for a particular customer's requirements. Typical factors that govern the type of paint formulation to be used are:
(a) Paint specification
(b) Production line speed
(c) Application equipment used
(d) Flash-off time before bake
(e) Oven capacity
(f) Post-paint requirements, such as post-forming, air-dry before packing, product abuse during assembly.

An example of the close relationship between paint performance and manufacturing parameters is as follows: If an appliance manufacturer builds a baking line based on 100 appliances per day and after five years the production requirement is 200 appliances per day, then the production line speed must be doubled to satisfy this need. If this increase in production was not envisaged in the original design, problems may arise which are apparently *paint deficiencies*. At 100 appliances per day line speed, the paint application and cure parameters could have been, for example:

(a) a 30 second dwell time in each metal pretreatment stage;
(b) a 1 minute dwell time in the dry-off oven;
(c) a 5 minute cool-down before paint application;
(d) paint application through a conventional disc at 400 mL/minute paint delivery;
(e) a 10 minute flash-off time before bake;
(f) a 20 minute bake at 175° C;
(g) a 10 minute cool-down period before unloading.

When the production rate is increased to 200 appliances per day, the following problems may arise:

(a) The dwell time in the pretreatment bath will drop to 15 seconds, which may not be enough to clean the metal thoroughly and apply the minimum film weight of iron or zinc phosphate. These shortcomings may result in craters if the surface is not clean or a lowering of properties (such as salt fog resistance) if the phosphate film weight is inadequate.
(b) The dry-off time will now be 30 seconds, which may not be sufficient to dry the article completely. This would result in uncoated metal surfaces and corrosion failure.
(c) The cooling time between pretreatment and application will now be 2.5 minutes; the metal at the time of application of the paint will be hotter, requiring a slower evaporating solvent to give the same appearance after application (for example, degree of orange peel).
(d) The required fluid delivery through the disc will now be 800 mL/minute, which exceeds the optimum delivery rate for a conventional disc (500–600 mL/minute) and will result in poor atomisation or 'slugging'.

(e) The flash-off time between application of the paint and the baking cycle is now only 5 minutes, which may cause 'solvent boil' in the oven.
(f) The oven bake time is now only 10 minutes at 175° F, which may cause a reduction in salt fog or detergent resistance. If the oven is capable of a higher temperature, then these properties may be maintained, but 'solvent boil' will again be aggravated.
(g) The cool-down time will be shortened to 5 minutes which leaves the parts hotter at the time of handling; if the paint film is significantly thermoplastic physical damage may occur.

Usually the change in production speed will not be as dramatic as that above, but will occur slowly over a number of years. The technical service representative must, however, be aware of such changes and the resultant demands on the paint system.

Obviously not all technical service problems are related to line-speed changes. Customer visits may be made for reasons such as adjustments due to ambient weather changes, modifications to products to allow for substrate variation, or familiarisation of personnel with the correct use of products. Typical changes made on line by technical service personnel are:

(a) adjustment of solvent balance to correct for ex . mple orange peel, atomisation, or solvent boil;
(b) additions to correct properties such as cratering, edge pull, pigment flotation, cure response or drying time;
(c) resin additions to adjust gloss, hardness, flexibility and related properties.

Typical 'on-line' tools usually carried by a technical service representative are listed in table 55.2.

The technical service representative, or chemist when fulfilling this role, must be aware of the capabilities and shortcomings of the product. Products can be adjusted only within certain limits: if the requirement is beyond the product capability, reformulation may be required. If this is not recognised because of inadequate product knowledge, then considerable time and money may be wasted.

TABLE 55.2
A technical service representative's kit for industrial servicing

Film-build gauge
Solvents (such as MEK to check the degree of cure)
Electrical resistance meter
Viscometer
Thermometer and a hygrometer for water-based or polyurethane coatings
Temperature recorder to plot oven temperature profiles (or temperature indicating strips)
Sample containers

56 COMPUTERS IN THE SURFACE COATINGS INDUSTRY

[*Editors Note:* This chapter provides a summary of the use of computers in the surface coatings industry. The subject alone merits a volume in its own right. References are included.—*JMW*]

56.1 INTRODUCTION

56.1.1 Background to the Use of Computers in the Paint Industry

With today's increasing emphasis on technology, greater use of computers to assist in obtaining solutions to problems, to automate parts of manufacturing processes, to do better colour matching, to develop lower-cost formulae, plus a myriad of other applications, must be increasingly expected.

Computers have been in use in some companies for up to twenty years. Their use has followed the normal commercial trend, generally starting with a central commercial batch-processing computer. Applications concentrated initially on invoicing/debtors/sales analyses and stock control. Then the trend has been to *formulae files* for mixing slips and costing purposes and control of raw materials and packaging, and to other accounting systems such as creditors' and general ledger.

The technical side of the business took second place initially, and as new generations of computer evolved and workloads increased, the problems of conversion to new and better computers also interfered with development.

In order to achieve good response (or turnaround times) for technical problems, external time-sharing services were often resorted to. However, these have the dual problems of continually escalating costs and the fact that the cost increases in direct proportion to the amount of usage.

During the 1970s the advent of *mini computers* and, subsequently, *microcomputers* provided a solution to these problems.

56.1.2 The Role of Mini Computers

The rapidly changing business scene has led to a rethinking in the use of computers. The dramatic increase in wage levels has combined with decreasing cost of computers, with the result that computers can be used economically in many places not possible a decade ago. The major effect of this is seen in the shift towards a new decentralized approach to the use of computers, particularly in Australia, with its high cost of communications and long distances. It is necessary for their effective use to set standards and monitor performance.

In the technical area, one can now justify the use of mini (and micro) computers both for research and for particular projects. These can be independent or have the capability of linking when required into a central computer system (as discussed later, the latter alternative is highly desirable).

56.1.3 Microprocessors for Manufacturing Process Control and Research

Microprocessors are, in this context, small 'computers', frequently using only one computing element 'chip', and usually dedicated to a specific application. They may be employed in resin plants to provide more consistent product quality, increased productivity and reduced labour.

Sufficient work has been done to show that microprocessors can make a large contribution to the smooth running of a resin plant. For quite modest expenditure, it is possible to achieve reduced resin plant labour, more consistent product quality and increased production (resulting from faster batch cycle times).

Some research has been done in this area in England,[1] and the use of microprocessors has started in the Australian paint industry. Two examples in use are as an interface between a spectrophotometer and a mini computer for colour matching work, and for data accumulation from a tensile tester. In view of their low cost there is little doubt that their use will spread very quickly.

With the development of the microprocessor, increasing numbers of research tools are available with inbuilt computer analysis. The areas include atomic absorption spectrophotometry, gas–liquid chromatography, high-performance liquid chromatography and nuclear magnetic resonance, where the combined machine and microprocessor enables almost instantaneous analysis of results. It is inevitable that in the future even the most simple of research tools, such as the pH meter, will be equipped with microprocessor abilities. Even now, laboratory balances with digital printout, automatic taring and zeroing and even memory capability are available. Because such devices save time (money) and effort, it is not surprising that this type of equipment, which was previously unavailable or too expensive, rapidly becomes essential.

56.1.4 The Question of Programming

The computer division of a company should set standards for systems and programming, provide the technical division with training in the application of those standards, and also provide appropriate facilities for the technical division to do its own programming for research and development projects unless appropriate manpower resources are available within the computer division at the time required. Such programming does not extend to operational systems to be used in the normal running of the business.

With the scarcity and cost of systems and programming resources, it is often hard to make a commercial case for R&D projects to be programmed by computer 'centres'. Another major difficulty often arises in the nature of the work, in that, being in a high technology area, there is often difficulty in obtaining the appropriately qualified person from the computer centre who can understand the problem sufficiently well to do the programming.

Most of computer centres' systems and programming staff are computer professionals who are most likely to be commercial in their business knowledge rather than technically orientated. R&D staff are likely to be familiar with computers and programming as a result of their own training, and therefore it would seem logical that many computer projects involving R&D work should be programmed by the technical division's own staff.

56.2 TECHNICAL APPLICATIONS

In this section, an attempt will be made to summarise the areas in the paint industry that can and do benefit from the use of computer techniques. Clearly, in this volume, it is not possible to cover all aspects in great detail, but it should alert the paint or resin manufacturer to the potential benefits available from these techniques.

56.2.1 Computers in Polymer Design

The term 'polymer design' covers a wide area of applications, including, among others, polyesters, alkyds, emulsions and polyamides. Programs have been written for manipulation of formulations for alkyds, emulsions and many others. One of the better documented areas in which computers are used is in the design and development of polyester resins, including alkyds. Several programs are available, each of which uses the same calculating basis to produce formulations with predicted constants.[2,3]

In polyester design, the formulator must avoid the possibility of gelation, or 100% reaction. T.C. Patton, in his book *Alkyd Resin Technology*[4], introduced the alkyd constants K and R to represent, respectively, the extent of reaction at theoretical gelation point, and the ratio of hydroxyl to acid equivalents. These concepts are discussed in chapter 5 (volume 1). Many of the computer programs available use these constants as the basis for formula manipulation.

The essential ingredients for a program of this nature include:

(a) a *file* containing the raw material data (physical and chemical constants); and
(b) the main calculating *program*.

Subsidiary programs are involved which, together with the above, comprise a linked suite of programs. Such adjuncts may include programs or files for:

(c) update of raw material data, especially costs;
(d) neat columnar print-out of the data base in an easily searched format; and
(e) storage and retrieval of standard formulations (*formula file data base*).

The data base for a program for polyester design contains the various data required for all expected raw materials. These data are extracted from the data base by the main program, depending on the initial input by the operator.

The following example indicates the flow chart for, and use of, a program, written by chemists, for the design of alkyd-type polyester resins. It is important to stress that this is an example only, indicative of one programming approach.

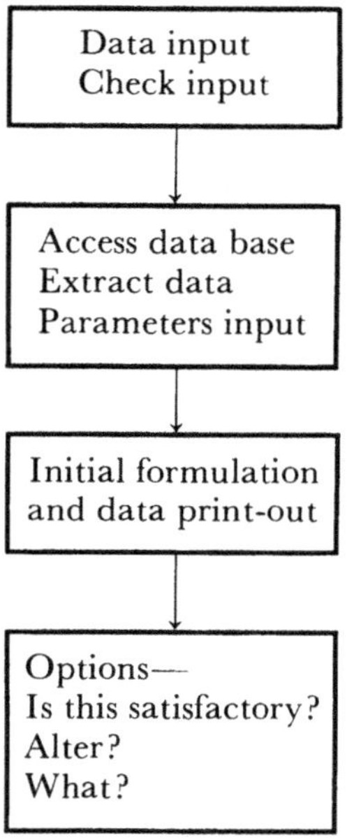

Data input may typically be 'raw material code, quantity'. The data are checked, on and/or after input, to ensure lack of typing errors, correct format.

Parameters, entered by the operator, may include request(s) for print-out of:
(a) K value (b) R value (c) Molecular weight predicted (d) Final—hydroxyl or acid values (e) Cost (f) Theoretical water loss (g) Yield

The options may include specification of K/R values, molecular weight, final acid value, final hydroxyl value, or possibly a range of values.

New data are now entered and used by the program for calculation of a new formulation based on the original raw materials, but with quantities adjusted to fit the values specified.

The following is an extract from a 'run' of a computerised polyester design program called PRPROG. Output from the computer is shown underlined, to distinguish it from operator input.

run prprog

PRPROG

1? soya, 1100
2? tofa, 700
3? rosin, 900
4? maleic, 157
5? phthalic, 1200
6? penta, 1317
7? end, 99

Here a series of inputs of 'raw material name' and 'quantities' are typed in. After checking, the computer will retrieve constants for each raw material from the data base.

CHECK INPUT—OK? Yes
NAME? 'soya polyester' This product now has a name.
OPTION? 1 First option.

The computer now produces the following data (not shown underlined for clarity):

Raw Material	Quantity	Equivalents
SOYABEAN OIL	15.69	0.107
T.O. FATTY ACID	38.51	0.133
ROSIN	12.84	0.038
MALEIC ANHYDRIDE	2.24	0.046
PHTHALIC ANHYDRIDE	17.12	0.231
PENTAERYTHRITOL	18.78	0.529
COST—C/KG	28	
REACTION WATER	5.17	
OIL LENGTH	55.8	
FATTY ACID LENGTH	53.5	
K VALUE	1.023	
MOL. WEIGHT	8593	
HYDROXYL VALUE	46	
ACID VALUE	0	

At this point, the computer program has been written to allow further changes:

OPTION? 2
HYDROXYL VALUE RANGE? 46, 76
MOL. WEIGHT RANGE? 8593, 9593

The program now prints a total of eight formulations with hydroxyl values of 46, 56, 66, 76, each at molecular weight 8593, 9593 and the relevant constants for each. The oil length, K value, and acid value are held constant. In this instance, Option 2 may be rerun a number of times to obtain a satisfactory formulation, followed by further options to input a new formulation, or to end the run.

The variations of this theme are considerable, some programs being more complex, others more orientated towards production of plant formulations, or costing/cost-reduction, but the broad basis is similar.

56.2.2 Computers in Product Formulation and Costing

One of the main problems besetting an industrial chemist is: 'What does it cost to make and will it do the job?' This is usually rapidly followed by: 'Can I make it more cheaply and still have it do the same job?'

Answering these questions involves, among other things, product formulation and costing. This may be a simple combined exercise, or quite complicated, depending on the nature of the product. Just as in polymer design, none of the calculations involved is particularly hard, but can be lengthy and tediously repetitive, and thus ideal for computerised treatments.

One application in this area arises because of changing raw material prices. A manufacturer needs regular updates on the costs of formulations, in order to fix satisfactory selling prices. This may involve the monthly calculation and print-out of hundreds or even thousands of separate formulation costings and generally requires a four-part computer program suite:

(a) a raw material file, containing the cost figures;
(b) a program for updating (a);
(c) a main formulation file, containing all standard formulations, perhaps in a 'code, quantity' format, plus other relevant data, such as yields, packaging type, production site, and production hours consumed; and
(d) the main running program, which accesses both data files and produces the updated costing print-outs.

There may be a number of subsidiary programs for:

(e) main formulation file updating;
(f) raw material cost-file listing; and
(g) main formulation file listing.

This type of suite is generally run in a number of steps. For example:

A specific, 'once-off' data file is set up by the operator, containing the codes of the raw materials for which the prices have changed, and the new prices, plus instructions for deletions, additions or changes to the main formulation file; this can be called a *transaction file*.

The main program is then run and does everything else:
(a) accesses the raw material file and the transaction file, and updates the former;
(b) accesses the main formulation file and updates it with additions, deletions or changes;
(c) runs the subsidiary programs to produce print-outs of the updated raw material and main formulation files;
(d) finally accesses both updated masterfiles, and produces the new costing print-outs for each product.

This type of operation is *non-interrogative*: the operator merely sets up the update or transaction file and runs the program. This is a classic *batch-mode* job, which may run overnight; providing the computer detects no errors in the transaction file, the whole process is automatic.

The off-line, or batch-job, approach is entirely suitable for formulations that are known to be correct and merely require some form of updating. However, the situation often arises where one or more formulas require alteration *before* a realistic costing can be obtained. This requires the use of an *interrogative*, or on-line program, which presents data to the operator and requests alterations. Such programs are often written by, or for, individual industrial concerns, as their needs vary widely.

Two examples follow of this type of application: the manipulation and costing of paint formulas; and emulsion formulations.

56.2.2.1 Paint Program

Computer programs designed for the *optimisation* of formulae, and calculation of various constants, are hardly new arrivals in the paint industry. Williams and Bacchetta[5] first described such a system in 1967, as did Briber[6] of General Electric (now Honeywell) in the same year. Two years later, the Philadelphia Society for Paint Technology (PSFPT) described a FORTRAN program[7] of considerable complexity using IBM equipment.

One of the first commercially available software packages offered to time-sharing customers of GE/Honeywell—and also certainly the first to employ the BASIC language—became available in 1970. The program was appropriately called PAINT$.[8] However, it was not until 1971 (although the package had been imported in early 1970) that Kemrez offered to Australian industry the opportunity to use the PSFPT package, largely as a sales aid. This program (or modifications of it) was subsequently adopted by a number of Australian companies; others used a GE package on time-sharing equipment.

Since then, most major Australian paint manufacturers, and many of their suppliers, have adopted computerised formulation processing techniques. Several companies can provide packages on time-sharing systems; software is also available from the larger mainframe and mini computer manufacturers.

A typical scheme is shown in the following example. Once again, it is stressed that this is only one of the many possible programming sequences.

(a) raw material data input by operator ('code number, quantity');
(b) data brought into the memory, from the masterfile, under program control;
(c) input by the operator of data not found on the masterfile, and update of latter if required;
(d) options presentation—any of:
  (i) brief formulation: code, % quantity, kg/1000L;
  (ii) density;
  (iii) non-volatile matter by mass;
  (iv) non-volatile matter by volume;
  (v) volatile solvent content;
  (vi) pigment volume concentration (PVC);
  (vii) pigment-to-binder ratio (PB);
  (viii) drier metal contents;
  (ix) brief costings-$/kg, $/L;
  (x) coverage cost in cents/sq. metre, 25μm dry film; and
  (xi) detailed costing breakdown.
(e) selection of all, none, or any combination of the above, followed by print-out of required data.
(f) selection of second-series options:
  (i) stop;
  (ii) input a new formulation;
  (iii) alter the PVC;
  (iv) alter the PB;
  (v) alter non-volatiles;
  (vi) replace one or more items in the original by new code/quantity combinations.
(g) returns to previous options.

This type of program is highly interrogative and requires many yes/no decisions by the operator. The second series options may be extremely complex, enabling the operator to produce, for example, PVC ladder formulations, while holding non-volatiles and/or selected formulation items constant.

See appendix 1 of this volume for details of the calculations required to produce constants and optimised formulas of paint systems.

More recent advances in the treatment of formulas on a mathematical basis have appeared in recent issues of the paint journals. The use of linear programming as an optimising tool appears promising[9,10] and mathematical approaches centred on the CPVC[11] and reduced PVC[12] also play an important role.

56.2.2.2 Emulsion Resin Design Program

In this case, the operation is similar to that for the paint program, but the options may typically include:

(a) alteration of monomer composition:
  (i) by percentage; or
  (ii) by specification of glass transition temperature (this presumes that one of the items on the 'raw material' data file is the T_g of the homopolymer, and that appropriate algorithms are available)
(b) alteration of non-volatile content;
(c) alteration of various component levels, including:
  (i) plasticisers;
  (ii) colloids;
  (iii) reducing agents;
  (iv) oxidising agents;
  (v) anionic surfactants; and
  (vi) non-ionic surfactants.

When the formulation is adjusted to the operator's satisfaction, the program may proceed to a costing stage. Data can be requested for:

(a) manufacturing vessel;
(b) plant site;
(c) labour hours;
(d) packaging;
(e) yields; and
(f) batch size.

The program finally produces a plant formulation, plus a detailed costing breakdown, and other relevant information such as calculated density, acid number, glass transition temperature, and so on.

These types of programs are invaluable to the formulator; the advantages of speed and accuracy are obvious.

56.2.3 Colour Matching by Computer

Colour matching is a highly specialised discipline, and its traditional problem has been its subjective nature, depending entirely upon the eyes of trained colour matchers. Pioneer work in this area was undertaken by Duncan[13,14] in 1940, but was hampered by the lack of adequate colorimetric and spectrophotometric techniques. Duncan's work was based on adaptations of the Kubelka-Munk theory[15] and led in 1957 to the design of programs for calculation of:

(a) the colours of pigment mixtures;
(b) the quantities of four given pigments required for a match to a given colour; and
(c) the chromaticity co-ordinates of binary mixtures of many pigments, to build up a background 'library'.

In 1962, Davidson and Hemminger developed the Colorant Mixture Computer (COMIC),[16] the operation of which was based on a simplified theory, using pigment absorption effects. COMIC was an analogue computer and thus restricted in its scope. Later programs of this type were mainly developed on digital computers, which are able to handle a wider range of colours.

Because of a number of factors, it is doubtful that the computer will ever completely replace trained colour matchers, as errors may arise in variations in illumination, substrate and fluorescence if present. However, a number of correcting matches can be made, which generally give satisfactory results. These concepts are covered further in chapter 46.

56.2.4 Information Retrieval

Information retrieval can be achieved by providing access to a computerised file of technical literature and patents, which are 'keyworded' for cross-referencing and retrieval. There is great potential benefit in retrieving promptly—and completely internally generating—such technical information as exposure results, technical reports, etc. Once a file of such information has been keyworded and created, one of the needs that arises is to produce and circulate a regularly updated index of keywords and summaries.

As computer storage costs continue to drop, more and more use will be made of computers for information retrieval and other applications where large-memory capacity is required.

56.2.5 Raw Material and Formula File Data Bases for Non-technical Applications

It is generally anticipated that all research projects will lead either to practical applications within company operations or to new or better products. From a planning point of view, it is desirable to ensure that all technical work stays close to the marketing/commercial environment; and that the requirements for eventual successful implementation are kept in mind from the outset and, where possible, the eventual environment is the one that is used for development. This is nowhere more critical than in the use of computers.

It is obvious that systems and programming resources are a scarce and valuable resource. Historically in most companies there has always been a much greater total workload requested than available resources. One of the biggest problems in EDP today is 'Why does it take so long to get this implemented?'.

Once a job is programmed it is likely to be used for anything from five to fifteen years. The cost of maintenance is significant. The key costs in deciding whether computerisation is viable are the one-time systems and programming cost and implementation cost (file creation, education/training, and parallel running). The running cost of most jobs is relatively small. The one-time development cost has to be recouped over a period of five to fifteen years, and the resulting programs affect the running of the business for that period. It is very costly to change a computer system once it has been implemented.

There is therefore a need for concern about the integration of systems design from the outset of all technical projects. Similarly the data base of information to be stored in computers that the technical division requires must be integrated from the outset with the production, purchasing and finance systems of the company for most effective results.

Two major areas have already been discussed, namely the formula file and the raw materials file. It is important that all areas of the business use the same information when decisions are made. Planning should therefore work towards one raw-material data base in each company, such that it can be used by the finance department for the payables system, the purchasing department for buying data, stock control personnel, and of course technical staff for developing and changing formulae. The technical division also has the responsibility for a raw-material register and codes.

Similarly the formula file is critical to the total operation of the business. It should be set up as a common data base of information to be used throughout the organisation, to avoid duplication of files and of input activities in creating and maintaining information.

The formula file is vital to costing, production (for checking availability of materials), purchasing (for determining material requirements), financial planning (for integration of financial business plans with production, purchasing division's stock control plans), and of course for the technical division's own use in developing and improving formulae.

The establishment of a computer formula file is essential to the better management of the paint business:

(a) to enable the impact of any proposed change (such as raw-material substitution) to be fully assessed (in terms of stocks, formula costs and impact on quality or effectiveness of products);
(b) to determine what materials are really required, compared to what actually is in stock;
(c) to assess the implications of cancelling a formula (what materials in stock are no longer required, advice to purchasing on what not to buy, a check to see what other factories could use the spare materials, and so on).

It is therefore critical that any proposed use of computers by the technical division of the company that involves formulae or raw-material information be developed on an integrated basis with the rest of the company. The initial creation of formulae cannot be done in a vacuum. It is wasteful of resources and increases the risk of errors to have to later re-input formulae into costing, production or finance systems.

It can be argued that the technical/R&D function should not have its own independent computers, because of the advantages of being totally integrated with the other functions of the business.

A specific goal should be to set up one computerised formula file for each company, accessible (with appropriate control) to all divisions. The technical division has a significant input to this key file, which is also needed by production, purchasing and finance. The technical or R & D division should be responsible for the initial creation of the basic formulae which are on this file. This one computer file can have tremendous impact on the whole business, by providing the tool to control stocks, prices and redundant materials.

A totally computerised raw-material data base containing edited specific information for access by formulating personnel—and which should include current costs to allow computation of minimum-cost formulations—is also a worthwhile aim.

This leads to two separate, but related computer processing needs within an organisation:

(a) The first is to be able to calculate the numerical attributes of a given trial formulation (including costs), as an aid to the formulator in determining his or her next experimental actions.
(b) The second is to be able to change one or more constituents of an existing operational formulation and, having done this, to produce amended mixing slip masters for all products affected by the change. A related situation could be a request to pro-rata the initial formulation quantities up to their '1000 litre' equivalents for production usage before a mixing slip is printed.

56.3 FUTURE TRENDS IN THE USE OF COMPUTERS

Experience with computers thus far suggests that computer applications will in the future be implemented to a larger extent by technical staff themselves rather than by computer professionals. The inefficiencies inherent in doing so will cost less with modern 'minis' and 'micros', as the computing power per dollar increases and as programming becomes easier.

This trend will continue, as all types of computer grow in use. Online terminal computer systems are essential in today's environment, and easy-to-use programming language is very clearly indicated with the ever-increasing shortage of skilled programming staff. Access to computer data bases containing details of all product formulae and all raw-material technical data, costs and stock information will gradually become more commonplace.

Apart from such inhouse company data bases, there are now data bases available to the public containing raw-material data and technical references.

REFERENCES

1 J S W Smith in *The Application of Microprocessors to the Synthetic Resin Industries* Paint Research Association U.K. 2 August 1979
2 D C Finney *J Paint Technology* 1971 **43** (556)
3 L J Lombardi and W R Tasker *J Aust Instit of Man* 1973 **51** 20
4 T C Patton *Alkyd Resin Technology* Interscience Wiley New York 1962
5 Williams and Bacchetta *J Paint Technology* 1967 **39** (508) 267
6 Briber and Sheerin *J Paint Technology* 1967 **39** (515) 728
7 PSEPT R G Alexander Chairman 'The Computer and Formulation Design' *J Paint Technology* 1969 **41** (528) 40
8 *General Electric Bulletin* PAINT$ program
9 P E Kavanagh *JOCCA* 1978 **61** 146
10 Wu Dao-Tsing *Application of Mathematical Modelling . . . in The Development of Coatings Formulation* (available on application from Du Pont)
11 A Ramig Jnr *J Paint Technology* 1975 **47** 602
12 Bierwagen and Hay *Prog Org Coatings* 1975 **3** 281
13 D R Duncan *Proc Phys Soc* 1940 **52** 390
14 D R Duncan *JOCCA* 1949 **32** 296
15 P Kubelka and F Munk *Z Tech Phys* 1931 **12** 593
16 D L Tilleard *Res Assoc Br Paint Colour and Varnish Man Bulletin* 1964 **13** (68) 223

57 STANDARDISATION, INSPECTION AND ACCREDITATION IN AUSTRALIA

57.1 INTRODUCTION

It should be recognised that a satisfactory standard of manufacture depends on the degree of conformity of a particular article or product with the design as described by a drawing, written specification or other standard. The article may be a can of paint, or it may be a motor vehicle or it might be a drawing pin. For the article to be acceptable, not only must it comply with the specification, but the specification itself must also be appropriate.

During the First World War, Australia's inability to assist England by providing munitions as well as primary products had been a source of disappointment to leaders of government and industry. Engineers and technologists generally saw that this inability to co-ordinate manufacturing processes was largely due to a lack of standardising facilities and that this labelled Australia as an undeveloped country.

One of the more progressive and significant steps in Australia's industrial development came in 1922 with the formation of the Australian Commonwealth Engineering Standards Association. This was followed in 1927 by the formation of the Australian Assoication of Simplified Practice, and these two bodies were amalgamated in 1929 to form Standards Association of Australia.

57.2 STANDARDS ASSOCIATION OF AUSTRALIA

SAA is the recognised national organisation for the promotion of industrial standardisation in Australia. The Association publishes standards that deal with raw materials, components and finished products; and with design, construction, testing, quality or performance.

The Association operates entirely on a non-profit basis. Its funds are provided by government grants, membership subscriptions and proceeds from the sale of publications. Companies, firms, organisations and individuals are eligible for subscribing membership of the Association. Such membership is distinct from committee membership, which is not contingent upon any direct financial contribution.

The headquarters office of the Association is located in Sydney and there is a major office also in Melbourne, with branch offices in Adelaide, Brisbane, Hobart, Perth and Newcastle.

SAA is governed by a Council representing a wide range of interests—governments and government departments, industry associations and professional bodies.

57.2.1 Committee Structure

The general pattern of SAA committee organisation is shown below:

Council

Standards Boards

Building | Electrical | Mechanical | Chemical | Consumer | (and others)

Technical committees

Gas cylinders | Boilers | Fasteners | (and others)

Sub-committees and working groups, as needed

57.2.2 Standards Boards

A Standards Board acts on behalf of Council in supervising the activities of technical committees within the sphere of industry concerned. Its major functions are to decide what projects should be undertaken; to allocate these projects to appropriate technical committees; to settle the terms of reference, constitution and chairmanship of committees; and to approve standards for publication.

57.2.3 Technical Committees

Technical committees are the basic groups responsible for the preparation of standards. Some committees are concerned with a single standard or a small group of standards; others may be responsible for a wide range of standards in a given field. The terms of reference of a technical committee are established by the Standards Board to which it is responsible.

Technical committees are formed on a representative basis by seeking nominations from associations or organisations, but not from individual firms. As far as possible the constitution of a committee is on an Australia-wide basis and provides for representation from all interested groups concerned with the subject, such as:

Users and purchasing bodies	(through trade associations and government departments)
Makers and suppliers	(through Chambers of Manufactures, etc.)
Regulatory or controlling authorities	(usually government departments)
Independent professional research bodies	(through professional institutions, CSIRO, universities, etc.)

Size of committees is usually a compromise between a reasonably broad basis of representation and the need to restrict membership to workable numbers.

Committees may form sub-committees, panels or working groups for convenience in drafting or for dealing with particular specialised aspects of their work. Apart from the principle that they are responsible to the main technical committee that appointed them, there is considerable flexibility in the mode of formation and operation of such subsidiary groups.

57.2.4 Preparation of Standards

Following approval of a project, the usual procedure is the formation of a representative technical committee, and then the committee members are responsible for establishing the content of the standard. The aim of technical committee work is to obtain and express a genuine consensus of expert opinion, in the form of a standard which is practical, realistic and acceptable to industry. The procedure for this is through meetings of the committee arranged by SAA. The normal stages in the preparation of an Australian standard are:

(a) preliminary draft;
(b) committee draft or drafts;
(c) draft for public review;
(d) draft for postal ballot.

The procedure after postal ballot depends on the result of the voting, particularly on the nature and extent of any negative votes. If necessary, comments arising from a ballot may be referred to the technical committee concerned.

Final approval of a draft for publication as an Australian standard is the responsibility of the relevant Standards Board. A full report on the voting and on any subsequent negotiations is submitted to the Standards Board, which gives careful attention to the reasons for any negative vote that may have been received.

Once published, a standard is always open to review to take account of experience in its use or of technological developments. Proposals for modification of a standard may be submitted at any time, by committee members or by anybody using the standard. Such proposals are referred to the technical committee and may lead either to the issue of amendments or to the preparation of a revised edition of the standard.

57.2.5 Application of Standards

SAA specifications, test methods and codes gain recognition as national standards from the fact that they are prepared and accepted by all interested parties represented on the drafting committees. As a general rule, standards derive authority from voluntary adoption based on their intrinsic merit.

Where, however, a standard is concerned with matters affecting safety of life or property, it may find compulsory application through reference in statutory regulations.

The voluntary adoption of standards may be effected in several ways. For producers, it is a matter of a decision to make products in compliance with a relevant standard and to use standard terminology, tests and specifications in quality control, in quotations and in marketing. For purchasers, it is a matter of citation of standard specifications as a condition of tendering and contract.

For standards such as those relating to drawing practice, symbols, glossaries and test methods, any organisation may issue internal instructions requiring that standard practices be followed. Another mode of application of standard specifications is through the use of the Australian Standards Mark.

57.2.6 The Australian Standards Mark

The Standards Association of Australia is the owner of the mark, which has been registered as a certification mark under the Commonwealth Trade Marks Act. It is available, under licence, to any manufacturer who can satisfy the Association that he or she is producing articles that comply with an appropriate published Australian standard, and who is prepared to enter into undertakings which include observance of general rules and of specific supervision and control requirements. More than 500 licences to use the Mark have been granted to manufacturers in the fourteen years that the Mark scheme has been in operation.

The Mark benefits the manufacturer and the purchaser. For the manufacturer, it is an assurance to customers that the goods meet a given standard; it provides protection against sub-standard products; and it reduces the need for special testing. For the purchaser, it provides an assurance of continuing production of goods to a standard; under an effective system of supervision and production control, it reduces the amount of check testing; and it provides an assurance that goods are fit for their purpose and will be reliable in service. In addition the national economy benefits through the encouragement of more effective production supervision and quality control in manufacturing industry, through the release of technical personnel from routine checking, and through the provision of an assurance of quality for exported products.

57.3 OTHER STANDARDISING ORGANISATIONS

The other major standardising authorities are largely governmental. These include federal bodies such as Departments of Defence, Housing and Construction, Transport, Health, Primary Industry, Telecom Australia and the Australian Motor Vehicle Certification Board. The various state bodies include Departments of Public Works, Government Stores, Railways and the various Electricity Authorities and Water Boards. There are others of course, but these will serve to illustrate the various ways in which governments prepare and use standards.

The Department of Defence is responsible for the publication of a range of specifications which are labelled DEF (AUST) specifications. These cover a tremendous range of products, and where possible they make reference to SAA specifications and test methods but also make extensive use of British Standards, DEF (U.K.), ASTM, U.S. Federal and U.S. Military specifications. Similarly, Navy, Army and Air Force specifications have a variety of sources.

Telecom Australia issues its own specifications, and this is understandable in view of the highly specialised nature of its operation.

Early in 1970 the Australian Transport Advisory Council established the Australian Motor Vehicle Certification Board. The object of this action was to simplify the procedures necessary to satisfy the individual state and territory Registering Authorities for compliance of a motor vehicle with the Australian Design Rules for motor vehicle safety.

The various state departments and instrumentalities issue their own specifications, but adopt the principle of using SAA specifications where possible.

57.4 INSPECTION

Inspection does not imply merely visual examination. Inspection incorporates the whole assessment of an article, including the testing and the more subjective elements, right through to the final decision to accept or reject.

The governmental inspection systems must play an important part in a discussion of this nature; in this chapter all that is intended is to have a brief look at some of the inspection systems that operate in order to illustrate the variety of systems available.

57.4.1 Service Inspection Systems

The three services—Army, Air Force and Navy—each have inspection branches which in some areas overlap. There is the Army Inspection Service, Directorate of Quality Assurance (known as DQA), and Naval Inspection. All three operate their own laboratories and have easy access to Department of Defence testing facilities. Departmental laboratories, however, are restricted to check testing, research and development.

Formal independent inspection of military stores commenced in 1888 with the inspection of ammunition produced by the Colonial Ammunition Company, by officers of the Victorian Permanent Artillery. The Australian Army Inspection Service was actually born in 1912. At that time, ammunition and weapons were the main interest; since that time it has grown, and its activities now cover the inspection of all Army stores.

It was during the 1950s that some direct responsibility was placed on manufacturers to control quality of their products. *Army (Aust.) 448* was issued in 1966, and this specification lays down inspection requirements and conditions for contractors supplying to the Army. Supervision of quality control procedures in manufacturers' plants is undertaken by the staff of the Army Inspection Services.

The activities of DQA are of much more recent origins and date from 1939. Today the Directorate operates an Approved Firms system. The DQA approval of companies is in three categories:

Part 1—Approval of manufacturing facilities
Part 2—Approval of warehousing facilities
Part 3—Approval of laboratory facilities

Part 1 approval involves an approval of a plant and its systems for control of quality. Part 3 approval involves operating a laboratory to the requirements for registration with the National Association of Testing Authorities. Frequently a particular plant will hold both Parts 1 and 3 approval.

Naval inspection is much more complex, in the sense that there are several naval inspecting authorities: Naval Ordnance Inspection, The Overseers and Superintendents of Inspection, Naval Air Engineering Overseeing Organisation, Inspectors of Victualling Stores, and Medical Branch Staff.

57.4.2 Other Governmental Inspection

Telecom Australia and the Department of Transport also operate Approved Firm schemes. The Department of Civil Aviation relies heavily on the approvals granted by DQA.

Telecom operates its own laboratories but also relies on manufacturers' facilities. Its scheme differs in points of detail from the others already mentioned, but it is similar and has the same basic philosophy—producing confidence in the goods supplied by a manufacturer.

In specialised areas, the defence services and Telecom may be obliged to accepts goods from non-approved firms. In these instances, testing has to be undertaken in departmental laboratories or other approved laboratories.

The state government departments are too numerous and their approach to the matter of inspection is too varied to consider in any great detail. Generally speaking, the states are not involved in the same amount of purchasing as the Commonwealth, and many of their activities are simply regulatory.

57.4.3 Government Paint Committee

The inspection system of greatest interest to the surface coatings industry is that operated by the Government Paint Committee (GPC). This committee has the responsibility of co-ordinating the specification, procurement, inspection, testing and use of paints and related materials by the participating Commonwealth and state government departments and instrumentalities.

The scheme serves the needs of Commonwealth departments (principally Defence, Housing and Construction, and Transport), the state government paint groups, and other government instrumentalities such as the Australian Atomic Energy Commission, British Phosphate Commissioners, Commonwealth Scientific and Industrial Research Organization (CSIRO), and the National Capital Development Commission.

The GPC paint approvals scheme ensures that paints of proven performance are identified and approved for use by its members. The GPC first issues recognition to a paint manufacturer, then grants approval to the manufacturer's products which have been shown to meet the relevant specifications. A List of Approved Products is issued to participating organisations, which then arrange for the purchase and use of paints to meet their needs. Inspection of supplies of approved paints is the responsibility of the purchaser.

The GPC grants recognition to a manufacturing unit when it is satisfied that the technical competence and facilities for manufacturing and testing are such as to ensure an on-going ability to control and maintain the quality of production within the specified limits.

After gaining recognition, a manufacturer is entitled to seek GPC approval for individual products. It is the responsibility of the manufacturer to test or have tested against the relevant GPC specification the products the manufacturer wishes to submit to the GPC for approval. Only material conforming to the requirements of the specification can be submitted. Under provisions safeguarding the approvals scheme, GPC may require check testing of the submitted sample. Once it is satisfied that the product meets the specification requirements, the GPC grants approval to the product and adds it to the List of Approved Products.

The List of Approved Products lists all products approved by the GPC and is amended from time to time as required. It is issued to the participating government departments and instrumentalities and is for their use only.

Specifications are issued by the Committee under the title 'Government Paint Committee Specification'. Each specification is for the manufacture, approval and supply of a type of paint; it shows appropriate usage conditions as well as the technical and test requirements. These specifications are largely based on those produced by the Standards Association of Australia (where appropriate), with additional requirements on composition or performance to meet the needs of the user organisation.

57.5 ACCREDITATION

Testing in Australia is conducted in laboratories operated by government authorities and instrumentalities, by companies operating manufacturing plants and by various public testing organisations.

Australia is in the fortunate position of having an organisation known as the National Association of Testing Authorities. This organisation provides a national testing service through a network of accredited laboratories throughout Australia. These laboratories undertake the complete range of testing of products and materials; the organisation offers to buyers of testing the assurance that testing is performed to a uniformly high standard.

57.5.1 National Association of Testing Authorities

In 1936, Federal Cabinet established a committee to advise the Government on extending the activities of the CSIR into Secondary Industries Testing and Research. This committee was known as the 1937 Secondary Industries Research Committee, and it saw that Australia had a problem of availability of testing facilities because of the sparsity of population and limited technological resources.

This Committee, in its report published in 1938, recommended among other things that Australia establish reference standards of measurement and associated calibration services. This proposal led to the establishment of National Standards Laboratory in 1939. The committee also recommended that Australia establish a comprehensive testing and certification service to meet normal industrial and commercial requirements.

During the Second World War the Department of Munitions Supply established a system of approval of laboratories called the Approved War Time Test House Scheme, and the successful operation of this scheme demonstrated that a co-ordinated testing service was feasible.

In 1945 CSIR again considered the recommendations of the 1937 Committee. After discussion between the various Commonwealth and state departments, the organisation now known as National Association of Testing Authorities was formed. It formally came into being in April, 1947.

The first laboratories to gain registration were Defence Research Laboratories, which are now known as Materials Research Laboratories. These were followed by the Western Australian Government Chemical Laboratory, and the first industrial laboratory to gain registration was the Cabarita NSW laboratory of British Australian Lead Manufacturers Pty Ltd (now Dulux Australia Limited).

NATA is an association of people and organisations that operate testing facilities. The basic philosophy of the Association is that membership is open to all operators of laboratories, provided only that the work they do is performed in a technically and ethically satisfactory manner. Technical competence and integrity are the essentials and are the basis for formal definition of the detailed requirements for membership.

The fundamental elements that determine a laboratory's abilities are:

(a) the quality of the staff;
(b) the availability of appropriate equipment that is satisfactorily calibrated and maintained;
(c) a high standard of laboratory practice, embracing sampling procedures, test methods used, supervision of staff, laboratory records and test reports;
(d) the accommodation available.

The Association is governed by a Council which consists of representatives of the members (that is, the registered laboratories), representatives of the Chamber of Manufactures, representatives of the Royal Australian Chemical Institute, the Australian Institute of Physics and the Institution of Engineers Australia, as well as representatives of the Commonwealth and state governments. NATA grants registration in nine fields of testing:

- Acoustic and vibration measurement
- Biological testing
- Chemical testing
- Electrical testing
- Heat and temperature measurement
- Mechanical testing
- Medical testing
- Metrology
- Nondestructive testing
- Optics and photometry

At 31 December 1983, there were nearly 1350 Australian laboratories holding NATA registration. Of these, 70% were industrial laboratories, 20% were operated by Commonwealth or state government departments or instrumentalities, 8% were commercial testing organisations, and the remainder were operated by universities or other educational institutions.

For each field of testing, the Council has appointed a Registration Advisory Committee which has the responsibility of supervising initial assessments of laboratories and maintaining surveillance of those already registered. Registration Advisory Committees co-opt other experts in particular fields to assist in assessment of laboratories. The system is therefore a voluntary one and relies on the co-operation of laboratory personnel, users of testing services and the technical community generally.

There are many competent laboratories operating in Australia that do not belong to the Association. However, it is generally accepted that membership of NATA provides evidence of competence in specific areas. All major inspecting authorities mentioned earlier rely on the network of laboratories provided by NATA to provide testing services associated with other elements of inspection. Practically all laboratories operated by such bodies are themselves also registered.

58 STATUTORY REQUIREMENTS OF THE PAINT INDUSTRY

The aim of this chapter is to outline the relevant Acts and regulations that apply to the paint industry. The subject will be examined in a broad sense only; the relevant regulations should be studied in detail before applying the principles to a particular case. Copies of these Acts and regulations are readily available from State and Commonwealth government printers.

58.1 POISONS

Probably the most widely known poison in paint is *lead*. Lead is a cumulative poison and may result in permanent disability or death.

An analysis of admissions to two Sydney hospitals over a twenty-year period revealed that ninety children had lead poisoning, and of these, seven died and 30 per cent were permanently affected. Eating old paint flakes containing lead was suspected as a major source of poisoning in these cases. Consequently there are now regulations controlling the use of lead and other poisons in paint.

58.1.1 The Uniform Paint Standard

The *Uniform Paint Standard*, as produced and periodically amended by the *National Health and Medical Research Council* (NHMRC), forms the basis for uniform legislation controlling poisons in paint. While it has no proper legal standing (the necessary legislation being the responsibility of the individual state parliaments), it does reflect quite accurately current medical opinion on the use of paints containing poisons.

This document is reproduced in Appendix 58.1 and an examination of it will indicate the severe restrictions placed on the use of lead-based paints. In addition, paints containing lead and other poisonous materials, such as mercury compounds and chlorinated solvents, must be labelled to indicate the hazards they present.

58.1.2 The Uniform Poisons Standard

There are a number of products that are generally manufactured and sold in association with paint products; these include thinners, catalysts, polishes and paint strippers. The *Uniform Poisons Standard* applies to these products. It is published by the NHMRC and is the basis for uniform legislation controlling the sale and use of products other than paints that contain poisons.

The Uniform Poisons Standard classifies poisons into eight groups depending on their hazard. These groups are known as *schedules*, and many of the poisons within Schedules 5 and 6 are commonly used in the paint industry. Schedule 5 covers substances or preparations of a hazardous nature which might be readily available to the public but which require caution in handling, use and storage. This schedule includes such substances as mineral turpentine, methylated spirit and methylene chloride. Schedule 6 covers substances or preparations of a poisonous nature which might be readily available to the public for domestic, agricultural, pastoral, horticultural, veterinary, photographic or industrial purposes or for the destruction of pests. It includes such substances as toluene, xylene, and isocyanates.

The standard also requires that products containing scheduled poisons are labelled and packaged appropriately.

58.1.2.1 Labelling

For Schedule 5 substances, the following warning must appear on the main face of the label:

WARNING

KEEP OUT OF REACH OF CHILDREN

For Schedule 6 poisons, the following warning must appear on the main face of the label:

POISON

NOT TO BE TAKEN

KEEP OUT OF REACH OF CHILDREN

READ SAFETY DIRECTIONS BEFORE OPENING

It is also mandatory for the following information to appear on the label:

The name and proportion (%, g/L, etc.) of each poison present in the product
The name and address of the manufacturer
Handling and first aid instructions appropriate to the poisons present in the product

58.1.2.2 Packaging

The immediate container in which a poison is sold shall:

(a) be impervious and non-reactive with the contents;
(b) be strong enough to prevent leakage or breakage during normal handling and transport;
(c) be securely closed and capable of secure reclosure (except for single-dose packs);
(d) have sufficient excess capacity (ullage) to prevent breakage because of the development of internal pressure under normal conditions of handling and transport; and
(e) be readily distinguishable from a container in which food or drink is sold.

This last requirement is achieved by the requirements:

(f) for Schedule 5 poisons, if the label is not permanently printed on to the container, by the container being embossed or permanently printed with NOT TO BE TAKEN or NOT TO BE USED AS A FOOD CONTAINER or POISON;
(g) for Schedule 6 poisons in cans, the marking must be POISON unless the label is lithographed. Plastic or glass containers must be the poisons bottle or jar specified in Australian Standard AS 2216.

58.2 HEALTH AND SAFETY

There is a great diversity of regulations that are designed to ensure the safety and well-being of employees and the users of paints.

58.2.1 Health and Welfare Provisions

Health and welfare provisions control the following items:

Abrasive blasting
Asbestos work
Atmosphere—discharge of impurities
Changing rooms
Drinking water
Electroplating
First aid and health facilities
Lead processing
Lighting
Maintaining cleanliness
Meal accommodation
Noise levels and protection from noise
Rest rooms
Sanitary accommodation
Seating in workplaces
Space per person
Spray painting
Ventilation
Washing accommodation

58.2.2 Safety Provisions

Safety provisions control the following items:

Access to and egress from buildings
Buildings and structures—safety
Use of compressed air
Entering of confined spaces
Dangerous enclosures
Electrical installations
Use of explosive powered tools
Falls of persons—protection
Fire safety
Use of flammable solutions for dipping
Floors
Flues and the guarding of heating appliances
Foundries
Fragile roofing materials
Safe use of harmful substances
Guarding of machinery
Manual handling of materials
Material stacks and containers
Personal protective equipment
Power presses—safety inspection
Refrigerated compartments
Testing closed vessels under pressure
Traffic control
Welding safety
Window cleaning
Working alone

To deal with all these subjects in depth is beyond the scope of this book, but the regulations relating to the protection of workers from atmospheric contamination in the workplace are worth describing in detail.

58.2.3 Atmospheric Contaminants in the Workplace

The NHMRC has published an *Approved Occupational Health Guide . . . Threshold Limit Values* (1981). It includes the full list of *threshold limit values* (TLVs) of the American Conference of Governmental Industrial Hygienists (ACGIH), with additional recommendations made by the NHMRC. While NHMRC occupational health guides have no automatic legal standing unless adopted by reference in state legislation, they must be regarded as expert advice. They would be regarded as reasonable standards in any civil action (e.g. worker's compensation claim). Without expert legal and technical advice to the contrary, they should be treated as if law. A summary of the major points follows.

The exposure of a worker to an airborne chemical should be controlled by reducing the concentration of the chemical in his or her breathing zone to a level at which he or she is protected from the adverse effects of the chemical. This recommendation lists levels (TLVs) for most of the chemicals found as contaminants of the air of the workplace.

The TLVs are intended as *guides* only in the control of health hazards and not as fine dividing lines between *safe* and *dangerous* concentrations. The data on which the standards are based are not accurate enough to warrant dispute about slight deviations (± 20 per cent) from the values given. Good occupational hygiene practice requires exposure to be maintained well below the standard, rather than at the standard.

In setting the TLVs, exposures are assumed to take place over an eight-hour work shift in the absence of severe physical stress, and to be followed by an exposure-free period of 16 hours or more during which detoxification occurs. When increased absorption of contaminants is promoted by working overtime, or by increased breathing accompanying hard physical labour, some adjustment of the TLV can be necessary. Similarly, adjustments could be required when exposure is accompanied by excessive heat, humidity or hyperbaric work.

The values are based on various criteria of adverse effect, including tissue changes, discomfort, and mild narcosis, which may increase accident risk. The values should not be regarded as a common denominator of toxicity in the calculation of so-called *toxicity ratios*, nor should they be considered as the sole criterion in the diagnosis of occupational disease.

Wide variations exist in individual susceptibility, and a small number of individuals may react at or even below the recommended standard.

The atmospheric contaminant concentration inhaled by an employee seldom remains constant during a working day. A typical day's exposure consists of concentration peaks superimposed on a lower-level background; the relative contribution of each needs to be decided in assessing the significance of variable exposures. The concentration peaks of fast-acting compounds such as irritants and narcotics can cause ill-effects in less than 15 minutes, and there are other compounds that can produce chronic or irreversible tissue changes after brief exposure to high concentrations. In both cases there is need to control peak exposures. Concentrations (prefixed by the letter 'C') should not be exceeded, even instantaneously. Short-term exposure limits (tabulated under the heading 'STEL') are allowed for periods of up to 15 minutes, spaced by at least one hour, and occurring no more than four times in an eight-hour shift.

Other listed values are time-weighted average (TWA) concentrations. To use these in assessing worker exposure, the varying concentrations to which an employee is exposed during the different activities of a full work shift have, theoretically, to be averaged according to the formula:

$$\text{Time-weighted average, } \bar{C} = \frac{(C_1 \times t_1) + (C_2 \times t_2) + \cdots + (C_n \times t_n)}{t_1 + t_2 + \cdots + t_n}$$

where t_1 is the time in hours spent in activity 1 in a contaminant concentration of C_1, t_2 is the time in hours spent in activity 2 in a contaminant concentration of C_2, and so on; and the sum $(t_1 + t_2 + \cdots + t_n)$ is close to eight hours.

In practice, application of the above formula should not be carried to extremes (for example, to an exposure of only one hour per day), and concentration peaks should not be neglected even though the time-weighted average is considered the appropriate parameter. The amount by which a time-weighted average can be exceeded for short periods without injury to health depends upon a number of factors, such as the nature of the contaminant, whether very high concentrations produce acute poisoning on brief exposure, whether the effects are cumulative, the frequency with which high concentrations occur, and the duration of such periods. As a guide, when C or STEL is not listed, peaks should not exceed three times the TWA limit. All factors should be taken into consideration in arriving at a decision as to whether a hazardous situation exists.

Atmospheric contaminant mixtures pose special problems of interpretation for which no simple rule is available. The physiological effects of two or more substances inhaled together might be exerted independently, additively, synergistically, or antagonistically. Specific consideration, in consultation with the literature, should be given to each mixture encountered.

The notation 'skin' placed after the listed compound indicates that it is capable of penetrating the intact human skin to the extent that the dose from skin absorption could approach or exceed the dose inhaled. The notation calls attention to the fact that under such conditions the TLV could be invalidated, unless skin absorption is prevented.

The values are intended for occupational hygiene application and for interpretation by persons trained in that discipline. They are not to be used or modified for use in the evaluation of health hazards caused by community air pollution. On this subject, reference can be made to the NHMRC recommended *National Emission Standards for Air Pollutants.*

The concentration of the contaminant present in the worker's breathing zone should be determined by sampling and analytical techniques so that the results yielded are capable of being interpreted in terms of the appropriate hygienic standard. Sampling, therefore, should be carefully planned.

Professional occupational hygiene judgement should always be exercised in the planning of a sampling programme in order to avoid error arising from poor selection of sampling technique, from inappropriate location of samplers and timing of samples, and from unnoticed variations in work procedures and processes.

Instrumentation suitable for measuring atmospheric vapours can range from relatively inexpensive Drager tubes through to sophisticated techniques involving gas chromatography, infra-red spectrophotometry, etc. The assessment of the health risks of dust must take into account the chance of it settling in the lungs. Coarser particles tend to impinge on surfaces in the nose and throat, and so do not reach the lungs. Very small particles may stay in suspension and be breathed out again. Ideal air sampling is from in front of the worker's face—that is, the air the worker is actually breathing—but the inconvenience to the worker and the cost are important factors against taking ideal measurements.

The threshold limit values of chemical substances are reviewed twice a year, and reference should therefore be made to the most recent publication.

58.3 FLAMMABLE MATERIALS

58.3.1 Fire Prevention and Protection

A fire requires the presence of air, fuel and an ignition source. The potential for all three occurring in a paint plant is relatively high. Extreme care is required to avoid any situation where a source of fuel and a source of ignition are available.

Flash point is defined as *the temperature of a solvent at which the solvent gives off vapour at a rate sufficient to form an ignitable mixture with the air at or near the surface of the liquid.* Ignition of a solvent vapour results in expansion of the gases and can be explosive in confined areas. This stresses the need for ventilation, which is one of the most important factors in keeping the fire hazard low. Probably the greatest potential source of ignition in the paint industry is static electricity. One of the major sources of this can be the break-up of an anhydrous liquid into small particles or a spray—toluene is particularly dangerous. Consequently in storage tanks the inlet pipes are extended to the bottom or sides of the tank to reduce the free fall. There is also a need to earth the tanks and mixers into which solvent is being emptied from a pipeline. Static electricity is more likely to build up when humidity is low. High moisture content in the air allows deposition of a moisture film on belts, floors and equipment, allowing drainage of static electricity to earth. This is said to occur at about 60 per cent relative humidity. Humidity has no effect, however, on static build-up of liquids broken into fine particles. Earthing of equipment is the only sure way to overcome the problem.

The use of non-sparking (non-conductive) equipment for cleaning floors and other equipment is essential, especially when handling nitrocellulose. Nitrocellulose is normally wetted with alcohol, which reduces its danger potential; however, when it is allowed to dry out, it becomes much more flammable and explosive. Any spills must be cleaned up and residues must be dissolved in solvent. The lids of all containers holding nitrocellulose must be retightened after use, and the rims of containers cleaned.

In the laboratory, it is good practice to follow these procedures:

(a) Soiled rags and paint waste must be removed daily. Soiled rags should be immersed in water.
(b) Containers of cleaning solvents must be kept closed. Laboratory stocks of flammable materials should be minimised by regular clean-up of storage areas.
(c) Adequate ventilation must be provided.
(d) Employees must know the location and use of fire extinguishers as well as their company's drill.

Effective means of extinguishing fires are available. Water can be used, mainly to absorb much of the heat and dilute the ignited material. It is most effective as a spray, but too much water or too strong a force of water can spread the fire. It can be used to blanket the fire if its density is lighter than that of the burning solvent or paint. By far the best types of extinguisher for flammable liquid fires are the carbon dioxide, dry chemical, or foam types.

58.3.2 Other Statutory Requirements

In all Australian states there is legislation separately controlling the *storage, conveyance, packaging and labelling of dangerous goods including gases, flammable liquids, poisons and corrosives*. The intent of each piece of legislation is basically the same, and the most important features are summarised below.

58.3.2.1 Storage

All stores must be licensed and constructed such that, in the case of a leakage, all materials are contained within the area.

Storage areas must be separated—dependent on the quantity of and hazard presented by the stored materials—from buildings in which people are working or meeting and from neighbouring properties. Appropriate signs and notices must be displayed in the storage area indicating the nature of the goods kept and the precautions to be taken. Fire protection facilities such as sprinklers must be provided.

58.3.2.2 Conveyance

The requirements for the conveyance of dangerous goods are based on the Australian Code for the Transport of Dangerous Goods by Road and Rail prepared by the Australian Department of Transport. The Code is being adopted into state legislation, with the intention of uniform legislation by January 1984.

The regulations stipulate that bulk tankers must be licensed, and that all vehicles must be appropriately constructed and maintained. Flammable liquids stored in drums and pails must be stacked in a specified manner. Trucks must carry signs, both front and rear, indicating that they are carrying flammable liquids. This sign is a red diamond marked with a symbol denoting fire and the word *flammable*. Information about safety precautions to be observed in the case of fire or accident must be carried in the cabin of the truk.

58.3.2.3 Labelling

Inner and outer packages containing dangerous goods must be appropriately labelled. There are nine classes of dangerous goods, and some classes are subdivided:

Class 1	Explosives
Class 2.1	Flammable gases
Class 2.2	Non-flammable gases
Class 2.3	Poisonous gases
Class 3	Flammable liquids
Class 4.1	Flammable solids
Class 4.2	Substances liable to spontaneous combustion
Class 4.3	Substances emitting flammable gas when wet
Class 5.1	Oxidising substances
Class 5.2	Organic peroxides
Class 6.1	Poisonous substances
Class 6.2	Infectious substances
Class 7	Radioactive substances
Class 8	Corrosives
Class 9	Miscellaneous dangerous substances

All of these classes and subclasses, except Class 9, have defined class labels, commonly referred to as 'diamonds'.

The requirements for labelling inner containers (i.e. containers not exposed during transport) are under review, but it seems certain that containers of capacity greater than specified limits will at least have to show the class label. For example, for flammable liquids with a flash point below 23° C, the threshold is 150 mL; for liquids with a flash point 23–61° C, 300 mL; for manufactured goods with at least 10% solids and flash point 23–61° C, 5 L (materials with flash point below 23° may be included in this last group, depending on viscosity). Many paints fall into this last group. Examples are shown in Appendix 58.2.

Outer containers (i.e. exposed during transport), again depending on quantity contained, must show class label(s), the correct technical name as defined and listed in the Code, the identification number (U.N. number), packaging group (see Section 58.4, below), and the name and address of the manufacturer, Australian agent, consignor or other person or body accepting responsibility for the goods.

58.4 PACKAGING

This is probably the most controversial requirement of recent legislation. Packages must comply with the recommendations prepared by the United Nations Committee of Experts on the Transport of Dangerous Goods. Recommendations have also been adopted for sea transport and, since 1 January 1983, for air transport.

Depending on the volumes involved and the hazard of the contents, there is the requirement for packages to pass specific drop, pressure, leakage and stacking tests, as well as general resistance to environment, chemical compatibility and similar requirements. In the Code a package is effectively defined as the smallest unit which may be transported without repacking and relabelling. A package may be a single container such as a drum—whether of single material (e.g. steel) or of composite construction (e.g. plastic-lined fibreboard)—or it may be a combination package where inner containers are packed in an outer container (e.g. cans or bottles in fibreboard container). Many packages in current use do not meet the requirements set by the Code.

Concessions have been given for the use, for a limited time, of containers complying with certain Australian Standards, but notice has been given that full compliance will be required in the future.

58.4.1 Consumer Protection

The requirements for labelling packages containing poisons and dangerous goods have already been discussed. There are also requirements to be fulfilled under the Australian Uniform Packaging regulations.

58.4.2 Identification of Packer or Marketer

Regulations still on the statute books in most Australian states require that the packer and the state in which the goods were packed must be identified on each package. The identification may be in plain language or by approved code number. To protect the commercial interests of manufacturers marketing their own products and also contract-packing for other marketers, the identity of packers and their codes is held confidentially by the weights and measures authorities. This has resulted, particularly for some house-brand products, in the situation where the consumer is given no indication of who is responsible for the goods. It is now acceptable to show what will be required by legislation—namely, the name and address of a person or company accepting responsibility for placing the goods on the market, providing that, at that address, information is available to the weights and measures inspectors identifying the place and identity of the packer.

58.4.3 Quantity Statement

Each package must carry a statement as to the quantity of material it contains. The quantity statement must appear on the main label(s) (it must be repeated on other faces of the package that may be used for display at the point of sale) in close proximity to whatever identifies the contents of the package. Different articles are sold by different units of measure. Paint is sold by volume, aerosols by mass, and resins by either mass or volume. The legal units of volume are the millilitre and litre. The legal units of mass are the gram and kilogram. Mass statements must include the word 'net'.

Expressions that qualify a unit of measurement such as 'bit litre' or 'full litre' are prohibited. Restrictions are placed on terms such as 'jumbo size' or 'economy size.'

Paint may be packed only in certain prescribed quantities, within a specified range. These are 250 mL, 500 mL, 1 litre, 2 litres, 4 litres, 6 litres and 10 litres. There is no prescription on size below 250 mL. Above 10 litres, increments must be in even litres.

Where two or more articles are sold in an outer package (e.g. paint and catalyst), weights and measures regulations could be satisfied with a total kit size, providing each inner package is marked with a quantity statement. However, if any of the articles comes under the poisons regulations, poisons warnings, the quantities of the hazardous articles and the proportion of the poison in them must be shown on the outer package.

A deficiency of up to 5 per cent of declared measure is allowed for a single package, provided that on checking 6 to 12 packages there is no average deficiency.

Tinting-base paint, which must be identified as such, is exempted from the size prescription so long as there is also a direction that it be a stated volume of tinter to produce a volume equal to one of the prescribed volumes.

Imported ready-for-sale articles must show the declaration of measure in Australian legal units and the country of origin but do not have to identify the actual packer. If further packaging is involved (e.g. sealing into a blister pack), the Australian company assuming responsibility must be identified. Quantity statements on the outer pack must comply with Australian metric requirements, although other equivalent units may be shown on the inner package.

58.4.4 Deceptive Packaging

Under the regulations a package is deceptive if any of the following features are in excess and are concealed:

(a) Unoccupied space as defined must not exceed 25 per cent of total volume.
(b) Bottom recesses and internal cavities must not exceed 10 per cent individually or in total.
(c) Wall thickness (the difference between the volumes of the inside and the outside of the container) must not exceed 25 per cent of total volume.
(d) Cap and lid volumes must not exceed 10 per cent of total volume.

Methods of measuring (and therefore the definition of) concealed volumes for each of these features are specified in the Standard Techniques of Measurement of Packages published by the Standing Committee on Packaging.

58.5 PROTECTION OF THE ENVIRONMENT

Legislation in this area is aimed at protecting the environment in which we live. By definition, the *environment* means all physical factors of the surrounding of human beings including the land, water, atmosphere, climate, sound, odours, tastes, the biological factors of animals and plants, and the social factors of aesthetics. Thus, the environment is protected against any substance (or pollutant) that:

(a) alters the quality of any segment or element of the receiving environment so as to adversely affect its beneficial use to mankind;
(b) is hazardous or potentially hazardous to health;
(c) imparts objectionable odours, radioactivity, noise, temperature change or physical, chemical or biological change to any segment or element of the environment.

This is achieved by a licensing system that requires individuals, companies and government authorities making a waste discharge to the environment to hold a licence. The licence specifies the conditions and limitations under which the discharge may be made without detrimental effect on the environment, and it requires its holder to institute a monitoring programme to ensure that the conditions are met.

In Australia, unlike other parts of the world such as California, there are as yet no air pollution regulations affecting the sale and use of paints. The states of Victoria and New South Wales have, however, given notice of their intention to enact such legislation as soon as possible. The strategy behind this legislation will be to reduce the emission of hydrocarbons to the atmosphere by promoting alternative technologies such as waterborne, high solids and powder coatings.

The authorities are currently negotiating on an industry-by-industry basis for companies to establish programs for pollution reduction. It is recognised that cost/effectiveness factors apply for the whole community, and high priority is being given to concentrations of activity involving high volumes of volatile solvents. The threat of across-the-board standards, and the consequent elimination of some sections of industry, is helpful in promoting a co-operative attitude.

APPENDIX 58.1 UNIFORM PAINT STANDARD (JUNE 1981)

1 Scope

1.1 These Regulations refer to imported paints and tinters as well as to locally produced products.
1.2 The provisions of these Regulations shall be read as being in addition to, and not in derogation of, the provisions of any other law of the State relating to the words, statements, particulars, or other matter required or permitted to be written on packages containing any type of paint.

2 Definitions

2.1 'Paint', without limiting the ordinary meaning, includes any substance used or intended to be used for application as a colouring or protective coating to any surface.
2.2 'Tinter' means any pigment or admixture of pigment with other substances, in powder or in liquid form, sold for the purpose of adding to paint in order to change the colour thereof.
2.3 'Poison' means any substance specified in the Schedules of the Poisons Regulations of the State or any regulations or amendment or substitution thereof.
2.4 'Non-Volatile Content' means that portion of the paint or tinter as determined by the method laid down in AS 1580, Method 301.1.
2.5 'Acid soluble substances' as defined in AS 1647, Part 3—1980.
2.6 'Toy' means an object or a number of objects manufactured and designed and/or labelled and/or marketed as a plaything for a child or children up to the age of 14 years.
2.7 'Accessible' means any part of a toy that can be contacted using the procedure and the articulate probe described in Appendix A of AS 1647, Part 2.
2.8 'Chemical toy' means a toy which contains two or more substances which when added together result in a chemical reaction, e.g. a chemistry set.
2.9 'Coating material' means a decorative, preservative or other coating applied to the surface of a toy; e.g. paint, varnish, lacquer, ink, metal.
2.10 'Graphic material' means the material deposited on another material by a graphic instrument during writing, drawing or marking; e.g. the core of a pencil (both 'lead' and coloured), crayon, chalk, finger paint, watercolour block.
2.11 'Intended' means designed, labelled or marketed.
2.12 'Printed material' means a toy in the form of a magazine or book containing printing and/or drawings; e.g. comic, magazine, rag book.

3 Prohibition

3.1 No person shall manufacture, sell or use any paint containing basic carbonate white lead except where State authorities have granted exemption for the manufacture of mirror backing paints.
3.2 The use of any Schedule 1 paint on:

- **3.2.1** the roof or other exterior portion of any house or other building or structure,
- **3.2.2** any fence, wall, post or gate,
- **3.2.3** any interior portion of any house or other building or structure intended for the use of human beings,
- **3.2.4** any household or other furniture or part thereof is prohibited.

3.3 No person shall manufacture, sell or use any of the following where the products do not comply with the Australian Standard AS 1647, Part 3—1980.

Finger colours, show card colours, poster paints, school pastels or crayons, toys, pencils, wallpaper, wraps or containers, or any surface which will come into contact with food.

4 Classification

4.1 Paint containing any material shown in Schedule 1 shall be classified as a Schedule 1 paint.
4.2 Paint containing any material shown in Schedule 2 shall be classified as a Schedule 2 paint.
4.3 Paint containing any material shown in Schedule 3 shall be classified as a Schedule 3 paint.

5 Labelling

5.1 Every package of paint or tinter for sale shall bear or have attached to it a label which shows:
- **5.1.1** the trade name or description,
- **5.1.2** the net weight or true volume of measure of the contents, and
- **5.1.3** the name and address of the vendor or manufacturer of the contents.

5.2 Subject to sub-regulation 5.3 of this regulation, the label of a package containing any Schedule 1 paint, Schedule 2 paint, or any tinter containing in excess of 10% lead shall contain:
- **5.2.1** the word 'WARNING', at the top of the label, in red capital letters with a face depth of not less than 5 mm, and
- **5.2.2** one or more of the following warnings according to the classification of the paint, in letters with a face depth of not less than 2.5 mm:
  - **5.2.2.1** in the case of Schedule 1 paint, in capital letters the statement 'KEEP AWAY FROM CHILDREN' followed by 'This paint presents a danger to young children from eating the dried paint, and should not be used on parts of buildings easily accessible to young children or on toys, furniture or any other article likely to come within the reach of young children'.
  - **5.2.2.2** in the case of Schedule 2 paint, in capital letters the statement 'KEEP AWAY FROM CHILDREN, PROVIDE ADEQUATE VENTILATION DURING APPLICATION' followed by 'Breathing in the vapour is dangerous. Do not use in the presence of a naked flame. Do not smoke'.
  - **5.2.2.3** in the case of any tinter containing in excess of 10% lead, in capital letters the statement 'KEEP AWAY FROM CHILDREN' followed by 'This tinter contains lead. Paint containing more than * mL of this tinter per litre should not be used on parts of buildings easily accessible to young children or on toys or furniture or any other articles likely to come within the reach of young children'.

5.3 Notwithstanding sub-regulation 5.2 of this regulation, the Director may approve of a smaller size of letters on a package than that prescribed by paragraph 5.2.1 or paragraph 5.2.2 of that sub-regulation where he is satisfied that the size of the letters prescribed by that paragraph is excessive in relation to the size of the package.

5.4 The label of a package containing any Schedule 1 paint or any tinter shall contain particulars with regard to:
- **5.4.1** every constituent of the paint or tinter that is specified in the first Schedule, and
- **5.4.2** the percentage of that constituent calculated on the non-volatile content of the paint or tinter except that this may be omitted for a tinter in the case of lead when the percentage is below 10.

6 Inspection, Sampling, Removal or Destruction and Penalties

The need for powers concerning inspection, sampling, ordering removal or destruction, and penalties should be noted and dealt with by each State in accordance with local conditions.

* The figure to be inserted by manufacturers is to comply with the levels indicated in Schedule 1 and 3 paints.

The First Schedule

Class I Paints

1 Lead and compounds of lead, exceeding 1.0 per cent calculated as Pb on the non-volatile content of the paint.
2 Arsenic and compounds of arsenic, exceeding 0.1 per cent calculated as As on the non-volatile content of the paint.
3 Antimony and compounds of antimony, exceeding 5.0 per cent calculated as Sb on the non-volatile content of the paint.
4 Cadmium and compounds of cadmium, exceeding 0.1 per cent calculated as Cd on the non-volatile content of the paint.
5 Selenium and compounds of selenium, exceeding 0.1 per cent calculated as Se on the non-volatile content of the paint.
6 Mercury and compounds of mercury, exceeding 0.1 per cent calculated as Hg on the non-volatile content of the paint.
7 Barium salts (except barium sulfate and barium metaborate) in paints containing more than 5 per cent barium calculated as a proportion of the non-volatile content of the paint.
8 Chromates and dichromates of alkali metals and ammonia in paints containing more than 5 per cent of chromium in these forms calculated as a proportion of the non-volatile content of the paint.
9 Tin organic compounds, being di-alkyl, tri-alkyl and tri-phenyl tin compounds where the alkyl group is methyl, ethyl, propyl or butyl not elsewhere specified in the Poisons List in paints containing more than 3 per cent of these compounds calculated as tin as a proportion of the non-volatile content of the paint.

The Second Schedule

Class II Paints

1 Benzene, exceeding 1.5 per cent by volume on the total volume of the paint.
2 Dichlorethane, exceeding 5 per cent by weight on the total weight of the paint.
3 Dichlorethylene, exceeding 10 per cent by weight on the total weight of the paint.
4 Dichlorbenzene, exceeding 5 per cent by weight on the total weight of the paint.
5 Dichlormethane, exceeding 5 per cent by weight on the total weight of the paint.
6 Methanol, exceeding 1 per cent by weight on the total weight of the paint.
7 Nitrobenzene, exceeding 1 per cent by weight on the total weight of the paint.
8 Pyridine, exceeding 2 per cent by weight of the total weight of the paint.
9 Trichlorethylene, exceeding 5 per cent by weight on the total weight of the paint.
10 Free organic isocyanates, exceeding 0.1 per cent by weight on the total weight of the paint.

The Third Schedule

Class III Paints

The standard for maximum permissible levels of toxic substances including paint, that may be present in toys, shall comply with the Australian Standard AS 1647, Part 3—1980.

APPENDIX 58.2 LABELLING REQUIREMENTS FOR FLAMMABLE LIQUIDS

TABLE A
Markings to be Displayed on Outer Packages

		Dangerous Goods Class Marking			Words	
Type of Liquid Column 1	Aggregate Quantity in Outer Package Column 2	Form Column 3	Colouring Column 4	Size (Minimum length of side) Column 5	Hazard Warning Column 6	Name of Liquid Column 7
Flammable liquid Class A Flash point less than 23° C	Not exceeding 150 mL	not required			not required	
	Exceeding 150 mL		Black lettering and symbol on red background	100 mm	HIGHLY FLAMMABLE	Trade name or technical name under which the liquid is sold
Flammable liquid Class B Flash point 23–61° C	Not exceeding 300 mL	not required			not required	
	Exceeding 300 mL		Black lettering and symbol on red background	100 mm	FLAMMABLE	Trade name or technical name under which the liquid is sold

TABLE B
Markings to be Displayed on Containers

		Dangerous Goods class Marking			Words	
Type of Liquid Column 1	Quantity in Container Column 2	Form Column 3	Colouring Column 4	Size (Minimum length of side) Column 5	Hazard Warning Column 6	Name of Liquid Column 7
Flammable liquid Class A Flash point less than 23° C	Not exceeding 50 mL	not required			not required	
	Exceeding 50 mL but not exceeding 150 mL	not required			HIGHLY FLAMMABLE	Trade name or technical name under which the liquid is sold
	Exceeding 150 mL but not exceeding 1 litre		Black lettering and symbol on red background	10 mm		
	Exceeding 1 litre but not exceeding 5 litres			20 mm		
	Exceeding 5 litres but not exceeding 10 litres			30 mm		
	Exceeding 10 litres but not exceeding 250 litres			100 mm		
Flammable liquid Class B Flash point 23°–61° C	Not exceeding 50 mL	not required			not required	
	Exceeding 50 mL but not exceeding 300 mL	not required			FLAMMABLE	Trade name or technical name under which the liquid is sold
	Exceeding 300 mL but not exceeding 1 litre		Black lettering and symbol on red background	10 mm		
	Exceeding 1 litre but not exceeding 5 litres			20 mm		
	Exceeding 5 litres but not exceeding 20 litres			30 mm		
	Exceeding 20 litres but not exceeding 250 litres			100 mm		

APPENDIX 1 PAINT CALCULATIONS

Standard mathematical procedures for calculating parameters most frequently used by the paint industry.

1. CALCULATIONS OF CONSTANTS

For the purposes of calculating the parameters shown in this section refer to appropriate chapter for definitions of *prime pigment(s)*, *total pigment* (including extenders), *solvent* (including that associated with the binder, if appropriate), *binder* (as solid binder), and *additives*.

1.1 Scaling to 100 parts Mass

For most calculations, a useful initial exercise is to *scale* the formula so that the mass total is 100 parts. The steps are:

(a) Sum the masses of the ingredient (which may be in any units so long as they are uniform) giving a total W_T.
(b) Multiply the masses of each component by a factor obtained by dividing 100 by W_T.
(c) The total of the new masses should be 100 exactly. Adjust if necessary the new mass of the predominant ingredient by adding or subtracting the difference between the new total and 100 exactly.
(d) Express the new masses as $W_I(1), \ldots, W_I(n)$.

Example

	ORIGINAL MASS	STEP 3—MULTIPLY EACH MASS BY FACTOR
Rutile titanium dioxide	350.0	$W_I(1) = 350.0 \times 0.0836 = 29.26$
Wetting agent	3.0	$W_I(2) = 3.0 \times 0.0836 = 0.25$
Alkyd resin	706.0	$W_I(3) = 706.0 \times 0.0836 = 59.02$
White Spirit	110.6	$W_I(4) = 110.6 \times 0.0836 = 9.25$
Mixed driers	24.5	$W_I(5) = 24.5 \times 0.0836 = 2.05$
Methyl ethyl ketoxime	2.0	$W_I(6) = 2.0 \times 0.0836 = 0.17$
	1196.1	100.00

STEP 1—SUM MASSES GIVING W_T

STEP 2: FACTOR $= \dfrac{100}{1196.1} = 0.0836$

Notes

W_T = total mass of the formula before scaling.

$W_I(1), \ldots, W_I(n)$ = ingredient masses for ingredients 1 to n.

1.2 Density

The *density* (preferred term to specific gravity) of a given formula is obtained by dividing the *mass* of the formula by the corresponding *volume*. Using metric units, it is expressed most frequently as kg/L (previously lb/gallon) at a defined temperature (usually 25° C).

The calculated density (ρ) is derived below, using the documented densities of the various components; data in the Raw Materials Index, or manufacturers' specifications, are useful sources. For practical purposes, the density can be measured directly, using pychometers ('weight/gallon cups' etc.).

The steps are:

a. Divide the mass of each ingredient ($W_I(1), \ldots, W_I(n)$) by its density ($\rho_I(1), \ldots, \rho_I(n)$).

b. This yields the respective volumes ($V_I(1), \ldots, V_I(n)$).

c. Sum the derived volumes giving a total volume V_T. Mathematically $V_T = \sum_{1}^{n} (V_I)$.

d. Calculate formula density, $\rho_T = \frac{\text{mass}}{\text{volume}} = \frac{100}{V_T}$.

Example

STEP 1—DIVIDE EACH MASS BY DENSITY

	SCALED MASS	DENSITY	
Rutile titanium dioxide	29.26	$D_I(1) = 4.07$	$V_I(1) = 29.26/4.07 = 7.19$
Wetting agent	0.25	$D_I(2) = 0.91$	$V_I(2) = 0.25/0.91 = 0.27$
Alkyd resin	59.02	$D_I(3) = 0.95$	$V_I(3) = 59.02/0.95 = 63.13$
White Spirit	9.25	$D_I(4) = 0.79$	$V_I(4) = 9.25/0.79 = 11.71$
Mixed driers	2.05	$D_I(5) = 1.08$	$V_I(5) = 2.05/1.08 = 1.90$
Methyl ethyl ketoxime	0.17	$D_I(6) = 0.83$	$V_I(6) = 0.17/0.83 = 0.20$
	W_T = 100.00		84.40

STEP 2—SUM VOLUMES GIVING V_T

STEP 3: Density, $\rho_T = \frac{W_T}{V_T} = \frac{100}{84.4} = 1.185$

Notes

W_T = total mass of the formula. If the scaling exercise has been first carried out, W_T = 100.

V_T = calculated total volume equivalent to 100 parts mass.

ρ_T = calculated density of the formula.

$V_I(1), \ldots, V_I(n)$ = volumes of ingredients 1 to n in a formula totalling 100 parts by mass.

$\rho_I(1), \ldots, \rho_I(n)$ = densities of ingredients 1 to n.

1.3 Bulking Value

The *bulking value* is the volume of unit mass of the formula, and is therefore the inverse of the density. Previously expressed as gallons/lb, the metric equivalent is litres/kilogram. Thus,

$$B_T = \frac{1}{\rho_T}$$

where B_T is the bulking value.

Bulking values are infrequently derived for formulae and are more useful for calculating the volumes of pigments, etc. They are not required for subsequent calculations.

Example
The bulking value of the formula in sections 1.1 and 1.2 is

$$\frac{1}{1.185} = \underline{0.844}$$

The bulking value of a titanium dioxide pigment of $\rho_I = 4.07$ is

$$\frac{1}{4.07} = \underline{0.246}$$

1.4 Non-Volatile Matter

The non-volatile fraction of a formula (or ingredient) is that part which will not evaporate under prescribed conditions of time and temperature. It is of significance to the surface coatings and related industries because it dictates the amount of material remaining after application and solvent loss. It is the preferred term to *solids content* as, strictly, some non-solid materials (such as plasticisers) are considered part of the final paint film.

Many of the significant parameters of paint formulae, such as the pigment/binder ratio and the pigment volume concentration, are calculated from the non-volatile fractions of the ingredients, using the data for each material.

The non-volatile matter can be calculated as a percentage of either the mass of the formula or, more usefully, of the volume.

1.4.1 Non-volatile Matter by Mass (NVM_T)

This is calculated using the 100 parts by mass scaled formula by the following steps:
(a) Obtain* the NVM_I for each of the ingredients (expressed as percentages by mass).
(b) Multiply the mass of each ingredient by the respective NVM_I and divide the result by 100.
(c) Sum the results to give the formula NVM_T expressed as a percentage.

ie $$NVM_T = \sum_1^n \left(\frac{W_I \times NVM_I}{100}\right)$$

Example

	SCALED MASS	STEP 1—OBTAIN NVM_I NVM_I (%)	STEP 2—MULTIPLY EACH MASS BY NVM_I/W_T
Rutile titanium dioxide	29.26	100.0	29.26 × 100/100 = 29.26
Wetting agent	0.25	100.0	0.25 × 100/100 = 0.25
Alkyd resin	59.02	70.0	59.02 × 70/100 = 41.31
White Spirit	9.25	0.0	9.25 × 0.0/100 = 0.00
Mixed driers	2.05	65.0	2.05 × 65/100 = 1.33
Methyl ethyl ketoxime	0.17	0.0	0.17 × 0.0/100 = 0.00
W_T =	100.00		STEP 3—SUM THE RESULTS → 72.15
	NVM_T =	72.15%	

Notes
NVM_T = Percentage non-volatile matter of the formula by mass.
$NVM_I(1), \ldots, NVM_I(n)$ = Percentage non-volatile matter of ingredients 1 to n by mass.

*The NVM_I for each ingredient can be found from manufacturers' specifications or data, or calculated (using similar procedures as above) if an 'ingredient' is a mixture or portion of a further formula, or measured using standard procedures.

1.4.2 Non-volatile Matter by Volume (NVV_T)

This is calculated using similar procedures as for NVM_T except that:

(a) The percentage of non-volatile material by volume for each ingredient is required.

(b) The volume of each formula ingredient (calculated from section 1.2 above, yielding $V_I(1), \ldots, V_I(n)$) is multiplied by the respective non-volatile by volume figure ($NVV_I(1), \ldots, NVV_I(n)$). Then

$$NVV_T = \sum_1^n \left(\frac{V_I \times NVV_I}{V_T} \right).$$

Example

		STEP 1—OBTAIN NVV_I	STEP 2—MULTIPLY EACH VOLUME BY NVV_I/V_T
	VOLUME[1]	NVV_I (%)	
Rutile titanium dioxide	7.19	100.0	7.19 × 100/84.4 = 8.52
Wetting agent	0.27	100.0	0.27 × 100/84.4 = 0.32
Alkyd resin	63.13	63.9[2]	63.13 × 63.9/84.4 = 47.80
White Spirit	11.71	0.0	11.71 × 0.0/84.4 = 0.00
Mixed driers	1.90	60.0	1.90 × 60.0/84.4 = 1.35
Methyl ethyl ketoxime	0.20	0.0	0.20 × 0.0/84.4 = 0.00
V_T =	84.40		STEP 3—SUM THE RESULTS → 57.99

$$NVV_T = 57.99\%$$

Notes

(1) Calculated in section 1.2, equivalent to $W_T = 100$

(2) Calculated using the formula on page 835, that is

$$NVV = \left[\frac{\dfrac{100}{0.95} - \left(\dfrac{100 - 70}{0.79} \right)}{\dfrac{100}{0.95}} \right] \times 100$$

$$= \left(\frac{105.26 - 37.97}{105.26} \right) \times 100$$

$$= 63.93\%$$

Notes

NVV_T = Percentage non-volatile matter of the formula by volume.

$NVV_I(1), \ldots, NVV_I(n)$ = Percentage non-volatile matter of ingredients 1 to n by volume.

(a) If V_T, the volume of the formula equivalent to 100 parts by mass, has not been previously calculated, it may be found from:

$$V_T = \frac{100}{\rho_T}$$

where ρ_T is the density of the formula.

(b) A significant problem is the determination of the ingredient NVV_I, and for solutions or dispersions of resins and other substances, it will require prior calculation using:
  (i) the formula of the mixture, totalling 100 parts by mass; and
  (ii) the densities of the solvent(s), and of the mixture.
  The term 'mixture' is used to describe in general the resin solution or similar formula ingredient.

The volume of the mixture, V_M, is given by:

$$V_M = \frac{100}{\rho_M}$$

where ρ_M is the density of the mixture.

Similarly, the total volume of the solvent(s) will be:

$$V_S = \sum_1^n \left(\frac{W_S}{\rho_S}\right)$$

where W_S = percentage mass of each solvent, and ρ_S = density of each solvent.

The remaining volume occupied by the non-volatile fraction, V_F, is given by:

$$V_F = V_M - V_S.$$

And the NVV_M of the mixture is:

$$NVV_M = \frac{V_F}{V_M} \times 100.$$

This calculation is considerably simplified when the mixture is a solution or dispersion in water: the V_S figure will be equivalent to W_S, that is, $(100 - NVM_M)$, where NVM_M is the non-volatile by mass of the mixture. Then

$$V_F = \left(\frac{100}{\rho_M}\right) - (100 - NVM_M)$$

and

$$NVV_M = \left[\frac{\dfrac{100}{\rho_M} - (100 - NVM_M)}{V_M}\right] \times 100.$$

However, as V_M has already been defined as $\dfrac{100}{\rho_M}$, the equation simplifies to

$$NVV_M = \left[\frac{100}{\rho_M} - (100 - NVM_M)\right] \times \rho_M.$$

Example. The NVV_M of an acrylic emulsion of density 1.12 and NVM_M of 52% is sought.

$$NVV_M = \left[\frac{100}{1.12} - (100 - 52)\right] \times 1.12$$

$$= (89.28 - 48) \times 1.12$$

$$= 46.2\%.$$

If required, the density of the non-volatile fraction, ρ_F, can be calculated from:

$$\rho_F = \frac{NVM_M}{\dfrac{100}{\rho_M} - \left(\dfrac{100 - NVM_M}{\rho_S}\right)}.$$

When there is only one solvent (other than water), the NVV_M can be calculated from:

$$NVV_M = \left[\frac{\dfrac{100}{\rho_M} - \left(\dfrac{100 - NVM_M}{\rho_S}\right)}{\dfrac{100}{\rho_M}}\right] \times 100$$

$$= 100 - \left(\frac{\rho_M}{\rho_S}\right)(100 - NVM_M).$$

(c) W_M = mass of mixture (forming part of a formula) = 100 parts.
V_M = volume of mixture equivalent to 100 parts by mass.
ρ_M = density of mixture.
NVM_M = percentage non-volatile matter of the mixture by mass.
NVV_M = percentage non-volatile matter of the mixture by volume.
$V_S(1), \ldots, V_S(n)$ = volumes of solvents 1 to n in the mixture.
$\rho_S(1), \ldots, \rho_S(n)$ = densities of solvents 1 to n in the mixture.
$W_S(1), \ldots, W_S(n)$ = masses of solvents 1 to n in the mixture (parts per hundred by mass).
V_F = volume of non-volatile fraction in the mixture.
ρ_F = density of non-volatile fraction in the mixture.

1.5 Pigment/Binder Ratio

This is a *ratio* (without units) of the sum of the *masses* of the pigments and extenders to the sum of the masses of the dry resin components. Conventionally, the pigment and extender masses are scaled such that the resin mass(es) equals 100 or, more usually, unity.

The pigment/binder ratio, sometimes abbreviated to 'P/B', is of little practical use in the surface coatings industry, except to allow comparisons between very similar formulae, because it ignores the effects of the densities of the ingredients. Pigment/binder ratios are often employed in sealant and adhesive formulations, where the quantities of extenders are relatively large and pigments of high density are absent or present in small quantities; they are found occasionally in alkyd paint formulae, probably for historical reasons.

P/B is calculated as follows, using the 'scaled weight' version of the formula (although the scaling exercise is not a necessary prerequisite).

(a) Sum the masses of all components identified as pigments and/or extenders:

$$W_{PE} = \sum_1^n (W_P(1), \ldots, W_P(n)) + \sum_1^n (W_E(1), \ldots, W_E(n)).$$

(b) Sum the masses of all components identified as resins. Note that these are *dry* masses, and hence derivation of these—as required for the non-volatile by mass calculation—is a necessary first step.

$$W_{RT} = \sum_1^n (W_R(1), \ldots, W_R(n)).$$

(c) The P/B is calculated:

$$PB = \frac{W_{PE}}{W_{RT}}$$

and expressed as 'PB/1', or alternatively as 'PB × 100/100'.

Example

	STEP 1—SUM MASSES OF DRY PIGMENTS	STEP 2—SUM MASSES OF DRY RESIN COMPONENTS
Rutile titanium dioxide	29.26	—
Wetting agent	—	—
Alkyd resin	—	41.31
White Spirit	—	—
Mixed driers	—	—
Methyl ethyl ketoxime	—	—
	$W_{PE} = 29.26$	$W_{RT} = 41.31$

STEP 3 ⇒ $PB = \frac{W_{PE}}{W_{RT}} = \frac{29.26}{41.31} = 0.71/1$

Note: Dry masses calculated from section 1.4.1

Notes

W_{PE} = total masses of all pigments and extenders in the formula.
W_{RT} = total dry masses of all resins in the formula.
PB = pigment/binder ratio.
$W_P(1), \ldots, W_P(n)$ = masses of pigments 1 to n in the formula.
$W_E(1), \ldots, W_E(n)$ = masses of extenders 1 to n in the formula.
$W_R(1), \ldots, W_R(n)$ = masses of dry resins 1 to n in the formula.

1.6 Pigment Volume Concentration (PVC)

The PVC is the sum of the *volumes* of the pigments and extenders compared to the sum of the *volumes* of all *solid* components in the formula, expressed as a *percentage*. It is a very widely adopted method of describing the degree of pigment binding in a dry paint film. Unlike the pigment/binder ratio,

(a) The PVC allows for wide differences in densities of the various ingredients; and
(b) It takes into account the effects of additives which, although present in relatively minor quantities, nevertheless increase the binding ability of the resin(s) present.

Note that, before the advent of computerised systems, the effect of additives was usually ignored, and the calculation simplified to consideration of the pigment (and extender) and resin volumes only.

The use of PVC is now so widespread that it is employed as a means of defining any given formula, and it frequently appears in specifications.

An extension of the PVC concept is the *critical pigment volume concentration* (CPVC), which is unique to each pigment and binder combination and is the point at which the resin matrix can no longer fill the voids between the pigment and extender particles. The CPVC can be determined experimentally using *PVC ladders* (see later) and recording the wet scrub resistance, for example, which decreases rapidly once the CPVC is exceeded. Other properties (such as opacity) may display the opposite effect.

A PVC ladder is a series of paints formulated to a range of PVC values, most frequently in 5 or 10% increments, and in which the same pigments, extenders and resin(s) are employed. Unless impractical, the same non-volatile (by volume) percentages should also be used.

Because the calculation of PVC requires manipulation of the dry volumes of the components, the calculations shown under section 1.2 should be first undertaken. The steps are:

(a) Sum the dry volumes of all components in the formula, i.e.

$$V_F = \sum_{1}^{n}\left(V_I \times \frac{NVV_I}{100}\right).$$

(Note the prior calculations necessary when resin solutions or dispersions are present—see section 1.4.3.)

(b) Identify and sum the volumes of all pigment and extender components in the formula

$$V_{PE} = \sum_{1}^{n}(V_P(1), \ldots, V_P(n)) + \sum_{1}^{n}(V_E(1), \ldots, V_E(n)).$$

(c) The PVC is calculated as follows:

$$PVC\ (\%) = \frac{V_{PE} \times 100}{V_F}.$$

Note that, if dispersions of pigments or extenders are present as ingredients, the non-volatile volumes must first be calculated.

Example

	STEP 1—SUM VOLUMES OF ALL DRY INGREDIENTS[1]	STEP 2—SUM VOLUMES OF ALL DRY PIGMENTS[1]
Rutile titanium dioxide	7.19 × 100/100 = 7.19	7.19 × 100/100 = 7.19
Wetting agent	0.27 × 100/100 = 0.27	—
Alkyd resin	63.13 × 63.9/100 = 40.34	—
White Spirit	—	—
Mixed driers	1.90 × 60/100 = 1.14	—
Methyl ethyl ketoxime	—	—
	V_F = 48.94	V_{PE} = 7.19

STEP 3 ⇨ $PVC = \frac{V_{PE} \times 100}{V_F} = \frac{7.19 \times 100}{48.94} = 14.69\%$

[1] Data calculated from figure in section 1.4.2, for example

$$V(1) = \frac{V_I(1) \times NVV_I(1)}{100}$$

Notes

V_F = total volume of all dry ingredients.
V_{PE} = total volume of pigments and extenders in the formula.
$V_P(1), \ldots, V_P(n)$ = dry volumes of pigments 1 to n in the formula.
$V_E(1), \ldots, V_E(n)$ = dry volumes of extenders 1 to n in the formula.

1.7 Drier Additions

Many resins, such as alkyd resins, require the addition of heavy metal driers to effect film cure, a subject covered in chapters 3 and 29 (volume 1). These driers are added as a certain proportion of the *metal* to the *solid resin*, and the rates of addition are normally specified by the resin manufacturer or are determined by the formulator for a particular end-use requirement. As an example, an alkyd resin requires 0.5% lead metal and 0.06% cobalt metal drier on the solids content.

Drier solutions are specified by the content of the metal in solution; 24% lead and 6% cobalt metals are standard concentrations when these driers are supplied in their naphthenate forms.

The calculations required to determine the quantity of drier required (W_D) are:

$$W_D = \frac{W_R \times \%\ \text{metal required}}{\%\ \text{metal in drier solution}}.$$

Notes

W_D = mass of drier solution.
W_R = mass of resin solid.
Nomograms are also available for deriving drier additions without the need for calculations.

Example. Calculate the quantity of 24% lead naphthenate such that 90 kg of alkyd resin (solids) are modified at the rate of 0.5% lead as metal.

$$W_D = \frac{90 \times 0.5}{24} = 1.88\ \text{kg}.$$

1.8 Calculation of Ball Mill Loadings

A frequent problem encountered by paint chemists is the determination of the quantity of a millbase for a ball mill of known capacity to ensure maxium grinding efficiency. Fairly conventional loadings are:

40/60 for mills with steel balls; and
50/60 for mills with porcelain balls.

The first figure in each case is the apparent volume occupied by the balls; the second is the optimum percentage volume occupied by the balls and millbase combined.

Assuming that 40% of the space occupied by the balls is voids (air), then a 40/60 *load ratio* results in:

$$20 + \left(\frac{40}{100} \times 40\right) = 36\%\ \text{of the mill volume occupied by millbase.}$$

Thus the quantity of millbase (W_{MB}) in kg required for a mill of V_M litres capacity is:

$$W_{MB} = 0.36 \times V_M \times \rho_{MB}\ \text{for steel balls,}$$

where ρ_{MB}, the density of the millbase, is given by:

$$\rho_{MB} = \frac{\sum W_{MB}}{\sum V_{MB}}$$

where $\sum W_{MB}$ is the sum of the masses of the millbase ingredients and $\sum V_{MB}$ is the sum of the respective volumes.

Similar calculations may be performed for other types of mill. The figures used will depend upon individual manufacturing practices and formulators. For example, an apparent media volume of 30% and voids of 36% might be nominated for a Bead mill.

The minimum loading of millbase is the quantity below which excessive wear will occur. The maximum loading for batch process milling is often set higher than the optimum in order to take advantage of the time available (for example, overnight grinding).

1.9 Formulation Costs

The cost of a formulation is the sum of the costs of the individual components, expressed usually as dollars or cents per kilogram or litre. *Raw material costs* are most commonly expressed as dollars/kg and occasionally as dollars/L.

Although the effects of air entrainment, volumetric reduction due to solvent combination and other non-additive factors can affect the cost of a formula, these are usually ignored. However, it is common to take into consideration *shrinkage* factors, which allow for the inevitable wastage of raw materials in the plant (evaporation of solvent and powder residues in bags are examples), and the loss of finished product due to filtration, tank residues and other processing activities. The factors are automatically compensated for if actual production *yields* are used.

The costing process is therefore:

(a) Scale the formula to 100 parts by mass.

(b) Calculate the *contribution* of each component as follows:

$$C_I = W_I \times C_{RM}.$$

(c) Sum the cost contributions to give the total formulation cost:

$$C_T = \sum (C_I/100).$$

(d) For volume costs:

$$C_V = C_T \times \rho_T.$$

(e) If initial costs for raw material(s) are in \$/L, convert to \$/kg using the density of the ingredient; the same expression is used for cents/L to cents/kg.

$$C_{RM} = \frac{C_{RMV}}{\rho_I}.$$

Notes

C_I = cost in dollars contributed by the raw material.
C_{RM} = cost of the raw material in dollars/kg.
C_T = cost of the formula in dollars/kg.
C_V = cost of the formula in dollars/L.
C_{RMV} = cost of the raw material in dollars/L.
ρ_T = density of the formulation.
ρ_I = density of an ingredient.

1.10 Paint Coverage

1.10.1 Film Thickness at a Nominated Coverage

It is often necessary to know the thickness of a coating which is obtained when a paint is applied at a given rate, in square metres/litre (m^2/L). The preferred paint thickness unit is micrometre (μm), replacing 'mils' (one-thousandth of an inch)*.

* 1 mil = 25.4 μm. The term 'thou' is also used for 'mils'.

The thickness of a paint coating is expressed by:

$$\text{Thickness} = \frac{\text{Volume}}{\text{Area}}.$$

Thus, for wet film thickness:

$$\text{Thickness}_{\text{wet}}(\mu\text{m}) = \frac{\text{Volume (L)}}{\text{Area (m}^2)} \times \frac{1000}{1}.$$

Example. 1 L of paint applied to 5 m^2 will therefore provide a wet coating 200 μm thick.

A more useful measure of paint performance is the derivation of *dry* film thickness. This is calculated in a similar manner, but taking the non-volatile by volume of the paint into account.

$$\text{Thickness}_{\text{dry}}(\mu\text{m}) = \frac{\text{Volume (L)}}{\text{Area (m}^2)} \times \frac{1000}{1} \times \frac{\text{NVV}_{\text{T}}}{100}.$$

Example. 6 L of paint of 42% NVV_T applied to 6.3 m^2 will yield a dry film thickness of 400 μm.

1.10.2. Amount of Paint Required to Cover a Specified Area at Specified Film Thickness

Calculations of this type are very useful for specification work and again may relate to wet or dry film thicknesses. From the above equation:

$$\frac{\text{Thickness } (\mu\text{m})}{1000} = \frac{\text{Volume (L)}}{\text{Area (m}^2)}.$$

Thus, for nominated wet film thickness:

$$\text{Volume (L)} = \frac{\text{Area (m}^2) \times \text{Thickness}_{\text{wet}}(\mu\text{m})}{1000}$$

and for dry film thickness:

$$\text{Volume (L)} = \frac{\text{Area (m}^2) \times \text{Thickness}_{\text{dry}}(\mu\text{m}) \times 100}{1000 \times \text{NVV}_{\text{T}}}.$$

For example, the volume of paint of NVV_T 42% required to cover 75 m^2 at a dry film thickness of 25 μm is 4.46 L.

1.10.3 Theoretical Spreading Rate (TSR)

Another parameter often required in specification work is the theoretical area (per litre) a paint will cover at a nominated dry film thickness. The NVV_T of the paint is required. Then,

$$\text{TSR} = \frac{\text{NVV}_{\text{T}} \times 10}{\text{Thickness}_{\text{dry}}(\mu\text{m})}\,\text{m}^2/\text{L}.$$

As the common method for expressing TSR is at 10 μm dry film thickness, then the TSR is numerically equivalent to the volume solids.

$$\text{TSR}_{(10\,\mu\text{m})} = \text{NVV}_{\text{T}}\text{m}^2/\text{L}.$$

1.11 Paint Coverage Costs

Whilst the raw material cost of a paint is clearly an important factor, a realistic approach to paint costs should be based on the cost to cover a specified area at a specified *dry* film thickness. The most usually quoted parameters are the cost (¢) to paint 1 m^2 at a dry film thickness of 25 μm, knowing the NVV_T (%) and cost ($/L) of the paint.

In section 1.10.2, the volume of paint required to coat a specified area at a nominated dry film thickness was given as:

$$\text{Volume (L)} = \frac{\text{Area (m}^2\text{)} \times \text{Thickness}_{\text{dry}}(\mu\text{m}) \times 100}{1000 \times NVV_T}.$$

Under the conditions specified, this simplifies to:

$$\text{Volume (L)} = \frac{2.5}{NVV_T}.$$

Thus the cost becomes:

$$\text{Cost (¢)} = \frac{\text{Cost (\$/litre)} \times 250}{NVV_T}.$$

Example. The cost at 25 μm dry film build per m^2 of a paint A of 33% NVV_T at $5.60/L is 42.42 cents. Similarly, the cost of a paint B of 53% NVV_T at $7.69/L is 36.27¢, demonstrating that, although paint B is more expensive on a volume basis, it is in fact cheaper to use.

Note. The standard parameters used before metrication were the cost (¢) to paint 1 square foot (ft^2) at a dry film thickness of 1/1000th inch ('mil') with reference to a base cost ($) of 1 gallon. Costs derived by this method can be converted to those calculated above by multiplying by 10.596 (assuming the same NVV_T).

2. OPTIMISATION

The preceding section examined the calculation of *constants*; that is, those fixed parameters derived mathematically from the physical characteristics of the ingredients in the formula.

Whilst the calculation of constants is an important activity associated with the formalisation of formulae, it is usual (and often more useful) to be able to change the formula to suit a required specification or a number of specifications. This process is called *optimisation.*

Optimisation in its various forms can range from relatively simple mathematical manipulation easily achieved using calculators, to extremely complex 'number-crunching' processes which in practice can only be carried out using computers (a subject discussed in chapter 56).

In this section, some relatively simple optimisation steps are outlined, for which computer techniques are not required. In most cases, the procedures will encompass the calculations outlined before (often more than once).

2.1 Cost Changes

There can be few formulating chemists who are not faced with the problem of reducing the cost of a formula at some stage of its development. Cost reduction usually involves one or more of the following:

(a) addition of a lower cost material,
(b) reduction in non-volatile content,

(c) substitution of one ingredient by a lower-cost alternative, and/or
(d) increase in PVC or P/B ratio.

A simple algorithm for adjusting the cost of a formula is shown below. The formula adjusts the cost and volume simultaneously, so no further calculations are necessary.

The cost of a formula, C_V, can be obtained by dividing the total cost in cents, C_F, by the total volume in litres, T_V.

$$C_V = \frac{C_F}{T_V}.$$

Thus C_V is in the units of cents/L.

The new required cost, C_N, is equal to the previous total cost plus the additional cost due to the addition or subtraction of some raw material (which may be already present) divided by the new total volume. Thus:

$$C_N = \frac{C_F + (C_{RM} \times W_{RM})}{T_V + (W_{RM}/\rho_{RM})}$$

C_N = new formula cost (in cents).
C_F = existing total formula cost (in cents).
C_{RM} = cost in cents/kg of the raw material to be added or removed.
W_{RM} = mass (in kg) of the raw material to be removed or added.
T_V = existing formula volume (in litres).
ρ_{RM} = density of the raw material to be added or removed.

Solving for W_{RM} produces the following:

$$W_{RM} = \frac{C_F - (C_N \times T_V)}{(C_N/\rho_{RM}) - C_{RM}}.$$

Example. The raw material cost of 1 L of paint is \$1.25/L and this has to be reduced to \$1.125/L by using a raw material costing \$2.00/kg and which has a density of 4.1.

$$W_{RM} = \frac{125 - (112.5 \times 1)}{(112.5/4.1) - 200} = \frac{12.5}{-172.56} = -0.072 \text{ kg, or } -72 \text{ g.}$$

In other words, 72 g of the raw material (or 14.4 cents worth) has to be removed from one litre of the formula to reduce the cost by 12.5 cents/L. The negative value for W_{RM} indicates that the raw material must be removed; a positive value would indicate that it should be added.

2.2 Adjustment of Non-volatile Content

Adjustment of the non-volatile content of a formula is achieved by the addition of a solvent to reduce the NV; by the addition of a solid material to increase the NV; or by the addition of a solution (or dispersion) to effect reduction or increase of the NV, depending on whether the NV of the solution is respectively lower or greater than that of the formula.

2.2.1 Reduction of Non-volatile Content

The steps are:
(a) Calculate the non-volatile by mass (NVM_T) of the formula, using the procedure described in section 1.4.1.
(b) Determine the total mass of the formula to be adjusted (W_F).

The mass of solvent to be added (W_S) is given by:

$$W_S = W_F\left(\frac{NVM_T}{Y}\right) - W_F$$

where Y is the *required NVM* of the formula.

A similar equation can be used for volume calculations, using volume terms throughout.

When the material to be added is a solution or dispersion and not a pure solvent, the steps are:
(a) Calculate the NVM_T of the formula, as above.
(b) Determine the total mass of the formula (W_F).
(c) Determine the NVM of the solution to be added (NVM_S).
Then the mass of solution to be added (W_S) is given by:

$$W_S = W_F\left(\frac{NVM_T - Y}{Y - NVM_S}\right)$$

where Y is the required NVM of the formula.

Again, a volume calculation can be derived.

2.2.2 Increase in Non-volatile Content

The equations required to effect an increase in non-volatile content are similar to those above, except that the ratios of the *volatile* components (as opposed to the non-volatile components) are required. Thus, when adding a solid material to increase the non-volatile content of a formula:

$$W_S = W_F\left(\frac{100 - NVM_T}{100 - Y}\right) - W_F.$$

W_S = mass of solid material to be added.
W_F = mass of formula to be adjusted.
NVM_T = non-volatile content by mass of the formula.
Y = required NVM of the formula.

When adjusting the non-volatile content of a formula upwards by adding a solution or dispersion of known non-volatile content, the following formula applies:

$$W_S = W_F\left(\frac{Y - NVM_T}{NVM_S - Y}\right).$$

W_S = mass of solution or dispersion to be added.
W_F = mass of the formula to be adjusted.
NVM_T = non-volatile content by mass of the formula to be adjusted.
NVM_S = non-volatile content by mass of the solution or dispersion to be added.
Y = required NVM of the formula.

2.3 More Complex Optimisation Routines

The paint formulator is occasionally required to modify paint compositions to suit particular specification requirements, but the calculations involved are more complex than can realistically be carried out without the use of computers. Examples are:
(a) reduction of prime pigment level with maintenance of current pigment volume concentration (PVC) and non-volatile content by volume;
(b) maintenance of prime pigment level on a volume basis with reduction of vehicle levels;
(c) adjustment of PVC using nominated extenders, pigments or groups of these; and
(d) ranges of formulae of increasing or decreasing PVC (PVC 'ladders') maintaining non-volatile

contents by mass or volume.

As an example of the calculation routines involved, consider the first optimisation, with respect to the change in prime pigment level with maintenance of PVC and NVV_T. The formula must first be scaled (see section 1.1), then the volumes of all dry components in the formula computed (see section 1.6).

$$PVC = \left(\frac{VOL_A + VOL_B}{VOL_A + VOL_B + VOL_C}\right) \times 100$$

where VOL_A = volume of pigment to be changed; VOL_B = total volume of all remaining pigments and extenders, i.e. $\sum(VOL_E(1), \ldots, VOL_E(n))$; and VOL_C = volume of all other ingredients in the formula on a dry basis.

Now, VOL_A, the volume of the pigment, is a function of its density ρ_A:

$$VOL_A = \frac{WT_A}{\rho_A}.$$

Assuming a 1 L unit, the increase or decrease in the mass of the pigment is:

$$WT_C = WT_A - WT_B$$

where WT_B = new mass required in g/L; and WT_C = difference in mass between the original and desired mass/L.

Thus the change in volume is:

$$V_X = \frac{WT_C}{\rho_A}.$$

To maintain the same PVC and NVV_T, a new 'remaining extender' volume, VOL_D, is calculated:

$$VOL_D = VOL_B + V_X$$

Thus all remaining extender volumes must be scaled by the VOL_D/VOL_B factor. Thus

$$\begin{aligned} NEW\ VOL_E(1) &= VOL_E(1) \times \frac{VOL_D}{VOL_B} \\ \vdots \qquad & \qquad \vdots \qquad\qquad \vdots \\ NEW\ VOL_E(n) &= VOL_E(n) \times \frac{VOL_D}{VOL_B} \end{aligned}$$

Each of the new extender volumes can be converted back to mass by multiplying by the respective density. Re-scaling of the formula is then necessary.

3. MISCELLANEOUS CALCULATIONS

The following techniques, whilst not directly dealing with the principal constants associated with paint formulae, will prove useful to the practical paint chemist.

3.1 Substitution of Extenders at Constant Gloss

It is well established that, past a certain level, as the PVC goes up, gloss goes down. It is often forgotten, however, that there is another factor working in the reduction of gloss: the specific oil absorption of the pigment(s) and/or extender(s).

The central idea is that a given mass of extender represents a certain volume; this volume may have a low vehicle demand or be like a sponge—large vehicle volume would be necessary

to saturate it. In the first case, the gloss will be little affected, whereas in the second, there will be a considerable effect on it. It should be remembered that extenders have a dual function:
(a) to impart specific properties to a coating;
(b) reduce cost.
The following formula permits the substitution of a given extender with another, maintaining the same gloss:

$$E = \frac{A_1 \times \rho_2}{A_2 \times \rho_1}$$

where
E = extender substitution factor (multiply the original extender mass by this factor to obtain the mass of the substituting extender);
A_1 = oil absorption of the substituted extender (%);
A_2 = oil absorption of the substituting extender (%);
ρ_1 = density of the substituted extender;
ρ_2 = density of the substituting extender.

3.2 Neutralisation of Water-reducible Resins by Amines

Many acid-containing resins are commonly neutrailised with amines to impart water solubility or compatibility (see volume 1, chapter 22, for detailed discussion). The amount of amine required to effect neutralisation can be calculated as follows:

$$W_A = \frac{W_R \times AN_R \times EW_A}{56\,100}$$

where
W_A = mass of amine required to neutralise W_R units of resin (non-volatile matter)
W_R = mass of non-volatile component of the resin to be neutralised;
AN_R = acid number of the non-volatile component of the resin;
EW_A = equivalent weight of the amine. For all common amines, this is equivalent to the molecular weight—see table A1.1.

TABLE 1
Common commercial amines

Amine	*Molecular Weight (equivalent to* EW_A*)*
Trimethylamine	59
Ammonia (28%)	60.7
Monoethanolamine	61
Diethylamine	73
Mono-*iso*-propanolamine	75
Morpholine	87
Dimethylethanolamine	89
Triethylamine	101
Diethanolamine	105
Diethylethanolamine	117
Di-*iso*-propanolamine	133
Triethanolamine	149
Tributylamine	185
Tri-*iso*-propanolamine	191

3.3 Epoxy Resin/Amine Hardener Calculations

For optimum curing of epoxy resins, one *epoxide equivalent* from the resin must be reacted with one *amine equivalent* from the amine hardener.

The epoxide equivalent (EEW) of most epoxy resins is known and provided in commercial data. It is calculated from:

$$\mathrm{EEW} = \frac{\mathrm{MW_E}}{\mathrm{N_{EG}}}$$

where $\mathrm{MW_E}$ = molecular weight of the resin; and $\mathrm{N_{EG}}$ = number of epoxide groups/molecule (very commonly 2).

The amine equivalent is, similarly, the molecular weight divided by the number of amine hydrogens per molecule:

$$\mathrm{AE} = \frac{\mathrm{MW_A}}{\mathrm{N_{AH}}}.$$

The amount of amine hardener required to neutralise 100 parts by mass of resin is therefore:

$$\mathrm{W_A} = \frac{\mathrm{AE} \times 100}{\mathrm{EEW}}.$$

Example. Calculate the level of TETA (triethylene tetramine), which has 6 'active hydrogen atoms' and a molecular weight of 146, which must be added to a liquid epoxy resin which has an epoxide equivalent of 190.

$$\text{The amine equivalent of TETA} = \frac{146}{6} = 24.3.$$

$$\text{Therefore, phr of TETA required} = \frac{100 \times 24.3}{190} = 12.8.$$

In this case, the recommendation would be the addition of 13 parts of TETA to 100 parts of the resin.

However, for polyamine and polyamide hardeners, neither the molecular weight nor the amine equivalent is known (or not disclosed). In such cases, the importance of the *amine value*, which can be measured directly by laboratory techniques, becomes apparent.

By definition, the *amine value* (AV) is the number of milligrams of potassium hydroxide equivalent to the amine alkalinity present in one gram of sample. The 'amine alkalinity' can be regarded as the number of 'active hydrogens'; they are considered alkaline because they exist in the amine group, which has a basic character.

To convert from milligrams of potassium hydroxide to the 'active hydrogen' concept, the amine value must be divided by 56.1 (the molecular weight of potassium hydroxide) and to convert from milligrams to grams, the amine value must be divided by 1000. To further bring the definition into line with the epoxide equivalent (i.e. the number of grams of resin *containing* one gram of 'active hydrogen'), the reciprocal of the modified amine value is taken. Therefore

$$\mathrm{AE} = \frac{56.1 \times 1000}{\mathrm{AV}}.$$

This can now be used in the above equation to find the ideal hardener/resin ratio.

Example. A polyamine has an amine value of 1500. Calculate the ideal mixing ratio of this resin with a liquid epoxy resin with an epoxide equivalent of 200.

Combining the equations above:

$$\text{pph of the polyamine} = \frac{180}{200} \times \frac{56.1 \times 1000}{1500} = 18.7$$

19 parts of the polyamine should be added to 100 parts of the liquid epoxy resin.

These methods are valid for primary and secondary amines. Tertiary amines react in a more catalytic fashion, which in effect entails a reaction between one amine group and more than one epoxide group. One large producer has decided, on an arbitrary basis, that for calculating ratios using tertiary amines, the 'amine equivalent' should be divided by three.

3.3.1 Notes

(a) In some American literature, the term 'weight per active hydrogen' is used, equivalent to 'amine equivalent'.

(b) Since epoxy resins are specified in EEW *ranges* rather than single values, the *average* EEW is used:

$$\frac{\text{lower limit} + \text{upper limit}}{2}$$

For optimum performance, it is advisable to compute the stoichiometric ratio first and use it as the base formula, and then experiment with varying excesses of hardener or epoxy resin. Depending on the specificity of the case, the level of amine can vary from -25% to $+25\%$ of the ideal ratio.

(c) As EEW goes up the amount of amine goes down.

(d) As EW_A goes up the amount of amine goes up.

3.4 Polyol/Isocyanate Calculations

Two-part polyurethane formulations are frequently the subject of paint development projects and, like epoxy/hardener systems, it is important to ensure optimum curing by balancing the *polyol* and *isocyanate* levels.

Published data indicate the isocyanate and hydroxyl concentrations in prepolymers for two-pack urethane coatings. Available terminal isocyanate groups are expressed as a percentage of the respective polymer *solution* mass; the hydroxyl concentrations are expressed as a *hydroxyl value* (HV) equivalent to the number of milligrams of potassium hydroxide per gram of resin (or solution of resin).

An equation for calculating stoichiometric relationships between the isocyanate-containing and hydroxyl-bearing components may be expressed as follows:

$$W_H = \frac{\%\ \text{NCO} \times 1333}{\text{HV}}$$

where W_H is the mass of hydroxyl-containing compound needed to react with 100 parts of isocyanate compound on a 1 to 1 basis (that is, one isocyanate group for every hydroxyl group).

The percent isocyanate (% NCO) refers to the prepolymer as furnished and does *not* involve the non-volatile content of the NCO-containing component.

Hydroxyl value (HV) indicates the amount of OH groups contained in any hydroxyl-bearing compound.

The factor of 1333 is a conversion number which converts the hydroxyl value into parts of hydroxyl-bearing compound required to react with a given amount of isocyanate compound. In this case, the isocyanate compound is kept constant as 100 parts.

Example. To determine the amount of a polyol of HV 326 mg KOH/g required to react with 100 parts of an isocyanate prepolymer (% NCO = 9.5) on a 1 to 1 basis:

$$W_H = \frac{9.5 \times 1333}{326} = 38.8 \text{ parts of the polyol per 100 parts of the isocyanate material.}$$

If an excess of the isocyanate or hydroxyl-bearing component is required, the desired ratio is simply multiplied as a factor of the isocyanate component.

Example. If a 1 : 1.2 OH/NCO ratio is required, the mass ratio would be 120 parts of the isocyanate material per 38.8 parts of the polyol.

It is frequently necessary to calculate, or recheck, the hydroxyl/isocyanate ratio in any given two-pack formulation. This ratio can be calculated from the following formula:

$$\text{ratio of hydroxyl/isocyanate} = 1/x$$

where

$$x = \frac{W_I \times \%\,NCO \times 1333}{W_H \times HV \times 100}$$

where
W_I = mass of isocyanate resin in the formulation;
% NCO = % isocyanate in that resin;
W_H = mass of hydroxyl-bearing resin in the formulation;
HV = hydroxyl value of that resin.
W_I and W_H refer to the figures shown in the formulation; non-volatile content figures are not taken into account.

Example. A formula is at hand containing two hydroxyl-bearing components. The formula (conventionally shown in a 'mixed' form) contains:

Polyol A 107 g/L	HV_A = 140 mg KOH/g
Polyol B 73 g/L	HV_B = 340 mg KOH/g
Isocyanate prepolymer 421 g/L	% NCO = 7.6%

The combined HV figure is first calculated:

$$\text{Total mass} = 107 + 73 = 180 \text{ g/L}$$

$$HV = \left(140 \times \frac{107}{180}\right) + \left(340 \times \frac{73}{180}\right) = 221 \text{ mg KOH/g.}$$

Then

$$x = \frac{421 \times 7.6 \times 1333}{170 \times 221 \times 100} = 1.13.$$

Thus the hydroxyl/isocyanate ratio in that formula is 1 : 1.13, a 13% excess of isocyanate.

Note. In practice, higher levels of the isocyanate component than would be expected on theoretical grounds are normally employed, because of side reactions taking place.

Formulators are advised to examine hydroxyl/isocyanate ratios between 1 : 1 and 1 : 1.4 with polyol resins, and between 1 : 1.5 and 1 : 2 when using selected alkyd resins as hydroxyl prepolymers.

APPENDIX 2 GLOSSARY OF TERMS

This glossary of terms is reproduced from Australian Standard 2310-1980 *Glossary of Paint and Paint Terms* with permission of the Standards Association of Australia, 80 Arthur Street, North Sydney, New South Wales, Australia.

Term	**Definition**
abrasive blast cleaning	A method of preparing surfaces before painting by the use of abrasive such as grit propelled either through nozzles by compressed air or from wheels by centrifugal force onto the surface.
accelerator	*See* catalyst.
acrylic resin	Synthetic resin resulting from the polymerisation or copolymerisation of various acrylic or acrylate monomers.
activator	*See* catalyst.
aeration	Incorporation of bubbles of air in paint during stirring, shaking or application.
ageing	Degeneration occurring in a coating during the passage of time and/or heating.
air drying	The formation of a solid paint film from a liquid paint film under natural ambient conditions.
airless spraying	Application of paint by means of equipment consisting of fluid pump, hose and spray nozzle to produce atomisation of the paint without the use of compressed air or other propellant. (*See also* spraying.)
alkyd resin	A synthetic resin made by condensation between a polyhydric alcohol such as glycerol, and a polybasic acid such as phthalic acid.
alligatoring	*See* crocodiling.
aluminium paint	A paint that includes particles or flakes of aluminium which form a silvery metallic finish.
anti-condensation paint	A paint formulated to minimise the effects of condensation of moisture under intermittent dry and humid conditions.
anti-corrosive paint	A paint formulated to protect metals from corrosion.
anti-fouling paint	A paint formulated to prevent the growth of barnacles and other sea water organisms on hulls of ships or other underwater surfaces.

Term	**Definition**
antique finish	A paint system giving the effect of old age.
anti-settling agent	An additive used to prevent or delay the settling or separation of pigments in paint during storage.
anti-skinning agent	An additive used to prevent the premature oxidation and the formation of an insoluble surface layer on paints containing drying oils before application as a thin film.
artificial weathering	The testing of coatings in which ageing is accelerated by exposure to ultraviolet radiation, moisture, etc.
baking	The process of curing a film of paint by heating above 100° C.
barrier coat	A coat used to isolate subsequent coats from the preceding coats or substrate to prevent adverse physical or chemical interaction.
binder	The non-volatile part of the medium.
bituminous paint	A black or dark coloured paint formulated with coal tar or bitumen as the binder.
bitty film	A film containing bits of skin, gel, flocculated material or foreign particles, which project above the surface of the film.
bleeding	Discolouration caused by migration of components from the underlying film.
blistering	Isolated convex deformation of a paint film in the form of blisters arising from the detachment of one or more of the coats.
blooming	The formation of a thin film on top of a paint film thereby reducing the lustre or veiling its depth of colour.
blowing	*See* popping.
blushing	The formation of milky opalescence in clear finishes caused by deposition of moisture from the atmosphere and/or precipitation of one or more of the solid constituents of the finish.
body	Used to indicate the consistency of a paint.
bodying	The increase of consistency of a paint.
boxing	Mixing paint by repeated pouring from one container to another.
breathing	The passage of moisture vapour through a paint film without the paint film exhibiting blistering, cracking or peeling.
bridging	The separation of a paint film from the substrate at internal corners or other depressions due to shrinkage of the film or the formation of paint film over a depression or crack.
bronzing	The formation in a paint film of a characteristic red or yellow metallic lustre that is visible only at certain angles of illumination.
brushability	The ease with which a paint can be uniformly applied with a paint brush.
brush blasting	*See* whip blasting.
brush marks	Lines of unevenness that remain in the dried paint film after brush application.
bubbling	The development or occlusion of bubbles in a wet paint film.

Term	Definition
build	Thickness of dried paint film.
burning-off	The softening of a paint film by a flame and scraping off while still soft.
caking	Hard settling. (*See* also settling.)
catalyst	A substance whose presence increases the rate of a chemical reaction. In some cases the catalyst functions by being consumed and regenerated; in other cases the catalyst seems not to enter the reaction and functions by surface characteristics of some kind. A negative catalyst (inhibitor) slows down a chemical reaction.
cement-based paint	A dry powder formulated with Portland cement and other materials, mixed with water just before use.
chalking	Change involving the release of one or more of the constituents of the film, in the form of loosely adherent fine powder.
checking	Breaks in the surface of a paint film which do not render the underlying surface visible when the film is viewed at a magnification of 10 ×. (*See also* crazing.)
—irregular pattern type	Checking in which the breaks are in no definite pattern.
—line type	Checking in which the breaks are, in general, in parallel lines.
—crow-foot type	Checking in which the breaks are in a series of three-pronged formations in which the prongs radiate from a point with an angle of approximately 120 degrees between prongs.
cheesy	The rather soft and mechanically weak condition of a dry-to-touch film but not a fully cured film.
chipping	The removal of paint and surface contaminants from a substrate by means of impact from a sharpened tool.
chlorinated rubber resin	Resin resulting from the action of chlorine on natural or synthetic rubber.
cissing	The recession of a wet paint film from a surface leaving small areas uncoated.
clouding	Opalescence caused by the precipitation of insoluble matter.
coat of paint	A continuous layer of dried paint film resulting from a single application of paint.
coating system	The number and types of coats of paint applied separately in a predetermined order.
cobwebbing	The formation of fine filaments of partly dried paint during the spray application of a fast drying paint.
cold curing	Chemical hardening without the application of heat.
colorant	A concentrated agent that can be added to paints to make a range of colours.
colour change	Any change in the colour of the film as a result of exposure other than that due to chalking and/or dirt collection.
colourfast	Resistant to changes in colour.

Term	**Definition**
compatibility (of products)	The ability of a product to mix with another without causing precipitation, coagulation, thickening, etc, of the resultant mixture.
compatibility (of a paint with the substrate)	The ability of a paint to be applied to a substrate or painted surface without causing undesirable effects.
consistency	The apparent viscosity of a paint when shearing forces of varying amounts are applied.
contrast ratio	An instrumental measure of the ability of a coat of paint to hide contrasts in colour of the underlying surface.
coverage	The spreading rate, expressed in square metres per litre.
covering power	*See* hiding power.
cracking	Formation of breaks in a paint film that expose the underlying surface.
—irregular pattern type	Cracking in which the breaks are in no definite pattern.
—line type	Cracking in which the breaks are, in general, in parallel lines.
—sigmoid type	Cracking in which the breaks are in relatively large curves which meet and/or intersect.
cratering	Residual effect of burst bubbles. (*See also* cissing.)
crazing	The formation of minute criss-cross cracks on the surface of a paint film. NOTE: Crazing resembles 'checking' but the cracks are deeper and broader and exhibit a polygonal pattern resembling crazy paving.
crinkling	*See* wrinkling.
critical pigment volume concentration (cpvc)	The pigment volume concentration at which there is a marked change in physical properties. NOTE: For solvent-borne paints it is considered the point at which the voids between the pigment particles are just filled with resin.
crocodiling	The formation of wide criss-cross cracks in a paint film.
crosslinking	Applied to polymer molecules, the setting up of chemical links between the molecular chains to form a three dimensional or network polymer, generally by covalent bonding. When extensive, as in most thermosetting resins, crosslinking makes one infusible molecule of all the linked chains. Crosslinking generally toughens and stiffens coatings and makes them insoluble.
crosslinking agent	A substance which will react chemically with the molecular chains of a thermoplastic resin and by linking them together create a more rigid structure resulting in a more or less infusible product.
curing	The process of condensation or polymerisation of paint by heat or chemical means, resulting in the full development of desirable properties.
curtain coating	An automated large scale technique of passing objects to be coated through a continuous curtain of paint.
curtaining	*See* sagging.
curtain spraying	The technique of passing objects to be coated through a curtain of sprayed paint.

Term	Definition
cutting-in	Careful painting of an edge, such as the wall colour at the ceiling line or at the edge of woodwork, to avoid spreading onto an adjacent area.
dilatancy	The fluid condition when paint consistency is increased by stirring or brushing.
diluent	A volatile liquid which, although not a true solvent for the non-volatile constituents of a paint medium, is miscible with and may be used in conjunction with a true solvent, without causing any deleterious effects.
dirt collection	The presence of matter adhering to the surface of or embedded in a film but not derived from the film.
discolouration	Any change in the colour of a film as a result of exposure, including that due to chalking and dirt collection.
drag	The resistance encountered when paint is spread by brush.
drier	A compound, usually organo-metallic and soluble in organic solvents and binders, which is added to paint to accelerate drying by catalytic oxidation.
dry to handle	A state during the drying or curing process when the paint film has hardened sufficiently for the object to be moved carefully without marring the film.
dry to sand	A state during the drying or curing process when the paint film can be sanded to remove imperfections without sticking or clogging of the abrasive paper or tearing of the film.
drying	The process of change in which a liquid film is converted to a solid film.
dry to recoat	The stage during the drying or curing process when the next coat can be applied without deleterious effects.
dry spray	A rough, powdery, non-coherent film produced when atomised paint dries before reaching the surface.
dry film thickness	The arithmetic mean of a series of dry film thickness measurements.
durability	The degree to which films of paint and paint materials withstand the destructive effect of the conditions to which they are subjected.
dust dry	A stage during the drying or curing process when particles of fine dust that settle on the surface do not stick to the paint film.
efflorescence	A deposit of salts that remains on the surface of masonry, brick or plaster after the evaporation of water.
electrophoretic deposition	A process of paint application in which an electric current is passed through an aqueous suspension of paint between the article to be coated and another electrode.
electrostatic painting	The application of paint by means of electrostatic attraction.
emulsion paint	*See* latex paint.
emulsion resin, emulsion	Stable dispersions of microscopic, insoluble resin particles in water.
epoxy resin	Synthetic resin utilising an epoxide (epoxy) group.
erosion	Attrition of the film by natural weathering which may expose the substrate.

Term	**Definition**
etch-primer	A primer containing a small amount of zinc chromate pigment, vinyl resin, phosphoric acid and solvent.
extender	An inorganic substance in powder form, usually white or slightly coloured with little opacity and having a refractive index usually less than 1.7, which is used in conjunction with pigments because of its physical or chemical properties.
exterior exposure	Direct exposure to the weather.
false body	*See* thixotropy.
fat edge, fatty edge	Accumulation of paint at the edge of a painted surface.
feathering, feather edging	The tapering of the edge of a film of paint by laying off with a comparatively dry brush.
feather sanding	The tapering of the edge of a dried paint film with abrasive paper.
filiform corrosion	A type of corrosion proceeding under a coat of paint, varnish or related product, in the form of threads, and generally starts from bare edges and local damage in the paint coat.
filler	A composition used for filling fine cracks and indentations to obtain a smooth finish preparatory to painting.
film (of paint)	*Synonymous with* coat of paint.
fineness of grind	The reading, in micrometres, obtained on a standard gauge under specified conditions of test indicating the depth of the gauge at which discrete solid particles in the product are readily discernible.
finish coat, finishing	The final coat of a paint system.
flat (finish)	A surface with a specular gloss reading not greater than 5 gloss units when the specular direction is 60 degrees. (*See also* specular gloss.)
flatting-down	Rubbing down a painted surface with fine abrasives to produce a smooth dull finish.
floating	Separation of pigment which occurs during drying, curing or storage which results in streaks or patchiness in the surface of the film and produces a variegated effect.
flooding	An extreme form of floating in which pigment floats to produce a uniform colour over the whole surface which is markedly different from that of a newly applied wet film. (*See also* floating *and* flow coating.)
flow	The ability of a paint to spread to a uniform thickness after application.
flow (flood) coating	A process of paint application in which the paint is poured or is allowed to flow over the object to be painted, the excess, if any, being allowed to drain off.
fly-off (flying)	The throwing-off of particles of paint from a paint roller.
full-gloss (finish)	A surface with a specular gloss reading above 85 gloss units when the specular direction is 60 degrees.
gelling	Deterioration of a paint or varnish by the partial or complete changing of the medium into a jelly-like condition.

Term	Definition
gel paints	Paints formulated to have a high degree of thixotropy. (*See also* thixotropy.)
general appearance (of a paint film)	The complete impression conveyed when a film is viewed normally from a distance of 3 m.
glaze	A translucent coating applied over a previous finish to enrich or modify the finish.
gloss	The visual impression created by the reflecting properties of a surface. (*See also* flat, low-gloss, semi-gloss, gloss (finish) and full-gloss.)
gloss (finish)	A surface with a specular gloss reading above 50 gloss units but not exceeding 85 gloss units when the specular direction is 60 degrees.
graining	The simulation of wood grain by the application of specially prepared colours or stains by the use of special graining tools or brushing techniques.
grain raising	The swelling and standing-up of wood fibre resulting from the absorption of water or solvent(s).
grinning through	The effect observed when a paint does not totally obscure the underlying surface.
grit blasting	*See* abrasive blast cleaning.
ground coat	The base coat in an antique or graining finish system that is applied before the graining, glaze or other finish coat.
hair cracks	Fine cracks in the top coat of a finishing system.
hammer finish	A finish similar in appearance to hammered metal obtained from specially formulated paints.
hard dry condition	The stage reached during a drying or curing process when a paint film has sufficient strength to withstand mechanical damage.
hardener	A cross-linking agent used to cure a resin or paint system.
hardness (of a film)	The property of a film to resist indentation, scratching, abrasion, etc.
hard settling	Accumulation of solids on the bottom of a container of paint which are difficult to reincorporate.
hiding power	*See* opacity.
high build coating (paint)	A paint which enables the application in one coat of a relatively thick film of paint without sagging or running.
holding primer	*See* shop primer.
holidays	Defects characterised by a film having areas of insufficient thickness even to the point where parts of the surface may remain uncoated.
hungry (surface)	A surface very absorbent of paint.
incorporation	The thorough mixing of a paint before use to ensure that no layering or sediment remains.
inhibitor	A material used in small proportion to slow a chemical reaction.
intumescent paint	A fire-retardant paint that when heated by a flame swells into a crust which insulates the substrate and retards substrate ignition.

Term	Definition
joint tape	A special paper or paper-faced cotton tape used over joints between wallboards to conceal the joint and provide a smooth surface for painting.
kalsomine	A dry powder from which paint is made by mixing with water. Consists essentially of calcium carbonate or clay and glue or casein.
knotting compound	A quick-drying composition used to paint knots or other resinous areas in joinery to prevent bleeding of resin through subsequent finish costs.
lacquer	A fast-drying clear or pigmented coating that dries solely by evaporation of solvent.
ladder (paint)	A pattern due to a miss in laying-off. (*See also* laying-off.)
lap	That part of a freshly applied coat which overlaps and blends with a previously applied coat which has not reached the hard dry condition.
latex	*See* emulsion
latex paint	Paint made with emulsion resin(s) as the binder.
laying-off	The final light strokes of a brush on a paint film which has been spread so as to even and smooth the film as much as possible.
leafing	The orientation of particles of flaky pigments to form a continuous sheet at the surface of the film.
levelling	The flowing-out of a paint film after application so as to produce a level surface.
life	The period of time during which a paint film continues to serve the purpose for which it was intended.
lifting	The softening, swelling and wrinkling of a dry coat by solvents in a subsequent coat being applied.
light bodied	A paint of low consistency.
livering	Early stage of gelling. Thickening of a paint material caused by a chemical reaction between the pigment and binder.
low-gloss (finish)	A surface with a specular gloss reading above 5 gloss units but not exceeding 20 gloss units when the specular direction is 60 degrees.
luminous fractional reflectance	The ratio of the luminous flux reflected from, to that incident on, a specimen for specified gloss angles.
lustreless	A surface finish practically free from gloss or side sheen. Usually defined as having less than 5 percent reflectance when viewed at 15 degrees to the surface.
marine varnish	Varnish specially formulated for immersion in water and exposure to marine atmospheres.
masking	Temporary covering of areas not to be painted.
masking-tape	A strip of paper or cloth similar to adhesive tape which can be used to temporarily cover areas not to be painted and then easily removed.

Term	Definition
mastic	A heavy-bodied, paste-like paint often applied with a trowel to produce a thick protective film.
medium	The total sum of the constituents of the liquid phase of the paint.
metallic finish	*See* aluminium paint and polychromatic finish.
metameric match	Close colour conformity under a particular illumination which changes to an appreciable colour difference under other light sources.
misses	*See* holidays.
mist coat	A very thin coat of paint or varnish which may be discontinuous, applied by spraying.
mudcracking	Visible irregular cracking in thick films of paint caused by shrinkage tension during drying.
nibs	Small pieces of foreign material, pieces of skin, coagulated medium, etc., which project above the surface of an applied film, usually a varnish.
nitrocellulose	A synthetic resin prepared by nitration of purified cellulose, cotton linters or wood pulp.
non-volatile content* by mass	The mass remaining after the removal of volatiles, expressed as a percentage of the total mass.
non-volatile content* by volume	Non-volatile content* of a paint, expressed as a percentage of the total volume.
non-volatile matter	The residue remaining after the removal of volatiles under specified conditions.
oil length	The percentage of oil in the binder.
oil stains	Stains containing drying oils, oleo-resinous varnishes or alkyd resins.
oleo-resinous	Generally refers to varnishes composed of vegetable drying oils in conjunction with hard resins, which may be either natural or synthetic.
opacity	The ability of a paint to obliterate the colour difference of a substrate.
orange peel	The pockmarked appearance of a sprayed film due to its failure to flow out to a level surface.
organosol	A dispersion of finely divided resin particles in an organic liquid, which may be wholly or partially volatile.
overspray	Sprayed paint which misses the surface to be coated.
paint	A product in liquid form which, when applied to a surface, forms a dry film having protective, decorative or other specific technical properties.
paint or varnish system	The sum of the various coats of paint or varnish which are to be applied to or which have been applied to a substrate.
paint remover	A compound that softens paint or varnish and permits the softened material to be scraped off, or hosed off with water.
peeling	Loss of adhesion resulting in detachment and curling out of the paint film.

* Non-volatile *matter* is term preferred in this book.

Term	**Definition**
phosphating	The treatment of metal surfaces by chemical solutions containing metal phosphates and phosphoric acid as the main ingredients, to form an adherent corrosion inhibiting layer which serves as a good base for paint.
pickling	A treatment for the removal of rust and millscale from steel by immersion in an acid solution.
pigment	A substance, generally in fine powder form, which is practically insoluble in media and which is physically dispersed in the binder to impart specific physical and chemical properties (optical, protective, decorative, etc).
pigment volume concentration	The volume of pigments in a paint expressed as a percentage of the total volume of the non-volatile matter.
pile	A fibrous surface produced on a roller covering in which the fibre stands up from the basic covering material.
pinholes	Minute holes in a dry film which form during application and drying of paint.
pitting	The formation of holes or pits by localised corrosion in a metal surface.
plastic paint	*See* latex paint.
plastisol	Dispersion of finely divided resin particles in a plasticiser or mixture of plasticisers which on heating softens and fuses the resin particles.
polychromatic finish	A finish which has a metallic lustre and gives an iridescent scintillating effect, the colour of which varies when viewed from different angles.
polyester	Synthetic resin resulting from the polycondensation of various polyacids and polyols.
polyurethane resin	Synthetic resin resulting from the reaction of polyfunctional isocyanates with compounds containing hydroxyl groups.
popping	A small bubble-like defect in a paint film resulting from the expansion on hydration of extraneous material in the plaster substrate. (*Also known as* blowing).
pot-life	The time period after the mixing of reactive components in a two-component paint system, during which the mixed paint can be used.
pre-construction primer	A fast-drying, abrasion-resistant primer for application to blast-cleaned steel plates and bars before cutting and welding into complex structures.
pressure pot	A pressure vessel containing paint and fitted with a compressed air supply to force paint to a spray gun.
pre-treatment (metal)	The chemical treatment of unpainted metal surfaces before painting.
primer coat, prime coat	The first coat of a painting system that helps bind subsequent coats to the substrate and which may inhibit its deterioration.
promoter	*See* catalyst.
putty	A plastic material with a high mineral filler content used for filling deep holes or wide gaps.

Term	Definition
reducer	*See* thinner.
reference film	A film of the material under test prepared in the same manner and at the same time as the test film used for comparing the condition of the test film after a period of exposure with its original condition.
relative dry hiding power	The ability of a paint to reduce the contrast of a black and white surface to which it is applied and allowed to dry. It is quantitatively expressed in terms of the proportional spreading rate of paint required to produce the same contrast reduction as that obtained with the paint chosen as standard.
resin	A natural or synthetic material used to bind pigments together, and to the substrate.
retarder	A slow-evaporating solvent used as a thinner to slow down the speed of drying of a paint or lacquer to improve the application properties or produce a better film.
rheology	The study of the flow properties of materials.
roller (paint)	A tool for the application of paint having a revolving cylinder covered with lambswool, fabric, foamed plastics or other material.
ropiness	Brush marks in a dry film caused by paint with poor levelling properties or by brushing paint after the film has begun to set.
runs	Paint film defects in the form of sagging paint in narrow ribbons flowing downwards on vertical surfaces from surface irregularities after the surrounding paint has set.
sagging	Excessive flow of paint on vertical surfaces causing imperfections with thick lower edges in the paint film.
sanding	The levelling of or the removal of imperfections from a surface by rubbing with abrasive papers or compounds.
sanding surfacer	A heavily pigmented undercoat used for building the surface to a smooth condition.
sand finish	*See* texture paint.
saponification	A defect resulting from attack on a binder by alkali.
satin finish	*See* semi-gloss.
scrubbability	*See* washability.
sealant	A permanently flexible material used to fill expansion joints and gaps in buildings so as to provide a weatherproof seal.
sealer	A product used to seal substrates to prevent materials from bleeding through to the surface, to prevent reaction of the substrate with incompatible top coats or to prevent undue absorption of the following coat into the substrate.
seeds	Undesirable particles or granules other than dust, found in a paint or varnish.
semi-gloss (finish)	A surface with a specular gloss reading above 20 gloss units but not exceeding 50 gloss units when the specular direction is 60 degrees.
settling	Separation of paint in a container in which the pigments and other dense insoluble materials accumulate and aggregate at the bottom.

Term	Definition
sheen	Reflection from a flat surface observed when the surface is viewed at an angle of 5 degrees or less.
shelf life	The period a paint may be stored in sealed unopened containers without the paint showing any significant deterioration in quality.
shellac	The product obtained by refining seedlac by heat processes or by both heat and solvent processes.
shop primer	A fast-drying, abrasion-resistant primer for application in the workshop to fabricated steel units.
side sheen	*See* sheen.
silicone resins	Resins based on polymers composed of silicon, carbon and hydrogen.
size	A water-thinned sealer or adhesive made from glue, casein or cellulose derivatives.
skinning	The formation of a tough, skin-like covering on liquid paints and varnishes when exposed to air.
solids content	*See* non-volatile content by mass *and* volume solids.
solvent	A liquid, single or blended, which is volatile under normal drying conditions and in which the binder is completely soluble.
spar varnish	*See* marine varnish.
spatter	Small particles or drips of liquid paint thrown or expelled from a paint brush or roller during the application of paint.
spectral match	Two or more materials which appear identical in colour under any visible light source.
specular gloss	The luminous fractional reflectance at the specular direction. (*See* luminous fractional reflectance).
spot priming	The priming of small areas of a previously painted surface where the substrate has been exposed.
spraying	A method of applying paint in which paint is atomised by compressed air or by high liquid paint pressure, the atomised paint being directed onto the surface being coated.
spreading rate	The area, in square metres, covered by 1 litre of paint.
stain	A solution or suspension of colouring matter in a vehicle designed primarily to be applied to create colour effects rather than to form a protective coating. Also used to describe a transparent coating that colours without completely obscuring the grain of the surface.
stainer	*See* colorant.
stippling	The process of producing a broken colour or pimpled texture by applying spots of a different colour or by disturbing the surface of the paint film before it has set by means of a brush, roller, sponge etc.
stoving	*See* baking.
streaking	The formation of irregular lines or streaks of various colours in a paint film caused by contamination of insufficient or improper incorporation of colorant.
substrate	The surface to which a coat of paint or varnish is applied.

Term	Definition
sulfide staining	Dark grey or black stains which occur on paint films caused by the reaction of compounds of lead and/or other metals with sulfur compounds.
surface dry condition	The stage during the drying or curing process when glass beads of 150μm to 250μm diameter can be dropped from a height of 150 mm onto the paint surface and be removed by brushing without visible damage to the film.
tack-free	The stage during the drying and curing process when the paint film is free from stickiness or tackiness under firm pressure.
tackiness	The degree of stickiness of a paint film after a given drying time.
tack-rag	A piece of loosely woven cloth that has been dipped into a varnish oil and wrung out. When the rag becomes tacky or sticky, it is wiped over a surface to remove small particles of dust.
tear-drops	Drops of paint which collect on the bottom edges of items painted by dipping.
terebine	*See* drier.
texture	The roughness or irregularity of a surface.
texture paint	A paint that can be manipulated by brush, roller, trowel or other tool to produce various types of rough, sandy or patterned effects.
thinner	A volatile liquid, single or blended, added to paint to facilitate application by lowering the viscosity.
thixotropy	The property of a paint whereby the consistency is reduced on brushing or stirring but increases again on standing.
tie-coat	A coat applied to a previous coat to improve the adhesion of subsequent coats.
tinter	*See* colorant.
tint-base	The basic paint to which colorants are added as required to make a wide range of colours.
topcoat	*See* finish coat.
total solids	*See* non-volatile content by mass *or* non-volatile content by volume.
touch-dry	The stage during the drying or curing process when the paint film no longer feels sticky when lightly touched.
touch-up	Spot painting to repair damaged or defective films to produce an even fault-free finish.
undercoat	An intermediate coat formulated to prepare a primed surface or other prepared surface for the finishing coat.
varnish	A non-pigmented paint which dries to a hard-gloss, semi-gloss or flat transparent film.
vehicle	*See* medium.
vinyl resin	Synthetic resin resulting from the polymerisation and/or copolymerisation of vinyl monomers.
volume solids	*See* non-volatile content by volume.

Term	Definition
wash primer	*See* etch primer.
washability	The ability of a paint film to withstand repeated washing under specified conditions without deterioration or change in appearance.
water spotting	Spotty appearance on a dry paint film caused by the drying out of droplets of water.
weathering	The exposure of paint films to the weather to determine their behaviour to natural elements and pollution.
whip blasting	Light abrasive blast cleaning.
wrinkling	The development of wrinkles in a paint film during drying.

USEFUL DATA AND CONVERSION TABLES

1. METRIC CONVERSION GUIDE

In the following table, conversion factors for units of measurement relevant to the surface coatings industry are provided.
For each conversion, the factor is shown in two forms:
(i) rounded to three significant figures, which is acceptable for many purposes
(ii) accurate to seven significant figures for more precise conversions; an asterisk (*) following a factor indicates an exact conversion.

To convert from	*to*	*multiply by factor (i)*	*or factor (ii)*
Linear equivalents			
angstrom (Å)	nanometre (nm)	0.100	$1.000\,000 \times 10^{-1}$*
mil	micrometer (μm)	25.4	$2.540\,000 \times 10^{1}$*
inch (in)	millimetre (mm)	25.4	$2.540\,000 \times 10^{1}$*
inch (in)	metre (m)	0.0254	$2.540\,000 \times 10^{-5}$*
foot (ft)	metre (m)	0.305	$3.048\,000 \times 10^{-1}$*
yard (yd)	metre (m)	0.914	$9.144\,000 \times 10^{-1}$*
mile	kilometre (km)	1.61	1.609 344*
international nautical mile (n mile)	kilometre (km)	1.85	1.852 000*
Area equivalents			
square inch (in^2)	square centimetre (cm^2)	6.45	6.451 600*
square foot (ft^2)	square metre (m^2)	0.0929	$9.290\,304 \times 10^{-2}$*
square yard (yd^2)	square metre (m^2)	0.836	$8.361\,274 \times 10^{-1}$
Volume equivalents			
cubic inch (in^3)	cubic centimetre (cm^3)	16.4	$1.638\,706 \times 10^{1}$
cubic foot (ft^3)	cubic metre (m^3)	0.0283	$2.831\,685 \times 10^{-2}$
cubic yard (yd^3)	cubic metre (m^3)	0.765	$7.645\,549 \times 10^{-1}$

To convert from	*to*	*multiply by factor (i) or*	*factor (ii)*
fluid ounce (fl oz)	millilitre (mL)	28.4	$2.841\,306 \times 10^{1}$
pint, Imperial (pt)	litre (L)	0.568	$5.682\,613 \times 10^{-1}$
gallon, Imperial (gal)	litre (L)	4.55	4.546 090*
gallon, US (gal, US)	litre (L)	3.79	3.785 412
Mass equivalents			
ounce, avoirdupois (oz)	gram (g)	28.3	$2.834\,952 \times 10^{1}$
pound (lb)	kilogram (kg)	0.454	$4.535\,924 \times 10^{-1}$
hundredweight (cwt)	kilogram (kg)	50.8	$5.080\,235 \times 10^{1}$
ton	tonne (t)	1.02	1.016 047
short ton	tonne (t)	0.907	$9.071\,847 \times 10^{-1}$
Density equivalents			
pound/gallon (lb/gal)	kilogram/litre (kg/L)	0.100	$9.977\,637 \times 10^{-2}$
pound/cubic foot (lb/ft^3)	kilogram/cubic metre (kg/m^3)	16.0	$1.601\,846 \times 10^{1}$
Force equivalents			
kip	kilonewton (kN)	4.45	4.448 222
pound-force (lbf)	newton (N)	4.45	4.448 222
kilogram-force (kgf)	newton (N)	9.81	9.806 650*
dyne (dyn)	millinewton (mN)	0.010	$1.000\,000 \times 10^{-2}$*
Pressure equivalents			
pounds/square inch (psi)	kilopascal (kPa)	6.89	6.894 757
atmosphere (atm)	megapascal (MPa)	0.101	$1.013\,250 \times 10^{-1}$*
inch of water (in H_2O)	kilopascal (kPa)	0.249	$2.486\,642 \times 10^{-1}$
inch of mercury (in Hg)	kilopascal (kPa)	3.39	3.386 384
centimetre of water (cm H_2O)	pascal (Pa)	97.9	$9.789\,039 \times 10^{1}$
centimetre of mercury (cm Hg)	kilopascal (kPa)	1.33	1.333 222

To convert from	*to*	*multiply by factor (i) or*	*factor (ii)*
millimetre of mercury (mm Hg)	kilopascal (kPa)	0.133	$1.333\,222 \times 10^{-1}$
torr	kilopascal (kPa)	0.133	$1.333\,224 \times 10^{-1}$
bar	megapascal (MPa)	0.100	$1.000\,000 \times 10^{-1}$*
millibar	kilopascal (kPa)	0.100	$1.000\,000 \times 10^{-1}$*
Work or energy equivalents			
British thermal unit (Btu)	kilojoule (kJ)	1.06	1.055 058
calorie (cal)	joule (J)	4.19	4.186 800*
Power equivalents			
horsepower (hp)	kilowatt (kW)	0.746	$7.456\,999 \times 10^{-1}$
Viscosity equivalents			
centipoise (cP)	millipascal second (mPa.s)	1.00	1.000 000*
centistokes (cSt)	square millimetre/second (mm^2.s^{-1})	1.00	1.000 000*
poise (P)	pascal second (Pa.s)	0.100	$1.000\,000 \times 10^{-1}$*
stokes (St)	square metre/second (m^2.s^{-1})	1×10^{-4}	$1.000\,000 \times 10^{-4}$*
Surface tension equivalents			
dyne/centimetre (dyn/cm)	millinewtons/metre (mN/m)	1.00	1.000 000*

* As determined from samples. Other V. M. & P. naphthas have flash points as low as 0° C

2. TEMPERATURE CONVERSIONS

Locate known temperature in centre column. If known temperature is in °C, read °F equivalent in right-hand column. If known temperature is in °F, read °C equivalent in left-hand column.

For temperatures not given in table, or to convert to other temperature scales, use the following:

Temperature Scale

	Water Boiling Point	Water Freezing Point	Absolute Zero
°F	212°F	32°F	−459
°C	100°C	0°C	−273
K	373 K	273 K	0
°R	80°R	0°R	−218
°Rank	671° Rank	491° Rank	0

F = Fahrenheit
C = Celsius or Centigrade
K = Kelvin
R = Reamur
Rank = Rankine

$°C = (F - 32)\frac{5}{9}$
$°F = \frac{9}{5}°C - 32$
$°R = 8° C$
$K = °C + 273$
$°Rank = °F + 459.67$

°C	T	°F	°C	T	°F	°C	T	°F
−273	**−459.4**		−17.8	**0**	32.0	10.0	**50**	122.0
−268	**−450**		−17.2	**1**	33.8	10.6	**51**	123.8
−262	**−440**		−16.7	**2**	35.6	11.1	**52**	125.6
−257	**−430**		−16.1	**3**	37.4	11.7	**53**	127.4
−251	**−420**		−15.6	**4**	39.2	12.2	**54**	129.2
−246	**−410**		−15.0	**5**	41.0	12.8	**55**	131.0
−240	**−400**		−14.4	**6**	42.8	13.3	**56**	132.8
−234	−390		−13.9	**7**	44.6	13.9	**57**	134.6
−229	**−380**		−13.3	**8**	46.4	14.4	**58**	136.4
−223	**−370**		−12.8	**9**	48.2	15.0	**59**	138.2
−218	**−360**		−12.2	**10**	50.0	15.6	**60**	140.0
−212	**−350**		−11.7	**11**	51.8	16.1	**61**	141.8
−207	**−340**		−11.1	**12**	53.6	16.7	**62**	143.6
−201	**−330**		−10.6	**13**	55.4	17.2	**63**	145.4
−196	**−320**		−10.0	**14**	57.2	17.8	**64**	147.2
−190	**−310**		−9.4	**15**	59.0	18.3	**65**	149.0
−184	**−300**		−8.9	**16**	60.8	18.9	**66**	150.8
−179	**−290**		−8.3	**17**	62.6	19.4	**67**	152.6
−173	**−280**		−7.8	**18**	64.4	20.0	**68**	154.4
−169	**−273**	−459.4	−7.2	**19**	66.2	20.6	**69**	156.2
−168	**−270**	−454	−6.7	**20**	68.0	21.1	**70**	158.0
−162	**−260**	−436	−6.1	**21**	69.8	21.7	**71**	159.8
−157	**−250**	−418	−5.6	**22**	71.6	22.2	**72**	161.6
−151	**−240**	−400	−5.0	**23**	73.4	22.8	**73**	163.4
−146	**−230**	−382	−4.4	**24**	75.2	23.3	**74**	165.2
−140	**−220**	−364	−3.9	**25**	77.0	23.9	**75**	167.0
−134	**−210**	−346	−3.3	**26**	78.8	24.4	**76**	168.8
−129	**−200**	−328	−2.8	**27**	80.6	25.0	**77**	170.6
−123	**−190**	−310	−2.2	**28**	82.4	25.6	**78**	172.4
−118	**−180**	−292	−1.7	**29**	84.2	26.1	**79**	174.2
−112	**−170**	−274	−1.1	**30**	86.0	26.7	**80**	176.0
−107	**−160**	−256	−0.6	**31**	87.8	27.2	**81**	177.8
−101	**−150**	−238	0.0	**32**	89.6	27.8	**82**	179.6
−96	**−140**	−220	0.6	**33**	91.4	28.3	**83**	181.4
−90	**−130**	−202	1.1	**34**	93.2	28.9	**84**	183.2
−84	**−120**	−184	1.7	**35**	95.0	29.4	**85**	185.0
−79	**−110**	−166	2.2	**36**	96.8	30.0	**86**	186.8
−73	**−100**	−148	2.8	**37**	98.6	30.6	**87**	188.6
−68	**−90**	−130	3.3	**38**	100.4	31.1	**88**	190.4
−62	**−80**	−112	3.9	**39**	102.2	31.7	**89**	192.2
−57	**−70**	−94	4.4	**40**	104.0	32.2	**90**	194.0
−51	**−60**	−76	5.0	**41**	105.8	32.8	**91**	19.8
−46	**−50**	−58	5.6	**42**	107.6	33.3	**92**	197.6
−40	−40	−40	6.1	**43**	109.4	33.9	**93**	199.4
−34	**−30**	−22	6.7	**44**	111.2	34.4	**94**	201.2
−29	**−20**	−4	7.2	**45**	113.0	35.0	**95**	203.0
−23	**−10**	14	7.8	**46**	114.8	35.6	**96**	204.8
−17.8	**0**	32	8.3	**47**	116.6	36.1	**97**	206.6
			8.9	**48**	118.4	36.7	**98**	208.4
			9.4	**49**	120.2	37.2	**99**	210.2
						37.8	**100**	212.0

°C	T	°F	°C	T	°F	°C	T	°F	°C	T	°F
38	**100**	212	260	**500**	932	538	**1000**	1832	816	**1500**	2732
43	**110**	230	266	**510**	950	543	**1010**	1850	821	**1510**	2750
49	**120**	248	281	**520**	968	549	**1020**	1868	827	**1520**	2768
54	**130**	266	277	**530**	986	554	**1030**	1886	832	**1530**	2786
60	**140**	284	282	**540**	1004	560	**1040**	1904	838	**1540**	2804
66	**150**	302	288	**550**	1022	566	**1050**	1922	843	**1550**	2822
71	**160**	320	293	**560**	1040	571	**1060**	1940	849	**1560**	2840
77	**170**	338	299	**570**	1058	577	**1070**	1958	854	**1570**	2858
82	**180**	356	304	**580**	1076	582	**1080**	1976	860	**1580**	2876
88	**190**	374	310	**590**	1094	588	**1090**	1994	866	**1590**	2894
93	**200**	392	316	**600**	1112	593	**1100**	2012	871	**1600**	2912
99	**210**	410	321	**610**	1130	599	**1110**	2030	877	**1610**	2930
100	**212**	413.6	327	**620**	**1148**	604	**1120**	2048	882	**1620**	2948
104	**220**	428	332	**630**	1166	610	**1130**	2066	888	**1630**	2966
110	**230**	446	338	**640**	1184	616	**1140**	2084	893	**1640**	2984
116	**240**	464	343	**650**	1202	621	**1150**	2102	899	**1650**	3002
121	**250**	482	349	**660**	1220	627	**1160**	2120	904	**1660**	3020
127	**260**	500	354	**670**	1238	632	**1170**	2138	910	**1670**	3038
132	**270**	518	360	**680**	1256	638	**1180**	2156	916	**1680**	3056
138	**280**	536	366	**690**	1274	643	**1190**	2174	921	**1690**	3074
143	**290**	554	371	**700**	1292	649	**1200**	2192	927	**1700**	3092
149	**300**	572	377	**710**	1310	654	**1210**	2210	932	**1710**	3110
154	**310**	590	382	**720**	1328	660	**1220**	2228	938	**1720**	3128
160	**320**	608	388	**730**	1346	666	**1230**	2246	943	**1730**	3146
166	**330**	626	393	**740**	1364	671	**1240**	2264	949	**1740**	3164
171	**340**	644	399	**750**	1382	677	**1250**	2282	954	**1750**	3182
177	**350**	662	404	**760**	1400	682	**1260**	2300	960	**1760**	3200
182	**360**	680	410	**770**	1418	688	**1270**	2318	966	**1770**	3218
188	**370**	698	416	**780**	**1436**	693	**1280**	2336	971	**1780**	3236
193	**380**	716	421	**790**	1454	699	**1290**	2354	977	**1790**	3254
199	**390**	734	427	**800**	1472	704	**1300**	2372	982	**1800**	3272
204	**400**	752	432	**810**	1490	710	**1310**	2390	988	**1810**	3290
210	**410**	770	**438**	**820**	1508	716	**1320**	2408	993	**1820**	3308
216	**420**	788	443	**830**	1526	721	**1330**	2426	999	**1830**	3326
221	**430**	806	449	**840**	1544	727	**1340**	2444	1004	**1840**	3344
227	**440**	824	454	**850**	1562	732	**1350**	2462	1010	**1850**	3362
232	**450**	842	460	**860**	1580	738	**1360**	2480	1016	**1860**	3380
238	**460**	860	466	**870**	1598	743	**1370**	2498	1021	**1870**	3398
243	**470**	878	471	**880**	1616	749	**1380**	2516	1027	**1880**	3416
249	**480**	896	477	**890**	1634	754	**1390**	2534	1032	**1890**	3434
254	**490**	914	482	**900**	1652	760	**1400**	2552	10o8	**1910**	3452
			488	**910**	1670	766	**1410**	2570	1043	**1910**	3470
			493	**920**	1688	771	**1420**	2588	1049	**1920**	3488
			499	**930**	1706	777	**1430**	2606	1054	**1930**	3506
			504	**940**	1724	782	**1440**	2624	1060	**1940**	3524
			510	**950**	1742	788	**1450**	2642	1066	**1950**	3542
			516	**960**	1760	793	**1460**	2660	1071	**1960**	3560
			521	**970**	1778	799	**1470**	2678	1077	**1970**	3578
			527	**980**	1796	804	**1480**	2696	1082	**1980**	3596
			532	**990**	1814	810	**1490**	2714	1088	**1990**	3614
			538	**1000**	1832				1093	**2000**	3632

3. COMMONLY USED SOLVENTS: TABLE OF PROPERTIES

	Boiling Initial (b.p.) °C	*Range End Point °C*	*Flash Point °C*	*Density (kg/L) at 20° C*	*Refractive Index at 20° C*
KETONES					
Acetone	55	57	−9	0.793	1.3591
Methyl acetone	54	70	−9	0.831	. . .
Methyl ethyl ketone	77	82	1	0.809	1.3791
Diethyl ketone	100	104	15	0.815	1.3927
Methyl propyl ketone	101	107	13	0.810	1.3895
Hexone (methyl isobutyl ketone)	112	118	23	0.802	1.3959
Methyl n-butyl ketone	114	137	32	0.818	1.4024
Mesityl oxide	117	139	32	0.855	1.4439
Diacetone alcohol	153	160	60	0.936	1.4204
Dipropyl ketone	138	144	40	0.816	1.4072
Methyl n-amyl ketone	147	153	54	0.818	1.4110
Diisobutyl ketone	164	169	60	0.808	. . .
Methyl n-hexyl ketone	169	173	71	0.820	1.4161
Cyclohexanone	130	173	54	0.946	1.4524
Methylcyclohexanone	114	173	57	0.921	. . .
Acetonyl acetone	188	193	85	0.973	1.449
Phorone	114	198	80	Crystals	1.4998
ETHERS					
Diethyl "Cellosolve"	119	125	38	0.849	. . .
Diethyl "Carbitol"	181	189	82	0.909	. . .
1, 4-Dioxane	98	102	18	1.034	1.422
Ethyl ether	. . .	. . .	−40	0.716	1.354 (17°)
Isopropyl ether	66	69	−9	0.724	1.368 (23°)
ESTERS					
Methyl acetate	53	55	−1	0.908	1.3619
Ethyl acetate	70	80	7	0.886	1.3727
Isopropyl acetate (95%)	84	94	2	0.870	1.3770
Isopropyl acetate (85%)	82	90	15	0.857	. . .
Ethyl propionate	90	118	15	0.891	1.3844
sec-Butyl acetate	105	127	31	0.858	1.389
Isobutyl acetate	114	118	31	0.870	1.3997
n-Butyl acetate	119	127	40	0.876	1.3951
Ethyl butyrate	107	131	29	0.880	1.3932
sec-Amyl acetate	121	144	44	0.862	. . .
Amyl acetate (Mixed isomers)	127	155	44	0.862	. . .
Butyl butyrate	152	170	52	0.874	1.4049
n-Butyl propionate	124	171	40	0.874	. . .
Methyl amyl acetate	140	147	40	0.857	1.4008
Hexyl (2-ethylbutyl) acetate	157	164	54	0.876	1.4103
sec-Hexyl acetate	129	158	43	0.861	. . .
Ethyl lactate	119	176	54	1.034	1.4118
Octyl (2-ethylhexyl) acetate	195	203	82	0.873	1.4300
Cyclohexanyl acetate	165	193	63	0.963	. . .
Isopropyl lactate	149	167	54	0.988	. . .
Butyl lactate	145	230	80	0.980	. . .
Glycol diacetate	184	191	96	1.107	1.415
Diglycol diacetate	238	251	135	1.116	. . .
Diethyl carbonate	87	127	21	0.957	1.3852
3-Methoxybutyl acetate	135	173	77	0.956	. . .
Methyl 'Cellosolve' acetate	143	145	60	1.005	1.4025
'Cellosolve' acetate	145	166	50	0.974	1.4030
Butyl 'Cellosolve' acetate	188	192	82	0.943	. . .

3. (continued)

	Boiling Initial (b.p.) °C	*Range End Point °C*	*Flash Point °C*	*Density (kg/L) at 20° C*	*Refractive Index at 20° C*
Methyl 'Carbitol' acetate	203	212	82	1.040	. . .
'Carbitol' acetate	211	220	110	1.011	1.4230
Butyl 'Carbitol' acetate	236	249	116	0.987	. . .
ETHER ALCOHOLS					
Methyl 'Cellosolve'	121	126	46	0.966	1.4028
'Cellosolve' (Trade Mark)	133	137	54	0.931	1.4080
Isopropyl 'Cellosolve'	140	143	46	0.906	. . .
Butyl 'Cellosolve'	163	172	74	0.902	1.4190
Methyl 'Carbitol'	190	194	93	1.035	1.4244
'Carbitol' (Trade Mark)	189	203	96	1.027	. . .
Butyl 'Carbitol'	220	231	116	0.955	1.4290
Benzyl 'Cellosolve'	254	258	130	1.070	. . .
ALCOHOLS					
Methanol (anhydrous)	64	66	18	0.793	1.329
Ethanol (anhydrous)	77	79	21	0.791	1.361
Isopropanol (anhydrous)	82	83	19	0.786	1.3776
sec-Butanol	96	103	31	0.808	1.397
Isobutanol	107	111	38	0.803	1.396
n-Butanol	116	119	44	0.811	1.3974
sec-Amyl alcohol	117	124	43	0.810	. . .
Amyl alcohol (mixed isomers)	121	139	44	0.814	. . .
Methyl amyl alcohol	131	132	45	0.807	1.4087
Hexyl (2-ethylbutyl) alcohol	144	156	57	0.833	1.4229
Octyl (2-ethylhexyl) alcohol	182	201	85	0.834	1.4300
Cyclohexanol	150	182	69	0.951	1.4656
Benzyl alcohol	199	204	140	1.047	1.5399
GLYCOLS					
Ethylene glycol	194	204	116	1.1154	1.431
Propylene glycol	183	200	110	1.038	1.433
Diethylene glycol	235	265	143	1.119	1.446
Triethylene glycol	275	305	166	1.125	. . .
HYDROCARBONS					
Benzene	78	81	−15	0.878	1.5014
Toluene	109	111	6	0.866	1.4962
Xylene	127	159	27	0.862	. . .
Hi-Flash naphtha	149	193	46	0.860	. . .
Petroleum naphtha	96	129	−1	0.735	. . .
V. M. & P. naphtha	168	210	49*	0.763	. . .
Mineral spirits	152	207	60	0.769	. . .
Cyclohexane	80	83	−14	0.779	1.4273
Tetrahydronaphthalene	196	217	82	0.972	1.5461
Decahydronaphthalene	186	194	57	0.885	. . .
'Solvesso' No. 1	94	139	7	0.796	. . .
'Solvesso' No. 2	134	185	43	0.847	. . .
'Solvesso' No. 3	179	214	69	0.877	. . .
'Safe-T-Esso'	157	217	63	0.874	. . .
CHLORINATED COMPOUNDS					
Ethylene dichloride	82	84	21	1.256	1.444
Monochlorobenzene	130	134	35	1.108	1.525
Orthodichlorobenzene	177	180	80	1.308	1.549
Dichloroethyl ether	172	178	85	1.222	1.4570

3. (continued)

	Boiling Initial (b.p.) °C	*Range End Point °C*	*Flash Point °C*	*Density (kg/L) at 20° C*	*Refractive Index at 20° C*
Carbon tetrachloride	76	77	....	1.597	1.461
Chloroform	60	62	....	1.492	1.447
Propylene dichloride	94	98	21	1.160	1.439
Trichlorethylene	85	87	....	1.466	1.474 (27°)
NAVAL STORES SOLVENTS					
Gum spirits of turpentine	154	201	43	0.868	...
Steam-distilled turpentine	155	173	46	0.857	...
Dipentene	172	200	63	0.853	1.473
'Hercosol'	176	262	66	0.912	...
Pine oil	99	226	85	0.925–0.935	...
Heavy Pine Oil	211	219	93	0.937–0.942	...
Fenchone	190	210	71	0.948	1.4647
FURANE DERIVATIVES					
Furfural	110	169	63	1.159	1.5261
Furfural acetate	180	186	85	1.119	1.4627
Furfural alcohol	152	220	80	1.137	1.4852
Tetrahydrofurfuryl alcohol	119	204	82	1.051	...

4. SURFACE TENSION DATA

Surface Tension of Organic Compounds in water

% = Mass % of solute
γ = Surface tension (mN/m)

Solute	*T* °C	% / γ							
Acetic acid	30	%	1.00	2.475	5.001	10.01	30.09	49.96	69.91
		γ	68.00	64.40	60.10	54.60	43.60	38.40	34.30
Acetone	25	%	5.00	10.00	20.00	50.00	75.00	95.00	100.00
		γ	55.50	48.90	41.10	30.40	26.80	24.20	23.00
Acetonitrile....	20	%	1.13	3.35	11.77	20.20	37.58	61.33	81.22
		γ	69.02	63.03	47.61	39.06	31.84	30.02	29.02
o-Aminobenzoic acid	25	%	12.35	22.36	30.45	37.44			
		γ	71.96	73.23	74.54	75.79			
m-Aminobenzoic acid	25	%	12.35	22.36	30.45	37.44			
		γ	73.30	74.59	76.16	77.89			
p-Aminobenzoic acid	25	%	12.35	22.36	30.45	37.44			
		γ	73.38	74.79	76.32	78.20			
Aminobutyric .. acid	25	%	4.96	9.34	13.43				
		γ	71.91	71.67	71.40				

Solute	T °C	% γ							
Ammonium lactate . . .	29	%	30.00	50.00	60.00	70.00	80.00	90.00	
		γ	35.40	34.40	35.40	35.60	38.20	44.50	
n-Butanol	30	%	0.04	0.41	9.53	80.44	86.05	94.20	97.40
		γ	69.33	60.38	26.97	23.69	23.47	23.29	22.25
n-Butyric acid	25	%	0.14	0.31	1.05	8.60	25.00	79.00	100.00
		γ	69.00	65.00	56.00	33.00	28.00	27.00	26.00
Dioxan	26	%	0.44	2.20	4.70	11.14	20.17	35.20	55.00
		γ	69.83	65.64	62.45	56.90	51.57	45.30	39.27
Dioxan	26	%	67.68	76.45	83.02	91.90	95.60	97.77	
		γ	36.95	35.80	35.00	33.95	33.60	33.10	
Formic acid . . .	30	%	1.00	5.00	10.00	25.00	50.00	75.00	100.00
		γ	70.07	66.20	62.78	56.29	49.50	43.40	36.51
Glycerol	18	%	5.00	10.00	20.00	30.00	50.00	85.00	100.00
		γ	72.90	72.90	72.40	72.00	70.00	66.00	63.00
Glycine	25	%	3.62	6.98	10.12	13.10			
		γ	72.54	73.11	73.74	74.18			
Hydrocinnamic . acid	21.5	%	7.02	12.62	18.39	26.09	31.06	38.25	47.93
		γ	69.08	66.49	63.63	59.25	56.14	52.96	47.24
Methyl acetate	25	%	0.66	1.29	2.29	3.56			
		γ	66.33	62.92	58.22	55.08			
Morpholine . . .	20	%	8.56	19.39	30.41	50.45	69.93	80.14	92.00
		γ	67.80	62.62	59.15	52.85	47.05	43.62	41.60
Potassium lactate	29	%	40.00	50.00	60.00	70.00			
		γ	66.40	66.40	65.40	63.40			
Phenol	20	%	0.024	0.047	0.118	0.417	0.941	3.76	5.62
		γ	72.60	72.20	71.30	66.50	61.10	46.00	42.30
n-Propanol	25	%	0.1	0.5	1.0	50.0	60.0	80.0	90.0
		γ	67.10	56.18	49.30	24.34	24.15	23.66	23.41
Propionic acid	25	%	1.91	5.84	9.80	21.70	49.80	73.90	100.00
		γ	60.00	49.00	44.00	36.00	32.00	30.00	26.00
Sodium lactate	29	%	1.00	10.00	30.00	40.00	50.00	60.00	70.00
		γ	70.40	69.60	68.50	64.80	45.40	56.70	60.70
Sucrose	25	%	10.00	20.00	30.00	40.00	55.00		
		γ	72.50	73.00	73.40	74.10	75.70		

Surface Tension of Methyl Alcohol in Water

% = Volume % of alcohol
γ = Surface tension (mN/m)

T°C	*%*	*7.5*	*10.00*	*25.0*	*50.0*	*60.0*	*80.0*	*90.0*	*100.0*
20	γ	60.90	59.04	46.38	35.31	32.95	27.26	25.36	22.65
30	γ	59.33	57.27	45.30	34.52	32.26	26.48	24.42	21.58
50	γ	56.19	55.01	43.24	32.95	30.79	25.01	22.55	19.52

Surface Tension of Ethyl Alcohol in Water

% = Volume % alcohol
γ = Surface tension (mN/m)

T°C	*%*	*5.00*	*10.00*	*24.00*	*34.00*	*48.00*	*60.00*	*72.00*	*80.00*	*96.00*
20	γ	. . .	. . .	. . .	33.24	30.10	27.56	26.28	24.91	23.04
40	γ	54.92	48.25	35.50	31.58	28.93	26.18	24.91	23.43	21.38
50	γ	53.35	46.77	34.32	30.70	28.24	25.50	24.12	22.56	20.40

Water Against Air

Temperature °C	*Surface tension mN/m*	*Temperature °C*	*Surface tension mN/m*	*Temperature °C*	*Surface tension mN/m*
−8	77.0	15	73.49	40	69.56
−5	76.4	18	73.05	50	67.91
0	75.6	20	72.75	60	66.18
5	74.9	25	71.97	70	64.4
10	74.22	30	71.18	80	62.6
				100	58.9

Interfacial Tension

Surface Tension at the interface between two liquids
(Each liquid saturated with the other)

Liquids	*Temperature °C*	*Interfacial tension mN/m*	*Liquids*	*Temperature °C*	*Interfacial tension mN/m*
Benzene—mercury	20	357	Water—heptylic acid	20	7.0
Ethyl ether—mercury	20	379	Water—n-hexane	20	51.1
Water—benzene	20	35.0	Water—mercury	20	375.0
Water—carbon tetrachloride	20	45.0	Water—n-octane	20	50.8
Water—ethyl ether	20	10.7	Water—n-octyl alcohol	20	8.5

5. REFRACTIVE INDICES

Liquids

Suitable media for refractometry; values at or near 20° C

Liquid	*D*
Water	1.333
Paraldehyde	1.405
Methyl acetate	1.450
Carbon tetrachloride	1.46
Glycerol	1.47
Liquid paraffin	1.48
Toluene	1.497
Benzene	1.501
Ethyl salicylate	1.523
Chlorobenzene	1.525
Methyl salicylate	1.538
Ethyl cinnamate	1.559
Benzyl benzoate	1.568
Aniline	1.586
Quinoline	1.627
α-Monobromonaphthalene	1.660
Mercury potassium iodide	1.717
Methylene iodide	1.737
Methylene iodide and sulfur (saturated)	1.78
Barium mercuric iodide aq	1.793
Potassium iodide and mercuric iodide aq	1.82*
Solution 35% by mass CH_2I_2; " 31% " SnI_4; " 16% " AsI_3; " 8% " SbI_3; " 10% " S	1.868
Hydrogen disulfide	1.885
Phosphorus in carbon disulfide	1.95*
Yellow phosphorus, 8 parts by mass; " sulfur 1 " " "; Methylene iodide 1 " " "	2.06
Mercuric iodide in aniline or quinoline	2.2*
Oil, paraffin	1.44
" olive	1.46
" turpentine	1.47
" cedar	1.516
" cloves	1.532
" cinnamon	1.601

* Maximum value obtainable.

Refractive index of water at 20° C for various wavelengths

Wave-length nm	1256	670.8	656.3	643.8	589.3	546.1	508.6	486.1	480.0	404.7	303.4	214.4
Refractive Index	1.3210	1.3308	1.3311	1.3314	1.3330	1.3345	1.3360	1.3371	1.3374	1.3428	1.3581	1.4032

Temperature coefficient is -8.0×10^{-5} per °C.

Optical Cement

Canada Balsam D = 1.530.

Plastic Materials

Material		D	Transmission range
Polystyrene	15° C	1.592	
	35° C	1.589	0.34μm to 2μm
	55° C	1.586	
Polycyclohexyl-methacrylate	15° C	1.507	
	35° C	1.504	
	55° C	1.501	
Polymethyl-methacrylate	20° C	1.491	0.34μm to 2μm

Temperature coefficient of refractive index for above three polymers is -14×10^{-5} per °C.

Refractive indices, dispersions and useful transmission range

Material	D	Temp. °C	Transmission range (μm)
Ice	1.31		
Sodium fluoride	1.3255		to 10
Lithium fluoride	1.3921	21	0.105 to 6
Strontium fluoride	1.43		to 11
Calcium fluoride (fluorite)	1.4339	15	0.125 to 9
Sodium aluminium sulfate (soda alum)	1.4388		
Potassium aluminium sulfate (potash alum)	1.4564		
Fused silica	1.4587		0.19 to 3.5
Ammonium aluminium sulfate (ammonia alum)	1.4594		
Barium fluoride	1.47		<0.2 to 13
Potassium chloride	1.4904	18	0.2 to 21
Sodium chloride, natural	1.5443	18	0.2 to 15
Sodium chloride, synthetic	1.5498		
Cadmium fluoride	1.55		to 10
Potassium bromide	1.5599	20	0.2 to 30
Topaz	1.63		
Potassium iodide	1.6666		0.25 to 31
Caesium bromide	1.6971	24	to 40
Magnesium aluminium oxide: natural spinel $MgO \cdot Al_2O_3$	1.715 to 1.723		
synthetic spinels $MgO \cdot 3Al_2O_3$	1.7266		0.2 to 6
synthetic spinels $MgO \cdot 5Al_2O_3$	1.729		
Grossularite (garnet)	1.735		
Magnesium oxide (periclase)	1.7373	20	0.22 to 7
Lead fluoride	1.76		to 11
Silver chloride	2.06*c*.		to 25
Zinc sulfide (blende or sphalerite)	2.370		

Refractive indices, dispersions and useful transmission range

Material		D	Temp. °C	Transmission range (μm)
Diamond		2.4195		
Thallium bromide-iodide (KRS 5)		2.629		0.5 to 40
Birefringent				
Potassium dihydrogen phosphate (KDP)	ord.	1.5095		
	ex.	1.4684		
Ammonium dihydrogen phosphate (ADP)	ord.	1.5242		<0.2 to 1.5
	ex.	1.4787		
Crystalline quartz	ord.	1.5443	18	0.18 to 3.5
	ex.	1.5534		
Sodium nitrate	ord.	1.5874		
	ex.	1.3361		
Calcspar	ord.	1.6584	18	0.2 to 2
	ex.	1.4864		
Aluminium oxide (corundum)	ord.	1.7686		0.2 to 6
	ex.	1.7604		
Titanium dioxide (rutile)	ord.	2.6		0.48 to 6
	ex.	2.9		

6. PHYSICAL PROPERTIES OF PIGMENTS

Opaque White Pigments

Pigment Name and Chemical Composition	Density t/m^3	Particle Characteristics	Refractive Index
White lead (basic sulfate) $PbSO_4 \cdot PbO$	6.46	. . .	Av. 1.93
Lithopone (regular), ZnS (28–30%), $BaSO_4$ (72–70%)	4.30	fine comp.	2.3 (ZnS)–1.64 ($BaSO_4$)
Zinc white (ordinary), ZnO	5.65	v. fine cryst.	2.02, 2.00
(ordinary), ZnO	. . .	spicules, founets	2.02, 2.00
White lead (basic carbonate), $2PbCO_3 \cdot Pb(OH)_2$	6.70	v. fine cryst.	1.94, 2.09
Antimony oxide, Sb_2O_3	5.75	v. fine cryst.	valentinite, 2.18, and 2.35 senarmonite, 2.09 (isot.)
Zirconium oxide (baddeleyite), ZrO_2	5.69	. . .	2.13, 2.20, 2.19 Av. 2.40
Titanium dioxide (anatase), TiO_2	3.9	min. round.	and 2.5
(rutile), TiO_2	4.2	round. or prism.	2.9, 2.6

Transparent White Pigments[†]

Pigment Name and Chemical Composition	*Density t/m³*	*Particle Characteristics*	*Refractive Index*
Diatomaceous earth, SiO_2	2.31	min. fossil forms	*n* mostly 1.435, some 1.40
Aluminum stearate, $Al(C_{18}H_{35}O_2)_3$	0.99	agg. of spher.	1.49
Pumice (volcanic glass), Na, K, Al, silicate	. . .	vesicular vitr. frag.	1.50
Aluminum hydrate, $Al(OH)_3$	2.45	v. fine amorph. part.	1.50–1.56
Gypsum, $CaSO_4 . 2H_2O$	2.36	fine cryst.	1.520, 1.530, 1.523
Silica (quartz), SiO_2	2.66	cryst. frag.	1.553, 1.544
(chalcedony), SiO_2	2.6	cryst. agg.	1.54
China clay (kaolinite), $Al_2O_3 . 2SiO_2 . 2H_2O$	2.60	fine, vermicular cryst.	1.558, 1.565, 1.564 (all ± .005)
Talc, $3MgO . 4SiO_2 . H_2O$	2.77	platy frag.	1.539, 1.589, 1.589
Mica (muscovite), $H_2KAl_3(SiO_4)_3$	2.89	platy frag.	1.563, 1.604, 1.599
Anhydrite, $CaSO_4$	2.93	cryst. frag.	1.570, 1.614, 1.575
Chalk (whiting), $CaCO_3$	2.70	hollow spherulites	Σc 1.510, Σc 1.645
Barytes (barite, nat.), $BaSO_4$	4.45	cryst. frag.	1.636, 1.648, 1.637
(blanc fixe, art.), $BaSO_4$	4.36	v. fine cryst. agg.	1.62–1.64
Barium carbonate, $BaCO_3$ (witherite)	4.3	. . .	1.529, 1.677, 1.676

[†] Also called "Extender" or "Inert" White Pigments.

Iron Oxide Pigments

Ochre, yellow (goethite), $Fe_2O_3 . H_2O$, clay, etc.	2.9–4.0	irr. spherulites	2.0 (isot. part); 2.05–2.31; 2.08–2.40
Sienna, raw (geothite), $Fe_2O_3 . H_2O$, clay, etc.	3.14	uneven spherulites	1.87–2.17 (mostly 2.06) (isot.)
Sienna, burnt, Fe_2O_3, clay, etc.	3.56	uneven, round. part.	1.85 (var.) (isot.)
Umber, raw, $Fe_2O_3 + MnO_2 + H_2O$, clay, etc.	3.20	uneven, round.	mostly 1.87–2.17
Umber, burnt, $Fe_2O_3 + MnO_2$, clay, etc.	3.64	uneven, round.	mostly 2.2–2.3
Iron oxide red (haematite), Fe_2O_3	5.2	min. cryst.	2.78, 3.01

7. VISCOSITY CONVERSION GUIDE

The following table relates efflux times and other measures of viscosity obtained by various viscometers commonly used in the surface coatings industry.

The values quoted are approximately true for Newtonian liquids of density 1.00 kg/L at 25° C. They are offered only as a guide and should not be used for converting test results from one scale to another.

Viscosity cSt	*AS 1580 Cup**	*Ford Cup #4*	*Zahn #2*	*Zahn #3*	*Zahn #4*	*Gardner Holdt*	*Krebs Stormer*
1						A-5	
6						A-4	
14			17			A-3	
22			19			A-2	
32			20			A-1	
50	20		22			A	
65	25	22	27			B	
85	32	27	34			C	
100	37	30	41			D	
125	45	36	49			E	
140	50	40	58			F	
165	60	46	66	18		G	
200	72	55	82	23		H	52
225	80	61		25		I	54
250	90	68		27	20	J	56
275	100	74		32	22	K	59
300	105	80		34	24	L	61
320	115	85		36	25	M	62
340	120	90		39	26	N	63
370		100		41	28	O	64
400				46	30	P	65
435				50	33	Q	66
470				52	34	R	67
500				57	37	S	68
550				63	40	T	69
630				68	44	U	71
885					64	V	78
1070						W	85
1290						X	95
1760						Y	100
2270						Z	105
2700						Z-1	114
3620						Z-2	129
4630						Z-3	136
6340						Z-4	
9850						Z-5	
14 800						Z-6	

*AS 1580 method 214.2; cup technically identical with BS 1733 and BS 3900: Part A6

8. HARDNESS CONVERSIONS

Mohs Hardness Scale

Hardness number	*Original scale*	*Modified scale*
1	Talc	Talc
2	Gypsum	Gypsum
3	Calcite	Calcite
4	Fluorite	Fluorite
5	Apatite	Apatite
6	Orthoclase	Orthoclase
7	Quartz	Vitreous silica
8	Topaz	Quartz or Stellite
9	Corundum	Topaz
10	Diamond	Garnet
11	....	Fused zirconia
12	....	Fused alumina
13	....	Silicon carbide
14	....	Boron carbide
15	....	Diamond

Hardness Conversion

Pencil Range	*Numerical Value*	*Approximate Sward Hardness*
5H-6H	<10	
4H-5H	9	
3H-4H	8	
2H-3H	7	
H-2H	6	
F-H	5	
HB-F	4	28
B-HB	3	26
2B-B	2	24
3B-2B	1	22
<3B	0	20

Results are reported as the range of hardness of consecutive pencils, the softer of which crumbles, and the harder of which penetrates the film.

Example: Pencil hardness HB-F is equivalent to a numerical value of.4.

9. PARTICLE SIZE DATA

Fineness of Grind: Conversion Table

HEGMAN & NS (North or National Std.)	0	1	2	3	4	5	6	7	8
mil (0.001 in)	4.0	3.5	3.0	2.5	2.0	1.5	1.0	.5	0
PC (Production Club)	0	1.25	2.50	3.75	5.00	6.25	7.50	8.75	10.00
micrometres*	100.0	87.5	75.0	62.5	50.0	37.5	25.0	12.5	0

* For simplicity, 1 mil = 25μm (instead of 25.4).

Particle Size Table: Sieve Information

SIEVE SIZE	*OPENING*		
U.S. mesh	*inches*	*micrometres*	*millimetres*
4	.187	4760	4.76
5	.157	4000	4.00
6	.132	3360	3.36
7	.111	2830	2.83
8	.0937	2380	2.38
10	.0787	2000	2.00
12	.0661	1680	1.68
14	.0555	1410	1.41
16	.0469	1190	1.19
18	.0394	1000	1.00
20	.0331	840	.84
25	.0280	710	.71
30	.0232	590	.59
35	.0197	500	.50
40	.0165	420	.42
45	.0138	350	.35
50	.0117	297	.297
60	.0098	250	.250
70	.0083	210	.210
80	.0070	177	.177
100	.0059	149	.149
120	.0049	125	.125
140	.0041	105	.105
170	.0035	88	.088
200	.0029	74	.074
230	.0024	62	.062
270	.0021	53	.053
325	.0017	44	.044
400	.0015	37	.037

10. PIGMENT APPLICATION:

Spreading Rates at Various Film Thicknesses

Spreading Rate

	US Gallon		*Imp. Gallon*		*Litre*		
mil	*ft²*	*m²*	*ft²*	*m²*	*μm*	*m²*	*ft²*
.25	6,416.0	596.0	7,706.0	715.8	6.3	159	1078
.50	3,208.0	298.0	3,853.0	357.9	12.5	80	861
.75	2,138.0	198.6	2,568.0	238.5	19	53	566
1.00	1,604.0	149.0	1,926.0	178.9	25	40	430
1.50	1,069.0	99.3	1,284.0	119.3	38	26	283
2.00	802.0	74.5	963.2	89.5	50	20	215
2.50	641.6	59.6	770.6	71.6	63	16	171
3.00	534.7	49.7	642.2	59.7	75	13.3	143
3.50	458.3	42.6	550.4	51.2	88	11.4	122
4.00	401.4	37.3	481.6	44.8	100	10.0	108
4.50	356.4	33.1	428.0	39.8	112	8.9	96
5.00	320.8	29.8	385.3	35.8	125	8.0	86
6.00	267.3	24.8	321.0	29.8	150	6.7	72
7.00	229.1	21.3	275.1	25.6	175	5.7	62
8.00	200.5	18.6	240.8	22.3	200	5.0	54
9.00	178.2	16.6	214.0	19.9	225	4.4	48
10.00	160.4	14.9	192.6	17.9	250	4.0	43

11. COLOUR CONVERSIONS (to the closest Gardner half-unit)

Gardner Standard 1963	US Rosin Standard	Hellige Varnish Comparator	Union Colour ASTM	Iodine Scale (g/100mL H_2O)	Potassium Dichromate Scale (g/100mL H_2SO_4)	P.R.S. Shellac Scale
1				.0013	.0039	3/8
2				.0019	.0048	1/2
			1			
3				.0024	.0071	5/8
4				.0031	.0112	3/4
			1½			7/8
5		1L		.0043	.0205	1
						1¼
6		1		.0072	.0322	1½
	X	2L				
7	WW	2	2	.0090	.0334	1¾
	WG					
8		3L		.0120	.0515	2
	N	3				
9			2½	.0204	.0780	2½
	M	4L				
10		4		.0403	.1640	3
	K		3			
11	I	5L		.0703	.2500	3½
			3½			
12	H	5		.1050	.3800	4
		6L	4			5
13	G	6		.1330	.5720	5½
			4½			6
14	F	7L		.1820	.7630	6¼
						6½
15		7	5	.2430	1.0410	6¾
						7
16		8L	6	.3440	1.2800	7¼
	E		7			7½
17		8		.4250	2.2200	8
			8			9
18	D	9L		.5400	3.0000	

12. PHYSICAL AND TOXICOLOGICAL DATA ON GASES AND VAPOURS

Vapour	*Rel. Density (Air = 1)*	*Flash Limit (% vol.) (20°C, 760 mm Hg)*		*Threshold of smell ppm*	*M.A.C. – ppm*
		Bottom	Top		
Acetone	2.00	1.6	13	450	1000
Acrylonitrile	1.83	3.0	17	-	20
Aliphatic hydrocarbon C-(50-200°C)	-	-	-	300	500 (USA)
n–Butanol	2.55	1.4	11.3	25	100
Carbon tetrachloride	5.32	-	-	70	10
Ethanol	1.59	3.1	19	-	1000
Formaldehyde	1.04	-	-	20	5
Hydrogen sulfide	1.19	4.3	45	0.1	10
Methanol	1.11	6.0	36.5	-	200
Mono-ethanolamine	-	-	-	-	3
Methylene chloride	3.93	13	18	-	500
Styrene monomer	3.6	1.1	8	25	100
Toluene	3.18	1.27	7.0	-	200
Trichlorethylene	4.62	-	-	-	100
Vinyl chloride	2.15	4.0	31	-	500
Xylene (ortho)	3.66	1.0	7.6	0.2	100
Xylene (meta)	3.66	1.1	7.0	-	100

* Figures extracted from AS 2430 Part 1–1982 "Classification of Hazardous Areas"

CONTRIBUTORS AND REVIEWERS

A Adomenas	Technical consultant.
L W Beatty	Technical Service Systems Officer, A C Hatrick Chemicals Pty Ltd.
R P Best	Director, Instrumental Colour Systems Ltd.
G J Biddle	Technical Manager, Protective Coatings and Marine, Taubmans Industries Ltd.
P R Biddles	General Manager, P T Dimet, Indonesia.
D M Bignell	Manager, Colours and Chemicals Pty Ltd.
I J Blainey	Technical Manager, Colours and Chemicals Pty Ltd.
R J Bolton	Product Manager, Liquid Colourants, Ciba Geigy Australia Ltd.
D D Bonney	Vice President, Technical and Industrial, Dutch Boy Philippines Inc.
G Buckley	Management Services Manager, BJN Paints.
L S Cash	Formerly Research Manager and Chief Chemist, Meggitt Ltd.
E V Collins	Consultant.
L C Cook	Marketing Manager, Colours and Chemicals Pty Ltd.
D Corless	Manager, Research and Development Laboratory, British Paints (Australia) Pty Ltd.
J B Craddock	Manager, Pacific Region Laboratory, Rohm and Haas Company.
D F Crossing	Chief Chemist, NSW, Berger Paints (Australia) Pty Ltd.
H V Davies	Technical Service Manager, Tioxide Australia Pty Ltd.
M R Domars	Chief Chemist, Regal Paints Pty Ltd.
R T Drew	Chief Chemist, Wattyl Pty Ltd.
R Drummond	Technical Manager, A C Hatrick Chemicals Pty Ltd.
K Duffy	Executive Director, Australian Paint Manufacturers Federation.
D H Duncan	Assistant Chemical Control Manager, A C Hatrick Chemicals Pty Ltd.
W T Elliott	Manager Research, BJN Paints.
D R Fellows-Smith	Chemical Safety Officer, Dulux Australia Ltd.
I D Fletcher	Regional Manager, Zerak Pty Ltd.
T D Flynn	Group Development Manager, Wattyl Ltd.
J H Foxton	Manager, Marketing Australasia, Pandrol Australia Pty Ltd.
K L Freeman	Teacher, Department of Technical and Further Education.
W W Gallagher	Senior Chemist, Paint Industries (Aust.) Pty Ltd.
J A Gilmour	Registrar, National Association of Testing Authorities.
W J Gutteridge	Market Development Manager, Chemical Specialities, Rohm and Haas Australia
J K Haken	Professor, Department of Polymer Science, University of New South Wales.
J Hale	Section Leader, Exploratory Research, A C Hatrick Chemicals Pty Ltd.
P F Hannaford	Senior Chemist, Laporte Australia Limited.
E J Harper	Technical Service Manager, A C Hatrick Chemicals Pty Ltd.
E J Hemmings	Manager, Protection/Marine Laboratory, British Paints (Australia) Pty Ltd.
B Hill	Chemical Superintendant, A C Hatrick Chemicals Pty Ltd.
M R Hughes	Research Manager, A C Hatrick Chemicals Pty Ltd.
O Jansma	Chief Chemist, Industrial Chemicals Laboratory, Shell Chemicals Australia Pty Ltd.
G W Jewell	Product Manager, Hoechst Australia Ltd.

G K Johnson	Manager, Dulux Powder Coatings Division.
W P Johnstone	National Technical Manager, Collie & Co.
R J Kennedy	Manager, Headquarters Automotive Laboratory, Berger Paints (Australia) Pty Ltd.
B Lathlean	Technical Manager, Meggitt Ltd.
B J Lourey	Technical Manager, British Paints (Aust.) Pty Ltd.
M R Luescher	Chief Chemist, Pigments and Additives Laboratory, Ciba Geigy Aust. Ltd.
D M Martin	Managing Director, Berger Paints (Australia) Pty Ltd.
P A Matthews	Manager, Analytical Laboratory, BJN Paints Pty Ltd.
T Maxwell	Development Manager, Building Paints, Taubmans Pty Ltd.
R G Mayrick	Research and Development Director, Mirotone Pty Ltd.
K McLaren	Director, Instrumental Colour Systems Ltd.
A J A Michie	Regulations Compliance Manager, Selleys Chemicals Company Pty Ltd.
A Millar	Technical Manager, Building Paints, Taubmans Pty Ltd.
B R Nagel	Works Chemist, NSW, Dulux Australia Ltd.
M V Powell	Technical Service and Development Manager, A C Hatrick Chemicals Pty Ltd.
R I Reti	Technical Service and Development Manager, D W Corning (Aust.) Pty Ltd.
G F Rhoades	Chief Chemist, Triton Paints (NSW) Pty Ltd.
A K Rileigh	Technical Director, Croda Polymers Group Pty Ltd.
A Robinson	Secretary, Government Paint Committee.
A S Salvatore	Development Manager, Selleys Chemical Company Pty Ltd.
J M Samios	Managing Director, Colorflo Dispersions Pty Ltd.
R Sendelbeck	Plant Supervisor, Borden Chemicals Pty Ltd.
G L Sepansky	Research Chemist, T A Taylor and Son Pty Ltd.
R A Simcock	Principal Development Officer, John Lysaght (Australia) Ltd.
A Sonego	Technical Manager, Quality Earths Division Steetley Industries Ltd.
R L Steele	Technical Service Specialist, Monsanto Australia Ltd.
H G Stephen	Technical Manager, Kemrey Chemicals.
D M D Stewart	Coatings Manager NSW, Rohm and Haas Australia Pty Ltd.
E W Talbot	Technical Manager, Automotive, Dulux Australia.
M Thomas	Market Development Manager, Coatings, Rohm and Haas Australia Pty Ltd.
I Tulloch	Senior Chemist, Paint Industries (Australia) Pty Ltd.
D B Turner	Technical Director, Selleys Chemical Company Pty Ltd.
Z Vachlas	Technical Manager Automotive, Dulux Australia Ltd.
T C Vindin	Manager, Industrial Laboratory, Berger Paints (Australia) Pty Ltd.
J M Waldie	Technical Manager, Selleys Chemical Company Pty Ltd.
T R Walsh	N.S.W. Sales Manager, Robert Bryce and Co Ltd.
R E Walton	Technical Services Manager, Pigments, Laporte Australia Limited.
I R Waples	Registration Manager, National Association of Testing Authorities, Australia.
B R Wearne	Chief Chemist, HMA Naval Dockyard Garden Island.
P J Wigney	Manager, Metalchem Division, Dulux Aust. Ltd.
R J Willis	Manager, Chemicals Department, Hardie Trading Ltd.
B J Wood	Technical Manager, Pascol Paints.
H H Wyatt	Formerly Chief Chemist, James Barnes Pty Ltd.
W Wyatt	Formerly Chief Chemist Plastics, Ciba Geigy Australia Ltd.
R C Young	Research and Development Chemist, Paint Industries (Aust.) Pty Ltd.

INDEX

Volume 1 pp 1–408
Volume 2 pp 409–899

GPSR Compliance

The European Union's (EU) General Product Safety Regulation (GPSR) is a set of rules that requires consumer products to be safe and our obligations to ensure this.

If you have any concerns about our products, you can contact us on ProductSafety@springernature.com

In case Publisher is established outside the EU, the EU authorized representative is:

Springer Nature Customer Service Center GmbH
Europaplatz 3
69115 Heidelberg, Germany

Batch number: 10824729

Printed by Printforce, the Netherlands